Tensors and Manifolds

Tensors and Manifolds
with Applications to Mechanics and Relativity

ROBERT H. WASSERMAN
Department of Mathematics
Michigan State University

New York Oxford
OXFORD UNIVERSITY PRESS
1992

For

Sara, Ziba, and Margaret

Oxford University Press

Oxford New York Toronto
Delhi Bombay Calcutta Madras Karachi
Petaling Jaya Singapore Hong Kong Tokyo
Nairobi Dar es Salaam Cape Town
Melbourne Auckland

and associated companies in
Berlin Ibadan

Published by Oxford University Press, Inc.
200 Madison Avenue, New York, New York 10016

Library of Congress Cataloging-in-Publication Data
Wasserman, Robert, 1923–
Tensors and manifolds : with applications to mechanics and relativity / by Robert H. Wasserman.
p. cm. Includes index.
ISBN 0-19-506561-1
1. Calculus of tensors. 2. Manifolds (Mathematics). 3. Spaces, Generalized. 4. Mechanics. 5. Relativity (Physics)
6. Mathematical physics. I. Title.
QC20.C28W37 1992 530.1'5563—dc20 91-19271 CIP

1 3 5 7 9 8 6 4 2

Printed in the United States of America
on acid-free paper

PREFACE

This book is based on courses taken by advanced undergraduate and beginning graduate students in mathematics and physics at Michigan State University.

The courses were intended to present an introduction to the expanse of modern mathematics and its application in modern physics. The book gives an introductory perspective to young students intending to go into a field of pure mathematics, and who, with the usual "pigeon-holed" graduate curriculum, will not get an overall perspective for several years, much less any idea of application. At the same time, it gives a glimpse of a variety of pure mathematics for applied mathematics and physics students who will have to be carefully selective of the pure mathematics courses they can fit into their curriculum.

Thus, in brief, I have attempted to fill the gap between the basic courses and the highly technical and specialized courses that both mathematics and physics students require in their advanced training, while simultaneously trying to promote, at this early stage, a better appreciation and understanding of each other's discipline.

A third objective is to try to harmonize the two aspects which appear at this level, variously described on the one hand as the "classical", "index", or "local" approach, and on the other hand as the "modern", "intrinsic", or "global" approach.

An underlying theme is an emphasis on mathematical structures. To model a physical phenomenon in general we use some kind of mathematical "space" on which various "physical properties" are defined. For example, in fluid mechanics we make a model in which the fluid consists of points or regions of a space with a certain structure ("ordinary Euclidean space"), and pressure, rate of strain, etc. are mappings into other spaces with certain structures. In general, to model a physical phenomenon, $\mathbb{R}^n$ won't suffice for the domains and images of our mappings—we have to start with manifolds. Moreover, these manifolds have to have additional structures the basic ingredients of which are tensor algebras.

We begin with the algebraic structures we will need and go briefly to $\mathbb{R}^n$ with these in Chapter 7. Manifolds are introduced in Chapter 9 and these structures come together as tensor fields on manifolds in Chapter 11. Chapters 12–14 cover the rudiments of analysis on manifolds, and Chapters

15–17 are devoted to geometry. Finally, a modern treatment of the major ideas of classical analytical mechanics is given in Chapters 18–19, and the remaining chapters are dedicated to an exposition of special and general relativity.

A few words about terminology and notation are needed. I have tried to stick as closely as possible to the most popular current usages, sometimes inserting parenthetically strongly competing alternatives. However, since our text borrows from many different branches of mathematics and physics, we require terminology and notation for a very large variety of concepts and their interrelations. Consequently, the usual problem of how to tread between high precision and readability occurs in aggravated form. Sometimes dropping some notation which is really needed for precision can make it easier to read a given discussion and get the main ideas. On the other hand, sometimes keeping extra terminology and/or notation makes it easier by reminding us of certain important distinctions which might otherwise be temporarily forgotten. A separate section on "Terminology and Notation" is included for convenient reference to some of the conventions used in this book.

Finally, with respect to general style, I have endeavored to steer a safe course between the Scylla of rigor, and the Charybdis of informality. By my not being too heavy-handed with some of the details (including taking some notational liberties as indicated above), hopefully the young student will be able to sail through this passage to enjoy a panorama of interesting mathematics and physics.

This book owes a great deal to the efforts of many classes of students who struggled with earlier classnote versions, and helped hone it into this final form. I was also fortunate to have the services of excellent typists, in particular, Cathy Friess who did the entire final version.

East Lansing, Michigan R. H. W.
November, 1991

CONTENTS

TERMINOLOGY AND NOTATION

Since we have to deal with a large variety of different kinds of mappings, I will use the term *function* to separate out those mappings whose values are real numbers. Also, I will use a dot, instead of parentheses, when indicating the value of a *linear* mapping, and a *transformation* or *operator* will be a mapping of a set into itself. Another notational practice I will employ, in order to emphasize the distinction between mappings, and bases and coordinate changes, is to put bars above symbols under a basis or coordinate change and to use different symbols under a mapping.

The following symbols are used generically for the indicated objects. The symbols in parentheses are used only on the infrequent occasions in which several of the same class of objects appear simultaneously. Some of the following symbols also appear in other contexts: 1. to conform to the standard notation in the literature (e.g.; c is the speed of light, g is a metric tensor field, and σ, μ and ε appear in Maxwell's equations); and 2. in very limitea contexts to fill in needed symbolism in particular arguments, examples, etc.

$a\ (b, c)$	real numbers or n-tuples of real numbers
$b(B)$	bilinear functions
$e_i\ (E_i)$	basis vectors
$f\ (g, h)$	(real-valued) functions
p, q	points of a manifold
t, u	parameters of curves or surfaces
$v, w\ (x)$	vectors
$A\ (B, C)$	tensors
K	tensor fields
M, N, P	manifolds
$V, W\ (X, Y, Z)$	vector spaces
$X, Y\ (Z)$	vector fields
$\mathscr{U}, \mathscr{V}\ (\mathscr{W})$	open sets
$\gamma, \lambda\ (\alpha\ \beta)$	curves
$\varepsilon^i\ (\Xi^i)$	basis 1-forms
ζ_{pi}	coordinate curves through p
$\mu^i\ (\nu^i, \xi^i)$	coordinate functions
$\sigma, \tau\ (\theta, \omega)$	1-forms, or 1-form fields
θ, ω	s-form fields
$\phi, \psi, \Phi, \Psi\ (\chi)$	mappings

I have taken the liberty to deviate from the usual notation in a couple of instances for what I believe to be good pedagogical reasons.

(i) Since notations like $(x^1, \ldots, x^n)$ and $(u^1, \ldots, u^n)$ are used ambiguously for both standard coordintes and points of $\mathbb{R}^n$ (perpetuating the notational abuse in wide currency which obscures the important distinction between a function and its values), I have avoided these in the first part of the manuscript. In particular, I use π^i for the natural coordinate functions (projections) on $\mathbb{R}^n$, a, or $(a^1, \ldots, a^n)$ for points of $\mathbb{R}^n$, and ϑ_{ai} for the natural coordinate curves of $\mathbb{R}^n$ through a. In later developments where additional technical superstructure complicates the discussion I sometimes revert to the standard notation.

(ii) For the derivative at a of a map ϕ, from $\mathbb{R}^m$ to $\mathbb{R}^n$, I use $D_a\phi$ instead of the usual $D\phi(a)$, which I believe better reflects the roles of a and ϕ. I then define $D\phi$ by $D\phi(a) = D_a\phi$. (This corresponds to the standard notation $Xf(p) = X_pf$ in Section 11.1.)

Tensors and Manifolds

1

VECTOR SPACES

Since tensor algebras are built from vector spaces we will recall some of the theory of the latter. We will review the basic properties of vector spaces, their representations and mappings, and mention a generalization.

1.1 Definitions, properties, and examples

Definition
If V is an abelian group with elements $v, w, x, \ldots$, if $a, b, c, \ldots$ are elements of a field, $\mathbb{K}$, and if a mapping $\mathbb{K} \times V \to V$ called "scalar multiplication" and denoted by

$$(a, v) \mapsto av \in V$$

is defined for all elements $a \in \mathbb{K}$ and all $v \in V$ such that

$$1v = v$$

$$a(v + w) = av + aw$$

$$(a + b)v = av + bv$$

$$(ab)v = a(bv)$$

then V is a *vector space.*

From these properties one can prove the additional properties

$$0v = 0 \qquad \text{for all} \qquad v \in V$$

$$a0 = 0 \qquad \text{for all} \qquad a \in \mathbb{K}$$

and, conversely, if $av = 0$ then either $a = 0$ or $v = 0$.

Definitions
Let S be any set (not necessarily finite) of elements in a vector space, V. Then the intersection of all subspaces of V which contain S and the set of all (finite) linear combinations of elements of S are the same. This set is a

subspace, $\langle S \rangle$, of V called *the linear closure of* S. If a subspace, W, of V is the linear closure of some set $S \subset V$ then we say S *spans*, or *generates* W, and S is *a set of generators* of W.

Definitions
Let $\{W_i\}$ be any set (not necessarily finite) of subspaces of a vector space, V. Then the intersection of all subspaces of V containing $\bigcup W_i$, and the set of all finite sums of the form $w_j + w_k + \cdots + w_p$ where $w_j \in W_j$, $w_k \in W_k, \ldots$, $w_p \in W_p$ are the same. This set is a subspace, $\sum W_i$, of V called *the sum of the subspaces*, W_i *of* V. $\sum W_i$ is *direct* if $W_j \cap \sum_{k \neq j} W_k = 0$ for all j. (Note, there is no restriction on the cardinality of the index set we are using.)

Definitions
If for all (finite) sums $\sum_i a_i v_i$ with $v_i \in S$ we have that $\sum_i a_i v_i = 0$ implies that $a_i = 0$ for all i, then S is *a linearly independent set*. This property is equivalent to $0 \notin S$ and every finite sum of subspaces of the form $\ell_v = \{av\colon a \in \mathbb{K}\}$ is direct. Finally, if W is a subspace of V, and if S spans W, and is a linearly independent set, then S *is a basis of* W. This property is equivalent to the property that for each $w \in W$ corresponding to each $v_i \in S$ there exists a unique element $a_i \in \mathbb{K}$ (with $a_i = 0$ for all but a finite number of i's) such that

$$w = \sum_i a_i v_i \tag{1.1}$$

Clearly, these definitions allow for the possibility that a vector space may contain an infinite linearly independent set, and an infinite basis.

EXAMPLES. (i) $\{(a^1, a^2, \ldots, a^n, \ldots)\colon a^i \in \mathbb{R}\}$ with "componentwise" addition and scalar multiplication with $\mathbb{R}$ is a vector space over $\mathbb{R}$. The set $S = \{(0, 0, \ldots, 1, 0, \ldots)\colon 1$ is in the ith place, $i = 1, 2, 3, \ldots\}$ is a linearly independent set but is not a basis.

(ii) $\{(a^1, a^2, \ldots, a^n, \ldots)\colon$ only a finite number of the a's are nonzero$\}$ with operations as in (i) is a vector space and now S is a basis.

(iii) The set of continuous real-valued functions on $[0, \pi]$ with the usual operations: $\{\sin nx\colon n = 1, 2, \ldots\}$ is a linearly independent set but not a basis.

(iv) The set of all ordered n-tuples of real numbers, $\mathbb{R}^n$, is an important example of a vector space having a finite basis. $\{(0, 0, \ldots, 1, \ldots, 0)\colon 1$ in the ith place, $i = 1, \ldots, n\}$ is called the *natural basis* of $\mathbb{R}^n$.

(v) As generalizations of (i)–(iv) we can consider the sets $\mathbb{R}^X$, the set of all (real-valued) functions on an arbitrary set, X, and $\mathbb{R}_0^X$ the set of all (real-valued) functions on an arbitrary set which vanish at all but a finite number of points. Again with pointwise addition and multiplication with $\mathbb{R}$ these are both vector spaces over $\mathbb{R}$. $\mathbb{R}_0^X$ is called *the free vector space over*

X. For each $x \in X$ we define the function f_x by

$$f_x(y) = \begin{cases} 1 & \text{when } y = x \\ 0 & \text{when } y \neq x \end{cases}$$

Then we have a 1–1 map $i: X \to \mathbb{R}^X$ (the canonical injection) defined by $x \mapsto f_x$. The image set of i, $i(X) \subset \mathbb{R}^X$ is a linearly independent set. It is a basis of $R_0^X \subset R^X$ but not of R^X. Every $f \in R_0^X$ has the unique representation

$$f = \sum_{x \in X} f(x) f_x$$

(Only a finite number of terms of this sum are nonzero). R_0^X will be important in the construction of tensor product spaces in Chapter 3.

Theorem 1.1. *If* **I** *is a linearly independent set, and* **G** *is a set of generators of a vector space, V, and* $\mathbf{I} \subset \mathbf{G}$, *then there exists a basis* **B** *such that* $\mathbf{I} \subset \mathbf{B} \subset \mathbf{G}$. *In other words, every linearly independent set can be extended to a basis, and every set of generators contains a basis.*

Proof. (1) If **G** is finite, we can let $\mathbf{G} = \{v_1, \ldots, v_n\}$ and proceed by induction on n. That is, let v_k be any element of **G** and consider the subspace, W, generated by $\mathbf{G} - v_k$. By the induction hypothesis, W has a basis **B** with $\mathbf{B} \subset \mathbf{G} - v_k$. If $v_k \in W$, then $W = V$ and we are done. If $v_k \notin W$, then $\mathbf{B} \cup v_k$ is a linearly independent set and spans V and is thus a basis of V.

(2) If **G** is infinite, we can use Zorn's Lemma, Problem 1.5. ■

In particular, since every (nontrivial) vector space has a linearly independent set and a set of generators with $\mathbf{I} \subset \mathbf{G}$, every vector space has a basis.

Note, however, that for many important spaces the bases are uncountable. In particular, every basis of the vector space of example (i) is uncountable (for method of proof, see Problem 1.12), every basis of the vector space of example (iii) is uncountable (it is an infinite-dimensional Banach space with norm $\max_{[0,\pi]} f$), and if X is an infinite set, then the bases of example (v) are uncountable.

In all the examples above, the field, $\mathbb{K}$, of scalars was $\mathbb{R}$, the field of real numbers. Many of the subsequent results are valid for an arbitrary field (of characteristic 0), in particular for the field of complex numbers, $\mathbb{C}$. It is sometimes useful to bring in complex vector spaces in connection with symplectic spaces (Section 5.4 and Chapter 18). Relations between real and complex vector spaces are indicated in Problems 1.7, 1.8, and 1.14. However, with rare exceptions all our vector spaces in the following will be real.

There is a generalization of the concept of a vector space which we will need when we get to vector fields in Chapter 11. In the definition of vector space we simply replace the field, $\mathbb{K}$, by a ring, $\mathbb{A}$ (with a unit). Such a structure is called an $\mathbb{A}$-*module*. All the definitions above in their various equivalent forms are valid for $\mathbb{A}$-modules. However, there are $\mathbb{A}$-modules that do not have a basis. For example, the module consisting of the elements $0, v_1, v_2, v_3$ over the integers, and with addition given by $v_i + v_i = 0$, $i = 1, 2, 3$ and $v_i + v_j = v_k$ for i, j, k all different, contains no linearly independent set, and hence has no basis. This example also shows that the property stated right after the definition of a vector space is not valid for modules.

An $\mathbb{A}$-module with a basis is called *a free module*. The $\mathbb{A}$-module we will be studying has a basis. There, $\mathbb{A}$ will be the algebra of (real-valued) functions on a certain set M; i.e., $\mathbb{R}^X$ with the structure of an algebra and with $X = M$.

Definition
If a vector space V has a linearly independent set of n elements and no linearly independent set of $n + 1$ elements, then *V has dimension n*, or $\dim V = n$.

Theorem 1.2. *If V has dimension n, then every linearly independent set of n elements is a basis, and every basis has exactly n elements.*

Proof. Problem 1.6.

(Note: Theorem 1.2 can be extended to the "infinite-dimensional" case. In particular, every two bases of V have the same cardinality (cf. Lang, 1965, p. 86).)

PROBLEM 1.1. In the definition of the linear closure of S we actually gave two different characterizations of this concept. Prove they are the same.

PROBLEM 1.2. The same as Problem 1.1 for the definition of the sum of subspaces.

PROBLEM 1.3. The same as Problem 1.1 for the definition of linearly independent set.

PROBLEM 1.4. The same as Problem 1.1 for the definition of basis.

PROBLEM 1.5. Prove Theorem 1.1 for the case **G** is an infinite set (cf., Greub, 1981, pp. 12–13).

PROBLEM 1.6. Prove Theorem 1.2.

PROBLEM 1.7. The set of all ordered n-tuples of complex numbers, $\mathbb{C}^n$, with "component-wise" addition and scalar multiplication with $\mathbb{C}$ is an

n-dimensional vector space with the natural basis $S = \{(0, \ldots, 1, 0, \ldots, 0)$: 1 is in the kth place, $k = 1, \ldots, n\}$ (cf., Example (iv)). Show that the *same set* of n-tuples is a real vector space with basis $S \cup \{(0, \ldots, i, 0, \ldots, 0)$: $i = \sqrt{-1}$ is in the kth place, $k = 1, \ldots, n\}$.

PROBLEM 1.8. Starting with a real vector space, V, we can construct a complex vector space, V_c, *the complexification of* V whose elements are the elements of $V \times V$ with "component-wise" addition and with scalar multiplication defined by $z(v, w) = (av - bw, aw + bv)$ where $z = a + ib$. Show that dim V = dim V_c (cf., Problem 1.14).

1.2 Representation of vector spaces

For an n-dimensional vector space, V^n (over $\mathbb{R}$), Theorem 1.2 implies that in the representation (1.1) we can choose the same finite set of basis elements v_i for all $v \in V^n$ so that (1.1) defines a 1–1 correspondence between elements $v \in V^n$ and ordered n-tuples $(a^1, \ldots, a^n) \in \mathbb{R}^n$. In particular, $v_i \mapsto (0, \ldots, 1, \ldots, 0)$ with 1 in the ith place. With different bases we have different correspondences. That is, once a basis is chosen we can represent any n-dimensional vector space by the particular space, $\mathbb{R}^n$, of example (iv) above.

If $\{e_1, \ldots, e_n\}$ and $\{\bar{e}_1, \ldots, \bar{e}_n\}$ are two bases for V^n, the relation between them can be written in the form

$$\bar{e}_j = \sum_i a^i_j e_i \tag{1.2}$$

For any $v \in V^n$ we have $v = \sum v^i e_i$ and $v = \sum \bar{v}^j \bar{e}_j$. Substituting (1.2) into the second representation for v, and then comparing the two expressions for v, we see that

$$\mathrm{v}^i = \sum_j \bar{v}^j a^i_j \tag{1.3}$$

Let

$$(a^i_j) = \begin{pmatrix} a^1_1 & a^1_2 & \cdots & a^1_n \\ a^2_1 & & & \\ \vdots & & & \\ a^n_1 & & \cdots & a^n_n \end{pmatrix}$$

Then (1.3) can be written in matrix notation as either

$$\begin{pmatrix} v^1 \\ \vdots \\ v^n \end{pmatrix} = (a^i_j) \begin{pmatrix} \bar{v}^1 \\ \vdots \\ \bar{v}^n \end{pmatrix}$$

or

$$\begin{pmatrix} \bar{v}^1 \\ \vdots \\ \bar{v}^n \end{pmatrix} = (a^i_j)^{-1} \begin{pmatrix} v^1 \\ \vdots \\ v^n \end{pmatrix} \tag{1.4}$$

Writing (1.2) in "matrix notation" as

$$\begin{pmatrix} \bar{e}_1 \\ \vdots \\ \bar{e}_n \end{pmatrix} = (a^i_j)^{\mathrm{tr}} \begin{pmatrix} e_1 \\ \vdots \\ e_n \end{pmatrix} \tag{1.5}$$

we see that the systems of basis vectors transform with the inverse transpose of the matrix of the transformation of the *components* of a vector. We call (a^i_j) *the change of basis* matrix.

Observe that we are using superscripts as well as subscripts in our notation. This enables us to use the "summation convention", which we will do from now on. That is, if in products, such as on the right side of (1.2), the same index occurs as a superscript and a subscript we will sum on that index omitting the Σ notation. Also, note that though in the matrix notation introduced above the superscript is the row index and the subscript is the column index this will not always be the case.

PROBLEM 1.9. Show that the matrix (a^i_j) defined above is nonsingular.

PROBLEM 1.10. If V^n is the direct sum of two subspaces and a change of basis is made in each of the subspaces, what is the form of the change of basis matrix for V^n?

1.3 Linear mappings

Definitions

If V and W are two vector spaces, a mapping $\phi\colon V \to W$ is a *linear mapping* if $\phi(v_1 + v_2) = \phi(v_1) + \phi(v_2)$ and $\phi(av) = a\phi(v)$ for all v_1, v_2, and v in V and a in the common field of V and W. For such a mapping we write $\phi(v) = \phi \cdot v$. $\phi(V)$ is the *image space* of V under ϕ, dim $\phi(V)$ is the *rank* of ϕ, and $\phi^{-1}(0)$ is the inverse image of $0 \in W$, or the *null space*, N_ϕ, of ϕ, or the *kernel*, ker ϕ, of ϕ. $\phi(V)$ is a subspace of W and $\phi^{-1}(0)$ is a subspace of V. If $W = \mathbb{R}$, then ϕ is called *a linear function*, or *a linear functional*, or *a linear form*, or *a covector*. If $W = V$ then ϕ is called a *linear transformation* or a *linear operator*. If, moreover, ϕ is 1–1 and onto, then ϕ is *an isomorphism.*

Again, we can generalize the preceding and much of the following to $\mathbb{A}$-modules.

We can construct many different linear mappings between two vector spaces (or, $\mathbb{A}$-modules) as the following theorem shows.

Theorem 1.3. *Given vector spaces V and W and a basis S of V, then there exists a unique linear mapping, $\phi: V \to W$ such that $\phi \cdot v_i = w_i$ where $v_i \in S$ and w_i are arbitrarily chosen elements of W.*

Proof. (i) Define ϕ by "extending the given conditions linearly", that is, let $\phi(v) = a^i w_i$ for $v = a^i v_i$. Then it is easy to verify the properties required by the definition. (ii) if ψ is another linear mapping such that $\psi \cdot v_i = w_i$, then by linearity $\psi \cdot v = a^i w_i$ for $v = a^i v_i$. Hence $\psi \cdot v = \phi \cdot v$ for all v so $\psi = \phi$.

Corollary. *If* $\dim V = \dim W$, *then V and W are isomorphic.*

The converse of this corollary is also true; if ϕ is an isomorphism of V and W then they have the same dimensions. More generally, we have the following result for the dimension of V.

Theorem 1.4. *If $\phi: V \to W$ is linear, then*

$$\dim V = \operatorname{rank} \phi + \dim \ker \phi$$

Proof. Let $\{e_1, \ldots, e_p\}$ be a basis of $\phi^{-1}(0)$, and let $\{e_1, \ldots, e_p, \bar{e}_1, \ldots, \bar{e}_q\}$ be a basis of V. (See Problem 1.5.) Then $\dim \phi^{-1}(0) = p$ and $\dim V = p + q$. Now we need only to show that $\{\phi \cdot \bar{e}_1, \ldots, \phi \cdot \bar{e}_q\}$ is a basis of $\phi(V)$, that is that $\dim \phi(V) = q$. A direct calculation shows that $\phi \cdot \bar{e}_1, \ldots, \phi \cdot \bar{e}_q$ is a linearly independent set and spans $\phi(V)$. ∎

If W is a subspace of V, and $v \in V$, we can form the subset $\{v + w: w \in W\}$. The set of all such subsets (as v ranges over V) is a *factor*, or *quotient space*, V/W. v_1 and v_2 are both in the same subset if and only if $v_1 - v_2 \in W$. The set (i.e., equivalence class) containing v is denoted by $[v]$. The dimension of V/W is called *the codimension of W*. The linear mapping $\pi: V \to V/W$ given by $v \mapsto [v]$ is onto, has the property $\pi^{-1}(0) \cong W$, and is called *the natural projection* of V onto V/W. From this and Theorem 1.4 we have $\operatorname{codim} W = \dim V - \dim W$.

Conversely, starting with a linear mapping, $\phi: V \to W$, we have $\phi^{-1}(0)$ and we can construct a factor space $V/\phi^{-1}(0)$ and an induced isomorphism $V/\phi^{-1}(0) \cong \phi(V)$. There is also the factor space $W/\phi(V)$ called *the cokernel of ϕ*, and $\operatorname{rank} \phi + \dim \operatorname{coker} \phi = \dim W$.

PROBLEM 1.11. Let V and W be vector spaces. (i) Define operations in $V \times W$ so that it becomes a vector space (called *the exterior direct sum of V and W* and denoted by $V \oplus W$). (ii) Show that the *projections* given by

$p_1: (v, w) \mapsto (v, 0)$ and $p_2: (v, w) \mapsto (0, w)$ are linear. (iii) Show that $V \oplus W = \ker p_1 \oplus (p_1(V \oplus W))$, where the first $\oplus$ on the right is the direct sum defined in Section 1.1.

PROBLEM 1.12. The set of all linear functions, $\mathscr{L}(V, \mathbb{R})$, on V is a vector space (this is a special case of Theorem 2.1). Use Theorem 1.3 to show that if V has a countably infinite basis, then the basis of $\mathscr{L}(V, \mathbb{R})$ is uncountably infinite.

(Hint: by Theorem 1.3, if V has a countably infinite basis, then every sequence of real numbers determines a unique element $f \in \mathscr{L}(V, \mathbb{R})$. For each positive number, r, let f_r be the element determined by $(1, r, r^2, r^3, \ldots)$. Consider the set of all such functions, and show that for any finite subset $\{f_1, \ldots, f_n\}$, $a^1 f_1 + a^2 f_2 + \cdots + a^n f_n = 0$ implies $a^1 = a^2 = \cdots = a^n = 0$.)

PROBLEM 1.13. If $(a, b) \in \mathbb{R}^2$, let $W = \{t(a, b): t \in \mathbb{R}\}$, a subset of the vector space $\mathbb{R}^2$. Give a geometrical interpretation of $\mathbb{R}^2/W$.

PROBLEM 1.14. Corresponding to each $a \in \mathbb{K}$ there is a linear operator ϕ on V given by $\phi: V \to aV$. In particular, for the space $\mathbb{C}^n$, corresponding to $i = \sqrt{-1}$ we have a linear operator, J with the property that $J^2 v = -v$. On the real vector space $\mathbb{R}^{2n}$, the linear operator J defined by

$$J: \begin{cases} e_k \mapsto -e_{k+n} & k = 1, \ldots, n \\ e_k \mapsto e_{k-n} & k = n+1, \ldots, 2n \end{cases}$$

where $\{e_1, \ldots, e_{2n}\}$ is the natural basis of $\mathbb{R}^{2n}$, has this property. Show that for any real vector space which admits such an operator, called *a complex structure*, the same set of elements can be made into a complex vector space by defining multiplication by a complex scalar by $(a + bi)v = av + bJv$ (cf., Problems 1.7 and 1.8).

PROBLEM 1.15. A set, S, on which there is defined a vector space, V, of transformations is called *an affine space of* V, if (i) $0 \in V$ is the identity mapping, if (ii) for $v \neq 0$ and $p \in S$, $v(p) \neq p$, if (iii) $u + v$ is the composition, and if (iv) for each ordered pair (p, q) of elements of S, there is an element $v \in V$ such that $v(p) = q$. Show that $(-v)(q) = p$ if $v(p) = q$. Show that (p, q) determines a unique vector, so we can write it as $\overrightarrow{pq}$ and call it a *free vector*, or *translation of* S. If $S = \{(a, b, c) \in \mathbb{R}^3: c = a + b + 1\}$ and V is the vector space $\mathbb{R}^2$, show that S is an affine space of $\mathbb{R}^2$ (cf., Problem 1.13).

1.4 Representation of linear mappings

Just as we have a 1–1 correspondence between vectors in V^n and elements of $\mathbb{R}^n$ once a basis is chosen in V^n, we can now, after choosing bases $\{e_1, \ldots, e_n\}$ in V^n and $\{E_1, \ldots, E_r\}$ in V^r, set up a 1–1 correspondence between the set of linear mappings from V^n to V^r and the set of r by n matrices. Thus, given a linear mapping, ϕ, then for each $i = 1, \ldots, n$, there is a unique set of components, ϕ_i^j, $j = 1, \ldots, r$, of the image of e_i under ϕ. That is,

$$\phi \cdot e_i = \phi_i^j E_j \tag{1.6}$$

If $w = \phi \cdot v$, then putting $v = v^i e_i$ and $w = w^j E_j$ we have $w^j E_j = w = \phi \cdot v^i e_i = v^i \phi \cdot e_i = v^i \phi_i^j E_j$, so

$$w^j = v^i \phi_i^j \begin{cases} i = 1, \ldots, n \\ j = 1, \ldots, r \end{cases} \tag{1.7}$$

or, in matrix notation, writing

$$(\phi_i^j) = \begin{pmatrix} \phi_1^1 & \phi_2^1 & \cdots & \phi_n^1 \\ \phi_1^2 & & & \\ \vdots & & & \\ \phi_1^r & \cdots\cdots & & \phi_n^r \end{pmatrix}$$

we have

$$\begin{pmatrix} w^1 \\ \vdots \\ w^r \end{pmatrix} = (\phi_i^j) \begin{pmatrix} v^1 \\ \vdots \\ v^n \end{pmatrix} \tag{1.8}$$

(Notice that the matrix of coefficients of the system (1.6), as it stands, is the transpose of (ϕ_i^j). Also, compare eqs. (1.6)–(1.8) respectively with eqs. (1.2)–(1.4).) In particular, if $r = 1$, then the mapping can be represented by a $1 \times n$ matrix, or a row vector. Thus, linear functions, or covectors, can be represented by row vectors.

Conversely, starting with an arbitrary r by n matrix, (ϕ_i^j), we can let the columns be the components, in the $\{E_i\}$ basis of V^r, of the images of the basis elements, e_i, of V^n; i.e., $e_i \mapsto \phi_i^1 E_1 + \phi_i^2 E_2 + \cdots + \phi_i^r E_r$. An arbitrary choice of these images determines a unique linear mapping by Theorem 1.3.

With the representation of vectors by elements of cartesian spaces and the representation of linear mappings by matrices, we see that the specific "concrete" examples of linear mappings of the form $\phi\colon \mathbb{R}^n \to \mathbb{R}^r$ given by

$$(v^1, \ldots, v^n) \mapsto (\phi_i^1 v^i, \ldots, \phi_i^r v^i) \tag{1.9}$$

or, in matrix notation

$$\begin{pmatrix} v^1 \\ \vdots \\ v^n \end{pmatrix} \mapsto \begin{pmatrix} \phi_1^1 & \cdots & \phi_n^1 \\ \vdots & & \vdots \\ \phi_1^r & \cdots & \phi_n^r \end{pmatrix} \begin{pmatrix} v^1 \\ \vdots \\ v^n \end{pmatrix}$$

are really all there are.

Just as a vector will have two different sets of components in two different bases, a linear mapping will have different matrices if different bases are chosen in V^n and V^r.

Definition
Two r by n matrices are *equivalent* if they represent the same linear mapping, ϕ, relative to different bases in V^n and V^r.

From matrix theory we know that two matrices are equivalent if and only if they have the same rank.

An important special case of the ideas above occurs when $V^r = V^n$, i.e., ϕ is a linear transformation. In this case, two n by n matrices representing ϕ are called *similar*, and matrices (ϕ_j^i) and (ψ_j^i) are similar if and only if there exists a nonsingular matrix, M such that $(\psi_j^i) = M^{-1}(\phi_j^i)M$. M is the change of basis matrix. For a linear transformation we also have the important concepts of eigenvectors, eigenvalues, invariant subspaces, etc. We will come back to these ideas later when we need them in Chapter 5.

PROBLEM 1.16. (i) Show that the correspondence $e_i \mapsto (0, \ldots, 1, \ldots, 0)$ between V^n and $\mathbb{R}^n$ described in Section 1.2 is an isomorphism. (ii) What is the form of (1.8) for this linear transformation if we choose in $\mathbb{R}^n$ (a) the natural basis, (b) an arbitrary basis $\{(a_{i1}, \ldots, a_{in}): i = 1, \ldots, n\}$?

2

MULTILINEAR MAPPINGS AND DUAL SPACES

We will discuss the important space, V^*, of linear functions on a vector space, V. We will describe isomorphisms between spaces of multilinear mappings, and, finally, we will focus on special properties of bilinear functions.

2.1 Vector spaces of linear mappings

In Section 1.3 we discussed briefly the idea of a linear mapping between vector spaces V and W. Now we consider the set of all linear mappings from V to W.

Theorem 2.1. *The set, $\mathscr{L}(V, W)$, of all linear mappings from V to W with operations $\phi + \psi$ and $a\phi$ defined by*

$$(\phi + \psi)\cdot v = \phi\cdot v + \psi\cdot v$$
$$(a\phi)\cdot v = a(\phi\cdot v)$$

is a vector space.

Proof. $\phi + \psi$ is a linear mapping since

$$\begin{aligned}(\phi + \psi)\cdot(av + bw) &= \phi\cdot(av + bw) + \psi\cdot(av + bw)\\ &= a\phi\cdot v + b\phi\cdot w + a\psi\cdot v + b\psi\cdot w\\ &= a(\phi + \psi)\cdot v + b(\phi + \psi)\cdot w.\end{aligned}$$

Similarly for $a\phi$. Each of the vector space properties of $\mathscr{L}(V, W)$ comes from the corresponding property of W. ■

Theorem 2.2. *With the standard definitions of addition and scalar multiplication, the set of r by n matrices is a vector space, and the 1–1 correspondence described in Section 1.4 between $\mathscr{L}(V^n, V^r)$ and the set of r by n matrices is an isomorphism.*

Proof. Problem 2.2.

Theorem 2.3. *Suppose $\{e_i\}$ and $\{E_i\}$ are bases of finite-dimensional spaces V and W respectively. Then the elements e^i_j of $\mathscr{L}(V, W)$ defined by $e^i_j\colon v \mapsto v^i E_j$ form a basis for $\mathscr{L}(V, W)$. Moreover, for $\phi \in \mathscr{L}(V, W)$,*

$$\phi = \phi^j_i e^i_j$$

where ϕ^i_j are given by $\phi \cdot e_k = \phi^j_k E_j$.

Proof. First of all, note that the definition of e^i_j is equivalent to $e^i_j\colon e_k \mapsto \delta^i_k E_j$ with e^i_j extended to V by linearity.

(i) $a^j_i e^i_j(e_k) = a^j_i \delta^i_k E_j = a^j_k E_j$ so $a^j_i e^i_j = 0$ implies that $a^j_k E_j = 0$ for all k, and since $\{E_i\}$ is linearly independent, $a^j_i = 0$.

(ii) Given ϕ in $\mathscr{L}(V, W)$, let $\phi \cdot e_i = \phi^j_i E_j$. Then $\phi^j_i e^i_j(e_k) = \phi^j_i \delta^i_k E_j = \phi^j_k E_j = \phi(e_k)$ so $\phi = \phi^j_i e^i_j$. ■

There are two important special cases of $\mathscr{L}(V, W)$; namely, $\mathscr{L}(\mathbb{R}, W)$ and $\mathscr{L}(V, \mathbb{R})$. We will devote our attention exclusively to the second case after giving one result for the first.

Theorem 2.4. *$\mathscr{L}(\mathbb{R}, W)$ is isomorphic with W.*

Proof. Note that every nonzero element of $\mathbb{R}$ is a basis of $\mathbb{R}$. In particular, 1 is the natural basis of $\mathbb{R}$. By Theorem 1.3 for each $w \in W$, there is an element $\bar{w}$ of $\mathscr{L}(\mathbb{R}, W)$ given by $\bar{w}\colon 1 \mapsto w$. Two different w's must clearly lead to two different mappings, so the correspondence $W \to \mathscr{L}(\mathbb{R}, W)$ is one-to-one. On the other hand, given any $\phi \in \mathscr{L}(\mathbb{R}, W)$, $\phi \cdot 1 \in W$ determines a mapping ψ given by $1 \mapsto \phi \cdot 1$. By the "uniqueness" part of Theorem 1.3 $\psi = \phi$ so the correspondence is onto. The linearity follows from the definition of the operations in $\mathscr{L}(\mathbb{R}, W)$. ■

This isomorphism will be invoked to make a certain "identification" when we study tangent maps in Section 7.3.

The space, $\mathscr{L}(V, \mathbb{R})$, of linear functions on V is also denoted by V^*. As noted in Section 1.3, its elements are also called linear forms (or linear functionals, or covectors).

As a concrete *example* of a space of linear forms we have $\mathscr{L}(\mathbb{R}^n, \mathbb{R})$, whose elements, f, are given by $f\colon (a^1, \ldots, a^n) \mapsto f_i a^i$, or, in matrix form,

$$\begin{pmatrix} a^1 \\ \vdots \\ a^n \end{pmatrix} \mapsto (f_1 \cdots f_n) \begin{pmatrix} a^1 \\ \vdots \\ a^n \end{pmatrix}$$

Clearly the elements of $\mathscr{L}(\mathbb{R}^n, \mathbb{R})$ are just special cases of the linear mappings, (1.9). As there, this example can be thought of as being quite general, in the sense that once we choose a basis, $\{e_i\}$, in any n-dimensional space, V^n, and a basis, $E \in R$, then $v \in V^n$ can be represented by $\begin{pmatrix} v^1 \\ \vdots \\ v^n \end{pmatrix}$, $f \in V^{n*}$ can be represented by a $1 \times n$ matrix (or row vector), $(f_1 \cdots f_n)$, where the f_i are given by $f \cdot e_i = f_i E$ (cf., eq. (1.6)), and

$$(f_1 \cdots f_n)\begin{pmatrix} v^1 \\ \vdots \\ v^n \end{pmatrix}$$

represents the image of v under f.

From the fact that V^{n*} is isomorphic to the set of $1 \times n$ matrices, it is clear the dim $V^{n*} = n$; i.e., dim $V^{n*} = \dim V^n$. It follows as a special case of Theorem 2.3 that the elements $\varepsilon^i \in V^{n*}$ defined by $\varepsilon^i \cdot e_k = \delta^i_k E$, $i, k = 1, \ldots, n$, $E \in R$, form a basis for V^{n*} and

$$f = f_i \varepsilon^i$$

for any f in V^{n*}.

Definition
If we choose $E = 1$, then the basis $\{\varepsilon_i\}$ of V^{n*} given by

$$\varepsilon^i \cdot e_k = \delta^i_k, \qquad i, k = 1, \ldots, n$$

is called *the dual basis* of $\{e_i\}$.

Since, when V is finite-dimensional, dim $V = \dim V^*$, it follows that we have a result for $\mathscr{L}(V, \mathbb{R})$ analogous to Theorem 2.4 for $\mathscr{L}(\mathbb{R}, W)$; namely, $\mathscr{L}(V, \mathbb{R})$ is isomorphic to V.

In general, however, if V is not finite-dimensional, V^* is not necessarily isomorphic with V. It is generally larger than V. See Problem 1.12. However, there is a subspace of V^* which is isomorphic with V.

Definition
Let $\{e_i\}$ be a basis of V. We denote the set $\{f \in V^*: f \cdot e_i = 0$ for all but a finite number of the e_i's$\}$ by V^*_0.

Theorem 2.5. *The set, $\{\varepsilon^i\}$, of elements of V^* defined by $\varepsilon^i \cdot e_k = \delta^i_k$ is a basis for V^*_0, and $V^*_0 \cong V$.*

Proof. (i) Consider any linear combination $a_i\varepsilon^i$. Then

$$(a_i\varepsilon^i)\cdot e_k = a_i(\varepsilon^i\cdot e_k) = a_k$$

so $a_i\varepsilon^i = 0$ implies that $a_k = 0$. (ii) Given f in V_0^*, let $f\cdot e_i = f_i$, and look at $f_i\varepsilon^i$. $(f_i\varepsilon^i)\cdot e_j = f_i\delta^i_j = f_j = f\cdot e_j$ for all e_j, so $f = f_i\varepsilon^i$. (i) and (ii) show that $\{\varepsilon^i\}$ is a basis for V_0^*. (iii) Notice that the condition $\varepsilon^i\cdot e_k = \delta^i_k$ defines a 1–1 correspondence between the basis sets $\{e_i\}$ and $\{\varepsilon^i\}$; namely, $e_i \mapsto \varepsilon^i$, the basis element of V_0^* which takes e_i to 1 and all the others to zero. If we extend this correspondence by linearity, then the resulting mapping is 1–1 and onto V_0^*.

Having constructed a new vector space V^* from V we can now ask about linear transformations on V^*; in particular, we can study the set $\mathscr{L}(V^*, \mathbb{R}) = V^{**}$.

Theorem 2.6. *Let v be any arbitrarily chosen (fixed) element of V. Then the map $\bar{v}$: $V^* \to \mathbb{R}$ defined by $\bar{v}$: $f \mapsto f\cdot v$ is in V^{**}.*

Proof. Problem 2.4.

Since for each $v \in V$ we have a $\bar{v} \in V^{**}$ by Theorem 2.6, that theorem gives us a mapping $\mathscr{I}$: $V \to V^{**}$. With the notation of Theorem 2.6, $\mathscr{I}(v) = \bar{v}$ and the map defined in that theorem can be written

$$\mathscr{I}(v)\cdot f = f\cdot v \tag{2.1}$$

Theorem 2.7. *The mapping $\mathscr{I}$: $V \to \mathscr{I}(V) \subset V^{**}$ is an isomorphism.*

Proof. (i)
$$\begin{aligned}\mathscr{I}(av_1 + bv_2)\cdot f &= f\cdot(av_1 + bv_2) && \text{by}(2.1)\\ &= af\cdot v_1 + bf\cdot v_2 && \text{by linearity of } f\\ &= a\mathscr{I}(v_1)\cdot f + b\mathscr{I}(v_2)\cdot f && \text{by (2.1)}\\ &= (a\mathscr{I}(v_1) + b\mathscr{I}(v_2))\cdot f\end{aligned}$$

by definition of operations in V^{**}. Hence $\mathscr{I}(av_1 + bv_2) = a\mathscr{I}(v_1) + b\mathscr{I}(v_2)$, so $\mathscr{I}$ is linear. (ii) $\mathscr{I}(v_1) = \mathscr{I}(v_2)$ implies $f\cdot v_1 = f\cdot v_2$ for all $f \in V^*$, so $v_1 = v_2$ and $\mathscr{I}$ is 1–1. (See Problem 2.7.) ■

Corollary. *If V is finite-dimensional, then V^{**} is isomorphic with V. Thus, if we identify V and V^{**} we can think of V as the space of linear functions on V^*. (See Section 2.2.)*

Definition
$\mathscr{I}$ is called the *natural injection* of V into V^{**}.

PROBLEM 2.1. (i) Prove Theorem 2.1 with $\mathscr{L}(V, W)$ replaced by W^V, the set of all maps from V to W. (ii) Prove Theorem 2.1 with $\mathscr{L}(V, W)$ replaced by W^X where X is an arbitrary set. (Compare with $\mathbb{R}^X$ in Section 1.1.)

PROBLEM 2.2. Prove Theorem 2.2.

PROBLEM 2.3. Suppose $\{e_i\}$ and $\{\bar{e}_i\}$ are bases of V, $i = 1, \ldots, n$, and suppose $\{\varepsilon^i\}$ and $\{\bar{\varepsilon}^i\}$ are corresponding dual bases of V^*. Show that if

$$(\bar{e}_1 \cdots \bar{e}_n) = (e_1 \cdots e_n)\begin{pmatrix} a^1_1 & \cdots & a^1_n \\ \vdots & & \\ a^n_i & \cdots & a^n_n \end{pmatrix}$$

$$\text{then} \quad \begin{pmatrix} \bar{\varepsilon}^1 \\ \vdots \\ \bar{\varepsilon}^n \end{pmatrix} = \begin{pmatrix} b^1_1 & \cdots & b^1_n \\ \vdots & & \\ b^n_1 & & b^n_b \end{pmatrix}\begin{pmatrix} \varepsilon^1 \\ \vdots \\ \varepsilon^n \end{pmatrix}$$

where $a^i_j b^j_k = \delta^i_k$. That is, the change of basis matrix (b^i_j) is the inverse of the change of basis matrix (a^i_j). Write the relation between the components, f_i and $\bar{f}_i$, of an element of V^* in the two bases in terms of (a^i_j).

PROBLEM 2.4. Write out the matrix of the linear transformation e^i_j in Theorem 2.3.

PROBLEM 2.5. In Theorem 2.5 we got a basis for a space of linear forms on a vector space which was permitted to be either finite- or infinite-dimensional. In Theorem 2.3 we got bases for spaces of mappings, and the proof given for Theorem 2.3 resembles that of Theorem 2.5. If in Theorem 2.3 V and/or W are infinite-dimensional we can define e^i_j as in the finite case. Precisely where will the proof fail if we try to prove Theorem 2.3 if V and/or W are infinite-dimensional?

PROBLEM 2.6. Prove Theorem 2.6.

PROBLEM 2.7. In the proof of part (ii) of Theorem 2.7 we stated that $f \cdot v_1 = f \cdot v_2$ for all $f \in V^*$ implies that $v_1 = v_2$. This is equivalent to the statement that if $v_1 \neq v_2$, then there is an f in V^* such that $f \cdot v_1 \neq f \cdot v_2$, or if $v \neq 0$, then there is an f in V^* such that $f \cdot v \neq 0$. Prove this.

PROBLEM 2.8. Prove that two matrices (ϕ^i_j) and (ψ^i_j) represent (with respect to two bases in V and their duals in V^*) the same linear mapping of V^n to V^{n*} if and only if there is a matrix M such that $(\psi^i_j) = M^{\text{tr}}(\phi^i_j)M$, and M is the change of basis matrix, (a^i_j). Two such matrices are called *congruent* (cf., definition of similar matrices in Section 1.4).

2.2 Vector spaces of multilinear mappings

Definition
Given vector spaces $V_1, V_2, \ldots, V_p, W$. A mapping $\phi: V_1 \times \cdots \times V_p \to W$ from the cartesian product $V_1 \times \cdots \times V_p$ to W is *multilinear* if it is linear in each argument.

EXAMPLES.

(i) We get an important example of a bilinear function ($p = 2$, $W = \mathbb{R}$) if we choose $V_1 = \mathbb{R}^n$, $V_2 = \mathbb{R}^m$. Then, given any a_{ij} $i = 1, \ldots, n, j = 1, \ldots, m$, the map $\phi: \mathbb{R}^n \times \mathbb{R}^m \to \mathbb{R}$ defined by

$$((v^1, \ldots, v^n), (w^1, \ldots, w^m)) \mapsto a_{ij}v^i w^j \tag{2.2}$$

is bilinear. Moreover, every bilinear function $\phi: \mathbb{R}^n \times \mathbb{R}^m \to \mathbb{R}$ can be written in this form. For, given ϕ, then

$$\phi: ((0, \ldots, \underset{i\text{th place}}{1}, \ldots, 0), (0, \ldots, \underset{j\text{th place}}{1}, \ldots, 0)) \mapsto a_{ij}$$

for some a_{ij} and then by bilinearity $\phi((v^1, \ldots, v^n), (w^1, \ldots, w^m)) = a_{ij}v^i w^j$. Clearly, these bilinear functions are generalizations of the dot product of vector analysis.

(ii) Another example of a bilinear function (in a certain sense both more and less general than the previous one) which we will see again in Chapter 3 is obtained if V_1 and V_2 are two vector spaces and we are given $f \in \mathscr{L}(V_1, \mathbb{R})$ and $g \in \mathscr{L}(V_2, \mathbb{R})$. Then $\phi: V_1 \times V_2 \to \mathbb{R}$ given by $(v_1, v_2) \mapsto (f \cdot v_1)(g \cdot v_2)$ is bilinear.

(iii) Finally, the "oriented volume" of a parallelepiped in "ordinary" space (a normed determinant function) is an example of a trilinear function.

The bilinear functions of example (i) are analogous to the "concrete" examples given by eq. (1.9) in the sense that every linear map is represented by the latter when bases are chosen in V and W, and every bilinear map is represented by the former when bases are chosen in V, W, and $\mathbb{R}$.

Theorem 2.8. *Given vector spaces $V_1, \ldots, V_p$, W and bases S_i of V_i, $i = 1, \ldots, p$, then there exists a unique multilinear mapping $\phi: V_1 \times \cdots \times V_p \to W$ such that*

$\phi(v_{k_1}, v_{k_2}, \ldots, v_{k_p}) = w_{k_1 \cdots k_p}$ where $v_{k_i} \in S_i$, $k_i = i1, \ldots, in_i$, $i = 1, \ldots, p$, $n_i =$ dim V_i, *and* $w_{k_1 \cdots k_p}$ *are arbitrary elements of* W.

Proof. See Theorem 1.3.

Theorem 2.9. *The set,* $\mathscr{L}(V_1, V_2, \ldots, V_p; \mathrm{W})$ *of all p-linear maps from* $V_1 \times \cdots \times V_p$ *to* W *(with the obvious definitions for the operations) is a vector space.*

Proof. Problem 2.9.

If the vector spaces $V_1, \ldots, V_p, W$ are finite-dimensional, then $\mathscr{L}(V_1, \ldots, V_p; W)$ has a basis in terms of those of $V_1, \ldots, V_p, W$ and the dimension of $\mathscr{L}(V_1, \ldots, V_P; W)$ is the product of the dimensions of $V_1, \ldots, V_p, W$. More explicitly, we have the following special case.

Theorem 2.10. *Suppose* $\{e_i\}$ *and* $\{E_i\}$ *are bases of* V *and* W *respectively. Then the functions* $f^{ij}\colon V \times W \to \mathbb{R}$ *defined by* $f^{ij}\colon (v, w) \mapsto v^i w^j$ *where* v^i *and* w^i *are the components of* v *and* w *in the chosen bases, form a basis for* $\mathscr{L}(V, W; \mathbb{R})$. *Moreover, for* $b \in \mathscr{L}(V, W; \mathbb{R})$

$$b = b_{ij} f^{ij}$$

where $b_{ij} = b(e_i, E_j)$.

Proof. First of all, note that the definition of f^{ij} is equivalent to $f^{ij}\colon (e_k, E_l) \mapsto \delta^i_k \delta^j_l$ with f^{ij} extended to $V \times W$ by bilinearity.

(i) $b_{ij} f^{ij}(e_k, E_l) = b_{ij} \delta^i_k \delta^j_l = b_{kl}$ so $b_{ij} f^{ij} = 0$ implies that $b_{kl} = 0$ for all k, l. Hence the f^{ij} form a linearly independent set.

(ii) If $b \in \mathscr{L}(V, W; \mathbb{R})$ then $b(v, w) = v^i w^j b(e_i, E_j) = b(e_i, E_j) f^{ij}(v, w)$ for all $v \in V$ and $w \in W$, so $b = b(e_i, E_j) f^{ij}$. ■

In the sequel we will be dealing with a variety of special cases of these spaces of multilinear mappings. It will be important to observe that these special cases are not all really different from one another—that certain spaces of mappings can be identified with certain others. What we mean by saying that two spaces can be "identified" with one another is that an isomorphism can be constructed between the two which does not require any choice of bases. Note that the isomorphisms established in Theorem 2.2 and Theorem 2.4 *do* require a choice of a basis. In the finite-dimensional case, isomorphisms between V and V^* require choosing bases, but the isomorphism $\mathscr{I}$ between V and V^{**} in the corollary of Theorem 2.7 does not. Hence we do not identify V and V^*, but we do identify V and V^{**}.

Definition
A linear mapping which is independent of the choice of bases is called *natural* (or *canonical*).

Theorem 2.11. *There is a natural isomorphism between* $\mathscr{L}(V_1, V_2; W)$ *and* $\mathscr{L}(V_1, \mathscr{L}(V_2, W))$.

Proof. Let b be an element of $\mathscr{L}(V_1, V_2; W)$. We define a map $\phi\colon \mathscr{L}(V_1, V_2; W) \to \mathscr{L}(V_1, \mathscr{L}(V_2, W))$ by $b \mapsto \phi(b) \in \mathscr{L}(V_1, \mathscr{L}(V_2, W))$ where $\phi(b)$ is the linear map whose values, $\phi(b)\cdot v_1 \in \mathscr{L}(V_2, W)$ are linear mappings given by

$$\phi(b)\cdot v_1\colon v_2 \mapsto b(v_1, v_2) \in W \tag{2.3}$$

That is, $\phi(b)$ is defined by its values, $\phi(b)\cdot v_1$, on V_1, each of which (i.e., for each fixed v_1) is a linear function from V_2 to W defined by (2.3).

We also write, for given b, and v_1, the partial map, $b(v_1, -)\colon v_2 \mapsto b(v_1, v_2)$, so with this notation $\phi(b)\cdot v_1$ is defined to be $b(v_1, -)$, that is, $\phi(b)\cdot v_1 = b(v_1, -)$.

We will prove that ϕ is 1–1 and onto. (The linearity of ϕ comes by following through the definitions.) Let $\mathscr{A} \in \mathscr{L}(V_1, \mathscr{L}(V_2, W))$. We define a map $\psi\colon \mathscr{L}(V_1, \mathscr{L}(V_2, W)) \to \mathscr{L}(V_1, V_2; W)$ by $\psi(\mathscr{A})\colon (v_1, v_2) \mapsto (\mathscr{A}\cdot v_1)\cdot v_2$ (Fig. 2.1). Now we form the compositions $\psi \circ \phi$ and $\phi \circ \psi$, and evaluate them on their respective domains.

(i) For all b, v_1, v_2

$$\begin{aligned} \psi(\phi(b))(v_1, v_2) &= (\phi(b)\cdot v_1)\cdot v_2 && \text{(by the definition of } \psi) \\ &= b(v_1, -)\cdot v_2 && \text{(by the definition of } \phi(b)) \\ &= b(v_1, v_2) \end{aligned}$$

Hence $\psi(\phi(b))(v_1, v_2) = b(v_1, v_2)$ for all v_1, v_2, and b, so $\psi(\phi(b)) = b$ for all

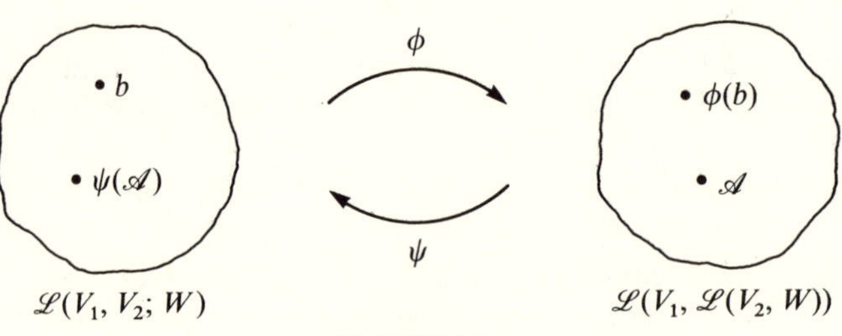

FIGURE 2.1.

b and $\psi \circ \phi =$ identity on $\mathscr{L}(V_1, V_2; W)$ which implies that ϕ is 1–1.

(ii) For all $\mathscr{A}$, v_1, v_2

$$
\begin{aligned}
(\phi(\psi(\mathscr{A}))\cdot v_1)\cdot v_2 &= \psi(\mathscr{A})(v_1, -)\cdot v_2 \qquad &\text{(by definition of } \phi(b))\\
&= \psi(\mathscr{A})(v_1, v_2)\\
&= (\mathscr{A}\cdot v_1)\cdot v_2 &\text{(by definition of } \psi)
\end{aligned}
$$

Again, since this is valid for all v_1, v_2 and $\mathscr{A}$, we have $\phi \circ \psi =$ identity on $\mathscr{L}(V_1, \mathscr{L}(V_2, W))$ which implies that ϕ is onto. ■

More generally, we have the following.

Theorem 2.12. *There is a natural isomorphism between* $\mathscr{L}(V_1, V_2, \ldots, V_p; W)$ *and* $\mathscr{L}(V_i, \mathscr{L}(V_1, \ldots, \hat{V}_i, \ldots, V_p; W))$. ($\hat{V}_i$ means that argument is to be omitted.)

Proof. Problem 2.13.

On the basis of these and other similar types of isomorphisms (see, for example, Problem 2.14), we will be able to describe tensor spaces in a variety of equivalent ways—which can be convenient and also confusing. Moreover, such isomorphisms are used for the construction of "vector-valued forms" which we will encounter in geometry, Chapter 16 (cf., Problem 3.5).

PROBLEM 2.9. Prove Theorem 2.9.

PROBLEM 2.10. Prove the general result stated below Theorem 2.9.

PROBLEM 2.11. We mentioned above Theorem 2.8 that every bilinear function is represented by the mapping (2.2) when bases are chosen. What are the coefficients of (2.2) for the f^{ij} of Theorem 2.10?

PROBLEM 2.12. Verify the statements in Theorem 2.9 that $\psi \circ \phi =$ identity on $\mathscr{L}(V_1, V_2; W)$ implies that ϕ is 1–1, and $\phi \circ \psi =$ identity on $\mathscr{L}(V_1$: $\mathscr{L}(V_2, W))$ implies that ϕ is onto.

PROBLEM 2.13. Prove Theorem 2.12.

PROBLEM 2.14. (i) Prove that there is a natural isomorphism between $\mathscr{L}(V_1, V_2, \ldots, V_p; W)$ and $\mathscr{L}(V_{i_1}, V_{i_2}, \ldots, V_{i_p}; W)$ where $i_1, \ldots, i_p$ is a permutation of $1, \ldots, p$. (ii) Prove that there is a natural isomorphism between $\mathscr{L}(V_1, \ldots, V_p; W)$ and $\mathscr{L}(V_{i_1}, \ldots, V_{i_s}; \mathscr{L}(V_{i_{s+1}}, \ldots, V_{i_p}; W))$ where $i_1, \ldots, i_p$ is a permutation of $1, \ldots, p$.

PROBLEM 2.15. The cartesian product $W_1 \times \cdots \times W_n$ of vector spaces with the operations

$$(w_1, \ldots, w_n) + (x_1, \ldots, x_n) = (w_1 + x_1, \ldots, w_n + x_n)$$
$$a(w_1, \ldots, w_n) = (aw_1, \ldots, aw_n)$$

is a vector space denoted by $W_1 \oplus W_2 \oplus \cdots \oplus W_n$ and called *the exterior direct sun* (see Problem 1.11). Notice that the sets $\tilde{W}_i = \{(0, \ldots, w_i, \ldots, 0)\}$ are subspaces of $W_1 \oplus \cdots \oplus W_n$ and $W_1 \oplus \cdots \oplus W_n$ is the (interior) direct sum of $\tilde{W}_i$ as defined in Section 1.1. Show that if $\dim V = n$, then $\mathscr{L}(V, W) \cong W \oplus \cdots \oplus W$, the direct sum of n copies of W. (See Theorem 2.4.)

2.3 Nondegenerate bilinear functions

Starting with a vector space V, we constructed a second vector space V^* each of whose elements is a mapping from V to $\mathbb{R}$. Instead of fixing an element of V^* and letting v vary in V to get a mapping with values in $\mathbb{R}$, we can, as we did in Theorem 2.6, fix an element $v \in V$ and let f vary in V^* and get a mapping $f \mapsto f \cdot v$ with values in $\mathbb{R}$. These two mappings appear in more symmetrical roles as partial mappings of a mapping $V^* \times V \to \mathbb{R}$ of the cartesian product of V^* and V to $\mathbb{R}$.

Definition
The function δ: $V^* \times V \to \mathbb{R}$ defined by $(f, v) \mapsto f \cdot v$ is called *the Kronecker delta*, or *the natural pairing of V^* and V into* $\mathbb{R}$. We write $\delta(f, v) = \langle f, v \rangle$, and $\delta = \langle -, - \rangle$; i.e., $\langle f, v \rangle = f \cdot v$.

The partial function $\langle f, - \rangle$: $V \to \mathbb{R}$ given by $v \mapsto f \cdot v$ is the same as f itself. The partial function $\langle -, v \rangle$: $V^* \to \mathbb{R}$ given by $f \mapsto f \cdot v$ is the function $\bar{v}$ in Theorem 2.6. It is easy to see that δ is a bilinear function, i.e., $\delta \in \mathscr{L}(V^*, V; \mathbb{R})$. By Theorem 2.10, if V is finite-dimensional and $\{e_i\}$ and $\{\varepsilon^i\}$ are dual bases, then f_i^j: $(\varepsilon^k, e_l) \mapsto \delta_i^k \delta_l^j$ define a basis for $\mathscr{L}(V^*, V; R)$, and

$$\delta = \delta_j^i f_i^j = f_1^1 + f_2^2 + \cdots + f_n^n$$

δ has a property which defines an important class of bilinear functions.

Definitions
A bilinear function b: $V \times W \to \mathbb{R}$ is (weakly) *nondegenerate* (or, *nonsingular*) if $b(v, w) = 0$ for all $v \in V$ implies $w = 0$, and $b(v, w) = 0$ for all $w \in W$ implies $v = 0$. If b is a nondegenerate bilinear function on $V \times W$, then *V and W are dual spaces with respect to b.*

The Kronecker delta is nondegenerate, for the fact that $\langle f, v \rangle = 0$ for all v implies $f = 0$ is just the definition of $f = 0$. The fact that $\langle f, v \rangle = 0$ for all f implies $v = 0$ was already used in Theorem 2.6. See Problem 2.7.

Thus, the natural pairing $\langle -, - \rangle$ of V^* and V is just one particular nondegenerate bilinear function, and V^* and V are dual with respect to $\langle -, - \rangle$.

Recall that in Section 2.2 we gave two examples of bilinear functions. In the first example, the bilinear function is nondegenerate and $\mathbb{R}^n$ and $\mathbb{R}^m$ are dual with respect to this function if and only if $m = n$, and the matrix (a_{ij}) is nonsingular.

By Theorem 2.11 to each bilinear function $b \in \mathscr{L}(V, W; \mathbb{R})$ there corresponds a unique linear map $\mathscr{A}_1 = \phi_1 \cdot b \in \mathscr{L}(V, \mathscr{L}(W; \mathbb{R}) \cong \mathscr{L}(V, W^*)$ where $\phi_1: \mathscr{L}(V, W; \mathbb{R}) \to \mathscr{L}(V, \mathscr{L}(W, \mathbb{R}))$. Since, by Problem 2.14, $\mathscr{L}(W, V; \mathbb{R}) \cong \mathscr{L}(V, W; \mathbb{R})$ corresponding to b there is also a unique linear map $\mathscr{A}_2 = \phi_2 \cdot b \in \mathscr{L}(W, \mathscr{L}(V, \mathbb{R})) \cong \mathscr{L}(W, V^*)$ where $\phi_2: \mathscr{L}(V, W; \mathbb{R}) \to \mathscr{L}(W, \mathscr{L}(V, \mathbb{R}))$. $\phi_1 \cdot b: V \to W^*$ is given by $v \mapsto b(v, -)$ and $\phi_2 \cdot b: W \to V^*$ is given by $w \mapsto b(-, w)$. In particular, if $W = V^*$ and $b = \delta$, then $\mathscr{A}_1 = \phi_1 \cdot b: V \to W^{**}$ is given by $v \mapsto \langle -, v \rangle = \bar{v}$, and $\mathscr{A}_2 = \phi_2 \cdot b: V^* \to V^*$ is given by $f \mapsto \langle f, - \rangle = f$.

If V and W are finite-dimensional and $\{e_i\}$ is a basis of V and $\{E_i\}$ is a basis of W, then $\phi_1 \cdot b: e_i \mapsto a_{ij}\Xi^j$ where $\{\Xi^j\}$ is the basis dual to $\{E_i\}$. But $\phi_1 \cdot b: e_i \mapsto b(e_i, -)$, so $a_{ij}\Xi^j \cdot E_k = b(e_i, E_k)$ and $a_{ik} = b(e_i, E_k) = b_{ik}$. That is, the ikth component of b in the $\{f^{ik}\}$ basis is the ikth element of the matrix of the linear map $\mathscr{A}_1$. Similarly, if $\phi_2 \cdot b: E_i \mapsto a_{ij}\varepsilon^j$, then $a_{ik} = b_{ki}$, which says that the ikth component of b in the $\{f^{ik}\}$ basis is the ikth element of the transpose of the matrix of the linear map $\mathscr{A}_2$.

PROBLEM 2.16. Prove the necessary and sufficient conditions stated for the nondegeneracy of the bilinear function $\mathbb{R}^n \times \mathbb{R}^m \to \mathbb{R}$ of the example given above.

PROBLEM 2.17. If V and W are dual with respect to b, then the maps $\mathscr{A}_1$ and $\mathscr{A}_2$ are injective.

PROBLEM 2.18. Choose a basis in V. (i) Express $\langle f, v \rangle$ in terms of this basis and its dual in V^*. (ii) What are the matrices, relative to these bases, of the linear maps $\phi_1 \cdot b$ and $\phi_2 \cdot b$ where b is the natural pairing of V and V^*?

PROBLEM 2.19. Give an example of a vector space with two nonisomorphic dual spaces.

2.4 Orthogonal complements and the transpose of a linear mapping

Definitions

Suppose b is a bilinear function on $V \times W$, and $S \subset V$ and $T \subset W$. $S^\perp = \{w \in W: b(v, w) = 0$ for all $v \in S\}$ is called *the orthogonal complement*, or *annihilator of* S *with respect to* b. ${}^\perp T = \{v \in V: b(v, w) = 0$ for all $w \in T\}$ is *the orthogonal complement*, or *annihilator of* T *with respect to* b.

Theorem 2.13. *For any set $S \subset V$, $S^\perp$ is a subspace of W, and for any set $T \subset W$, ${}^\perp T$ is a subspace of V.*

Proof. *Problem 2.20.*

Theorem 2.14. $N_1 = {}^\perp W$ *and* $N_2 = V^\perp$ *where* $N_1 \subset V$ *is the null space of* $\phi_1 \cdot b$ *and* $N_2 \subset W$ *is the null space of* $\phi_2 \cdot b$.

Proof. *Problem 2.21.*

Corollary. *b is nondegenerate if and only if $N_1 = 0$ and $N_2 = 0$.*

Theorem 2.15. *If finite-dimensional vector spaces V and W are dual with respect to b, then $W \cong V^*$ and $V \cong W^*$.*

Proof. Since $N_1 = N_2 = 0$ the linear maps $\phi_1 \cdot b$ and $\phi_2 \cdot b$ are 1–1. Hence $\dim V \leq \dim W^* = \dim W \leq \dim V^* = \dim V$. So $\dim V = \dim W^*$ and $\dim W = \dim V^*$ and the maps are onto. ■

Corollary. *If V and W are dual with respect to b, then* $\dim V = \dim W$.

On the basis of Theorem 2.15 we frequently say that V^* is *the* dual space of V, and $V^{**} = V$ is *the* dual space of V^*.

(Note that in the infinite-dimensional case the conclusion of Theorem 2.15 is not necessarily valid. If it is, we say that b is strongly nondegenerate.)

Theorem 2.16. *If $S \subset V$ and $T \subset W$, $\langle S \rangle \subset {}^\perp(S^\perp)$ and $\langle T \rangle \subset ({}^\perp T)^\perp$.*

Proof. If $v \in \langle S \rangle$, then for all $w \in S^\perp$, $b(v, w) = a_i b(v^i, w) = 0$ so $v \in {}^\perp(S^\perp)$. Similarly for the second part. ■

We can strengthen this result, and get other interesting relations in the case $W = V^*$ and, $b = \delta$. Hence, *the orthogonal complement will be with respect to δ from now on.*

Theorem 2.17. $\langle S \rangle = {}^\perp(S^\perp)$.

Proof. From Theorem 2.16 we know $\langle S \rangle \subset {}^\perp(S^\perp)$. Now suppose $\tilde{v} \in {}^\perp(S^\perp)$ and $\tilde{v} \notin \langle S \rangle$. We can choose the $\tilde{v}$ to be one of the basis elements of V by Theorem 1.1. Let $\tilde{f}$ be the element of V^* which maps $\tilde{v}$ to 1, and the other basis elements to zero. $\tilde{f}$ exists and is unique by Theorem 1.3. In particular, for all $v \in \langle S \rangle$, $\langle \tilde{f}, v \rangle = 0$. That is, $\tilde{f} \in \langle S \rangle^\perp$. But $S \subset \langle S \rangle$, so $\tilde{f} \in S^\perp$. Hence, for all $v \in {}^\perp(S^\perp)$, $\langle \tilde{f}, v \rangle = 0$. In particular, since by assumption $\tilde{v} \in {}^\perp(S^\perp)$, we must have $\langle \tilde{f}, \tilde{v} \rangle = 0$, but this contradicts $\tilde{f} \cdot \tilde{v} = 1$, so ${}^\perp(S)^\perp \subset \langle S \rangle$. ■

Theorem 2.18. *If v is finite-dimensional and S is a subspace of V then* $\dim S^{\perp} = \dim V - \dim S$. *($\perp$ is with respect to δ).*

Proof. *Problem 2.23.*

Now suppose we have two vector spaces V and W and a linear mapping $\Phi: V \to W$. Φ determines a certain linear mapping from W^* to V^*.

Definition
If $\Phi: V \to W$, then the *transpose*, or *dual of* Φ is the map $\Phi^*: W^* \to V^*$ defined by $\Phi^*: w^* \mapsto w^* \circ \Phi$ where $w^* \in W^*$.

We can think of Φ^* as taking functions defined on W and "pulling them back" to functions defined on V.

Theorem 2.19. *Φ^* is linear.*

Proof. We evaluate $\Phi^*(aw_1^* + bw_2^*)$ at an arbitrary $v \in V$. Thus,

$$\begin{aligned}
\Phi^*(aw_1^* + bw_2^*)\cdot v &= ((aw_1^* + bw_2^*)\circ\Phi)\cdot v \qquad \text{by definitions of } \Phi^* \\
&= (aw_1^* + bw_2^*)\cdot(\Phi\cdot v) = a(w_1^*\cdot(\Phi\cdot v)) + b(w_2^*\cdot(\Phi\cdot v)) \\
&= a(\Phi^*(w_1^*)\cdot v) + b(\Phi^*(w_2^*)\cdot v) \qquad \text{by definition of } \Phi^* \\
&= (a\Phi^*(w_1^*) + b\Phi^*(w_2^*))\cdot v \\
&\qquad\qquad \text{by definition of operations in } V^*
\end{aligned}$$

So $\Phi^*(aw_1^* + bw_2^*) = a\Phi^*(w_1^*) + b\Phi^*(w_2^*)$. ■

We had an example of a mapping and its transpose at the end of Section 2.3. More precisely, if $\mathscr{A}_1 = \phi_1 \cdot b: V \to W^*$ and $\mathscr{A}_2 = \phi_2 \cdot b: W \to V^*$, then $\mathscr{A}_2 = \mathscr{A}_1^*|_W$. That is, $(\mathscr{A}_2 \cdot w)\cdot v = (\mathscr{A}_1^* \cdot w^{**})\cdot v$ when $w^{**} = w$ (by the isomorphism $\mathscr{I}$). This is seen by rewriting the right side: $(\mathscr{A}_1^* \cdot w^{**})\cdot v = [w^{**}\circ(\phi_1 \cdot b)](v) = w^{**}\cdot w^* = w^* \cdot w$ (when $w^{**} = w$) $= (\phi_1 \cdot b)\cdot v \cdot w = (\phi_2 \cdot b)\cdot w \cdot v$.

Theorem 2.20. *If $\Phi_1: V \to W$ and if $\Phi_2 = \Phi_1^*$, then $\Phi_1 = \Phi_2^*|_V$.*

Proof. *Problem 2.24.*

Theorem 2.21. *Φ^* is the transpose of Φ if and only if*

$$\langle \Phi^* \cdot w^*, v\rangle = \langle w^*, \Phi \cdot v\rangle \tag{2.4}$$

for all $v \in V$ and $w^* \in W^*$.

Proof. If we evaluate $\Phi^* \cdot w^* = w^* \circ \Phi$ at v we get (2.4) immediately. ■

Theorem 2.22. *If* $\Phi: V \to W$ *and* Φ^* *is its transpose, then*

(i) $\ker \Phi^* = (\Phi(V))^{\perp}$
(ii) $\Phi(V) = {}^{\perp}(\ker \Phi^*)$
(iii) $\ker \Phi = {}^{\perp}(\Phi^*(W^*))$
(iv) $\Phi^*(W^*) = \ker (\Phi)^{\perp}$

See Fig. 2.2.

Proof. (i) If $w^* \in \ker \Phi^*$, then $\Phi^* \cdot w^* = 0$ and by eq. (2.4) $\langle w^*, \Phi \cdot v \rangle = 0$ for all $v \in V$, so $w^* \in \Phi(V)^{\perp}$. On the other hand, if $w^* \in \Phi(V)^{\perp}$, then $\langle w^*, \Phi \cdot v \rangle = 0$ for all $v \in V$. By eq. (2.4) $\langle \Phi^* \cdot w^*, v \rangle = 0$ for all $v \in V$. Since $\langle -, - \rangle$ is nondegenerate $\Phi^* \cdot w^* = 0$ so $W^* \in \ker \Phi^*$.

(ii) By Theorem 2.17 $\Phi(V) = {}^{\perp}(\Phi(V)^{\perp})$ which is ${}^{\perp}(\ker \Phi^*)$ by part (i).

(iii) and (iv): *Problem 2.26.* ■

Corollary. Φ *is onto if and only if* Φ^* *is* 1–1. Φ *is* 1–1 *if and only if* Φ^* *is onto.*

Some of the results above are the algebraic abstractions of important results in other areas of mathematics. Thus, for example, Theorem 2.22(ii) has the interpretation that a nonhomogeneous system of equations has a solution if and only if the nonhomogeneous part is orthogonal to every solution of the adjoint homogeneous system. This is known a the "Fredholm Alternative" in the theory of integral equations, or, more generally, in Hilbert Space theory.

PROBLEM 2.20. Prove Theorem 2.13.

PROBLEM 2.21. Prove Theorem 2.14.

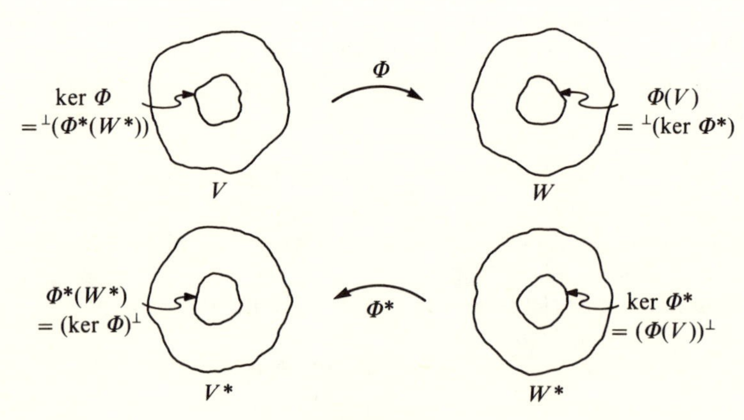

FIGURE 2.2.

PROBLEM 2.22. Prove that $\mathscr{I}({}^{\perp}S^*) = \mathscr{I}(V) \cap {}^{\perp}(S^*)$ for $S^* \subset V^*$.

PROBLEM 2.23. Prove Theorem 2.18.

PROBLEM 2.24. Prove Theorem 2.20.

PROBLEM 2.25. In the finite-dimensional case with given bases, write (2.4) in matrix notation and find the relation between the matrices of Φ and Φ^*.

PROBLEM 2.26. Prove parts (iii) and (iv) of Theorem 2.22.

PROBLEM 2.27. Using parts (iii) and (iv) of Theorem 2.22 prove that if $T \subset W$, then $\langle T \rangle = ({}^{\perp}T)^{\perp}$.

PROBLEM 2.28. We can generalize the concept of the dual of a mapping as follows. If V_1 and W_1 are dual with respect to b_1 and V_2 and W_2 are dual with respect to b_2, and Φ: $V_1 \to V_2$, then we define Φ^* by $b_1(\Phi^* \cdot w_2, v_1) = b_2(\Phi \cdot v_1, w_2)$. Prove that Φ^* satisfies parts (i) and (iii) of Theorem 2.22.

PROBLEM 2.29. (Abstract homology, cf., Whitney, 1957 or Greub, 1981). Suppose $W = V$ and Φ is a linear transformation on V with the property $\Phi^2 = 0$. (Φ is called *a differential operator*). Show (i) $\Phi(V) \subset \ker \Phi$, (ii) $\Phi^{*2} = 0$, (iii) $\Phi^*(V^*) \subset \ker \Phi^*$. Finally, the factor spaces $\ker \Phi / \Phi(V)$ and $\ker \Phi^* / \Phi^*(V^*)$, called the *homology* and *cohomology* spaces of V, respectively, are isomorphic. These ideas will arise in a more specific context in Chapter 12.

3

TENSOR PRODUCT SPACES

We will describe the space of bilinear functions on a pair of finite-dimensional vector spaces as the tensor product space of the duals of those two vector spaces. We then consider tensor product spaces of more than two vector spaces. Finally we define tensor product spaces in the general case so that they reduce to spaces of multilinear functions in finite dimensions. From now on we will denote vectors by $v, w, \ldots$, which may have subscripts, and linear functions by Greek letters $\sigma, \tau, \ldots$, which may have superscripts.

3.1 The tensor product of two finite-dimensional vector spaces

Consider the set of bilinear functions $\mathscr{L}(V, W; \mathbb{R})$, with V and W finite-dimensional. Notice that this set has certain elements of the following type. Let $\sigma \in V^*$ and $\tau \in W^*$. Then with each fixed pair $(\sigma, \tau) \in V^* \times W^*$ we have the function $V \times W \to \mathbb{R}$ given by

$$(v, w) \mapsto (\sigma \cdot v)(\tau \cdot w) \tag{3.1}$$

By the linearity of σ and τ this function is bilinear, so it is in $\mathscr{L}(V, W; \mathbb{R})$. (See example (ii) in Section 2.2.) Since the values of this function are products of the values of σ and τ, we use the product notation $\sigma \otimes \tau$ for this function. That is, $\sigma \otimes \tau \colon V \times W \to \mathbb{R}$ is the element of $\mathscr{L}(V, W; \mathbb{R})$ given by (3.1).

For an example, consider the case where $V = \mathbb{R}^n$ and $W = \mathbb{R}^m$. Then $\sigma \in \mathbb{R}^{n*}$ will have the form $\sigma \colon (v^1, \ldots, v^n) \mapsto \sigma_i v^i$, and $\tau \in \mathbb{R}^{m*}$ will have the form $\tau \colon (w^1, \ldots, w^m) \mapsto \tau_i w^i$. (See below Theorem 2.4.) According to (3.1) $\sigma \otimes \tau$ will be given by $\sigma \otimes \tau \colon ((v^1, \ldots, v^n), (w^1, \ldots, w^m)) \mapsto \sigma_i \tau_j v^i w^j$. Note that this is a special case of example (i) in Section 2.2. Note also, that with bases in V and W every element in $\mathscr{L}(V, W; \mathbb{R})$ of the form $\sigma \otimes \tau$ can be represented by such examples.

Now, for each pair $(\sigma, \tau) \in V^* \times W^*$ we have an element $\sigma \otimes \tau$ of $\mathscr{L}(V, W; \mathbb{R})$. Thus, $\mathscr{L}(V, W; \mathbb{R})$ contains a set of "products", the image set of a mapping

$$\otimes \colon V^* \times W^* \to \mathscr{L}(V, W; \mathbb{R})$$

given by

$$(\sigma, \tau) \mapsto \sigma \otimes \tau \tag{3.2}$$

One easily verifies that $\otimes$ is bilinear. Thus, $\sigma \otimes \tau$ plays two roles: it is a bilinear function given by (3.1), and it is the image of a bilinear map, $\otimes$, given by (3.2).

The image set of $\otimes$ is not a subspace of $\mathscr{L}(V, W; \mathbb{R})$ and, in particular, not $\mathscr{L}(V, W; \mathbb{R})$, since $\sigma^1 \otimes \tau^1 + \sigma^2 \otimes \tau^2$ is not always of the form $\sigma \otimes \tau$. However, when V and W are finite-dimensional spaces, if $\{\varepsilon^i\}$ is a basis of V^* and $\{\Xi^i\}$ is a basis of W^*, then $\{\varepsilon^i \otimes \Xi^j\} = \{f^{ij}\}$ is a basis of $\mathscr{L}(V, W; \mathbb{R})$ (cf., Theorem 2.10), so the set of products does span $\mathscr{L}(V, W; \mathbb{R})$. This accounts for the following terminology and notation.

Definitions
If V and W are finite-dimensional spaces, $\mathscr{L}(V, W; \mathbb{R})$ is called *the tensor* (or *Kronecker*) *product* of V^* and W^*. We write $\mathscr{L}(V, W; \mathbb{R}) = V^* \otimes W^*$. Similarly, since in $\mathscr{L}(V^*, W^*; \mathbb{R})$ we have the "product" elements

$$v \otimes \mathrm{w}\colon (\sigma, \tau) \mapsto (\sigma \cdot v)(\tau \cdot w),$$

(cf., corollary of Theorem 2.7), $\mathscr{L}(V^*, W^*: \mathbb{R})$ is the *tensor* (or *Kronecker*) *product* of V and W, and we write $\mathscr{L}(V^*, W^*; \mathbb{R}) = V \otimes W$. (Note that we are using the notation $\otimes$ in three different ways.)

$\mathscr{L}(V, V^*; \mathbb{R}) = V^* \otimes V$ is an example of a tensor product space. So is $\mathscr{L}(V^*, V; \mathbb{R}) = V \otimes V^*$. The natural pairing, $\langle -, - \rangle$ (Section 2.3), is an element of $V \otimes V^*$.

Theorem 3.1. *Given two finite-dimensional vector spaces V and W. Form $V \otimes W = \mathscr{L}(V^*, W^*; \mathbb{R})$ and the map $\otimes\colon V \times W \to V \otimes W$ with $\otimes(v, w)$ defined as above. The pair $(V \otimes W, \otimes)$ has the following properties:*

(i) *$\otimes$ is bilinear, and $\otimes(V \times W)$ spans $V \otimes W$*

(ii) *If Z is any vector space, and b is any bilinear map, $b\colon V \times W \to Z$, then there exists a unique* linear *map, $\phi\colon V \otimes W \to Z$, onto $\langle b(V \times W) \rangle$, the linear closure of the range of b, and such that $b = \phi \circ \otimes$; i.e., any bilinear map, b, has a unique factorization into the product of $\otimes$ and a linear map. (The fact that every bilinear map, b, can be factored into a linear map and a particular unique map is referred to as* the universal property of bilinear maps.)

Proof. (i) These two properties of $\otimes\colon V \times W \to V \otimes W$ are precisely analogous to those of the mapping given by eq. (3.2). (ii) Let $\{v_i \otimes w_j\}$ be a basis of $V \otimes W$. Define ϕ by mapping $v_i \otimes w_j \mapsto b(v_i, w_j)$ and then extending this correspondence to all of $V \otimes W$ by linearity. Every $z \in \langle b(V \times W) \rangle$ can be written $z = a^{ij} b(v_i, w_j)$. But this is the image of

$a^{ij}v_i \otimes w_j$ under ϕ, so ϕ is onto $\langle b(V \times W)\rangle$. If $\phi_1 \circ \otimes = \phi_2 \circ \otimes$, then $(\phi_1 - \phi_2)\cdot \otimes(v, w) = 0$ for all v, w. But, since $\{\otimes(v, w) = v \otimes w : v \in V, w \in W\}$ spans $V \otimes W$, $(\phi_1 - \phi_2)\cdot A = 0$ for all $A \in V \otimes W$, which implies that $\phi_1 = \phi_2$. ■

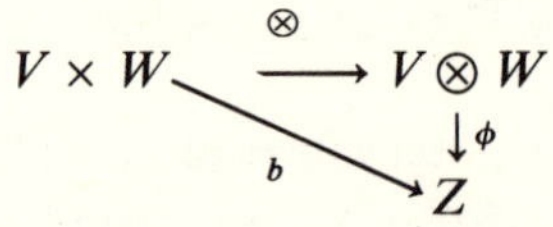

$$\begin{array}{ccc} V \times W & \xrightarrow{\otimes} & V \otimes W \\ & \searrow^{b} & \downarrow \phi \\ & & Z \end{array}$$

Corollary. *For a bilinear map, b, in Theorem 3.1 (ii) with the additional property that the dimension of the linear closure* $\langle b(V \times W)\rangle$ *of its range is* $\dim V \otimes W$, *the corresponding linear map,* ϕ, *is an isomorphism of* $V \otimes W$ *with* $\langle b(V \times W)\rangle$.

Proof. By the theorem, ϕ is onto $\langle b(V \times W)\rangle$. This plus the fact that $\dim \langle b(V \times W)\rangle = \dim V \otimes W$ makes ϕ an isomorphism. ■

Corollary. *Given spaces V, W, and Z, there is a natural isomorphism between the space* $\mathscr{L}(V, W; Z)$ *of bilinear mappings and the space* $\mathscr{L}(V \otimes W, Z)$ *of linear mappings.*

Proof. Problem 3.4.

At the beginning of this section we saw that if $\{\varepsilon^i\}$ is a basis of V^* and $\{\Xi^i\}$ is a basis of W^*, then $\{\varepsilon^i \otimes \Xi^j\}$ form a basis of $V^* \otimes W^*$, so for any $A \in V^* \otimes W^*$ we can write

$$A = A_{ij}\varepsilon^i \otimes \Xi^j \tag{3.3}$$

and A has values

$$A(v, w) = A_{ij}v^i w^j \tag{3.4}$$

Similarly $\{e_i \otimes E_j\}$ form a basis of $V \otimes W$, so for any $A \in V \otimes W$,

$$A = A^{ij}e_i \otimes E_j \tag{3.5}$$

and A has values

$$A(\sigma, \tau) = A^{ij}\sigma_i \tau_j \tag{3.6}$$

PROBLEM 3.1. Write the explicit form of the product $v \otimes w \in V \otimes W$ as a bilinear function if $V = \mathbb{R}^n$ and $W = \mathbb{R}^m$.

PROBLEM 3.2. Give a specific example in which $\sigma^1 \otimes \tau^1 + \sigma^2 \otimes \tau^2$ in $V^* \otimes W^*$ is not a product.

PROBLEM 3.3. Prove part (i) of Theorem 3.1.

PROBLEM 3.4. Prove the second corollary of Theorem 3.1.

PROBLEM 3.5. Prove $\mathscr{L}(V, \ldots, V, V^*; \mathbb{R}) \cong \mathscr{L}(V \otimes \cdots \otimes V; \mathscr{L}(V^*, \mathscr{R}))$ (cf., vector-valued forms, Sections 4.1 and 16.3).

3.2 Generalizations, isomorphisms, and a characterization

We can generalize the development of the last section and construct tensor products of any finite number of vector spaces. For example, we note that with every triple $(\sigma, \tau, \omega) \in V^* \times W^* \times X^*$ we have an element of $\mathscr{L}(V, W, X; \mathbb{R})$

$$(v, w, x) \mapsto (\sigma \cdot v)(\tau \cdot w)(\omega \cdot x)$$

and we call this function $\sigma \otimes \tau \otimes \omega$. That is, we have a map $\otimes$: $V^* \times W^* \times X^* \to \mathscr{L}(V, W, X; \mathbb{R})$ whose images are of the form $\sigma \otimes \tau \otimes \omega$. We have a theorem analogous to Theorem 3.1. In particular, $\otimes$ is trilinear and $\otimes(V^* \times W^* \times X^*)$ spans $\mathscr{L}(V, W, X; \mathbb{R})$, which we now denote by $V^* \otimes W^* \otimes X^*$. Also, any trilinear map has a unique factorization into the product of $\otimes$ and a linear map.

We can generalize eqs. (3.3)–(3.6). In particular, if $\{\varepsilon_1^i\}$, $\{\varepsilon_2^i\}$, and $\{\varepsilon_3^i\}$ are bases respectively of V^*, W^*, and X^*, then $\{\varepsilon_1^i \otimes \varepsilon_2^j \otimes \varepsilon_3^k\}$ is a basis of $V^* \otimes W^* \otimes X^*$ and for any $A \in V^* \otimes W^* \otimes X^*$,

$$A = A_{ijk} \varepsilon_1^i \otimes \varepsilon_2^j \otimes \varepsilon_3^k \tag{3.7}$$

and A has values

$$A(v, w, x) = A_{ijk} v^i w^j x^k \tag{3.8}$$

Generally, given m vector spaces $v_1, \ldots, v_m$, we can construct $V_1 \otimes \cdots \otimes V_m$. If $\{e_{i_1}^1\}, \ldots, \{e_{i_m}^m\}$ are bases of $V_1, \ldots, V_m$ ($i_k = 1, \ldots, n_k$ where $n_k = \dim V_k$), then $\{e_{i_1}^1 \otimes \cdots \otimes e_{i_m}^m\}$ is a basis of $V_1 \otimes \cdots \otimes V_m$ and for any $A \in V_1 \otimes \cdots \otimes V_m$,

$$A = A^{i_1 \cdots i_m} e_{i_1}^1 \otimes \cdots \otimes e_{i_m}^m \tag{3.9}$$

and A has values

$$A(\sigma^1, \ldots, \sigma^m) = A^{i_1 \cdots i_m} \sigma_{i_1}^1 \cdots \sigma_{i_m}^m \tag{3.10}$$

Note that $\dim V_1 \otimes \cdots \otimes V_m = n_1 \cdots n_m$. This is a special case of the general result stated above Theorem 2.10. (See Problem 2.10.)

Since we can construct the tensor product of any two spaces, we can form the tensor product of tensor products. In general, the operations of taking tensor products and/or duals can be iterated, resulting in an apparently bewildering proliferation of vector spaces. The following theorems help to keep things under control.

Theorem 3.2. *The vector spaces* $V \otimes W \otimes X$, $(V \otimes W) \otimes X$, *and* $V \otimes (W \otimes X)$ *are naturally isomorphic.*

Proof. A bilinear map $b: (V \otimes W) \times X \to V \otimes W \otimes X$ is determined by $(v \otimes w, x) \mapsto v \otimes w \otimes x$. (For each fixed $v \otimes w$ this prescription determines a linear map from X to $V \otimes W \otimes X$, and for each fixed x, by Theorem 3.1, since $(v, w) \mapsto v \otimes w \otimes x$ is bilinear, it determines a linear map from $V \otimes W$ to $V \otimes W \otimes X$.) Then by Theorem 3.1 we have the linear map $\phi: (v \otimes w) \otimes x \mapsto v \otimes w \otimes x$

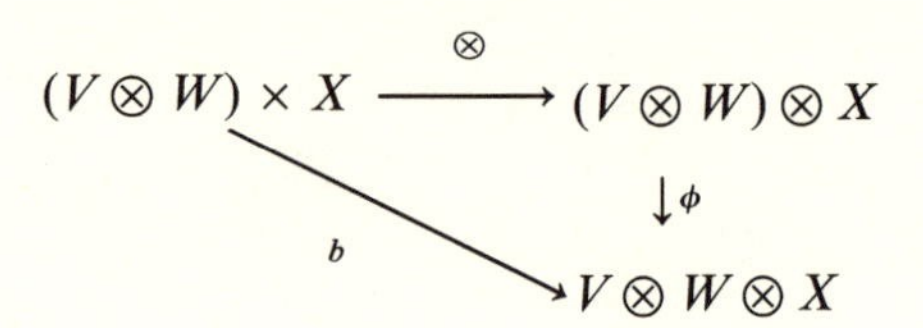

From the trilinear map $t: V \times W \times X \to (V \otimes W) \otimes X$ given by $(v, w, x) \mapsto (v \otimes w) \otimes x$ we have the liner map $\psi: v \otimes w \otimes x \mapsto (v \otimes w) \otimes x$

$$\begin{array}{ccc} V \times W \times X & \xrightarrow{\otimes} & V \otimes W \otimes X \\ & {\scriptstyle t} \searrow & \downarrow \psi \\ & & (V \otimes W) \otimes X \end{array}$$

But $\phi \circ \psi$ and $\psi \circ \phi$ are both identities so ϕ is an isomorphism. A similar argument shows that $V \otimes (W \otimes X)$ and $V \otimes W \otimes X$ are isomorphic. ∎

Theorem 3.3. $V \otimes W$ *and* $W \otimes V$ *are naturally isomorphic.*

Proof. Problem 3.7.

We saw that $\otimes$ is bilinear, and, in particular, distributive with respect to $+$. We might be tempted to infer from Theorems 3.2 and 3.3 that $\otimes$ is also associative and commutative. However, in general, these properties are not even defined for $\otimes$ and, when they are, they are not generally true. (See Problem 3.8.) When we define a closely related mapping in Section 4.3 this situation will be partially rectified. (See Theorem 4.6.)

Theorem 3.4. *$(V \otimes W)^*$ is naturally isomorphic to $V^* \otimes W^*$.*

Proof. The result follows immediately from the second corollary of Theorem 3.1 for the case $Z = \mathbb{R}$. ∎

Corollary. *$(V^* \otimes W^*)^*$ is naturally isomorphic to $V \otimes W$.*

Theorem 3.4 implies that $V \otimes W$ and $V^* \otimes W^*$ are dual spaces with respect to the natural pairing. It is an example of "duality" in tensor product spaces. (See Section 4.1.) It is important to note that Theorem 3.4 is not necessarily valid if V and W are not both finite-dimensional.

We have seen that when V and W are finite-dimensional, the pair $(V \otimes W, \otimes)$ has the properties listed in Theorem 3.1, which with its corollaries, in turn, lead to the isomorphisms described above. Now suppose we abstract this situation a bit; suppose V, W, and X are any vector spaces, and $\mathscr{T}$ is any map $V \times W \to X$, and suppose the pair $(X, \mathscr{T})$ satisfies the properties listed in Theorem 3.1. It turns out that $(X, \mathscr{T})$ is really no more general than $(V \otimes W, \otimes)$; that is, the properties of Theorem 3.1 essentially uniquely determine the pair $(X, \mathscr{T})$.

Theorem 3.5. *If $(X, \mathscr{T})$ has the properties of Theorem 3.1, then $X \cong V \otimes W$ and $\mathscr{T} = \phi \circ \otimes$ where ϕ is the isomorphism $V \otimes W \to X$.*

Proof. Letting $\mathscr{T}$ have the role of b in Theorem 3.1 we have $\mathscr{T} = \phi \circ \otimes$ where ϕ: $V \otimes W \to X$. Then letting $\otimes$ have the role of b in Theorem 3.1 we have $\otimes = \psi \circ \mathscr{T}$ where ψ: $X \to V \otimes W$. Combining these two equalities we get $\mathscr{T} = \phi \circ \psi \circ \mathscr{T}$ and $\otimes = \psi \circ \phi \circ \otimes$. Since $\mathscr{T}(V \times W)$ spans X, $\phi \circ \psi$ is the identity on X, so ϕ is onto. Since $\otimes(V \times W)$ spans $V \otimes W$, $\psi \circ \phi$ is the identity on $V \otimes W$ so ϕ is 1–1. Hence ϕ is an isomorphism. ∎

On the basis of Theorem 3.5 we can now say that given two finite-dimensional vector spaces, there is one and only one tensor product space with the properties of Theorem 3.1, and that tensor product space is the space $V \otimes W$ defined in Section 3.1. The major significance, however, of the characterization of Theorem 3.1 is that it can be extended to any two vector spaces, not necessarily finite-dimensional. The uniqueness requirement is shown exactly as in the proof of Theorem 3.5, as in that proof the nature of $V \otimes W$ is irrelevant. The existence requirement will be satisfied by construction, in the next section. Clearly, the constructed space will have to reduce to $V \otimes W = \mathscr{L}(V^*, W^*; \mathbb{R})$ in the finite-dimensional case.

PROBLEM 3.6. The proof of Theorem 3.2 requires only that the tensor products satisfy Theorem 3.1 and its generalizations. Hence it is valid for arbitrary vector spaces (not necessarily finite-dimensional). Give a simpler proof of Theorem 3.2 by using Theorem 3.4. (Note, however, that Theorem 3.4 is only valid in the finite-dimensional case.)

PROBLEM 3.7. Prove Theorem 3.3 for arbitrary vector spaces (not necessarily finite-dimensional).

PROBLEM 3.8. If V and W are the same, then $V \otimes W$ and $W \otimes V$ are the same space, but in general $v \otimes w \neq w \otimes v$.

PROBLEM 3.9. Generalize Theorem 3.4 to $(V_1 \otimes V_2 \otimes \cdots \otimes V_p)^* \cong V_1^* \otimes V_2^* \otimes \cdots \otimes V_p^*$.

3.3 Tensor products of infinite-dimensional vector spaces

In Section 3.1 we defined tensor products of finite-dimensional vector spaces. In the general case where V and/or W may not be finite-dimensional we still have the maps $\otimes\colon V^* \times W^* \to \mathscr{L}(V, W; \mathbb{R})$ given by eq. (3.2) and $\otimes\colon V \times W \to \mathscr{L}(V^*, W^*; \mathbb{R})$ in Theorem 3.1. However, now the sets $\otimes(V^* \times W^*)$ and $\otimes(V \times W)$ do not necessarily span the spaces $\mathscr{L}(V, W; \mathbb{R})$ and $\mathscr{L}(V^*, W^*; \mathbb{R})$. These latter spaces may be too large.

It turns out that there are a space and a mapping which satisfy Theorem 3.1. As we have already pointed out, this can be the only such pair. We will denote it by $(V \otimes W, \otimes)$. That is, the tensor product of V and W, $(V \otimes W, \otimes)$, will be defined by the properties of Theorem 3.1.

Given V and W to get a space $V \otimes W$, and a mapping, $\otimes$ which satisfy Theorem 3.1, we start out by considering the vector space, $\mathbb{R}_0^{V \times W}$, of all real-valued functions with domain $V \times W$ which vanish at all but a finite number of points. As we saw in Section 1.1, there is a 1–1 correspondence between elements $(v, w) \in V \times W$ and functions $f_{(v,w)} \in \mathbb{R}_0^{V \times W}$ which are 1 on (v, w) and 0 otherwise, and $\{f_{(v,w)}\colon (v, w) \in V \times W\}$ is a basis of $\mathbb{R}_0^{V \times W}$. We can think of $V \times W \subset \mathbb{R}_0^{V \times W}$, and we will use the notation (v, w) for $f_{(v,w)}$.

Now consider the set of all elements of $\mathbb{R}_0^{V \times W}$ of the form

$$(a^1 v_1 + a^2 v_2, b^1 w_1 + b^2 w_2) - a^1 b^1 (v_1, w_1) - a^1 b^2 (v_1, w_2) - a^2 b^1 (v_2, w_1) - a^2 b^2 (v_2, w_2)$$

with $v_1, v_2 \in V$, $w_1, w_2 \in W$ and $a^1, a^2, b^1, b^2 \in \mathbb{R}$. They generate a subspace, N. Then we put

$$V \otimes W = \mathbb{R}_0^{V \times W} / N$$

That is, $V \otimes W$ is a set of equivalence classes of functions from $V \times W$ to $\mathbb{R}$. The map

$$\otimes = \Pi|_{V \times W}$$

is the restriction to $V \times W$ of the projection $\mathbb{R}_0^{V \times W} \to \mathbb{R}_0^{V \times W}/N$.

Theorem 3.6. *The pair* $(V \otimes W, \otimes)$ *defined above has the properties of Theorem 3.1.*

Proof. (i) Since Π is linear and maps all of N to 0,

$$\begin{aligned}\Pi(a^1 v_1 + a^2 v_2, b^1 w_1 + b^2 w_2) = a^1 b^1 \Pi(v_1, w_1) + a^1 b^2 \Pi(v_1, w_2) \\ + a^2 b^1 \Pi(v_2, w_1) + a^2 b_2 \Pi(v_2, w_2)\end{aligned}$$

so $\otimes(a^1 v_1 + a^2 v_2, b^1 w_1 + b^2 w_2) = a^1 b^1 \otimes (v_1, w_2) + \cdots$ for all $v_1, v_2 \in V$, $w_1, w_2 \in W$, and $a^1, a^2, b^1, b^2 \in \mathbb{R}$. That is, $\otimes$ is bilinear. Also, since Π is linear and $V \times W$ spans $\mathbb{R}_0^{V \times W}$, $\otimes(V \times W) = \Pi(V \times W)$ spans $V \otimes W$.

(ii) Now suppose we have $b\colon V \times W \to Z$. We define a linear map $\Phi\colon \mathbb{R}_0^{V \times W} \to Z$ by $a^{ij}(v_i, w_j) \mapsto a^{ij} b(v_i, w_j)$. Then we define a map $\phi\colon V \otimes W \to Z$ by $\phi\colon V \otimes W \to Z$ by $\phi\colon [f] \mapsto \Phi(f)$ where $f \in \mathbb{R}_0^{V \times W}$ and $[f]$ is the equivalence class of f. It remains only to check that ϕ is well-defined, and that ϕ satisfies the four properties of part (ii) of Theorem 3.1; namely, uniqueness, linearity, surjectivity, and factoring of b. We leave these details for the problems at the end of this section. ■

The first corollary of Theorem 3.1 is not meaningful in the infinite-dimensional case. The second corollary is valid in general. Theorem 3.2 and Theorem 3.3—"associativity" and "commutativity" are valid in general.

Theorem 3.4 is not valid in general. From the second corollary of Theorem 3.1 with $Z = \mathbb{R}$ we get $(V \otimes W)^* \cong \mathscr{L}(V, W; \mathbb{R})$, and, as we noted above, $\mathscr{L}(V, W; \mathbb{R})$ can be larger than $V^* \otimes W^*$ in the infinite-dimensional case, so $(V \otimes W)^*$ can be larger than $V^* \otimes W^*$.

We can generalize Theorem 3.6 as follows.

Theorem 3.7. *Given n vector spaces $V_1, \ldots, V_n$, there exists an essentially unique vector space $V_1 \otimes \cdots \otimes V_n$ and mapping*

$$\otimes\colon V_1 \times \cdots \times V_n \to V_1 \otimes \cdots \otimes V_n$$

with the properties

(i) $\otimes$ *is n-linear, and $\otimes(V_1 \times \cdots \times V_n)$ spans $V_1 \otimes \cdots \otimes V_n$.*
(ii) *If Z is any vector space, and Ψ is any n-linear mapping,*

$\Psi: V_1 \times \cdots \times V_n \to Z$, then there exists a unique linear map $\phi: V_1 \otimes \cdots \otimes V_n \to Z$, onto $\langle \Psi(V_1 \times \cdots \times V_n) \rangle$, and such that $\Psi = \phi \circ \otimes$.

Proof. Problem 3.14.

Thus far, as we have gone along, we have tried to focus some attention on infinite-dimensional vector spaces as well as finite-dimensional ones and point out some similarities and some differences. We did this because, while "classical" tensor analysis is restricted to finite dimensions, the infinite-dimensional spaces are becoming important in applications; for example, applications in which it is required to do tensor analysis on Banach or Frechet manifolds. (See, for example, Marsden, 1981.) However, *in the following chapters we will assume, unless otherwise explicitly stated, that all our vector spaces are finite-dimensional.*

PROBLEM 3.10. Prove that the map ϕ in Theorem 3.6 is well-defined.

PROBLEM 3.11. Prove that the map ϕ in Theorem 3.6 has the four required properties.

PROBLEM 3.12. Denote the image of (v, w) under $\otimes$ by $v \otimes w$, With each element $v \otimes w$ of $V \otimes W$ we can associate a bilinear function $b_{v \otimes w}$ given by $(\sigma, \tau) \mapsto (\sigma \cdot v)(\tau \cdot w)$. (i) Prove that this assignment defines an isomorphism of $V \otimes W$ with a subspace, Z, of $\mathscr{L}(V^*, W^*; \mathbb{R})$. (ii) Prove that if V and W are finite-dimensional, then $Z = \mathscr{L}(V^*, W^*; \mathbb{R})$.

PROBLEM 3.13. We can get a direct proof of $V^* \otimes W^* \subset (V \otimes W)^*$ as a nice application of Theorem 3.1. (Hint: Apply Theorem 3.1 for $V \otimes W \otimes V^* \otimes W^*$, choose b (not bilinear, but 4-linear) and show that for each $A \in V^* \otimes W^*$ there is a linear function $V \otimes W \to \mathbb{R}$).

PROBLEM 3.14. Prove Theorem 3.7.

4

TENSORS

We restrict our attention to tensor products of a single vector space, V, with itself and with products of V^*. We look, in these special cases, at the isomorphisms, and representations in terms of components we discussed in Chapters 2 and 3. Finally, we will describe several important mappings of these spaces.

4.1 Definitions and alternative interpretations

We can form a variety of tensor products all based on a single given space, V. Thus, we can form

$$V_0^r = \underbrace{V \otimes \cdots \otimes V}_{r \text{ factors}} \qquad \text{the contravariant tensor product spaces}$$

$$V_s^0 = \underbrace{V^* \otimes \cdots \otimes V^*}_{s \text{ factors}} \qquad \text{the covariant tensor product spaces}$$

and

$$V_s^r = \underbrace{V \otimes \cdots \otimes V}_{r \text{ factors}} \otimes \underbrace{V^* \otimes \cdots \otimes V^*}_{s \text{ factors}} \qquad \text{the mixed tensor product spaces,}$$

Recall that when V is finite-dimensional,

$$\underbrace{V \otimes \cdots \otimes V}_{r} \otimes \underbrace{V^* \otimes \cdots \otimes V^*}_{s} \cong \mathscr{L}(\underbrace{V^*, \ldots, V^*}_{r}, \underbrace{V, \ldots, V}_{s}; \mathbb{R})$$

We will restrict ourselves to this case in the sequel (unless otherwise stated).

We could also form products containing both V's and V^*'s with V's and V^*'s in different orders from those in V_s^r. However, Theorem 3.2 and Theorem 3.3 give us some justification for ignoring these. In particular, we have $V_s^r \otimes V_q^p \cong V_{s+q}^{r+p}$. So we will now restrict ourselves to the tensor product spaces V_s^r, including the special cases V_0^r, V_s^0 and the space $V_0^0 \equiv \mathbb{R}$. Note that dim $V_s^r = n^{r+s}$ if dim $V = n$.

Definitions
A tensor of type (r, s) is an element of V^r_s. It is *contravariant of degree* (*order*, or *rank*) r, and *covariant of degree* (*order*, or *rank*) s. If $r = 1$ and $s = 0$, an element of V^1_0 $(=\mathscr{L}(V^*, \mathbb{R}) = V)$ is a *vector*. If $r = 0$ and $s = 1$ an element of V^0_1 $(=\mathscr{L}(V, \mathbb{R}) = V^*)$ is a *covector* or *1-form* (or, *linear form*, *function*, or *functional*). An element of V^r_s of the form $v_1 \otimes v_2 \otimes \cdots \otimes v_r \otimes \sigma^1 \otimes \cdots \otimes \sigma^s$ is called *decomposable*.

In addition to restricting the possible types of tensor product spaces based on a single vector space, V, to the spaces V^r_s, the natural isomorphisms, Theorems 3.2–3.4, of the last chapter give us a certain "duality" in the set of spaces V^r_s when V is finite-dimensional. That is, in addition to the duality of V and V^*, and, from Theorem 3.4 and its Corollary, the duality of V^2_0 and V^0_2, we have

$$(\underbrace{V \otimes \cdots \otimes V}_{r} \otimes \underbrace{V^* \otimes \cdots \otimes V^*}_{s})^* \cong \underbrace{V^* \otimes \cdots \otimes V^*}_{r} \otimes \underbrace{V \otimes \cdots \otimes V}_{s}$$
$$\cong \underbrace{V \otimes \cdots \otimes V}_{s} \otimes \underbrace{V^* \otimes \cdots \otimes V^*}_{r}$$

so $(V^r_s)^* \cong V^s_r$, and V^r_s and V^s_r are dual spaces with respect to the natural pairing. That is, for $A \in V^r_s$ and $B^* \in V^s_r = (V^r_s)^*$ we have the natural pairing $(A, B^*) \mapsto \langle B^*, A \rangle$.

Recall that, in addition to the various natural isomorphisms between tensor product spaces which we observed in the last chapter, we had, in Section 2.2, Theorem 2.11 and 2.12, and Problem 2.14, natural isomorphisms between certain spaces of multilinear mappings. These latter give us alternative interpretations for the spaces V^r_s which are especially important for small values of r and s. We have the following explicit results for V^1_1, V^2_0, and V^0_2.

Theorem 4.1.

(i) $V^2_0 \cong \mathscr{L}(V^*, V)$
(ii) $V^1_1 \cong \mathscr{L}(V, V) \cong \mathscr{L}(V^*, V^*)$
(iii) $V^0_2 \cong \mathscr{L}(V, V^*)$

Proof. (i) In Theorem 2.11 put $V_1 = V_2 = V$ and $W = \mathbb{R}$ and by the corollary of Theorem 2.7 we can put $\mathscr{L}(V^*, \mathbb{R}) = V$. (ii) and (iii): Problem 4.1. ■

Corollary. *A tensor of type* (2, 0), (1, 1) *or* (0, 2) *is nondegenerate if and only if the corresponding linear mappings are isomorphisms* (cf., Corollary of Theorem 2.14 and Theorem 2.15).

For larger values of r and s the isomorphisms, and hence the alternative interpretations, proliferate rapidly. As one illustration, consider the space $V_2^1 = V \otimes V^* \otimes V^*$. We have the following examples.

(i) $V \otimes V^* \otimes V^* \cong \mathscr{L}(V, V \otimes V^*)$; that is, according to Theorem 2.12, $\mathscr{L}(V^*, V, V; \mathbb{R}) \cong \mathscr{L}(V, \mathscr{L}(V^*, V; \mathbb{R}))$. If $A \in V \otimes V^* \otimes V^*$, then the isomorphism, ϕ, is given by $\phi\colon A \mapsto \phi(A) \in \mathscr{L}(V, \mathscr{L}(V^*, V; \mathbb{R}))$ where $\phi(A)\cdot v \in \mathscr{L}(V^*, V; \mathbb{R})$ is given by

$$(\phi(A)\cdot v)(\sigma, w) = A(\sigma, v, w) \tag{4.1}$$

(ii) $V \otimes V^* \otimes V^* \cong \mathscr{L}(V, V; V)$; that is according to Problem 2.14 and Theorem 2.12, $\mathscr{L}(V^*, V, V; \mathbb{R}) \cong \mathscr{L}(V, V; \mathscr{L}(V^*, \mathbb{R}))$. If $A \in V \otimes V^* \otimes V^*$, then the isomorphism, ϕ, is given by $\phi\colon A \mapsto \phi(A) \in \mathscr{L}(V, V; \mathscr{L}(V^*, \mathbb{R}))$ where $\phi(A)(v, w) \in \mathscr{L}(V^*, \mathbb{R})$ is given by

$$(\phi(A)(v, w))\cdot \sigma = A(\sigma, v, w) \tag{4.2}$$

In example (i) we have "(1, 1) tensor-valued 1-forms" (Section 16.1) and example (ii) includes "vector-valued 2-forms" (Section 16.3).

PROBLEM 4.1. Prove parts (ii) and (iii) of Theorem 4.1. Exhibit the isomorphisms.

PROBLEM 4.2. Describe the linear transformations corresponding to $\langle -, - \rangle$ ($\langle -, - \rangle \in V_1^1$).

PROBLEM 4.3. Show that the elements of $\mathscr{L}(V, V)$ and $\mathscr{L}(V^*, V^*)$ corresponding to a given tensor of type (1, 1) are dual linear transformations.

PROBLEM 4.4. List all the remaining possible interpretations of $V \otimes V^* \otimes V^*$ and describe the corresponding isomorphisms.

4.2 The components of tensors

In Section 3.2 we saw that if $\{e_i^1\}, \ldots, \{e_i^n\}$ are bases of $V_1, \ldots, V_n$, then $\{e_{i_1}^1 \otimes \cdots \otimes e_{i_n}^n\}$ is a basis of the tensor product space $V_1 \otimes \cdots \otimes V_n$, so that an element of $V_1 \otimes \cdots \otimes V_n$ can be "expanded" in terms of its *components* $A^{i_1 \cdots i_n}$, in this basis, as in (3.9).

As a special case of (3.9), any tensor $A \in V_s^r$ can be written

$$A = A^{i_1 \cdots i_r}_{j_1 \cdots j_s} e_{i_1} \otimes \cdots \otimes e_{i_r} \otimes \varepsilon^{j_1} \otimes \cdots \otimes \varepsilon^{j_s} \tag{4.3}$$

where $\{e_i\}$ is a basis of V and $\{\varepsilon^k\}$ is its dual basis. As a special case of (3.10), the values of A are

$$A(\sigma^1, \ldots, \sigma^r, v_1, \ldots, v_s) = A^{i_1 \cdots i_r}_{j_1 \cdots j_s} \sigma^1_{i_1} \cdots \sigma^r_{i_r} v^{j_1}_1 \cdots v^{j_s}_s \tag{4.4}$$

It is sometimes useful to describe the isomorphisms described in Section 4.1 in terms of the components of the tensors when a basis is chosen. Thus, for example, in the given illustration for tensors of type $(1, 2)$ we have for interpretation (i), $A = A^i_{jk} e_i \otimes \varepsilon^j \otimes \varepsilon^k$ and $(\phi \cdot A) \cdot e_m = \tilde{A}^i_{mk} e_i \otimes \varepsilon^k$ where $\{\varepsilon^i\}$ is the dual basis in V^*. If we evaluate A at $(\varepsilon^l, e_m, e_n)$ and evaluate $(\phi \cdot A) \cdot e_m$ at (ε^l, e_n) and then, according to (4.1), equate the results, we get $A^l_{mn} = \tilde{A}^l_{mn}$. That is, the components of A are the elements of the matrix of the linear transformation $\mathscr{L}(V, V \otimes V^*)$. Finally, if $e^{lm}_n \colon e_p \mapsto \delta^l_p e_n \otimes \varepsilon^m$ is the basis of $\mathscr{L}(V, V \otimes V^*)$ given by Theorem 2.3 we can write

$$\phi \cdot e_i \otimes \varepsilon^j \otimes \varepsilon^k = \phi^{njk}_{ilm} e^{lm}_n$$

and evaluating both sides on e_p we get

$$(\phi \cdot e_i \otimes \varepsilon^j \otimes \varepsilon^k) \cdot e_p = \phi^{njk}_{ilm} \delta^l_p e_n \otimes \varepsilon^m$$

Evaluating both sides of this last equation, in turn, on (ε^s, e_t) we obtain by eq. (4.1) $e_i \otimes \varepsilon^j \otimes \varepsilon^k(\varepsilon^s, e_p, e_t) = \phi^{njk}_{ilm} \delta^l_p \delta^s_n \delta^m_t$ so $\phi^{sjk}_{ipt} = \delta^s_i \delta^j_p \delta^k_t$.

Once we choose a basis in V, and hence in V^r_s, we have exactly the same situation as in Section 1.2. That is, we have an isomorphism between V^r_s and the set of components $A^{i_1 \cdots i_r}_{j_1 \cdots j_s}$; $A^{i_1 \cdots i_r}_{j_1 \cdots j_s} + B^{i_1 \cdots i_r}_{j_1 \cdots j_s}$ are the components of $A + B$ and for $a \in \mathbb{R}$, $aA^{i_1 \cdots i_r}_{j_1 \cdots j_s}$ are the components of aA. See Problem 1.16. This justifies the approach of "classical" tensor analysis where it is customary to work with the components of tensors rather with the tensors themselves.

It is important to know how the components of a tensor transform when we change bases in V. As in eq. (1.2), $\bar{e}_i = a^j_i e_j$ and according to Problem 2.3, $\bar{\varepsilon}^i = b^i_j \varepsilon^j$ where $a^i_j b^j_k = \delta^i_k$. We write $A \in V^r_s$ in terms of each basis, just as we did for any vector, v, in Section 1.1;

$$A = {}^{i_1 \cdots i_r}_{j_1 \cdots j_s} e_{i_1} \otimes \cdots \otimes e_{i_r} \otimes \varepsilon^{j_1} \otimes \cdots \otimes \varepsilon^{j_s}$$

and

$$A = \bar{A}^{p_1 \cdots p_r}_{q_1 \cdots q_s} \bar{e}_{p_1} \otimes \cdots \otimes \bar{e}_{p_r} \otimes \bar{\varepsilon}^{q_1} \otimes \cdots \otimes \bar{\varepsilon}^{q_s}$$

and substitute into the first the expression for the e's and ε's in terms of the $\bar{e}$'s and $\bar{\varepsilon}$'s. Comparing the resulting expression for A with the second one above, we get our result:

$$\bar{A}^{p_1 \cdots p_r}_{q_1 \cdots q_s} = A^{i_1 \cdots i_r}_{j_1 \cdots j_s} a^{j_1}_{q_1} \cdots a^{j_s}_{q_s} b^{p_1}_{i_1} \cdots b^{p_r}_{i_r} \tag{4.5}$$

Compare this with eq. (1.4).

Theorem 4.2. *Two sets of numbers, $A^{i_1 \cdots i_r}_{j_1 \cdots j_s}$, and $\bar{A}^{i_1 \cdots i_r}_{j_1 \cdots j_s}$, are components using different bases $\{e_i\}$ and $\{\bar{e}_i\}$ of the same tensor if and only if they are related by (4.5) where (a^i_j) is a nonsingular matrix and $a^i_j b^j_k = \delta^i_k$.*

Proof. Problem 4.8.

Theorem 4.2 characterizes a tensor in terms of its components. In "classical" treatments this characterization is taken as the definition. In specific applications one almost always works with components. Then we must, however, be careful to be sure that our results do not depend on the particular choice of basis.

PROBLEM 4.5. Choose a basis for V, and write the specific form of (4.3) and (4.4) for the tensor $\langle -, - \rangle$ in V^1_1.

PROBLEM 4.6. Describe the nondegeneracy of tensors of type (1, 1), (2, 0), and (0, 2) in terms of their components.

PROBLEM 4.7. Interpret the components of $A \in V \otimes V^* \otimes V^*$ as certain coefficients of elements of $\mathscr{L}(V^*, V; V^*)$. (See example below eq. (4.4).)

PROBLEM 4.8. What are the matrices of the isomorphisms of Theorem 4.1(ii) and (iii)?

PROBLEM 4.9. Prove Theorem 4.2.

PROBLEM 4.10. Let $\{e_1, e_2, e_3\}$ be a basis for V, and let $A \in V^1_1$ be $A = A^i_j e_i \otimes \varepsilon^j$ where the components A^i_j arranged in a matrix, (A^i_j) with row index i, are

$$\begin{pmatrix} 3 & 0 & 1 \\ 0 & 3 & -1 \\ 1 & -1 & 0 \end{pmatrix}$$

Let $\bar{e}_1 = e_1 + e_2$, $\bar{e}_2 = 2e_2$, $\bar{e}_3 = -e_2 + e_3$ be a new basis, and let $v = -e_1 + 2e_3$ and $\tau = 5\varepsilon^1 - 2\varepsilon^2 + \varepsilon^3$. (i) Evaluate A at (τ, v) and evaluate the two corresponding linear maps at τ and v, respectively. (ii) Find $\bar{\varepsilon}^i$ in terms of ε^i. (iii) Find the expression for v and τ in the new basis. (iv) Find the components of A in the new basis. (v) Show $\det(A^i_j) = \det(\bar{A}^i_j)$ and $\operatorname{tr}(A^i_j) = \operatorname{tr}(\bar{A}^i_j)$. (vi) Do the computation of part (i) in the new basis.

PROBLEM 4.11. (i) Suppose $A \in V^1_1$ has components δ^i_j for some basis of V. Show that then A has components δ^i_j for every basis of V, and $A = \langle -, - \rangle$.

(ii) If A is a tensor of type (1, 1) and A has the same components in every basis, show that A is a multiple of $\langle -, - \rangle$; i.e., the components of A are $A^i_j = a\delta^i_j$ for some $a \in \mathbb{R}$.

(Problem 4.11 gives the existence and "uniqueness" of (1, 1) tensors having the same components in every basis.)

PROBLEM 4.12. If A is a tensor of type (r, s), and A has the same components in every basis, show that either $A = 0$, or $r = s$.

PROBLEM 4.13. If A is a tensor of type (2, 2), and A has the same components in every basis, show that

$$A^{ij}_{pq} = a(\delta^i_p\delta^j_q + \delta^i_q\delta^j_p) + b(\delta^i_p\delta^j_q - \delta^i_q\delta^j_p) \tag{4.6}$$

See Problem 5.14. (There is the related topic of *isotropic tensors*, important in continuum mechanics, cf., Aris, 1962, p. 30, and in relativity, Section 24.2. There, only orthonormal bases are used.)

PROBLEM 4.14. If the following given set of numbers is the set of components of A for some basis of V, find the set of components of A for another basis of V.

(i) $\delta_{ij} = 1$ when $i = j$
$= 0$ otherwise

(ii) $\delta^{i_1 \cdots i_r}_{j_1 \cdots j_r} = 1$ when $j_1 \cdots j_r$ are distinct integers between 1 and $p \geq r$, and $i_1 \cdots i_r$ is an even permutation of $j_1 \cdots j_r$
$= -1$ when $j_1 \cdots j_r$ are distinct and $i_1 \cdots i_r$ is an odd permutation of $j_1 \cdots j_r$
$= 0$ otherwise

(Hint: $\delta^{i_1 \cdots i_r}_{j_1 \cdots j_r} = \sum_\pi \operatorname{sgn} \pi \delta^{i_{\pi(1)}}_{j_1} \cdots \delta^{i_{\pi(r)}}_{j_r}$ where π is a permutation of the integers $1, \ldots, r$.)

PROBLEM 4.15. Let $\varepsilon^{i_1 \cdots i_r} = \delta^{i_1 \cdots i_r}_{1 \cdots r}$, which is defined as in Problem 4.14(ii) except that $i_1, \ldots, i_r$ can only take on the values $1, \ldots, r$. Show that $\bar{\varepsilon}^{j_1 \cdots j_r} = \det(a^i_j)\varepsilon^{i_1 \cdots i_r} b^{j_1}_{i_1} \cdots b^{j_r}_{i_r}$ have the same values as the $\varepsilon^{i_1 \cdots i_r}$ ((a^i_j) is the change of basis matrix and $a^i_j b^j_k = \delta^i_k$.) Two sets of numbers related this way are the components in their respective bases of *an* rth *order tensor density of weight 1.* $\varepsilon^{i_1\cdots i_r}$ (or $\varepsilon_{j_1\cdots j_s}$ defined the same way) is called *a permutation symbol*. (For a general discussion of relative tensors see Synge and Schild (1966). This concept is useful in integration theory in Chapter 12.)

4.3 Mappings of the spaces V^r_s

There are several types of mappings that can be defined for the tensor product spaces V^r_s. Some of the following ones can be generalized, but they are basic and most important in subsequent developments.

1. We saw in Section 2.4 that a given linear mapping $\Phi: V \to W$ "induces" another linear mapping, $\Phi^*: W^* \to V^*$, the dual, or transpose of Φ. Now that we have a whole hierarchy of spaces, V^r_s, built on V, we can generalize this result.

Definitions
If $\phi: V \to W$ is a linear mapping, then (i) *the* rth *power of* ϕ is the linear mapping $\phi_r: V^r_0 \to W^r_0$ with the property

$$\phi_r(v_1 \otimes \cdots \otimes v_r) = \phi \cdot v_1 \otimes \cdots \otimes \phi \cdot v_r \tag{4.7}$$

and (ii) *the* sth *power of* ϕ^* is the linear mapping ϕ^s. $W^0_s \to V^0_s$ with the property

$$\phi^s(\tau^1 \otimes \cdots \otimes \tau^s) = \phi^* \cdot \tau^1 \otimes \cdots \otimes \phi^* \cdot \tau^s \tag{4.8}$$

The existence of unique maps with the given properties comes from the generalization of Theorem 3.1.

We also have the following alternative description of these induced mappings.

Theorem 4.3. (i) $\phi_r: V^r_0 \to W^r_0$ *has the property described by eq.* (4.7) *if and only if for* $A \in V^r_0$, *the values of* $\phi_r \cdot A$ *are given by*

$$(\phi_r \cdot A)(\sigma^1, \ldots, \sigma^r) = A(\phi^* \cdot \sigma^1, \ldots, \phi^* \cdot \sigma^r) \tag{4.9}$$

where σ^i *are in* W^*.

(ii) $\phi^s: W^0_s \to V^0_s$ *has the property described by eq.* (4.8) *if and only if for* $A \in W^0_s$, *the values of* $\phi^s \cdot A$ *are given by*

$$(\phi^s \cdot A)(v_1, \ldots, v_s) = A(\phi \cdot v_1, \ldots, \phi \cdot v_s) \tag{4.10}$$

where v_i *are in* V.

Proof. (i) (4.9) defines $\phi_r \cdot A$ as a multilinear function on $W^* \times \cdots \times W^*$ since ϕ^* is linear and A is multilinear; that is $\phi_r \cdot A \in W^r_0$. Now, if $\phi_r \cdot A$ satisfies (4.9) for all A $\in V^r_0$ then, in particular,

$$\begin{aligned}
&(\phi_r \cdot (v_1 \otimes \cdots \otimes v_r))(\sigma^1, \ldots, \sigma^r) \\
&\quad = (v_1 \otimes \cdots \otimes v_r)(\phi^* \cdot \sigma^1, \ldots, \phi^* \cdot \sigma^r) = \langle \phi^* \cdot \sigma^1, v_1 \rangle \cdots \langle \phi^*, \sigma^r, v_r \rangle \\
&\quad = \{\sigma^1, \phi \cdot v_1 \rangle \cdots \langle \sigma^r, \phi \cdot v_r \rangle \qquad \text{by Theorem 2.21} \\
&\quad = (\phi \cdot v_1 \otimes \cdots \otimes \phi \cdot v_r)(\sigma^1, \ldots, \sigma^r)
\end{aligned}$$

So $\phi_r \cdot (v_1 \otimes \cdots \otimes v_r) = \phi \cdot v_1 \otimes \cdots \otimes \phi \cdot v_r$. Thus, if ϕ_r satisfies (4.9) it satisfies (4.7).

For the converse, let $A = a^i(v_{i1} \otimes \cdots \otimes v_{ir})$. Then

$$\phi_r \cdot A = a^i \phi_r \cdot (v_{i1} \otimes \cdots \otimes v_{ir}) = a^i(\phi \cdot v_{i1} \otimes \cdots \otimes \phi \cdot v_{ir})$$

by (4.7). Then

$$\begin{aligned}
\phi_r \cdot A(\sigma^1, \ldots, \sigma^r) &= a^i \langle \sigma^1, \phi \cdot v_{i1} \rangle \cdots \langle \sigma^r, \phi \cdot v_{ir} \rangle \\
&= a^i \langle \phi^* \cdot \sigma^1, v_{i1} \rangle \cdots \langle \phi^r \cdot \sigma^r, v_{ir} \rangle = A(\phi^* \cdot \sigma^1, \ldots, \phi^* \cdot \sigma^r).
\end{aligned}$$

Thus, if ϕ_r satisfies (4.7) it satisfies (4.9).

(ii) *Problem 4.16.* ∎

We can define another generalization of the case $r = 2$.

Definition

If ϕ: $V \to W$ and ψ: $X \to Y$ are linear, then *the tensor product of* ϕ *and* ψ is the linear mapping $\phi \otimes \psi$: $V \otimes X \to W \otimes Y$ with the property

$$\phi \otimes \psi(v \otimes x) = \phi \cdot v \otimes \psi \cdot x \tag{4.11}$$

In the special case $W = Y = \mathbb{R}$ we have a problem with notation, since in this case $\phi \otimes \psi$ has already been defined. In Section 3.1 $\phi \otimes \psi$ was defined as an element of $V^* \otimes X^*$. Here it is in $(V \otimes X)^*$. However, these two functions, with different domains, are corresponding elements in the isomorphism of Theorem 3.4 and have the same values, so we keep the same notation for both, relying on the context to make things clear. Thus, we have

$$\phi \otimes \psi(v, x) = \langle \phi \otimes \psi, v \otimes x \rangle \tag{4.12}$$

where on the left side $\phi \otimes \psi$ stands for a bilinear function belonging to

$V^* \otimes X^*$ and on the right side $\phi \otimes \psi$ is the corresponding linear function belonging to $(V \otimes X)^*$.

More generally, we define a linear mapping

$$\phi^1 \otimes \cdots \otimes \phi^p\colon V_1 \otimes \cdots \otimes V_p \to W_1 \otimes \cdots \otimes W_p$$

by

$$\phi^1 \otimes \cdots \otimes \phi^p(v_1 \otimes \cdots \otimes v_p) = \phi^1 \cdot v_1 \otimes \cdots \otimes \phi^p \cdot v_p \tag{4.13}$$

where $\phi^i\colon V_i \to W_i$, and in the special case $W_1 = \cdots W_p = \mathbb{R}$ we have

$$\phi^1 \otimes \cdots \otimes \phi^p(v_1, \ldots, v_p) = \langle \phi^1 \otimes \cdots \otimes \phi^p, v_1 \otimes \cdots \otimes v_p \rangle \tag{4.14}$$

where on the left side $\phi^1 \otimes \cdots \otimes \phi^p$ stands for a multilinear function belonging to $V_1^* \otimes \cdots \otimes V_p^*$ and on the right side $\phi^1 \otimes \cdots \otimes \phi^p$ is the corresponding linear function belonging to $(V_1 \otimes \cdots \otimes V_p)^*$. Note that this result, from a generalization of Theorem 3.4, gives an explicit description of the pairings which, in the special cases discussed in Section 4.1, gives the dualities described there.

2. There are two other important general types of mappings; a mapping $V_s^r \times V_q^p \to V_{s+q}^{r+p}$ and a mapping $V_q^p \to V_{q-1}^{p-1}$. These are frequently combined to yield other important bilinear mappings.

Theorem 4.4. *There exists one and only one bilinear map* $V_0^p \times V_0^q \to V_0^{p+q}$ *such that for* $v_1 \otimes \cdots \otimes v_p \in V_0^p$ *and* $w_1 \otimes \cdots \otimes w_q \in V_0^q$,

$$(v_1 \otimes \cdots \otimes v_p, w_1 \otimes \cdots \otimes w_q) \mapsto v_1 \otimes \cdots \otimes v_p \otimes w_1 \otimes \cdots \otimes w_q.$$

Proof. Let j be the natural isomorphism $j\colon V_0^p \otimes V_0^q \to V_0^{p+q}$ (see Theorem 3.2). In particular,

$$j\colon (v_1 \otimes \cdots \otimes v_p) \otimes (w_1 \otimes \cdots \otimes w_q) \mapsto v_1 \otimes \cdots \otimes w_q.$$

So $j \circ \otimes$ is a bilinear map with the required property. By the second corollary of Theorem 3.1 there is only one such map. ∎

Definition

$AB = j \circ \otimes(A, B) \in V_0^{p+q}$ is *the* (*outer*) *product* of the tensors $A \in V_0^p$ and $B \in V_0^q$ and the map $j \circ \otimes$ is called *tensor multiplication.*

Theorem 4.5. *If* $A \in V_0^p$, *and* $B \in V_0^q$ *then* (*i*) *The values of AB are*

$$AB(\sigma^1, \ldots, \sigma^{p+q}) = A(\sigma^1, \ldots, \sigma^p)B(\sigma^{p+1}, \ldots, \sigma^{p+q}).$$

(ii) *The components of AB (relative to a basis of V_0^{p+q}) are $A^{i_1 \cdots i_p}B^{j_1 \cdots j_q}$.*

Proof. (i) For $A = v_1 \otimes \cdots \otimes v_p$ and $B = w_1 \otimes \cdots \otimes w_q$,

$$\begin{aligned} AB(\sigma^1, \ldots, \sigma^{p+q}) & \\ &= j \circ \otimes((v_1 \otimes \cdots \otimes v_p)(w_1 \otimes \cdots \otimes w_q))(\sigma^1, \ldots, \sigma^{p+q}) \\ &= (v_1 \otimes \cdots \otimes w_q)(\sigma^1, \ldots, \sigma^{p+q}) \\ &= (v_1 \otimes \cdots \otimes v_p)(\sigma^1, \ldots, \sigma^p)(w_1 \otimes \cdots \otimes w_q)(\sigma^{p+1}, \ldots, \sigma^{p+q}) \\ &= A(\sigma^1, \ldots, \sigma^p)B(\sigma^{p+1}, \ldots, \sigma^{p+q}). \end{aligned}$$

For linear combinations of elements of this form, the result follows from the bilinearity of $j \circ \otimes$.

(ii) *Problem 4.19.* ■

Note that because of the natural isomorphism j: $V_0^p \otimes V_0^q \to V_0^{p+q}$, most writers do not bother to make a distinction between AB and $A \otimes B$. However, $A \otimes B$, a product of *vectors*, is in $V_0^p \otimes V_0^q$ and defined (in Chapter 3) on pairs $(\alpha, \beta) \in V_0^{p*} \times V_0^{q*}$, while AB, a product of *tensors*, is in V_0^{p+q} and defined on elements of $V^* \times \cdots \times V^*$. Thus, AB and $A \otimes B$ are different as mappings, since they have different domains, even though their values on "corresponding arguments" are the same—see Problem 4.21(ii).

The definition of the product of two tensors and Theorems 4.4 and 4.5 for V_0^p and V_0^q can be generalized to V_s^r and V_q^p with only slight complication of notation.

Theorem 4.6. *Multiplication of tensors is bilinear and associative, but not commutative.*

Proof. Problem 4.20.

Definition

The contraction, C_μ^λ, is the unique linear mapping $V_q^p \to V_{q-1}^{p-1}$ with the property

$$v_1 \otimes \cdots \otimes v_p \otimes \sigma^1 \otimes \cdots \otimes \sigma^q \mapsto \langle \sigma^\mu, v_\lambda \rangle v_1 \otimes \cdots \otimes \hat{v}_\lambda \otimes \cdots \otimes v_p \otimes \sigma^1 \otimes \cdots \otimes \hat{\sigma}^\mu \otimes \cdots \otimes \sigma^q \quad (4.15)$$

where the "hats" above v_λ and σ^μ on the right mean that those factors are to be omitted.

It is not immediately obvious that there does indeed exist a unique linear mapping with this property. This comes from the generalization of Theorem 3.1 for V_s^r when we choose V_{q-1}^{p-1} for the vector space Z and choose

$$(v_1, \ldots, v_p, \sigma^1, \ldots, \sigma^q) \mapsto \langle \sigma^\mu, v_\lambda \rangle v_1 \otimes \cdots \otimes \hat{v}_\lambda \otimes \cdots \otimes v_p \otimes \sigma^1 \otimes \cdots \otimes \hat{\sigma}^\mu \otimes \cdots \otimes \sigma^q \in V_{q-1}^{p-1}$$

for the multilinear map in part (ii) of that theorem.

As an example C_1^2: $V_1^2 \to V$ takes $v \otimes w \otimes \sigma \mapsto \langle \sigma, w \rangle v$, and, in particular, takes $e_i \otimes e_j \otimes \varepsilon^k \mapsto \langle \varepsilon^k, e_j \rangle e_i = \delta_j^k e_i$. Hence

$$A_k^{ij} e_i \otimes e_j \otimes \varepsilon^k \mapsto A_k^{ij} \delta_j^k e_i = A_k^{ik} e_i.$$

This example illustrates the general result that if A has components $A_{j_1 \cdots j_r}^{i_1 \cdots i_r}$, then $C_\mu^\lambda \cdot A$ has components $A_{j_1 \cdots j_{\mu-1} k \cdots j_s}^{i_1 \cdots i_{\lambda-1} k \cdots i_r}$ (sum on k). Clearly, if r and s are large enough, one can perform several successive contractions. In particular, if $A \in V_1^1$, then $C_1^1 \cdot A = \text{trace } A$. See Problem 4.10(v). From this result for the components of $C_\mu^\lambda \cdot A$ we get

$$C_\mu^\lambda \cdot A(\sigma^1, \ldots, \hat{\sigma}^\lambda, \ldots, \sigma^r, v_1, \ldots, \hat{v}_\mu, \ldots, v_s) = \sum_i A(\sigma^1, \ldots, \varepsilon^i, \ldots, \sigma^r, v_1, \ldots, e_i, \ldots, v_s) \quad (4.16)$$

Finally, in certain cases, given two tensors A and B we can form the product and then contract. Thus, for example if, in a basis, $A = A_k^{ij} e_i \otimes e_j \otimes \varepsilon^k \in V_1^2$, and $B = B_{lmn}^q e_q \otimes \varepsilon^l \otimes \varepsilon^m \otimes \varepsilon^n \in V_3^1$, then

$$C_3^2 \cdot AB = A_k^{ij} B_{ljn}^q e_i \otimes e_q \otimes \varepsilon^k \otimes \varepsilon^l \otimes \varepsilon^n \in V_3^2.$$

Important special cases of this operation will be described in Section 6.3.

In particular, the evaluation of a tensor, A, on the arguments $\sigma^1, \ldots, \sigma^r$, $v_1, \ldots, v_s$ can be performed by forming the product

$$A \otimes \sigma^1 \otimes \cdots \otimes \sigma^r \otimes v_1 \otimes \cdots \otimes v_s$$

and then contracting successively $r + s$ times.

In Theorem 4.2 we had a characterization of tensors in terms of components which is both of central importance in the classical theory and also very useful in doing actual calculations. Now, having introduced products and contractions we can describe other similar results known classically as *quotient rules.*

Theorem 4.7. *Two sets of numbers* $A^{i_1 \cdots i_r}_{j_1 \cdots j_s}$ *and* $\bar{A}^{i_1 \cdots i_r}_{j_1 \cdots j_s}$ *are components of the same tensor relative to bases* $\{e_i\}$ *and* $\{\bar{e}_i\}$, *respectively, if and only if for all tensors* $B \in V^p_q$, $A^{i_1 \cdots k \cdots i_r}_{j_1 \cdots j_s} B^{i_{r+1} \cdots i_{r+p}}_{j_{s+1} \cdots k \cdots j_{s+q}}$ *and* $\bar{A}^{i_1 \cdots k \cdots i_r}_{j_1 \cdots j_s} \bar{B}^{i_{r+1} \cdots i_{r+p}}_{j_{s+1} \cdots k \cdots j_{s+q}}$ *are components of the same tensor relative to these bases.*

Proof. **If**: To keep the notation under control, we consider a special case. Suppose $A^{ij}_k B_{il}$ are components of a tensor of type (1, 2) for all tensors B of type (0, 2). Then

$$A^{ij}_k B_{il} e_j \otimes \varepsilon^k \otimes \varepsilon^l = \bar{A}^{pm}_n \bar{B}_{pq} \bar{e}_m \otimes \bar{\varepsilon}^n \otimes \bar{\varepsilon}^q \tag{4.17}$$

We put $\bar{e}_m = a^j_m e_j$, $\bar{\varepsilon}^n = b^m_k \varepsilon^k$, and $\bar{\varepsilon}^q = b^q_l \varepsilon^l$ where $a^i_j b^j_k = \delta^i_k$, and, by Theorem 4.2, $\bar{B}_{pq} = B_{ir} a^i_p a^r_q$ into eq. (4.17). Then

$$A^{ij}_k B_{il} e_j \otimes e^k \otimes \varepsilon^l = \bar{A}^{pm}_n B_{ir} a^i_p a^r_q a^j_m e_j \otimes b^n_k \varepsilon^k \otimes b^q_l \varepsilon^l$$

or

$$A^{ij}_k B_{il} = \bar{A}^{pm}_n B_{ir} a^i_p a^r_q a^j_m b^n_k b^q_l \qquad \text{for all } j, k, l$$

Since $a^r_q \cdot b^q_l = \delta^r_l$ and $B_{ir} \delta^r_l = B_{il}$, and B_{il} can be chosen arbitrarily, we get

$$A^{ij}_k = \bar{A}^{pm}_n a^i_p a^j_m b^n_k$$

which is equivalent to the criterion eq. (4.5) of Theorem 4.2.

Only if: Apply Theorem 4.2 to A and B, multiply, and contract. ■

Note that instead of requiring all tensors $B \in V^p_q$ for the condition in Theorem 4.7 we need only a linearly independent set of pq elements of V^p_q. Also, it should be clear that any other contractions of AB or sequence of contractions could replace those indicated in Theorem 4.7.

PROBLEM 4.16. Prove part (ii) of Theorem 4.3.

PROBLEM 4.17. If $\phi: V \to W$ and $\psi: W \to X$, then $(\psi \circ \phi)_r = \psi_r \circ \phi_r$ and $(\psi \circ \phi)^s = \psi^s \circ \phi^s$.

PROBLEM 4.18. If $\phi: V \to W$ is an isomorphism, there are linear maps $\phi_s^{-1}: V^0_s \to W^0_s$ with $\sigma^1 \otimes \cdots \otimes \sigma^s \mapsto \phi^{-1*} \cdot \sigma^1 \otimes \cdots \otimes \phi^{-1*} \cdot \sigma^s$ and $\phi^{-1r}: W^r_0 \to V^r_0$ with $w_1 \otimes \cdots \otimes w_r \mapsto \phi^{-1} \cdot w_1 \otimes \cdots \otimes \phi^{-1} \cdot w_r$. Define maps $\phi_{r,s}: V^r_s \to W^r_s$ and $\phi^{r,s}: W^r_s \to V^r_s$ and describe all four of these in terms of the values of their images as in eqs. (4.9) and (4.10).

PROBLEM 4.19. Prove part (ii) of Theorem 4.5.

PROBLEM 4.20. Prove Theorem 4.6.

PROBLEM 4.21. (i) Generalize eq. (4.14) to

$$B^*(v_1, \ldots, v_p) = \langle B^*, v_1 \otimes \cdots \otimes v_p \rangle$$

where B^* on the left is in $V_1^* \otimes \cdots \otimes V_p^*$ and B^* on the right is in $(V_1 \otimes \cdots \otimes V_p)^*$.

(ii) Show that if A is in V_p^0 and, equivalently, in V_0^{p*}, and if B is in V_q^0 and equivalently in V_0^{q*}, then

$$A \otimes B(v_1 \otimes \cdots \otimes v_p, v_{p+1} \otimes \cdots \otimes \mathrm{v}_{p+q}) = AB(v_1, \ldots, v_{p+q})$$

PROBLEM 4.22. (i) Define linear maps $V_q^p \to V_{q-r}^{p-r}$ for $p, q \geq r$.

(ii) Define a nondegenerate bilinear function on $V_s^r \times V_r^s$ and thus conclude that V_s^r and V_r^s are dual spaces.

PROBLEM 4.23. For $A \in V_0^p$ and $B^* \in (V_0^p)^*$, $\langle B^*, A \rangle = C_p^p \cdots C_1^1 \cdot B^*A$.

5

SYMMETRIC AND SKEW-SYMMETRIC TENSORS

We will examine two classes of subspaces of the covariant tensor product spaces, V_s^0; the symmetric subspaces, $S^s(V^*)$, and the skew-symmetric (alternating) subspaces, $\Lambda^s(V^*)$. In particular, we will describe bases of these spaces. Finally, we will look at special properties of $S^2(V^*)$ and $\Lambda^2(V^*)$. We simply mention here that obvious corresponding results are valid for contravariant tensors.

5.1 Symmetry and skew-symmetry

Recall that in Problem 3.8 we noted that in general the two functions $v \otimes w$ and $w \otimes v$ in $V \otimes V$ are not the same. That is, $\langle \sigma, v \rangle \langle \tau, w \rangle \neq \langle \sigma, w \rangle \langle \tau, v \rangle$ for all σ and τ. Or, equivalently, $v \otimes w(\sigma, \tau) \neq v \otimes w(\tau, \sigma)$ for all σ, τ.

Definition
If the values of a tensor remain unchanged when two of its covariant arguments or two of its contravariant arguments are transposed, then the tensor is *symmetric in these two arguments.*

For example, if

$$A(\tau^1, \tau^2, \tau^3, \ldots, \tau^r, w_1, w_2, \ldots, w_s) = A(\tau^1, \tau^3, \tau^2, \ldots, \tau^r, w_1, w_2, \ldots, w_s)$$

for all values of all the τ's and w's then A is symmetric in its second and third covariant arguments. Notice that it makes no sense to transpose a covariant and contravariant argument.

Definition
If the values of a tensor change sign when two of its covariant arguments or two of its contravariant arguments are transposed, then the tensor is *skew-symmetric* (or *antisymmetric*, or *alternating*) in these two arguments.

Theorem 5.1. *A is skew-symmetric in two (covariant, or contravariant) arguments if and only if whenever these two arguments have the same value, the value of A is zero.*

Proof. We will do the proof for the first two covariant arguments of A to simplify the notation. (i) Suppose $A(\sigma^1, \sigma^2, \ldots) = -A(\sigma^2, \sigma^1, \ldots)$ for all σ^1 and σ^2. Then, when $\sigma^1 = \sigma^2$, $A(\sigma^2, \sigma^2, \ldots) = -A(\sigma^2, \sigma^2, \ldots)$ so that $A(\sigma^2, \sigma^2, \ldots) = 0$. (ii) If $A(\sigma, \sigma, \ldots) = 0$ for all σ, then $A(\sigma + \tau, \sigma + \tau, \ldots) = 0$ for all σ and τ. Expanding the left side we get $A(\sigma, \tau, \ldots) + A(\tau, \sigma, \ldots) = 0$ for all σ, τ and so $A(\sigma, \tau, \ldots) = -A(\tau, \sigma, \ldots)$. ■

Theorem 5.2. *A is symmetric in two (covariant, or contravariant) arguments if and only if the components $A^{ij\cdots}_{pq\cdots}$ are symmetric in their corresponding two (covariant, or contravariant) indices. A is skew-symmetric in two (covariant, or contravariant) arguments if and only if the components $A^{ij\cdots}_{pq\cdots}$ are skew-symmetric in their corresponding two (covariant, or contravariant) indices. Hence, in particular, the property of the components of a tensor being symmetric or skew-symmetric in two (covariant, or contravariant) indices is independent of the choice of a basis.*

Proof. Problem 5.1.

There is another way of describing the symmetry and skew-symmetry in two arguments of a tensor. We will discuss the symmetric case and leave the skew-symmetric case for a problem. Again, to simplify the notation we will work with the first two arguments.

If $v_1 \otimes v_2 \otimes \cdots \in V^r_s$, let $T: V^r_s \to V^r_s$ be such that $v_1 \otimes v_2 \otimes \cdots \mapsto v_2 \otimes v_1 \otimes \cdots$ where the dots on each side of the arrow are supposed to represent the same thing. By Theorem 3.1, or its generalization, T defines an automorphism of V^r_s.

Theorem 5.3. *If $A \in V^r_s$, and T is as given above, then $T \cdot A = A$ if and only if A is symmetric in its first two arguments.*

Proof. The proof follows from the following lemma.

Lemma. $T \cdot A(\tau^1, \tau^2, \ldots) = A(\tau^2, \tau^1, \ldots)$.

Proof of lemma. A is a linear combination of decomposable elements, $v_{i1} \otimes v_{i2} \otimes \cdots$. Write $A = a^i(v_{i1} \otimes v_{i2} \otimes \cdots)$. Then

$$T \cdot A = a^i T \cdot (v_{i1} \otimes v_{i2} \otimes \cdots) = a^i (v_{i2} \otimes v_{i1} \otimes \cdots).$$

So

$$\begin{aligned} T \cdot A(\tau^1, \tau^2, \ldots) &= a^i(v_{i2} \otimes v_{i1} \otimes \cdots)(\tau^1, \tau^2, \ldots) = a^i \langle \tau^1, v_{i2} \rangle \langle \tau^2, v_{i1} \rangle \cdots \\ &= a^i \langle \tau^2, v_{i1} \rangle \langle \tau^1, v_{i2} \rangle \cdots = a^i (v_{i1} \otimes v_{i2} \otimes \cdots)(\tau^2, \tau^1, \ldots) \\ &= A(\tau^2, \tau^1, \ldots). \end{aligned}$$ ■

PROBLEM 5.1. Prove Theorem 5.2.

PROBLEM 5.2. (i) If a tensor of type (3, 0) is symmetric in two arguments, and skew-symmetric in two arguments then it is the zero tensor. (ii) Generalize (i).

PROBLEM 5.3. (i) If the components of a tensor of type (1, 1) are symmetric for every basis, then it is a scalar multiple of $\langle -, - \rangle$. (ii) If the components of a tensor of type (1, 1) are skew-symmetric for every basis, then it is the zero tensor. These results indicate that the concepts of symmetry and skew-symmetry with respect to a covariant and a contravariant argument (index) are essentially vacuous.

PROBLEM 5.4. Define an automorphism, T, of V_s^r and prove that $T \cdot A = -A$ if and only if A is skew-symmetric in its first two arguments.

PROBLEM 5.5. Suppose dim $V = 3$, and A is a tensor of type (0, 3) whose components have the symmetries $A_{ijk} + A_{jki} + A_{kij} = 0$ and $A_{ijk} = -A_{ikj}$. How many components of A are independent? Choose an independent set, and express the remaining components in terms of them.

PROBLEM 5.6. Suppose A is a tensor of type (0, 4) having the following symmetries:

(a) $A_{ijkl} = -A_{jikl}$
(b) $A_{ijkl} = -A_{ijlk}$
(c) $A_{ijkl} + A_{iklj} + A_{iljk} = 0$

Then

(i) $A_{ijkl} = A_{klij}$.
(ii) $A(v, w, v, w) = 0$ for all $v, w \in V$ implies $A = 0$. (Hence, if B and C are two tensors having the given symmetries and $B(v, w, v, w) = C(v, w, v, w)$ for all $v, w \in V$, then $B = C$.)
(iii) Let b be a symmetric tensor of type (0, 2). Then $A_{ijkl} = b_{ik}b_{jl} - b_{il}b_{jk}$ has symmetries (a), (b), and (c).

PROBLEM 5.7. If dim $V = n$, then a (0, 4) tensor with the symmetries of Problem 5.6 has $n^2(n^2 - 1)/12$ distinct nonvanishing components. (Hint: Find the number of distinct nonvanishing components with 2, 3, and 4 distinct indices and add.)

5.2 The symmetric subspace of V_s^0

In the following discussion we will be restricting our attention to the spaces V_s^0 of covariant tensors. It will be evident that a completely parallel development is valid for contravariant tensors.

Definition
A tensor $A \in V_s^0$ is *symmetric* if it is symmetric in all pairs of its arguments. This is equivalent to the property $A(v_1, \ldots, v_s) = A(v_{\pi(1)}, \ldots, v_{\pi(s)})$ for all permutations, π, of $1, 2, \ldots, s$.

Note that all zero and first order tensors are symmetric.

Theorem 5.4. *A is symmetric if and only if* $A_{j_1 \cdots j_s} = A_{j_{\pi(1)} \cdots j_{\pi(s)}}$ *for all permutations,* π, of $1, 2, \ldots, s$.

We will let π also denote the automorphism of V_s^0 determined by $\sigma^1 \otimes \sigma^2 \otimes \cdots \otimes \sigma^s \mapsto \sigma^{\pi(1)} \otimes \sigma^{\pi(2)} \otimes \cdots \otimes \sigma^{\pi(s)}$. We get the following from Theorem 5.3.

Theorem 5.5. *A is symmetric if and only if* $\pi \cdot A = A$ *for all permutations,* π, of $1, 2, \ldots, s$.

Proof. Problem 5.8.

Theorem 5.6. *The set,* $\mathrm{S}^s(V^*)$, of all symmetric tensors of V_s^0 is a subspace of V_s^0.

There are several other ways to describe the subspace of symmetric tensors of V_s^0.

Definitions
The linear transformation

$$\mathfrak{S}\colon V_s^0 \to \mathrm{S}^s(V^*)$$

defined by

$$A \mapsto \frac{1}{s!} \sum_{\pi} \pi \cdot A \tag{5.1}$$

where $\sum_\pi$ is the sum over permutations of $1, \ldots, s$ is called *the symmetrization operator on* V_s^0. $\mathfrak{S} \cdot A$ is called *the symmetric part of A*.

It is clear that $\mathfrak{S}$ given by (5.1) is linear, and it has values in $\mathrm{S}^s(V^*)$ since for any permutation, $\tilde{\pi}$, $\tilde{\pi} \cdot (\sum_\pi \pi \cdot A) = \sum_\pi \pi \cdot A$.

Theorem 5.7. *$\mathfrak{S}\cdot A = \frac{1}{s!}\sum_\pi \pi\cdot A$ (i.e., (5.1)) if and only if*

$$\mathfrak{S}\cdot A(v_1, \ldots, v_s) = \frac{1}{s!}\sum_\pi A(v_{\pi(1)}, \ldots, v_{\pi(s)}) \tag{5.2}$$

Proof. The proof is based on the following lemma.

Lemma. *For any permutation, π,*

$$\pi\cdot A(v_1, \ldots, v_s) = A(v_{\pi^{-1}(1)}, \ldots, v_{\pi^{-1}(s)})$$

Proof of lemma. Let $A = a_i(\sigma^{i1}\otimes\cdots\otimes\sigma^{is})$. Then

$$\pi\cdot A = a_i(\sigma^{i\pi(1)}\otimes\cdots\otimes\sigma^{i\pi(s)}).$$

So

$$\begin{aligned}\pi\cdot A(v_1, \ldots, v_s) &= a_i(\sigma^{i\pi(1)}\otimes\cdots\otimes\sigma^{i\pi(s)})(v_1, \ldots, v_s)\\ &= a_i\langle\sigma^{i\pi(1)}, v_1\rangle\langle\sigma^{i\pi(2)}, v_2\rangle\cdots\langle\sigma^{i\pi(k)}, v_k\rangle\cdots\langle\sigma^{i\pi(s)}, v_s\rangle.\end{aligned}$$

Now, for some k, $\pi(k) = 1$, $k = \pi^{-1}(1)$, and $\langle\sigma^{i\pi(k)}, v_k\rangle = \langle\sigma^{i1}, v_{\pi^{-1}\{1\}}\rangle$. Another factor will be $\langle\sigma^{i2}, v_{\pi^{-1}(2)}\rangle$, and so on. So

$$\pi\cdot A(v_1, \ldots, v_s) = a_i\langle\sigma^{i1}, v_{\pi^{-1}(1)}\rangle\cdots\langle\sigma^{is}, v_{\pi^{-1}(s)}\rangle = A(v_{\pi^{-1}(1)}, \ldots, v_{\pi^{-1}(s)}).$$

■

Proof of theorem. Now suppose $\mathfrak{S}\cdot A = \frac{1}{s!}\sum_\pi \pi\cdot A$. Then

$$\begin{aligned}\mathfrak{S}\cdot A(v_1, \ldots, v_s) &= \left(\frac{1}{s!}\sum_\pi \pi\cdot A\right)(v_1, \ldots, v_s) = \frac{1}{s!}\sum_\pi \pi\cdot A(v_1, \ldots, v_s)\\ &= \frac{1}{s!}\sum_\pi A(v_{\pi^{-1}(1)}, \ldots, v_{\pi^{-1}(s)}) \qquad \text{(by the lemma)}\\ &= \frac{1}{s!}\sum_\pi A(v_{\pi(1)}, \ldots, v_{\pi(s)})\end{aligned}$$

All the steps of this argument are reversible, so the required equivalence is established. ■

We saw that $\mathfrak{S}$ maps $V_s^0(V^*)$ into $\mathrm{S}^s(V^*)$. Actually, it is onto $\mathrm{S}^s(V^*)$, since if A is any element of $\mathrm{S}^s(V^*)$ then $\mathfrak{S}$ maps $A \in V_s^0$ into itself, so

$$\mathrm{S}^s(V^*) = \mathfrak{S}(V_s^0)$$

Moreover, $\mathfrak{S}^2 = \mathfrak{S}$; i.e., $\mathfrak{S}$ is a projection operator on V_s^0. So

$$V_s^0 = \ker \mathfrak{S} \oplus \mathrm{S}^s(V^*)$$

(See Problem 1.9.)

There are alternative description of ker $\mathfrak{S}$, and hence of $\mathrm{S}^s(V^*)$.

Theorem 5.8. *The linear closure of the set of all elements of V_s^0 which are skew-symmetric in two arguments is the same as the linear closure of the set of all elements of V_s^0 of the form*

$$\sigma^1 \otimes \cdots \otimes \sigma^i \otimes \cdots \otimes \sigma^j \otimes \cdots \otimes \sigma^s - \sigma^1 \otimes \cdots \otimes \sigma^j \otimes \cdots \otimes \sigma^i \otimes \cdots \otimes \sigma^s.$$

Proof. Problem 5.9.

Theorem 5.9. $N_s = \ker \mathfrak{S}$ *where N_s is the subspace of V_s^0 described in* Theorem 5.8.

Proof. (i) Let A be an element of N_s of the form in Theorem 5.8. $\mathfrak{S}\cdot A$ is $1/s!$ times a sum of permutations of these elements. For each permutation, π, with $i \mapsto \pi(i)$ and $j \mapsto \pi(j)$, there is another permutation, $\tilde{\pi}$, which is the same as π except that $\tilde{\pi}: i \mapsto \pi(j)$ and $j \mapsto \pi(i)$. The terms in $\mathfrak{S}\cdot A$ corresponding to such pairs cancel each other, so for such elements $\mathfrak{S}\cdot A = 0$. Since every element of N_s is a linear combination of these elements, $\mathfrak{S}\cdot A = 0$ for all $A \in N_s$.

(ii) First we show that for $A \in V_s^0$, $\pi\cdot A - A \in N_s$. For $A = \sigma^1 \otimes \cdots \otimes \sigma^s$ and π a transposition this is clearly true. Assume it is true for $A = \sigma^1 \otimes \cdots \otimes \sigma^s$ and π a product of m transpositions, and proceed by induction. Since π is linear, it follows that $\pi\cdot A - A \in N_s$ for any $A \in V_s^0$. Now,

$$\mathfrak{S}\cdot A - A = \frac{1}{s!}\sum_{\pi} (\pi\cdot A - A) \in N_s.$$

So, if $A \in \ker \mathfrak{S}$, then $A \in N_s$. ∎

Corollary. (i) $\mathrm{S}^s(V^*) \cong V_s^0/N_s$.

(ii) $V_s^0 = N_s \oplus \mathrm{S}^s(V^*)$.

To summarize, we have described a certain subspace of V_s^0 in three ways: (i) the subset of V_s^0 stable (invariant) under symmetry automorphisms of V_s^0; (ii) the image of the symmetrization operator; and (iii) a quotient (factor) space.

We can express the components of $\mathfrak{S}\cdot A$ in a given basis of V_s^0 in terms of those of A. We can write

$$A(v_{\pi(1)}, \ldots, v_{\pi(s)}) = A_{i_{\pi(1)}\cdots i_{\pi(s)}} v_{\pi(1)}^{i_{\pi(1)}} \cdots v_{\pi(s)}^{i_{\pi(s)}}$$

so, from eq. (5.2),

$$\begin{aligned} \mathfrak{S}\cdot V(v_1, \ldots, v_s) &= \frac{1}{s!}\sum_\pi v_{\pi(1)}^{i_{\pi(1)}} \cdots v_{\pi(s)}^{i_{\pi(s)}} A_{i_{\pi(1)}\cdots i_{\pi(s)}} \\ &= \frac{1}{s!} v_1^{i_1}\cdots v_s^{i_s} \sum_\pi A_{i_{\pi(1)}\cdots i_{\pi(s)}} \\ &= \frac{1}{s!}\sum_\pi A_{i_{\pi(1)}\cdots i_{\pi(s)}} \varepsilon^{i_1}\otimes\cdots\otimes\varepsilon^{i_s}(v_1, \ldots, v_s) \end{aligned}$$

Thus,

$$\mathfrak{S}\cdot A = \frac{1}{s!}\sum_\pi A_{i_{\pi(1)}\cdots i_{\pi(s)}} \varepsilon^{i_1}\otimes\cdots\otimes\varepsilon^{i_s} \tag{5.3}$$

and the components of $\mathfrak{S}\cdot A$ in the basis $\{\varepsilon^{i_1}\otimes\cdots\otimes\varepsilon^{i_s}; i_1, \ldots, i_s = 1, \ldots, n\}$ of V_s^0 are $(1/s!)\sum_\pi A_{i_{\pi(1)}\cdots i_{\pi(s)}}$. Notice that there are n^s terms in (5.3) and the coefficients of the terms in which the values of $i_1, \ldots, i_s$ are simply permuted are all the same.

We can also write the symmetric part of A *in a basis of* $\mathrm{S}^s(V^*)$ with components in terms of the components of A in a basis of V_s^0. From (5.1), the definition of $\mathfrak{S}$, for any $A = A_{i_1\cdots i_s}\varepsilon^{i_1}\otimes\cdots\otimes\varepsilon^{i_s}\in V_s^0$,

$$\mathfrak{S}\cdot A = \frac{1}{s!} A_{i_1\cdots i_s}\sum_\pi \varepsilon^{i_{\pi(1)}}\otimes\cdots\otimes\varepsilon^{i_{\pi(s)}} \tag{5.4}$$

for all $A\in V_s^0$. Note that the factors $\sum_\pi \varepsilon^{i_{\pi(1)}}\otimes\cdots\otimes\varepsilon^{i_{\pi(s)}}$ are in $\mathrm{S}^s(V^*)$. In (5.4), $i_1, \ldots, i_s$ each take on values $1, \ldots, n$ so there will be n^s terms. We will now separate these into a certain number of bunches.

(i) Look at one term in which the values of $i_1, \ldots, i_s$ are distinct. There will be a total of $s!$ terms with $i_1, \ldots, i_s$ having these particular values, but permuted. For example, for $s = 3$, we might have $i_1 = 3$, $i_2 = 1$ and $i_3 = 4$. Then we also have $i_1 = 4$, $i_2 = 1$, $i_3 = 3$ and four more terms with indices 1, 3 and 4. For all these $s!$ terms the second factor $\sum_\pi \varepsilon^{i_{\pi(1)}}\otimes\cdots\otimes\varepsilon^{i_{\pi(s)}}$ in

(5.4) is the same, so we can write the sum of these terms as

$$\frac{1}{s!}(A_{i_1\cdots i_s} + A_{i_2 i_1\cdots i_s} + \cdots)(\varepsilon^{i_1}\otimes\cdots\otimes\varepsilon^{i_s} + \varepsilon^{i_2}\otimes\cdots\otimes\varepsilon^{i_s} + \cdots)$$

where now the values of $i_1, \ldots, i_s$ are fixed (no sum). Since the two factors of this expression are unchanged if we permute the $i_1, \ldots, i_s$, we can choose $i_1 < i_2 < \cdots < i_s$ and write this sum as

$$\sum_0 = \frac{1}{s!}\sum_\pi A_{i_{\pi(1)}\cdots i_{\pi(s)}}\sum_\pi \varepsilon^{i_{\pi(1)}}\otimes\cdots\otimes\varepsilon^{i_{\pi(s)}}, \qquad i_1 < i_2 < \cdots < i_s$$

(ii) If two of the values of the indices $i_1, \ldots, i_s$ are the same, there will be $s!/2$ terms of (5.4) with these values of the indices, and again each term will have the same factor $\sum_\pi \varepsilon^{i_{\pi(1)}}\otimes\cdots\otimes\varepsilon^{i_{\pi(s)}}$. Now in this common factor, the same term appears twice, so we can write

$$\sum_\pi \varepsilon^{i_{\pi(1)}}\otimes\cdots\otimes\varepsilon^{i_{\pi(s)}} = 2\sum_{\pi *}\varepsilon^{i_{\pi(1)}}\otimes\cdots\otimes\varepsilon^{i_{\pi(s)}}$$

where $\sum_{\pi *}$ means that we add up only half the terms—all the distinct ones. Further, in $\sum_\pi A_{i_{\pi(1)}\cdots i_{\pi(s)}}$ each of the $s!/2$ terms with the common factor $\sum_{\pi *}\varepsilon^{i_{\pi(1)}}\otimes\cdots\otimes\varepsilon^{i_{\pi(s)}}$ appears twice, so if we add up all the terms in which $i_1, \ldots, i_s$ have these values we get

$$\frac{1}{s!}\frac{\sum_\pi A_{i_{\pi(1)}\cdots i_{\pi(s)}}}{2}\cdot 2\sum_{\pi *}\varepsilon^{i_{\pi(1)}}\otimes\cdots\otimes\varepsilon^{i_{\pi(s)}}$$

$$= \frac{1}{s!}\sum_\pi A_{i_{\pi(1)}\cdots i_{\pi(s)}}\sum_{\pi *}\varepsilon^{i_{\pi(1)}}\otimes\cdots\otimes\varepsilon^{i_{\pi(1)}}$$

Again, since the two factors of this expression are unchanged, if we permute the $i_1, \ldots, i_s$ we can choose $i_1 \le i_2 \le \cdots \le i_s$, and write this sum as

$$\sum_1 = \frac{1}{s!}\sum_\pi A_{i_{\pi(1)}\cdots i_{\pi(s)}}\sum_{\pi *}\varepsilon^{i_{\pi(1)}}\otimes\cdots\otimes\varepsilon^{i_{\pi(s)}} \qquad i_1 \le i_2 \le \cdots \le i_s$$

(iii) If the set $i_1, \ldots, i_s$ has a value repeated p times and another value repeated q times, there will be $s!/p!q!$ terms in (5.4) with these values of the indices, and each term will have the same factor

$$\sum_{\pi} \varepsilon^{i_{\pi(1)}} \otimes \cdots \otimes \varepsilon^{i_{\pi(s)}} = p!q! \sum_{\pi*} \varepsilon^{i_{\pi(1)}} \otimes \cdots \otimes \varepsilon^{i_{\pi(s)}}$$

since the $s!$ permutations of $i_1, \ldots, i_s$ yield only $s!/p!q!$ *distinct* terms. The sum of these terms will be

$$\sum_{pq} = \frac{1}{s!} \frac{\sum_{\pi} A_{i_{\pi(1)} \cdots i_{\pi(s)}}}{p!q!} \sum_{\pi} \varepsilon^{i_{\pi(1)}} \otimes \cdots \otimes \varepsilon^{i_{\pi(s)}}$$

$$= \frac{1}{s!} \sum_{\pi} A_{i_{\pi(1)} \cdots i_{\pi(s)}} \sum_{\pi*} \varepsilon^{i_{\pi(1)}} \otimes \cdots \otimes \varepsilon^{i_{\pi(s)}}, \qquad i_1 \le i_2 \le \cdots \le i_s$$

Again $\sum_{\pi*}$ means we add up only distinct terms.

(iv) For any fixed set of values of $i_1, \ldots, i_s$ with no matter how many repetitions, we can draw out of (5.4) a set of terms whose sum is given by the same expression as in (i), (ii), and (iii). Thus, letting $i_1, \ldots, i_s$ range over all fixed sets of values, the sum of the terms of (5.4) will be the sum of sums of the form $\sum_0, \sum_1, \sum_{pq}, \ldots$. That is, we get

$$\mathfrak{S} \cdot A = \frac{1}{s!} \sum_{\pi} A_{i_{\pi(1)} \cdots i_{\pi(s)}} \sum_{\pi*} \varepsilon^{i_{\pi(1)}} \otimes \cdots \otimes \varepsilon^{i_{\pi(s)}}, \qquad i_1 \le i_2 \le \cdots \le i_s \quad (5.5)$$

Notice that the coefficients in (5.5) are exactly the same as those of (5.3). However, now they are not repeated as they were in (5.3).

Theorem 5.10. $\{\sum_{\pi*} \varepsilon^{i_{\pi(1)}} \otimes \cdots \otimes \varepsilon^{i_{\pi(s)}};\ i_1 \le i_2 \le \cdots \le i_s\}$ *forms a basis of* $\mathsf{S}^s(V^*)$, and (5.5) is the representation of the symmetric part of A in that basis (in terms of the components of A in the V_s^0 basis.)

Proof. Problem 5.11.

Corollary. *If* $\dim V = n$, *then* $\dim \mathsf{S}^s(V^*) = \binom{n+s-1}{s}$.

Proof. The number of terms in (5.5) is the number of ways of choosing s integers between 1 and n in nondecreasing order. ∎

Finally, if A is symmetric, (5.5) becomes

$$A = A_{i_1 \cdots i_s} \sum_{\pi*} \varepsilon^{i_{\pi(1)}} \otimes \cdots \otimes \varepsilon^{i_{\pi(s)}}, \qquad i_1 \le \cdots \le i_s \quad (5.6)$$

Note that the components in this basis are simply a subset of the components of A in the V_s^0 basis.

Let us illustrate the formulas (5.3), (5.5), (5.6) above for the specific example in which V^* has dimension 4, and $s = 3$; i.e., the covariant tensors of rank 3 built on a vector space of dimension 4.

If we choose a basis $\{\varepsilon^1, \varepsilon^2, \varepsilon^3, \varepsilon^4\}$ in V^*, then for any $A \in V_3^0$ we can write $A = A_{ijk}\varepsilon^i \otimes \varepsilon^j \otimes \varepsilon^k$, and for this example (5.3) reads

$$\mathfrak{S}\cdot A = \frac{1}{3!}[A_{ijk} + A_{jki} + A_{kij} + A_{jik} + A_{ikj} + A_{kji}]\varepsilon^i \otimes \varepsilon^j \otimes \varepsilon^k \quad (5.7)$$

There are four sets of *distinct* values of i, j, k for each set of which there are six terms of (5.7) with the same coefficients. There are 12 sets of values of i, j, k, for each set of which there are three terms, and when $i = j = k$ there are four more terms.

For this example, eq. (5.5) is

$$\mathfrak{S}\cdot A = \frac{1}{3!}(A_{ijk} + A_{jik} + \cdots)(\varepsilon^i \otimes \varepsilon^j \otimes e^k + \varepsilon^j \otimes \varepsilon^i \otimes \varepsilon^k + \cdots) \quad (5.8)$$

The subspace $S^3(V^*)$ of symmetric tensors in V_3^0 has dimension $\binom{4+3-1}{3} = 6!/3!3! = 20$ according to the last corollary. We can check this directly in this case by explicitly writing all possible triples of integers between 1 and 4 in nondecreasing order. So eq. (5.8) has 20 terms. If, in particular, A is symmetric, then the coefficients simplify and we get

$$\begin{aligned}
A = {} & A_{123}(\varepsilon^1 \otimes \varepsilon^2 \otimes \varepsilon^3 + \varepsilon^2 \otimes \varepsilon^3 \otimes \varepsilon^1 + 4\,\text{more terms}) \\
& + A_{124}(\varepsilon^1 \otimes \varepsilon^2 \otimes \varepsilon^4 + \varepsilon^2 \otimes \varepsilon^4 \otimes \varepsilon^1 + 4\,\text{more terms}) \\
& \quad \vdots \qquad\qquad\qquad\qquad\qquad\qquad \left.\right\} \text{4 terms} \\
& + A_{112}(\varepsilon^1 \otimes \varepsilon^1 \otimes \varepsilon^2 + \varepsilon^1 \otimes \varepsilon^2 \otimes \varepsilon^1 + \otimes\varepsilon^2 \otimes \varepsilon^1 \otimes \varepsilon^1) \\
& + A_{113}(\varepsilon^1 \otimes \varepsilon^1 \otimes \varepsilon^3 + \varepsilon^1 \otimes \varepsilon^3 \otimes \varepsilon^1 + \otimes\varepsilon^3 \otimes \varepsilon^1 \otimes \varepsilon^1) \\
& \quad \vdots \qquad\qquad\qquad\qquad\qquad\qquad \left.\right\} \text{12 terms} \\
& + A_{111}(\varepsilon^1 \otimes \varepsilon^1 \otimes \varepsilon^1) \\
& + A_{222}(\varepsilon^2 \otimes \varepsilon^2 \otimes \varepsilon^2) \\
& \quad \vdots \qquad\qquad\qquad\qquad\qquad\qquad \left.\right\} \text{4 terms}
\end{aligned}$$

PROBLEM 5.8. Prove Theorem 5.5.

PROBLEM 5.9. Prove Theorem 5.8.

PROBLEM 5.10. Prove $\mathfrak{S}$ commutes with the mapping ϕ_r of Section 4.3. In particular, symmetry is preserved under linear mappings.

Problem 5.11. Prove Theorem 5.10.

5.3 The skew-symmetric (alternating) subspace of V_s^0

The following treatment closely parallels that of Section 5.2. Again we restrict our attention to the covariant tensors V_s^0. Obvious analogs exist for the spaces V_0^r.

Definitions
A tensor $A \in V_s^0$ is *skew-symmetric* (or *alternating*) if it is skew-symmetric in all pairs of its arguments. This is equivalent to the property $A(v_1, \ldots, v_s) = (\operatorname{sgn} \pi) A(v_{\pi(1)}, \ldots, v_{\pi(s)})$ for all permutations π of $1, 2, \ldots, s$. A skew-symmetric tensor of type $(0, s)$ is called *an s-form.* (A skew-symmetric tensor of type $(r, 0)$ is called *an r-vector.*)

Note that all zero and first order tensors are skew-symmetric. Note also, that by Theorem 5.1, in contrast to the symmetric case, the only skew-symmetric tensors of order greater than dim V are the zero tensors.

We will let π also denote the automorphism of V_s^0 determined by $\sigma^1 \otimes \sigma^2 \otimes \cdots \otimes \sigma^s \mapsto \sigma^{\pi(1)} \otimes \sigma^{\pi(2)} \otimes \cdots \otimes \sigma^{\pi(s)}$.

Theorem 5.11. *A is skew-symmetric if and only if $\pi \cdot A = (\operatorname{sgn} \pi) A$ for all permutations, π, of $1, \ldots, s$.*

Proof. Problem 5.12.

Corollary. *A is skew-symmetric if and only if $A_{i_1}, \ldots, {}_{i_s} = (\operatorname{sgn} \pi) \mathrm{A}_{i_{\pi(1)} \cdots i_{\pi(s)}}$* for all permutations, π, of $1, 2, \ldots, s$.

Theorem 5.12. *The set, $\Lambda^s(V^*)$, of all skew-symmetric tensors of V_s^0 is a* subspace of V_s^0.

There are several other ways to describe the subspace $\Lambda^s(V^*)$ of skew-symmetric tensors of V_s^0.

Definitions
The linear transformation

$$\mathfrak{A}: V_s^0 \to \Lambda^s(V^*)$$

defined by

$$A \mapsto \frac{1}{s!} \sum_{\pi} (\operatorname{sgn} \pi) \pi \cdot A \tag{5.9}$$

is called *the alternating operator on* V_s^0 and $\mathfrak{A} \cdot A$ is called *the skew-symmetric part* of A.

The following three theorems are analogs of Theorems 5.7–5.9.

Theorem 5.13. $\mathfrak{A} \cdot A = \frac{1}{s!} \sum_\pi (\text{sgn } \pi)\pi \cdot A$ *if and only if*

$$\mathfrak{A} \cdot A(v_1, \dots, v_s) = \frac{1}{s!} \sum_\pi (\text{sgn } \pi) A(v_{\pi(1)}, \dots, v_{\pi(s)}) \tag{5.10}$$

Corollary. *If* $A = \sigma^1 \otimes \cdots \otimes \sigma^s$, *then* $\mathfrak{A} \cdot A(v_1, \dots, v_s) = \frac{1}{s!} \det(\sigma^i \cdot v_j)$.

Theorem 5.14. *The linear closure of the set of all elements of* V_s^0 *which are symmetric in two arguments is the same as the linear closure of the set of all elements of* V_s^0 *of the form*

$$\sigma^1 \otimes \cdots \otimes \sigma^i \otimes \cdots \otimes \sigma^j \otimes \cdots \otimes \sigma^s + \sigma^1 \otimes \cdots \otimes \sigma^j \otimes \cdots \otimes \sigma^i \otimes \cdots \otimes \sigma^s$$

and the same as the linear closure of the set of decomposable elements with two or more factors the same.

Theorem 5.15. $N_A = \ker \mathfrak{A}$ *where* N_a *is the subspace of* V_s^0 *described in Theorem 5.14.*

Corollary. (i) $\Lambda^s(V^*) = V_s^0 / N_A$.
(ii) $V_s^0 = N_A \oplus \Lambda^s(V^*)$.

To summarize, we have described a certain subspace of V_s^0 in three ways: (i) the subset of V_s^0 stable (invariant) under skew-symmetry automorphisms of V_s^0; (ii) the image of the alternating operator; and (iii) a quotient (factor) space.

We have representations of $\mathfrak{A} \cdot A$ and A when A is skew-symmetric corresponding to those for the symmetric case. Thus, corresponding to (5.3), we can express the components of $\mathfrak{A} \cdot A$ in terms of those of A by

$$\mathfrak{A} \cdot A = \frac{1}{s!} \sum_\pi (\text{sgn } \pi) A_{i_{\pi(1)} \cdots i_{\pi(s)}} \varepsilon^{i_1} \otimes \cdots \otimes \varepsilon^{i_s} \tag{5.11}$$

As in the corresponding formula for the symmetric part of A, there are n^s terms in (5.11). However, now whenever two of the indices $i_1, \dots, i_s$ have the same value, the coefficient vanishes.

Using

$$\mathfrak{A}\cdot A=\frac{1}{s!}A_{i_1\cdots i_s}\sum_{\pi}(\text{sgn }\pi)\varepsilon^{i_{\pi(1)}}\otimes\cdots\otimes\varepsilon^{i_{\pi(s)}} \tag{5.12}$$

which we get from the definition, (5.9), of $\mathfrak{A}$, we can write the skew-symmetric tensor $\mathfrak{A}\cdot A$ *in a basis of* $\Lambda^s(V^*)$ in terms of its components in a basis of V_s^0. Note that the factors $\sum_\pi(\text{sgn }\pi)\varepsilon^{i_{\pi(1)}}\otimes\cdots\otimes\varepsilon^{i_{\pi(s)}}$ in (5.12), which are in $\Lambda^s(V^*)$, vanish whenever two of the indices $i_1,\ldots,i_s$ have the same value. In particular, if $s>n$, then $\mathfrak{A}\cdot A=0$. For every set of distinct values of $i_1,\ldots,i_s$, half the $s!$ factors $\sum_\pi(\text{sgn }\pi)\varepsilon^{i_{\pi(1)}}\otimes\cdots\otimes\varepsilon^{i_{\pi(s)}}$ with these values have the value

$$\sum_{\pi}(\text{sgn }\pi)\varepsilon^{i_{\pi(1)}}\otimes\cdots\otimes\varepsilon^{i_{\pi(s)}},$$

and half have the value

$$-\sum_{\pi}(\text{sgn }\pi)\varepsilon^{i_{\pi(1)}}\otimes\cdots\otimes\varepsilon^{i_{\pi(s)}}.$$

Hence

$$\begin{aligned}\mathfrak{A}\cdot A&=\sum_{\pi_e}A_{i_{\pi(1)}\cdots i_{\pi(s)}}\sum_{\pi}(\text{sgn }\pi)\varepsilon^{i_{\pi(1)}}\otimes\cdots\otimes\varepsilon^{i_{\pi(s)}}\\&\quad+\sum_{\pi_o}A_{i_{\pi(1)}\cdots i_{\pi(s)}}\left(-\sum_{\pi}(\text{sgn }\pi)\varepsilon^{i_{\pi(1)}}\otimes\cdots\otimes\varepsilon^{i_{\pi(s)}}\right),\end{aligned} \tag{5.13}$$

where π_e in the coefficient of the first term is an even permutation of $1,\ldots,s$, and π_o in the coefficient of the second term is an odd permutation of $1,\ldots,s$. Moreover, for each set of distinct values of $i_1,\ldots,i_s$ we can choose the order $i_1<i_2<\cdots<i_s$. For either this order is an even permutation of the given order, in which case (5.13) is valid for $i_1<\cdots<i_s$, or, if this order is an odd permutation of the given order, a minus sign must be introduced in the second factors of (5.13) and π_e and π_o must be interchanged. The result is

$$\mathfrak{A}\cdot A=\frac{1}{s!}\sum_{\pi}(\text{sgn }\pi)A_{i_{\pi(1)}\cdots i_{\pi(s)}}\sum_{\pi}(\text{sgn }\pi)\varepsilon^{i_{\pi(1)}}\otimes\cdots\otimes\varepsilon^{i_{\pi(s)}},\qquad i_1<i_2<\cdots<i_s \tag{5.14}$$

(cf., eq. (5.5)).

Theorem 5.16. *The set* $\{\sum_\pi(\text{sgn }\pi)\varepsilon^{i_{\pi(1)}}\otimes\cdots\otimes\varepsilon^{i_{\pi(s)}};\ i_1<\cdots<i_s\}$ *of tensors appearing in (5.14) form a basis of* $\Lambda^s(V^*)$.

Corollary. *If* $\dim V=n$, *then* $\dim\Lambda^s(V^*)=\binom{n}{s}$. *In particular, if* $s=\dim V$, *then $\Lambda^s(V^*)$ is 1-dimensional and spanned by $\varepsilon^1 \otimes \cdots \otimes \varepsilon^s$. If $s > n$, then $\Lambda^s(V^*)$ has only the zero element.*

Finally, if A is skew-symmetric (5.14) reduces to

$$A = A_{i_1 \cdots i_s} \sum_\pi (\text{sgn } \pi)\varepsilon^{i_{\pi(1)}} \otimes \cdots \otimes \varepsilon^{i_{\pi(s)}}, \qquad i_1 < i_2 < \cdots < i_s \quad (5.15)$$

We can simplify eqs. (5.14) and (5.15) if we introduce the following definition and notation.

Definition
For $\sigma^i \in V^*$, $\mathfrak{A} \cdot \sigma^1 \otimes \cdots \otimes \sigma^s$ is called *the exterior* (or *wedge*, or *skew-symmetric*) *product* of $\sigma^1, \sigma^2, \ldots, \sigma^s$, and we write

$$\mathfrak{A} \cdot \sigma^1 \otimes \cdots \otimes \sigma^s = \sigma^1 \wedge \sigma^2 \wedge \cdots \wedge \sigma^s \qquad (5.16)$$

(Similarly, for $v_i \in V$ and $v_1 \otimes \cdots \otimes v_r \in V_0^r$ we write

$$\mathfrak{A} \cdot v_1 \otimes \cdots \otimes v_r = v_1 \wedge \cdots \wedge v_r)$$

Then

$$\sum_\pi (\text{sgn } \pi)\sigma^{\pi(1)} \otimes \cdots \otimes \sigma^{\pi(s)} = s!\sigma^1 \wedge \cdots \wedge \sigma^s$$

and

$$\sum_\pi (\text{sgn } \pi)\sigma^{i_{\pi(1)}} \otimes \cdots \otimes \sigma^{i_{\pi(s)}} = s!\sigma^{i_1} \wedge \cdots \wedge \sigma^{i_s}.$$

In particular, by Theorem 5.16, $\{\varepsilon^{i_1} \wedge \cdots \wedge \varepsilon^{i_s}: i_1 < \cdots < i_s\}$ is a basis of $\Lambda^s(V^*)$.

Using the wedge product notation we can write (5.14) as

$$\mathfrak{A} \cdot A = \left(\sum_\pi (\text{sgn } \pi)\mathrm{A}_{i_{\pi(1)} \cdots i_{\pi(s)}}\right)\varepsilon^{i_1} \wedge \cdots \wedge \varepsilon^{i_s}, \qquad i_1 < \cdots < i_s \quad (5.17)$$

If A is skew-symmetric, and we use this notation then both (5.15) and (5.17) reduce to

$$A = s!A_{i_1 \cdots i_s}\varepsilon^{i_1} \wedge \cdots \wedge \varepsilon^{i_s}, \qquad i_1 < \cdots < i_s \qquad (5.18)$$

Comparing (5.17) with (5.14) and (5.18) with (5.15), we get that the components of a tensor in $\Lambda^s(V^*)$ with respect to the wedge basis are $s!$ times the components of Theorem 5.16.

(Note that the expression for $\mathfrak{A}\cdot A$ given by eq. (5.11) also simplifies if we use the wedge product notation. Thus, $\mathfrak{A}\cdot A = A_{i_1\cdots i_s}\varepsilon^{i_1}\wedge\cdots\wedge\varepsilon^{i_s}$. However, here $i_1, \ldots, i_s$ all run from 1 to n, and $\mathfrak{A}\cdot A$ is expressed in terms of a linearly dependent set and not a basis.)

Finally, it should be noted that our definition (5.16) of the exterior product is only one of two prevalent definitions in the literature which differ by a factor. According to the other definition, $\sigma^1\wedge\cdots\wedge\sigma^s = s!\mathfrak{A}\cdot\sigma^1\otimes\cdots\otimes\sigma^s$. With this definition the wedge product basis is the same as the basis of Theorem 5.16 and the components of $\mathfrak{A}\cdot A$ and A are just as they were in eqs. (5.14) and (5.15).

PROBLEM 5.12. Prove Theorem 5.11.

PROBLEM 5.13. For $\Lambda^2(V^*)$:

(i) If $\dim V = 3$, every element of $\Lambda^2(V^*)$ is decomposable.

(ii) If $\dim V = 4$, not every element of $\Lambda^2(V^*)$ is decomposable. (Decomposable elements of $\Lambda^s(V^*)$ have the form $\sigma^1\wedge\cdots\wedge\sigma^s$ (cf., Section 4.1).

PROBLEM 5.14. From the results of Chapters 2 and 3 there is natural isomorphism $\mathscr{L}(V_2^0, V_2^0) \cong V_2^2$, so for $s = 2$, $\mathfrak{A}\in V_2^2$. If $\mathfrak{A}^{ij}_{pq}$ are components of $\mathfrak{A}$, show that in every basis $\mathfrak{A}^{ij}_{pq} = \delta^i_p\delta^j_q - \delta^i_q\delta^j_p$. (See Section 4.2 and Problem 4.13.)

PROBLEM 5.15. (i) If σ, τ, ω are in V^*, express the values and the components of $\sigma\wedge\tau\wedge\omega$ in terms of their values and components in a basis of V^*. (ii) Show that $\sigma^1\wedge\cdots\wedge\sigma^s(v_1, \ldots, v_s) = (1/s!)\det(\langle\sigma^i, v_j\rangle)$.

PROBLEM 5.16. Write out the formulas (5.11), (5.14), and (5.15) for $\mathfrak{A}\cdot A$, and for a skew-symmetric tensor for the special case of covariant tensors of rank 3 with $\dim V = 4$.

PROBLEM 5.17. For the decomposable elements $\sigma^1\otimes\cdots\otimes\sigma^s\in V_s^0$ we define *a symmetric product* by $\sigma^1\circledast\cdots\circledast\sigma^s = \mathfrak{S}\cdot\sigma^1\otimes\cdots\otimes\sigma^s$. Then a symmetric tensor can be written as a homogeneous polynomial in $\sigma^1, \ldots, \sigma^s$. Write the coefficients in terms of the $A_{i_1\cdots i_s}$ of eq. (5.6).

5.4 Some special properties of $\mathrm{S}^2(V^*)$ and $\Lambda^2(V^*)$

We saw, in Section 4.1, that the space of second order covariant tensors, $V^*\otimes V^*$, is isomorphic to the space of linear mappings, $\mathscr{L}(V, V^*)$, from V

to V^*. Specifically, we have an isomorphism given by $\phi_1: b \mapsto \mathscr{A}_1$ where $b \in V^* \otimes V^*$ and $\mathscr{A}_1 \in \mathscr{L}(V, V^*)$ is given by $\mathscr{A}_1: v \mapsto b(v, -)$, and we also have an isomorphism $\phi_2: b \mapsto \mathscr{A}_2$ where $\mathscr{A}_2: v \mapsto b(-, v)$ (cf., Theorem 2.11.)

Also, since the elements of $V^* \otimes V^*$ are bilinear functions we have the concept of nondegeneracy (nonsingularity) and the result of Section 4.1 that a tensor of type (0, 2) (i.e., an element of $V^* \otimes V^*$)) is nondegenerate iff $\mathscr{A}_1$ and $\mathscr{A}_2$ are nonsingular. Corresponding properties are, of course, valid for the symmetric and skew-symmetric subspaces, $\mathrm{S}^2(V^*)$ and $\Lambda^2(V^*)$, of V_2^0. These two subspaces play prominent roles in geometry and mechanics, respectively, so we need to examine their properties a bit further.

(i) *Properties of* $\mathrm{S}^2(V^*)$. First of all we note that for $b \in \mathrm{S}^2(V^*)$, ϕ_1 and ϕ_2 are the same. For such a b, we write $\mathscr{A}_1 = \mathscr{A}_2 = b^\flat$ and the null space of the common linear mapping $b^\flat: V \to V^*$ is called *the null space of b*. It is the set $\{v \in V: b(v, w) = 0 \text{ for all } w \in V\}$ (cf., Theorem 2.14).

Next we look at the restriction of b to the diagonal elements of $V \times V$.

Definition
Given b (symmetric), the map, Q, from V to $\mathbb{R}$ given by $Q: v \mapsto b(v, v)$ is the *quadratic function (form) associated with b.*

If $Q(v) = b(v, v)$, then $Q(v + w) = b(v + w, v + w)$, from which we get

$$b(v, w) = \tfrac{1}{2}[Q(v + w) - Q(v) - Q(w)] \qquad \text{(the cosine law)} \qquad (5.19)$$

which shows that the association $b \mapsto Q$ is 1–1. Alternatively, using the expansion for $Q(v - w)$ in addition to that above for $Q(v + w)$ we get

$$b(v, w) = \frac{Q(v + w) - Q(v - w)}{4} \qquad \text{(the polarization identity).}$$

The mapping $Q \mapsto b$ is called *the polarization of Q*.

Definitions
b is *positive definite* if $Q(v) > 0$ for all $v \neq 0$. b is *positive semidefinite* if $Q(v) \geq 0$ for all v. There are corresponding definitions with "positive" replaced by "negative." If b is not definite there can be *null vectors*, or *lightlike vectors* $v \neq 0$ for which $Q(v) = 0$. The *null cone of b* is the set of all null vectors of b. In the *indefinite* (neither positive semidefinite, nor negative semidefinite) case, vectors for which $Q(v) < 0$ are called *timelike*, and those for which $Q(v) > 0$ are called *spacelike*.

Theorem 5.17. (The Cauchy–Schwarz inequality) *If b is semidefinite, then it satisfies*

$$[b(v, w)]^2 \le b(v, v)b(w, w)$$

for all $v, w \in V$.

Proof. For all real a, $b(av + w, av + w) = a^2 b(v, v) + 2ab(v, w) + b(w, w)$. This expression is either always nonnegative, or always nonpositive. In either case as a quadratic in a, its discriminant, $[b(v, w)]^2 - b(v, v)b(w, w)$, must be either negative or zero. (In contrast to the usual case, now $w = av$ is sufficient, but not necessary for equality.) ∎

Corollary. *If b is semidefinite and not definite, then b is degenerate.*

Theorem 5.18. *b is definite if and only if b is semidefinite and nondegenerate.*

Proof. Problem 5.19.

Theorem 5.19. *b is definite if and only if it has no null vectors.*

Proof. The only case not simply a matter of definition is when b is indefinite. Then, if $b(v, v) < 0$ and $b(w, w) > 0$, there is a number, $0 < a < 1$, such that for $z = av + (1 - a)w$, $b(z, z) = 0$. ∎

For examples of the various possibilities for a symmetric tensor of type $(0, 2)$, let $V = \mathbb{R}^n$ then with respect to the natural basis of $\mathbb{R}^n$, b is given by a set of numbers b_{ij} (Section 2.2). With these arranged in a matrix, (the matrix of $b^\flat$) consider the following examples.

(1) $(b_{ij}) = \begin{pmatrix} 1 & 0 \\ 0 & 1 \end{pmatrix}$ b is nondegenerate, positive definite

(2) $(b_{ij}) = \begin{pmatrix} 0 & 1 \\ 1 & 0 \end{pmatrix}$ b is nondegenerate, indefinite

(3) $(b_{ij}) = \begin{pmatrix} -1 & \frac{1}{2} \\ \frac{1}{2} & -1 \end{pmatrix}$ b is nondegenerate, negative definite

(4) $(b_{ij}) = \begin{pmatrix} -1 & 1 \\ 1 & -1 \end{pmatrix}$ b is degenerate, negative semidefinite

(5) $(b_{ij}) = \begin{pmatrix} 1 & 0 & 0 \\ 0 & -1 & 0 \\ 0 & 0 & 0 \end{pmatrix}$ b is degenerate, indefinite

We have been looking at the space $S^2(V^*)$ and classifying its elements according to the possible values of $Q(v)$. Now we switch our point of view and pick one element out of $S^2(V^*)$ and see what this does for V. First of all with a given $b \in S^2(V^*)$ we can generalize the concepts of length and orthogonality of vectors in V.

Definitions
A vector space, V, for which a nondegenerate element of $S^2(V^*)$ is chosen is called *a scalar (inner) product vector space. The length (magnitude) of v*, $\|v\|$, is $|Q(v)|^{\frac{1}{2}}$ $(=|b(v, v)|^{\frac{1}{2}})$ *v and w are orthogonal* if $b(v, w) = 0$. Since, in the semidefinite case, we have the Cauchy–Schwarz inequality, we can in that case define *the angle between v and w* by $\cos\vartheta = |b(v, w)|/(\|v\|\,\|w\|)$ when neither $\|v\|$ nor $\|w\|$ are zero.

With a given b, we can decompose V into a direct sum of subspaces: $V = V_+ \oplus V_- \oplus V_0$. V_0 is the null space of b. b is positive definite on V_+ and not on any larger subspace of V, and b is negative definite on V_- and not on any larger subspace of V. The choice of V_+ is not unique, though the dimension is. Similarly for V_-. See Greub (1981, p. 265, ff).

Theorem 5.20. *Given b we can choose an orthonormal basis for V.*

Proof. (i) The case $V = V_+$. (The Gram–Schmidt process.) Let $\{e_1, \ldots, e_n\}$ be a basis of V_+. We construct an orthogonal basis $\{\bar{e}_1, \ldots, \bar{e}_n\}$ as follows. Let $\bar{e}_1 = e_1$. Let $\bar{e}_2 = e_2$—the projection of e_2 on $\bar{e}_1$,

$$= e_2 - \|e_2\| \frac{b(e_2, \bar{e}_1)}{\|e_2\|\,\|\bar{e}_1\|} \frac{\bar{e}_1}{\|\bar{e}_1\|} = e_2 - \frac{b(e_2, \bar{e}_1)}{b(\bar{e}_1, \bar{e}_1)} \bar{e}_1$$

Let $\bar{e}_3 = e_3$—the projection of e_3 *on the plane of* $(\bar{e}_1, \bar{e}_2)$,

$$= e_3 - \frac{b(e_3, \bar{e}_2)}{b(\bar{e}_2, \bar{e}_2)} \bar{e}_2 - \frac{b(e_3, \bar{e}_1)}{b(\bar{e}_1, \bar{e}_1)} \bar{e}_1 \qquad \text{and so on}$$

These vectors will be orthogonal. Now divide each $\bar{e}_i$ by its magnitude.

(ii) For the general case, $V_+ \oplus V_- \oplus V_0$, we can construct a basis for V_+ and a basis for V_- as in (i), and any basis of V_0 will be orthonormal. Together these will form an orthonormal basis of $V_+ \oplus V_- \oplus V_0$ (loc. cit.). ■

In terms of an orthonormal basis of V, b has the form

$$b = \varepsilon^1 \otimes \varepsilon^1 + \cdots + \varepsilon^r \otimes \varepsilon^r - \varepsilon^{r+1} \otimes \varepsilon^{r+1} - \cdots - \varepsilon^m \otimes \varepsilon^m \qquad (5.20)$$

The matrix, in this basis, of the linear mapping $b^\flat\colon V \to V^*$ corresponding to b is

$$\begin{pmatrix} I_r & 0 & 0 \\ 0 & -I_s & 0 \\ 0 & 0 & 0 \end{pmatrix}$$

where I_r and I_s are respectively $r \times r$ and $s \times s$ identity matrices. In matrix language, Theorem 5.17 says that every real symmetric matrix is congruent to a diagonal matrix with 1's, -1's, and 0's. See Problem 2.8.

To illustrate these results, we again consider the space $V = \mathbb{R}^n$, and suppose b is given by its components in the natural bases as in the examples above. The natural basis is not an orthonormal basis unless b has the form (5.20). In example (1), b has the form (5.20) so the natural basis is orthonormal. In example (2), there is a basis for which $b = \varepsilon^1 \otimes \varepsilon^1 - \varepsilon^2 \otimes \varepsilon^2$. In example (3), we can get $b = -\varepsilon^1 \otimes \varepsilon^1 - \varepsilon^2 \otimes \varepsilon^2$. In example (4) we can get $b = -\varepsilon^1 \otimes \varepsilon^1$. And in example (5) the natural basis is again orthonormal. Going one step further for example (3), since

$$\begin{pmatrix} 0 & 1 \\ 2(\frac{1}{3})^{\frac{1}{2}} & (\frac{1}{3})^{\frac{1}{2}} \end{pmatrix} \begin{pmatrix} -1 & \frac{1}{2} \\ \frac{1}{2} & -1 \end{pmatrix} \begin{pmatrix} 0 & 2(\frac{1}{3})^{\frac{1}{2}} \\ 1 & (\frac{1}{3})^{\frac{1}{2}} \end{pmatrix} = \begin{pmatrix} -1 & 0 \\ 0 & -1 \end{pmatrix}$$

$$\begin{pmatrix} -1 & \frac{1}{2} \\ \frac{1}{2} & -1 \end{pmatrix} \quad \text{and} \quad \begin{pmatrix} -1 & 0 \\ 0 & -1 \end{pmatrix}$$

are congruent. Alternatively, according to (4.5),

$$\begin{pmatrix} 0 & 2(\frac{1}{3})^{\frac{1}{2}} \\ 1 & (\frac{1}{3})^{\frac{1}{2}} \end{pmatrix}$$

is the change of basis matrix, (a^i_j), so using (1.3) we find that for this b, $\{(0, 1), (2(\frac{1}{3})^{\frac{1}{2}}, (\frac{1}{3})^{\frac{1}{2}})\}$ is an orthonormal basis for $\mathbb{R}^2$.

Definitions

In a decomposition of V by b, the dimension of V_- is called *the index* of b. *A Euclidean vector space*, E^n_0, is a scalar (inner) product vector space with index 0. *A Lorentzian vector space*, E^n_1 is a scalar product vector space with index 1 (or $n - 1$). An affine space (Problem 1.15) whose vector space is E^n_0 is called *a Euclidean affine space*, and is denoted by $\mathscr{E}^n_0$. An affine space whose vector space is E^n_1 is *a Lorentzian affine space*, $\mathscr{E}^n_1$.

In Lorentzian vector spaces, V can be decomposed as the direct sum of 1-dimensional timelike subspaces, and $n-1$-dimensional spacelike subspaces. Lorentzian vector spaces form the basis of Einstein's theory of relativity, whose "bizarre" geometry comes from the fact that Theorem 5.17 is no longer valid.

Theorem 5.21. (**Backward Cauchy–Schwarz inequality**) *If v and w are timelike vectors of a Lorentzian vector space, then*

$$[b(v, w)]^2 \geq b(v, v)b(w, w)$$

Proof. We can write $w = av + z$ where $z \in V_+$. Then

$$b(w, w) = a^2 b(v, v) + b(z, z)$$

and $b(v, w) = ab(v, v)$. So

$$\begin{aligned} [b(v, w)]^2 = a^2[b(v, v)]^2 &= b(v, v)[b(w, w) - b(z, z)] \\ &= b(v, v)b(w, w) - b(v, v)b(z, z). \end{aligned}$$

Since the last term is positive, we have our result. ■

Two timelike vectors of a Lorentzian space cannot be orthogonal (Problem 5.23), so either $b(v, w) > 0$ or $b(v, w) < 0$. In the latter case we say that v and w are *forward-facing* or *in the same time-cone*. In that case we can define a function, ϑ, of v and w, by

$$\cosh \vartheta = \frac{-b(v, w)}{\|v\| \, \|w\|}$$

called *the hyperbolic angle between v and w.*

Theorem 5.22. *If v and w are timelike, then*

(i) $b(v, w) < 0 \Rightarrow v + w$ *is timelike and forward-facing.*
(ii) $b(v, w) > 0 \Rightarrow v + w$ *is spacelike.*

Proof. (i) $b(v + w, v + w) = b(v, v) + 2b(v, w) + b(w, w) < 0$ and $b(v + w, v) = b(v, v) + b(v, w) < 0$.

(ii) Apply the backward Cauchy–Schwarz inequality to the expansion in part (i). ■

Finally, we will be interested in mappings of inner product spaces which preserve their structures. If $\phi: V \to W$ is a liner mapping, b is a symmetric bilinear function in $\mathsf{S}^2(V^*)$, and B is a bilinear function in $\mathsf{S}^2(W^*)$ then ϕ is an *isometry* if

$$B(\phi \cdot v_1, \phi \cdot v_2) = b(v_1, v_2) \tag{5.21}$$

Using the notation of eq. (4.10), we can write this as

$$\phi^2 \cdot B = b$$

and we say b is *the pull-back of B*.

Frequently, the term isometry is reserved for the case where V and W have the same dimension. In particular, if ϕ is a linear transformation on a space V with inner product b, then ϕ is called *an orthogonal transformation*. If b is represented in an orthogonal basis by

$$\begin{pmatrix} I_r & & \\ & -I_s & \\ & & 0 \end{pmatrix}$$

then ϕ is an orthogonal transformation iff the matrix (ϕ_{ij}) of ϕ in that basis satisfies

$$(\phi_{ij})^{\mathrm{tr}} \begin{pmatrix} I_r & & \\ & -I_s & \\ & & 0 \end{pmatrix} (\phi_{ij}) = \begin{pmatrix} I_r & & \\ & -I_s & \\ & & 0 \end{pmatrix} \tag{5.22}$$

As an example, if $V = \mathbb{R}^2$ and b is given by

$$\begin{pmatrix} 1 & 0 \\ 0 & -1 \end{pmatrix},$$

then the transformations represented by

$$\begin{pmatrix} \cosh \vartheta & \sinh \vartheta \\ \sinh \vartheta & \cosh \vartheta \end{pmatrix}$$

for all real ϑ are orthogonal.

(ii) *Properties of* $\Lambda^2(V^*)$. First of all, for $b \in \Lambda^2(V^*)$, $\phi_2 \cdot b = -\phi_1 \cdot b$, and we write $b^\flat = \phi_1 \cdot b$. Again the null space of b is the null space of $b^\flat$ and b is nondegenerate iff its null space is just zero.

Now, it is clear that we cannot emulate for skew-symmetric tensors the classification of elements in terms of the values of $b(v, v)$ as we did for $S^2(V^*)$, since $b(v, v) = 0$ now for all v. However, given an element, b, of $\Lambda^2(V^*)$ we can get a result corresponding to Theorem 5.20; i.e., we can find a basis of V in terms of which b has a "canonical" representation.

Theorem 5.23. *Given $b \in \Lambda^2(V^*)$, we can choose a basis $\{\Xi^1, \dots, \Xi^k\}$ of the range of the linear map $b^\flat$ such that*

$$b = \Xi^1 \wedge \Xi^2 + \Xi^3 \wedge \Xi^4 + \cdots + \Xi^{2p-1} \wedge \Xi^{2p} \tag{5.23}$$

with $2p = k$.

Proof. If $\{\varepsilon^i\}$ is a basis of V^*, then

$$\begin{aligned} b &= a_{12}\varepsilon^1 \wedge \varepsilon^2 + a_{13}\varepsilon^1 \wedge \varepsilon^3 + \cdots + a_{1n}\varepsilon^1 \wedge \varepsilon^n \\ &\quad + a_{23}\varepsilon^2 \wedge \varepsilon^3 + \cdots + a_{2n}\varepsilon^2 \wedge \varepsilon^n + \sum_{2<i<j} a_{ij}\varepsilon^i \wedge \varepsilon^j \\ &= \varepsilon^1 \wedge \Xi^1 + \varepsilon^2 \wedge \Sigma^2 + \sum_{2<i<j} a_{ij}\varepsilon^i \wedge \varepsilon^j \end{aligned} \tag{5.24}$$

where

$$\begin{aligned} \Xi^1 &= a_{12}\varepsilon^2 + \cdots + a_{1n}\varepsilon^n = a_{12}\varepsilon^2 + \Sigma^1 \\ \Sigma^2 &= a_{23}\varepsilon^3 + \cdots + a_{2n}\varepsilon^n \end{aligned} \tag{5.25}$$

We can assume $a_{12} \neq 0$, so from (5.25),

$$\varepsilon^2 = \frac{\Xi^1}{a_{12}} - \frac{\Sigma^1}{a_{12}} \tag{5.26}$$

Substituting (5.26) into (5.24) we get

$$\begin{aligned} b &= \varepsilon^1 \wedge \Xi^1 + \left(\frac{\Xi^1}{a_{12}} - \frac{\Sigma^1}{a_{12}}\right) \wedge \Sigma^2 + \sum_{2<i<j} a_{ij}\varepsilon^i \wedge \varepsilon^j \\ &= \Xi^1 \wedge \left(\frac{\Sigma^2}{a_{12}} - \varepsilon^1\right) - \frac{\Sigma^1 \wedge \Sigma^2}{a_{12}} + \sum_{2<i<j} a_{ij}\varepsilon^i \wedge \varepsilon^j \end{aligned}$$

so

$$b = \Xi^1 \wedge \Xi^2 + \sum_{2<i<j} \tilde{a}_{ij}\varepsilon^i \wedge \varepsilon^j \tag{5.27}$$

where

$$\Xi^2 = \frac{\Sigma^2}{a_{12}} - \varepsilon^1 = \frac{a_{23}\varepsilon^3 + \cdots + a_{2n}\varepsilon^n}{a_{12}} - \varepsilon^1$$

Let $\{e_i\}$ be the basis of V dual to $\{\varepsilon^i\}$ then,

$$\Xi^1(e_i) = a_{1k}\varepsilon^k(e_i) = a_{1i} \qquad \text{and} \qquad b(e_1, e_i) = \frac{a_{1i}}{2}$$

for $i \neq 1$ and $\Xi^1(e_1) = b(e_1, e_1) = 0$ so $\Xi^1(e_i) = b(2e_1, e_i)$ for all e_i, and hence, $\Xi^1 = b(2e_1, -)$. That is, Ξ^1 is the image of $2e_1$ under the linear map $b^\flat$, so Ξ^1 is in the range of $b^\flat$. Similarly, for Ξ^2. Clearly, Ξ^1 and Ξ^2 are linearly independent.

Now, either there is a nonzero term in the sum on the right side of (5.27), or all $\tilde{a}_{ij} = 0$, In the first case we can split $\sum_{2<i<j} \tilde{a}_{ij}\varepsilon^i \wedge \varepsilon^j$ into a term $\Xi^3 \wedge \Xi^4$ plus a sum containing two fewer ε's (by the same method we used on b). In any case we will eventually get $b = \Xi^1 \wedge \Xi^2 + \Xi^3 \wedge \Xi^4 + \cdots + \Xi^{2p-1} \wedge \Sigma^{2p}$ where the Ξ's are all linearly independent and are all in the image of $b^\flat$.

Now it only remains to show that $\{\Sigma^1, \ldots, \Xi^{2p}\}$ is a basis of the range of $b^\flat$. Extend $\{\Xi^1, \ldots, \Xi^{2p}\}$ to a basis of V^* and let $\{\bar{e}_1, \ldots, \bar{e}_n\}$ be the dual basis of V. If i is odd and $\leq 2p - 1$, then $b(\bar{e}_i, -)\cdot\bar{e}_j = b(\bar{e}_i, \bar{e}_j) = (\Xi^{i+1}/2)\bar{e}_j$ for all j, so $b^\flat(\bar{e}_i) = b(\bar{e}_i, -) = \Xi^{i+1}/2$ for i odd and $\leq 2p - 1$. Similarly $b^\flat(\bar{e}_i) = -\Xi^{i-1}/2)$ for i even and $\leq 2p$, and $b^\flat(\bar{e}_i) = 0$ for $i > 2p$. Hence $b^\flat$ takes every $v \in V$ into the space $\langle \Xi^1, \ldots, \Xi^{2p} \rangle$. Since all Ξ^i are in the range of $b^\flat$ and $\{\Xi^1, \ldots, \Xi^{2p}\}$ is a linearly independent set, it is a basis of the range of $b^\flat$. ■

Corollary. *The rank of $b \in \Lambda^2(V^*)$ (defined to be the rank of $b^\flat$) is even, and b can be nondegenerate only if* dim V *is even.*

Definition

A basis for V with respect to which b is written as eq. (5.23) is *a symplectic basis.* If b is nondegenerate, V is called *a symplectic vector space.*

Instead of (5.23), it is more common to write b in terms of the basis of Theorem 5.23 using a different indexing:

$$b = \varepsilon^1 \wedge \varepsilon^{1+p} + \varepsilon^2 \wedge \varepsilon^{2+p} + \cdots + \varepsilon^p \wedge \varepsilon^{2p} \tag{5.28}$$

The matrix in this basis of the linear mapping $b^\flat$, corresponding to b, is $\frac{1}{2}J$,

where

$$J = \begin{pmatrix} 0 & I & 0 \\ -I & 0 & 0 \\ 0 & 0 & 0 \end{pmatrix} \tag{5.29}$$

and I is the $p \times p$ identity matrix. Note that $J^2: V \to V$ and $J^2 = -id$. (See Problem 1.14.) Theorem 5.23, in matrix terms, says that every skew-symmetric matrix is congruent to a matrix of type (5.29).

By means of $b \in \Lambda^2(V^*)$ we can identify certain subspaces of V just as we did for elements of $S^2(V^*)$. Thus, for $W \subset V$, let $W^\perp$ be the orthogonal complement of W with respect to b, then from $\dim b(W, -) = \dim V - \dim W^\perp$ and $\dim W = \dim (W \cap V^\perp) + \dim b(W, -)$ (Theorem 1.4) we get

$$\dim W + \dim W^\perp = \dim V + \dim (W \cap V^\perp) \tag{5.30}$$

We can use (5.30) to prove the following result.

Theorem 5.24. *If* $W \subset W^\perp$, *then* $W = W^\perp$ *iff* $\dim W = \frac{1}{2}(\dim V + \dim V^\perp)$.

Proof. Problem 5.27.

Definition
A subspace W of V is (i) *Lagrangian* if $W = W^\perp$; (ii) *isotropic* if $W \subset W^\perp$; and (iii) *coisotropic* if $W^\perp \subset W$.

From eq. (5.30) we get that if W is isotropic (coisotropic) then $\dim W \leq (\geq) \frac{1}{2}(\dim V + \dim V^\perp)$.

It can be shown that every isotropic subspace can be enlarged to a Lagrangian subspace, and with every coisotropic subspace, W, we can associate a symplectic vector space, $W/W^\perp$.

We can illustrate these ideas in $\mathbb{R}^n$ (if the space is to be symplectic, n must be even). Thus, in $\mathbb{R}^2$, if W is the set of pairs $\{p = (a_1 t, a_2 t): t \in \mathbb{R}\}$, then $W^\perp = \{q: b(q, p) = 0 \, \forall \, p \in W\}$ is W, so W is Lagrangian.

In addition to $\mathbb{R}^{2n}$, there are other important examples of symplectic vector spaces

1. Given V, the direct sum $V \oplus V^*$ with

$$b(v + \sigma, w + \tau) = \langle \sigma, v \rangle - \langle \tau, w \rangle$$

is a symplectic space, and conversely, every symplectic space can be represented as a direct sum.

2. In Problem 1.14 we saw that for some $2n$-dimensional vector spaces (e.g., $\mathbb{R}^{2n}$) there is a complex structure, J. Suppose V also has a *Hermitian inner product*, that is, an inner product, b, such that $b(Jv, Jw) = b(v, w)$. Then with $B(v, w) = b(v, Jw)$, V is a symplectic vector space. Straightforward calculations show that B is skew-symmetric, nondegenerate and $B(Jv, Jw) = B(v, w)$.
3. Symplectic vector spaces abound in mathematical physics. (cf., Guillemin and Sternberg, 1984). In particular, the "tangent spaces" of the "phase space" of a mechanical system have a natural symplectic structure. We will elucidate this remark in detail in Chapter 18 when we study mechanics. We have to introduce other ideas, such as manifolds, before we can explain further.

Finally, we have the concept of a *symplectic mapping*, analogous to the concept of isometry introduced at the end of subsection (i); that is, a linear mapping $\phi: V \to W$ which preserves the structures $b \in \Lambda^2(V^*)$ and $B \in \Lambda^2(W^*)$ according to eq. (5.21) just as before. The matrix form of the condition for a linear transformation is the same as before (eq. (5.22)) with

$$\begin{pmatrix} I_r & & \\ & -I_s & \\ & & 0 \end{pmatrix}$$

replaced by J, eq. (5.29).

(This may be a good point at which to insert some parenthetical remarks of a more general, but less precise nature. We have already encountered, and will continue to encounter, situations which could be said to fit into the vague general context of having a set, which (i) has a certain "structure" defined on it, and (ii) "undergoes changes." A set with only one of these things does not define much. We have a defined concept if we have both (i) and (ii) and they are linked by "the changes preserve a given structure," or "the structures are preserved under a given change." These two opposite ways of looking at the same situation occur frequently. In the present case, when we defined symplectic transformations we focused on the transformations which preserve a given structure, and when we defined Hermitian inner products we focused on the structures which are preserved under a given transformation. The formal relationship, $b(Jv, Jw) = b(v, w)$ is the same for both problems, but the point of view is different.)

PROBLEM 5.18. (i) Show that Q satisfies the *parallelogram property*: $Q(v + w) + Q(v - w) = 2(Q(v) + Q(w))$ and $Q(v) = Q(-v)$. (ii) If $Q: V \to \mathbb{R}$ satisfies the parallelogram property, then b defined by (5.19) is bilinear and symmetric. (See Greub, 1981, p. 262.)

PROBLEM 5.19. Prove Theorem 5.18.

PROBLEM 5.20. Show there exist null vectors not in the null space iff b is indefinite.

PROBLEM 5.21. If b is semidefinite, we have the triangle (Minkowski) inequality:

$$(b(v + w, v + w))^{\frac{1}{2}} \leq (b(v, v))^{\frac{1}{2}} + (b(w, w))^{\frac{1}{2}}$$

PROBLEM 5.22. (i) We gave examples of elements of $S^2(V^*)$ having all possible combinations of properties except one. Exemplify this last possibility. (ii) Find subspaces V_+, V_-, and V_0 for each example.

PROBLEM 5.23. Show that two timelike vectors of a Lorentzian vector space cannot be orthogonal.

PROBLEM 5.24. For forward-facing timelike vectors v and w of a Lorentzian vector space, we have the backward triangle inequality

$$[b(v + w, v + w)]^{\frac{1}{2}} \geq [b(v, v)]^{\frac{1}{2}} + [b(w, w)]^{\frac{1}{2}}$$

PROBLEM 5.25. Suppose P is a hyperplane of a Lorentzian vector space, E_1^n and n is a vector in E_1^n such that $b(n, t) = 0$ for all $t \in P$. Then

$$b(n, n) < 0 \Leftrightarrow P \text{ is a Euclidean vector space}$$

$$b(n, n) > 0 \Leftrightarrow P \text{ is a Lorentzian vector space}$$

$$b(n, n) = 0 \Leftrightarrow P = V_+ \otimes V_0 \text{ where } V_0 \text{ is 1-dimensional.}$$

(Hint: Choose a basis for E_1^n such that $n = n^0 e_0 + n^1 e_1$ where $e_0 \in V_-$ and $e_1 \in V_+$.)

PROBLEM 5.26. If $b \in S^2(V^*)$ we can define a mapping $B: \Lambda^p(V^*) \times \Lambda^p(V^*) \to \mathbb{R}$ by $\langle \tau^1 \wedge \cdots \wedge \tau^p, \sigma^1 \wedge \cdots \wedge \sigma^p \rangle \mapsto \det(b(\tau^i, \sigma^j))$. Show that B is a nondegenerate element of $S^2(\Lambda^p(V^*))$; that is, B is a scalar (inner) product on $\Lambda^p(V^*)$.

PROBLEM 5.27. Prove Theorem 5.24.

PROBLEM 5.28. Classify subspaces of $\mathbb{R}^4$ considered as a symplectic vector space.

6

EXTERIOR (GRASSMANN) ALGEBRA

We will imbed collections of the spaces V_s^r into larger structures containing them. We will then focus primarily on the exterior, or Grassmann, algebras, ΛV, and ΛV^*.

6.1 Tensor algebras

Each of the spaces V_s^r, V_0^r, $\mathsf{S}^r(V)$, $\Lambda^s(V^*)$, etc., is a vector space, and so we can add tensors within each of these spaces and multiply by scalars, and get elements within the same space. But we do not add tensors from two different spaces.

On the other hand, according to Section 4.3, we have a multiplication for tensors from any two tensor product spaces. As we saw there, this operation in the set $\bigcup_p V_0^p$ (or, more generally in $\bigcup_{p,q} V_q^p$) is bilinear and associative, but unlike addition and scalar multiplication, which form standard algebraic structures; i.e., vector spaces, the operation of tensor multiplication in $\bigcup_p V_0^p$ (or $\bigcup_{p,q} V_q^p$) does not. Moreover, it would be desirable to have addition and multiplication of tensors defined on the same set. We can accomplish this by essentially imbedding all our spaces in one large set which can be given the structure of an algebra.

There are several tensor algebras. To be specific we will look at *the contravariant tensor algebra.* That is, we look at the spaces $\mathbb{R}$, V, $V_0^2, \ldots, V_0^p, \ldots$. (Forget about the spaces $V_0^p \otimes V_0^q$, etc.). Now, form the (weak) direct sum of these spaces; that is, the vector space consisting of all sequences $(A_0, A_1, A_2, \ldots, A_p, \ldots)$ where $A_0 \in \mathbb{R}$, $A_1 \in V, \ldots, A_p \in V_0^p, \ldots$ and all but a finite number of A's are zero. This is a vector space, $\oplus V_0^r$, where addition and scalar multiplication are defined "componentwise".

Finally, for $V = (A_0, A_1, \ldots)$ and $\mathbf{W} = (B_0, B_1, \ldots)$ in $\oplus V_0^r$ we define multiplication in $\oplus V_0^r$ by

$$\mathbf{VW} = \left(A_0B_0, A_0B_1 + A_1B_0, \ldots, \sum_{i+j=p} A_iB_j, \ldots \right) \tag{6.1}$$

where A_iB_j is the product defined in Section 4.3.

Note that for the special cases $\mathbf{V} = (0, \ldots, A, \ldots, 0, \ldots)$ and $\mathbf{W} = (0, \ldots, B, \ldots, 0, \ldots)$ where all components except the pth component of $\mathbf{V}$ and the qth component of $\mathbf{W}$ are zero, $\mathbf{VW} = (0, \ldots, AB, \ldots, 0, \ldots)$ so AB is a special case of (6.1).

Theorem 6.1. *With the operation of multiplication defined above, $\oplus V_0^r$ is an associative algebra; i.e., a vector space (or, more generally, an $\mathbb{A}$-module) with an associative bilinear multiplication.*

Proof. Problem 6.1.

Evidently a tensor algebra, $\oplus V_s^0$, can also be made out of the spaces $\mathbb{R}$, V^*, V_2^0, $V_3^0, \ldots$, and one can be made out of all of the V_s^r. There are also *the symmetric algebras* $\mathsf{S}V$ and $\mathsf{S}V^*$, built out of the symmetric subspaces $\mathsf{S}^r(V) \subset V_0^r$ or out of the $\mathsf{S}^s(V^*) \subset V_s^0$. Finally, there are *the exterior algebras* ΛV and ΛV^* (or Grassmann algebras), formed from the spaces $\Lambda^r(V) \subset V_0^r$ of r-vectors, or from the spaces $\Lambda^s(V^*) \subset V_s^0$ of s-forms. We will now focus our attention on the exterior algebras.

PROBLEM 6.1. Prove Theorem 6.1.

6.2 Definition and properties of the exterior product

Since the exterior algebra ΛV^* is constructed from the direct sum of the spaces $\Lambda^s(V^*)$, its elements are $n + 1$-tuples $(A_0, A_1, \ldots, A_n)$ where $A_0 \in \mathbb{R}$, $A_1 \in V^*, \ldots, A_n \in \Lambda^n(V^*)$. We have to define products so they have this form. We can make a definition like (6.1) if we first define a product in $\bigcup_s \Lambda^s(V^*)$ where $0 \leq s \leq \dim V^*$.

Definition
For $A \in \Lambda^p(V^*)$ and $B \in \Lambda^q(V^*)$ *the exterior* (or *wedge*, or *skew-symmetric*) *product of A and B* is the element $A \wedge B \in \Lambda^{p+q}(V^*)$ defined by

$$A \wedge B = \mathfrak{A} \cdot AB \tag{6.2}$$

Definition
A p, q shuffle is a permutation π^*, of $1, 2, \ldots, p + q$ such that

$$\pi^*(1) < \cdots < \pi^*(p)$$

and

$$\pi^*(p + 1) < \cdots < \pi^*(p + q).$$

Theorem 6.2. (i) *The values of* $A \wedge B$ *are* $A \wedge B(v_1, \ldots, v_{p+q})$

$$= \frac{p!q!}{(p+q)!} \sum_* (\operatorname{sgn} \pi) A(v_{\pi(1)}, \ldots, v_{\pi(p)}) B(v_{\pi(p+1)}, \ldots, v_{\pi(p+q)})$$

where $\sum_*$ *is the sum over* p, q *shuffles.*

(ii) *The components of* $A \wedge B$ *in a wedge product basis in* $\Lambda^{p+q}(V^*)$ *are*

$$\sum_* (\operatorname{sgn} \pi) A_{i_{\pi(1)} \cdots i_{\pi(p)}} B_{i_{\pi(p+1)} \cdots i_{\pi(p+q)}}$$

where $A_{i_{\pi(1)} \cdots i_{\pi(p)}}$ *and* $B_{i_{\pi(p+1)} \cdots i_{\pi(p+q)}}$ *are the components of* A *and* B*, respectively, in wedge product bases.* (Compare Theorem 4.5.)

Proof. (i) $A \wedge B(v_1, \ldots, v_{p+q}) = \mathfrak{A} \cdot AB(v_1, \ldots, v_{p+q})$

$$= \frac{1}{(p+q)!} \sum_\pi (\operatorname{sgn} \pi) AB(v_{\pi(1)}, \ldots, v_{\pi(p+q)})$$

$$= \frac{1}{(p+q)!} \sum_\pi (\operatorname{sgn} \pi) A(v_{\pi(1)}, \ldots, v_{\pi(p)}) B(v_{\pi(p+1)}, \ldots, v_{\pi(p+q)})$$

$$= \frac{1}{(p+q)!} [p!q!\, A(v_1, \ldots, v_p) B(v_{p+1}, \ldots, v_{p+q})$$

$+ (-1)^p p!q!\, A(v_2, \ldots, v_{p+1}) B(v_1, v_{p+2}, \ldots, v_{p+q}) +$ terms in which at least one argument of A is interchanged with one argument of B, and in each term the indices are ordered as in a shuffle]

(ii) From part (i), $(p+q)! A \wedge B(e_{i_1}, \ldots, e_{i_{p+q}})$

$$= \sum_* (\operatorname{sgn} \pi) p!\, A(e_{i_{\pi(1)}}, \ldots, e_{i_{\pi(p)}}) q!\, B(e_{i_{\pi(p+1)}}, \ldots, e_{i(\pi(p+q)})$$

For $i_1 < \cdots < i_{p+q}$, the numbers on the left side are the coefficients of $\varepsilon^{i_1} \wedge \cdots \wedge \varepsilon^{i_{p+q}}$ for $A \wedge B$ in eq. (5.18); that is, they are the components of $A \wedge B$ in a wedge product basis. Since $i_{\pi(1)} < \cdots < i_{\pi(p)}$, $p!\, A(e_{i_{\pi(1)}}, \ldots, e_{i_{\pi(p)}})$ is the coefficient of $\varepsilon^{i_{\pi(1)}} \wedge \cdots \wedge \varepsilon^{i_{\pi(p)}}$ for A in eq. (5.18); that is, it is the component of A in a wedge product basis. Similarly for

$$q!\, B(e_{i_{\pi(p+1)}}, \ldots, e_{i_{\pi(p+q)}}).$$

■

To illustrate, suppose $A \in \Lambda^1(V^*) = V^*$ and $B \in \Lambda^2(V^*)$, then
(i) the values of $A \wedge B$ are

$$
\begin{aligned}
A \wedge B(v, w, x) &= \mathfrak{A} \cdot AB(v, w, x) \\
&= \frac{1}{3!}(AB(v, w, x) + AB(w, x, v) + AB(x, v, w) - AB(v, x, w) \\
&\quad - AB(x, w, v) - AB(w, v, x)) \\
&= \frac{1}{3!}[(A(v)B(w, x) - A(v)B(x, w)) \\
&\quad + (A(w)B(x, v) - A(w)B(v, x)) \\
&\quad + (A(x)B(v, w) - A(x)B(w, v))] \\
&= \frac{1}{3}[A(v)B(w, x) - A(w)B(v, x) + A(x)B(v, w)]
\end{aligned}
$$

(ii) the components of $A \wedge B$ are

$$\sum_{*} (\text{sgn } \pi) A_{i_{\pi(1)}} \mathrm{B}_{i_{\pi(1)} i_{\pi(3)}} = A_i B_{jk} - A_j B_{ik} + A_k B_{ij}$$

Recall that in Section 5.3 we have already introduced exterior products in V and in V^*. We will see shortly that the two definitions are consistent. Just as in Section 5.3, there is a commonly used alternative to the definition (6.2), namely,

$$A \wedge B = \frac{(p+q)!}{p!q!} \mathfrak{A} \cdot AB$$

By introducing the factor $(p+q)!/p!q!$ in the definition we eliminate factors in some subsequent formulas such as in Theorem 6.2(i).

Theorem 6.3. *Exterior multiplication is bilinear.*

Proof. Problem 6.3.

Theorem 6.4. *Exterior multiplication is associative.*

Proof. We wish to show that for $A \in V_p^0$, $B \in V_q^0$ and $C \in V_r^0$

(i) $\mathfrak{A}\cdot(\mathfrak{A}\cdot AB)C = \mathfrak{A}\cdot ABC$, and
(ii) $\mathfrak{A}\cdot A(\mathfrak{A}\cdot BC) = \mathfrak{A}\cdot ABC$.

The proof of (i) follows from the following lemma with $t = p + q$ and $u = p + q + r$, and from the linearity of $\mathfrak{A}$.

Lemma. *For any set* $\{\sigma^1, \ldots, \sigma^u\}$ *of* u *(not necessarily distinct) elements of* V^*,

$$\mathfrak{A}\cdot(\mathfrak{A}\cdot\sigma^1\otimes\cdots\otimes\sigma^t)(\sigma^{t+1}\otimes\cdots\otimes\sigma^u) = \mathfrak{A}\cdot\sigma^1\otimes\cdots\otimes\sigma^u \tag{6.3}$$

Proof. The left-hand side is

$$\mathfrak{A}\cdot\frac{1}{t!}\left(\sum_\pi (\operatorname{sgn}\pi)\sigma^{\pi(1)}\otimes\cdots\otimes\sigma^{\pi(t)}\right)(\sigma^{t+1}\otimes\cdots\otimes\sigma^u)$$

$$= \frac{1}{t!}\sum_\pi [(\operatorname{sgn}\pi)\mathfrak{A}(\sigma^{\pi(1)}\otimes\cdots\otimes\sigma^{\pi(t)}\otimes\sigma^{t+1}\otimes\cdots\otimes\sigma^u)$$

by the distributivity of multiplication in $\oplus V_p^0$, and the linearity of $\mathfrak{A}$

$$= \frac{1}{t!}\sum_\pi (\operatorname{sgn}\pi)\left(\frac{1}{u!}\sum_\lambda (\operatorname{sgn}\lambda)\sigma^{\lambda\pi(1)}\otimes\cdots\otimes\sigma^{\lambda\pi(t)}\otimes\sigma^{\lambda(t+1)}\otimes\cdots\otimes\sigma^{\lambda(u)}\right)$$

where λ is a permutation of $1, \ldots, u$.

Now for each of the $t!/2$ even π's,

$$\sum_\lambda (\operatorname{sgn}\lambda)\sigma^{\lambda\pi(1)}\otimes\cdots\otimes\sigma^{\lambda(u)} = \sum_\lambda (\operatorname{sgn}\lambda)\sigma^{\lambda(1)}\otimes\cdots\otimes\sigma^{\lambda(u)}$$

so for each of the $t!/2$ even π's the term under $\sum_\pi$ is the same, namely,

$$\frac{1}{u!}\sum_\lambda (\operatorname{sgn}\lambda)\sigma^{\lambda(1)}\otimes\cdots\otimes\sigma^{\lambda(u)}.$$

For each of the $t!/2$ odd π's,

$$\sum_\lambda (\operatorname{sgn}\lambda)\sigma^{\lambda\pi(1)}\otimes\cdots\otimes\sigma^{\lambda(u)} = -\sum_\lambda (\operatorname{sgn}\lambda)\sigma^{\lambda(1)}\otimes\cdots\otimes\sigma^{\lambda(u)}$$

and for each of these π's, $\operatorname{sgn}\pi = -1$, so again, for each of these π's the term under $\sum_\pi$ is the same as for the even π's. Thus all $t!$ terms under $\sum_\pi$ are the same so the left side of (6.3) is $(1/u!)\sum_\lambda (\operatorname{sgn}\lambda)\sigma^{\lambda(1)}\otimes\cdots\otimes\sigma^{\lambda(u)}$ which is precisely the right side of (6.3). ■

Corollary. $\sigma^1 \wedge \cdots \wedge \sigma^s = \mathfrak{A}\cdot\sigma^1 \otimes \cdots \otimes \sigma^s$.

Proof. (i) For $s = 3$,

$$\begin{aligned}\mathfrak{A}\cdot(\sigma^1 \otimes \sigma^2 \otimes \sigma^3) &= \mathfrak{A}\cdot[(\mathfrak{A}\cdot(\sigma^1 \otimes \sigma^2))\sigma^3] = (\mathfrak{A}\cdot(\sigma^1 \otimes \sigma^2)) \wedge \sigma^3 \\ &= (\mathfrak{A}\cdot(\sigma^1\sigma^2)) \wedge \sigma^3 = (\sigma^1 \wedge \sigma^2) \wedge \sigma^3 \\ &= \sigma^1 \wedge \sigma^2 \wedge \sigma^3,\end{aligned}$$

(ii)

$$\begin{aligned}\mathfrak{A}\cdot[(\sigma^1 \otimes \cdots \otimes \sigma^s \otimes \sigma^{s+1}) &= \mathfrak{A}\cdot[(\mathfrak{A}\cdot(\sigma^1 \otimes \cdots \otimes \sigma^s))\sigma^{s+1}] \\ &= \mathfrak{A}\cdot[(\sigma^1 \wedge \cdots \wedge \sigma^s)\sigma^{s+1}]\end{aligned}$$

by induction hypothesis, and this last expression is $\sigma^1 \wedge \cdots \wedge \sigma^{s+1}$. ■

This corollary shows that definition (6.2) is consistent with definition (5.16).

Theorem 6.5. *If* $A \in \Lambda^p(V^*)$ *and* $B \in \Lambda^q(V^*)$, *then* $A \wedge B = (-1)^{pq}B \wedge A$. (*Anticommutativity.*)

Proof. For elements of the form

$$\sigma^1 \wedge \cdots \wedge \sigma^p \in \Lambda^p(V^*)$$

and

$$\tau^1 \wedge \cdots \wedge \tau^q \in \Lambda^q(V^*)$$

we have

$$\begin{aligned}(\sigma^1 \wedge \cdots \wedge \sigma^p) \wedge (\tau^1 \wedge \cdots \wedge \tau^q) &= \sigma^1 \wedge \cdots \wedge \sigma^p \wedge \tau^1 \wedge \cdots \wedge \tau^q \\ &= (-1)^{pq}\tau^1 \wedge \cdots \wedge \tau^q \wedge \sigma^1 \wedge \cdots \wedge \sigma^p\end{aligned}$$

and A and B are linear combinations of such elements. ■

Now that we have defined a product in $\bigcup_s \Lambda^s(V^*)$ we can construct *the exterior algebra of* V^*, ΛV^*, by forming the direct sum of the spaces $\Lambda^s(V^*)$, and then defining a multiplication as in (6.1) with A_iB_j replaced by $A_i \wedge B_j$. Note that $\dim \Lambda V^* = \sum_s \dim \Lambda^s(V^*) = 2^n$.

PROBLEM 6.2. (i) Show there are $\binom{p+q}{q}$ terms in $\sum_*$, each one of which corresponds to a p, q shuffle. (ii) Write out the values and the components of $A \wedge B$ as in the illustration, if $A \in \Lambda^2(V^*)$ and $B \in \Lambda^3(V^*)$.

PROBLEM 6.3. Prove Theorem 6.3.

PROBLEM 6.4. Illustrate Theorems 6.3–6.5 with specific examples.

PROBLEM 6.5. In Corollary (i) of Theorem 5.15 we saw that $\Lambda^s(V^*)$ can be described as a quotient space. For each s we have an alternation operator and a kernel. The direct sum of these kernels is an ideal in the algebra $\oplus V_s^0$. ΛV^* is isomorphic to the quotient of $\oplus V_s^0$ by this ideal. (Greub, 1981, p. 102, ff.)

6.3 Some more properties of the exterior product

Theorem 6.6. *The set $\{v_1, \ldots, v_p\}$ of elements of V is a linearly dependent set if and only if $v_1 \wedge \cdots \wedge v_p = 0$.*

Proof. Problem 6.6.

Recall (Problem 5.13) that elements of $\Lambda^r(V)$ of the form $v_1 \wedge \cdots \wedge v_r$ are called decomposable.

Theorem 6.7. *There is a 1–1 correspondence between r dimensional subspaces W of V and 1-dimensional subspaces, Z, of $\Lambda^r(V)$ consisting of decomposable elements. Moreover, for $w \in W$ and $A \in Z$, $w \wedge A = 0$.*

Proof. Given W, let $\{e_1, \ldots, e_r\}$ be a basis of W and let $Z = \Lambda^r(W) = \{ae_1 \wedge \cdots \wedge e_r\}$. W and Z have the required properties. On the other hand, every 1-dimensional subspace of $\Lambda^r(V)$ consisting of decomposable elements has the form $Z = \{av_1 \wedge \cdots \wedge v_r\}$. If $W = \langle\{v_1, \ldots, v_r\}\rangle$ then $Z = \Lambda^r(W)$ and W and Z have the required property. Finally, $\bar{Z} = Z \Rightarrow e_1 \wedge \cdots \wedge e_r = a\bar{e}_1 \wedge \cdots \wedge \bar{e}_r \Rightarrow e_1 \wedge \cdots \wedge e_r \wedge \bar{e}_i = 0$ for all $\bar{e}_i \Rightarrow \bar{e}_i$ is a linear combination of $e_1, \ldots, e_r$ for all $\bar{e}_i \Rightarrow W = \bar{W}$. ■

Theorem 6.8. *Suppose $v \in V$ and $A \in \Lambda^r(V)$. Then there is a $B \in \Lambda^{r-1}(V)$ such that $A = v \wedge B$ (i.e., A has v as a factor) if and only if $v \wedge A = 0$.*

Proof. **Only if:** $A = v \wedge B$ implies $v \wedge A = 0$.
If: For $A \in \Lambda^r(V)$ we can write

$$\begin{aligned} A &= C^{i_1 \cdots i_r} e_{i_1} \wedge \cdots \wedge e_{i_r} \qquad i_1 < \cdots < i_r \\ &= C^{1 i_2 \cdots i_r} e_1 \wedge \cdots \wedge e_{i_r} + C^{i_1 \cdots i_r} e_{i_1} \wedge \cdots \wedge e_{i_r}, \qquad \text{with } i_1 > 1 \end{aligned}$$

Now choose a basis in V with e_1 the given vector v. Then $v \wedge A = 0$ implies $C^{i_1 \cdots i_r} = 0$ when $i_1 > 1$ and $A = e_1 \wedge C^{1 i_2 \cdots i_r} e_{i_2} \wedge \cdots \wedge e_{i_r} = v \wedge B$ with $B \in \Lambda^{r-1}(V)$. ■

Theorem 6.9. (Cartan's lemma.) *Let $\{e_i\}$ be a basis of V, and suppose v_i, $i = 1, \ldots, p$ are elements of V such that $\sum_{i=1}^{p} e_i \wedge v_i = 0$. Then $v_i = A_i^j e_j$ where $A_i^j = A_j^i$ for $i, j = 1, \ldots, p$ and $A_i^j = 0$ for $j = p + 1, \ldots, n$.*

Proof. *Problem 6.8.*

There are some important properties of exterior algebras corresponding to ones we have seen before for tensor algebras, in Section 4.3.

1. If $\phi: V \to W$ is a linear mapping, then there are induced linear maps in the tensor algebra corresponding to those we described in Section 4.3. That is, we have $\phi_r: \Lambda^r(V) \to \Lambda^r(W)$ given by

$$\phi_r: v_1 \wedge \cdots \wedge v_r \mapsto \phi \cdot v_1 \wedge \cdots \wedge \phi \cdot v_r \tag{6.4}$$

and $\phi^s: \Lambda^s(W^*) \to \Lambda^s(V^*)$ given by

$$\phi^s: \tau^1 \wedge \cdots \wedge \tau^s \mapsto \phi^* \cdot \tau^1 \wedge \cdots \wedge \phi^* \cdot \tau^s \tag{6.5}$$

In the special case when $W = V$, and $r = n = \dim V$, $\phi_n: \Lambda^n(V) \to \Lambda^n(V)$. If $\{e_1, \ldots, e_n\}$ is a basis for V, then $e_1 \wedge \cdots \wedge e_n$ is a basis for $\Lambda^r(V)$ and

$$\begin{aligned}\phi_n: e_1 \wedge \cdots \wedge e_n &\mapsto \phi_1^{i_1} e_{i_1} \wedge \cdots \wedge \phi_n^{i_n} e_{i_n} \\ &= \sum_{\pi} (\operatorname{sgn} \pi) \phi_1^{\pi(1)} \cdots \phi_n^{\pi(n)} e_1 \wedge \cdots \wedge e_n \\ &= \det(\phi_j^i) e_1 \wedge \cdots \wedge e_n\end{aligned}$$

So for any $A \in \Lambda^n(V)$

$$\phi_n \cdot A = \det(\phi_j^i) A \tag{6.6}$$

That is, ϕ_n is simply multiplication by $\det(\phi_j^i)$.

As interesting by-products, we get the results for determinants that (i) since ϕ_n is independent of the basis, $\det(\phi_j^i)$ is an invariant and we can talk about $\det \phi$, *the determinant of the linear transformation* ϕ, and (ii) from the result $(\phi \circ \psi)_r = \phi_r \circ \psi_r$ for compositions, we get that

$$\det(\phi_j^i)(\psi_k^j) = \det(\phi_j^i) \cdot \det(\psi_k^j).$$

(Compare this with the direct proof of this result.)

2. Recall that in Section 4.3, from the definition of the tensor product of two linear mappings ϕ and ψ we obtained the result that (in the special case $W = Y = \mathbb{R}$) $\phi \otimes \psi$, as a tensor product of linear functions, was the linear function that corresponded to the bilinear function $\phi \otimes \psi$ in the isomorphism $(V \otimes X)^* \cong V^* \otimes X^*$ and that the pairing $\langle \phi \otimes \psi, v \otimes x \rangle$ induced by the duality is given by $\langle \phi \otimes \psi, v \otimes x \rangle = \phi \otimes \psi(v, x)$ (see eq. (4.12)).

We can get corresponding results in exterior algebra, but we cannot get these by following the procedure of Section 4.3. (In particular, we cannot define a general exterior product of linear mappings by the pattern in Section 4.3, since if $\phi \cdot v$ and $\psi \cdot x$ are in different spaces $\phi \cdot v \wedge \psi \cdot x$ is not defined.) Rather, we define, for σ^1 and σ^2 in V^*, a linear function $\sigma^1 \wedge \sigma^2$ on $\Lambda^2(V)$ by

$$\sigma^1 \wedge \sigma^2 = \mathfrak{A} \cdot \sigma^1 \otimes \sigma^2 = \tfrac{1}{2}(\sigma^1 \otimes \sigma^2 - \sigma^2 \otimes \sigma^1)$$

where $\sigma^1 \otimes \sigma^2$ and $\sigma^2 \otimes \sigma^1$ are the linear functions defined in eq. (4.11). Then using eq. (4.12)

$$\langle \sigma^1 \wedge \sigma^2, v_1 \wedge v_2 \rangle = \sigma^1 \wedge \sigma^2(v_1, v_2)$$

More generally, we define

$$\sigma^1 \wedge \cdots \wedge \sigma^p = \mathfrak{A} \cdot \sigma^1 \otimes \cdots \otimes \sigma^p$$

where $\sigma^1 \otimes \cdots \otimes \sigma^p \in V^{p*}$. Then

$$\langle \sigma^1 \wedge \cdots \wedge \sigma^p, v_1 \wedge \cdots \wedge v_p \rangle = \sigma^1 \wedge \cdots \wedge \sigma^p(v_1, \ldots, v_p) \tag{6.7}$$

and by linearity

$$\langle A, v_1 \wedge \cdots \wedge v_p \rangle = A(v_1, \ldots, v_p) \tag{6.8}$$

where A on the left is in $\Lambda^p(V)^*$ and A on the right is in $\Lambda^p(V^*)$. Equation (6.8) gives the isomorphism between $\Lambda^p(V)^*$ and $\Lambda^p(V^*)$, and hence, the pairing of $\Lambda^p(V)$ and $\Lambda^p(V^*)$.

Note that in the special case, eq. (6.7), from Problem 5.15 we have the formula

$$\langle \sigma^1 \wedge \cdots \wedge \sigma^p, v_1 \wedge \cdots \wedge v_p \rangle = \frac{1}{p!} \det(\langle \sigma^i, v_j \rangle) \tag{6.9}$$

Finally, there are two additional important mappings in exterior algebra.

1. The duality in exterior algebras described above enables us to define a pairing of r-vectors and s-forms, $s \geq r$. This is an important special case of the mapping described in Section 4.3, consisting of the composition of tensor multiplication followed by contractions.

Definition
The (left) interior product is a bilinear map

$$\rfloor: \Lambda^r(V) \times \Lambda^s(V^*) \to \Lambda^{s-r}(V^*) \qquad \text{for } s \geq r \text{ given by } (A, B^*) \mapsto A \rfloor B^*$$

where $A \rfloor B^*$ is defined by

$$\langle A \rfloor B^*, C \rangle = \langle B^*, A \wedge C \rangle \tag{6.10}$$

for all $C \in \Lambda^{s-r}(V)$ when $s \geq r$. (Note that when $s = r$, $\Lambda^{s-r}(V) = \mathbb{R}$, C is a real number, $A \wedge C = CA$, and (6.10) reduces to $A \rfloor B^* = \langle B^*, A \rangle$).

We can illustrate this definition schematically by arranging the spaces involved in a linear order. Thus,

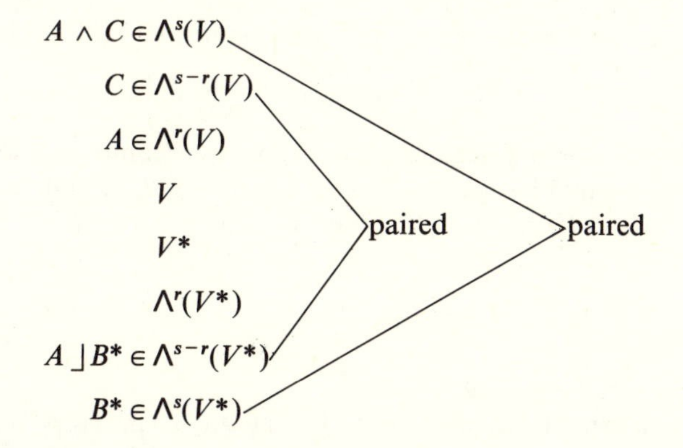

The right side of (6.10) is a pairing of an s-vector and an s-form, and the left side of (6.10) is a pairing of an $(s - r)$-vector and an $(s - r)$-form.

Definition
Given $A \in \Lambda^r(V)$, we define a linear map $A\rfloor: \Lambda^s(V^*) \to \Lambda^{s-r}(V^*)$ for $s \geq r$ by $A\rfloor \cdot B^* = A \rfloor B^*$.

Comparing eq. (6.10) with eq. (2.4) we see that the mapping $A\rfloor$ is the transpose of the mapping $A \wedge$. The mapping $sA\rfloor: \Lambda^s(V^*) \to \Lambda^{s-1}(V^*)$ for the case $r = 1$ (i.e., $A \in V$) is denoted by i_A and called *interior multiplication by A*. This case will be examined further in Chapter 11, and will be important when we get to exterior and Lie derivatives.

Definition
If we fix the second argument, B^*, in $\lrcorner$ then we get a mapping

$$\lrcorner B^*: \Lambda^r(V) \to \Lambda^{s-r}(V^*)$$

(from r-vectors to $s-r$-forms).

In the special case when $s = \dim V$, we get isomorphisms which map decomposable elements to decomposable elements, and the image of a decomposable element corresponding to a subspace W of V of dimension r (see Theorem 6.7) has as its corresponding subspace in V^* the orthogonal complement of W. Corresponding elements in such an isomorphism are called *dual tensors* in the classical literature (Synge and Schild, 1966, p. 247).

In the special case $r = 1$, ker $\lrcorner B^*$ is called *the associated subspace of* B^*. It is the orthogonal complement of the subspace of V^* generated by $A \lrcorner B^*$ where $A \in \Lambda^{s-1}(V)$. It will appear again in Section 14.3 when we discuss characteristics of forms.

2. We saw in Section 5.4, that given $b \in S^2(V^*)$, a symmetric second-order covariant tensor, then V has an orthonormal basis with respect to b.

Definition
Let $b \in S^2(V^*)$ be nondegenerate and positive definite and let $\{e_1, \ldots, e_n\}$ be an orthonormal basis of V. The set of $p, n-p$ shuffles defines a mapping of a basis of $\Lambda^p(V)$ to $\Lambda^{n-p}(V)$ by prescribing

$$*: e_{\pi(1)} \wedge \cdots \wedge e_{\pi(p)} \mapsto (\operatorname{sgn} \pi) e_{\pi(p+1)} \wedge \cdots \wedge e_{\pi(n)}$$

where π is a $p, n-p$ shuffle.

We can extend $*$ to all of $\Lambda^p(V)$ by linearity so that $*: \Lambda^p(V) \to \Lambda^{n-p}(V)$. We define such a map for each p, so finally $*: \Lambda V \to \Lambda V$. $*$ is called *the (Hodge) star operator* on ΛV.

(In Problem 6.16 we will define a mapping $\Lambda^p(V) \to \Lambda^{n-p}(V)$. That definition will require a choice of $B^* \in \Lambda^n(V^*)$ and a choice of a nondegenerate $b \in S^2(V^*)$, but does not involve any choice of bases. That definition will generalize the one above which, in particular, shows that the definition of the star operator above does not actually depend on the particular choice of orthonormal basis of V.)

As an illustration, consider the case when $\dim V = 3$. Then, if $v = v^i e_i$ and $w = w^i e_i$ we get $v \wedge w = v^i w^j e_i \wedge e_j$ and $*(v \wedge w) = v^i w^j * (e_i \wedge e_j) = (v^1 w^2 - v^2 w^1) e_3 + (v^2 w^3 - v^3 w^2) e_1 + (v^3 w^1 - v^1 w^3) e_2$. Thus, when $\dim V = 3$, $*(v \wedge w) = v \times w$, the cross-product of vector analysis.

Theorem 6.10. $b(v, w) = *(v \wedge *w)$

Proof. Problem 6.14.

(Note that in vector analysis we write $b(v, w) = v \cdot w$; that is, Theorem 6.10 gives a formula for the dot product.)

Theorem 6.11. $*\cdot* = (-1)^{p(n-p)} id$ where id is the identity map on $\Lambda^p(V)$.

Proof. Problem 6.15.

PROBLEM 6.6. Prove Theorem 6.6.

PROBLEM 6.7. Prove that for $v, w \in V$, $v \wedge w \neq 0$, and $A \in \Lambda^r(V)$, $A = v \wedge w \wedge B$ for $B \in \Lambda^{r-2}(V)$ if and only if $v \wedge A = 0$ and $w \wedge A = 0$ (cf. *Theorem 6.8.*) Generalize.

PROBLEM 6.8. Prove Theorem 6.9.

PROBLEM 6.9. Show that the mappings given by (6.4) and (6.5) are the restrictions to $\Lambda^r(V)$ and $\Lambda^s(V^*)$, respectively, of the mappings ϕ_r and ϕ^s given by (4.7) and (4.8). In particular, they have the composition properties of Problem 4.17.

PROBLEM 6.10. Show (i) that there is a 1–1 correspondence between alternating multilinear functions on the cartesian product of r factors of V and alternating linear functions on V_0^r; (ii) that there is a 1–1 correspondence between alternating linear functions on V_0^r and linear functions on $\Lambda^r(V)$; and (iii) that (i) and (ii) give the isomorphism given by eq. (6.8).

PROBLEM 6.11. If $A \in \Lambda^r(V)$ and $B^* \in \Lambda^r(V^*)$, then

$$A \rfloor B^* = C_r^r \cdots C_1^1 \cdot B^* A.$$

See Problem 4.23.

PROBLEM 6.12. If $\{e_i\}$ and $\{\varepsilon^i\}$ are dual bases of V and V^*, respectively, then

$$e_{i_1} \wedge \cdots \wedge e_{i_r} \rfloor \varepsilon^{j_1} \wedge \cdots \wedge \varepsilon^{j_s}$$
$$= \begin{cases} \dfrac{(s-r)!}{s!} (\operatorname{sgn} \pi) \varepsilon^{k_1} \wedge \cdots \wedge \varepsilon^{k_{s-r}} & \text{if } \{i_1, \ldots, i_r\} \subset \{j_1, \ldots, j_s\} \\ 0 & \text{otherwise} \end{cases}$$

Here the k's are the elements of $\{j_1, \ldots, j_s\}$ not in $\{i_1, \ldots, i_r\}$ and π is the permutation $j_1, \ldots, j_s \to i_1, \ldots, i_r, k_1, \ldots, k_{s-r}$.

PROBLEM 6.13.

(i) $$e_i \rfloor \varepsilon^{j_1} \wedge \cdots \wedge \varepsilon^{j_s} = \frac{1}{s} \sum_{k=1}^{s} (\operatorname{sgn} \pi) \delta_i^{j_k} \varepsilon^{j_1} \wedge \cdots \wedge \hat{\varepsilon}^{j_k} \wedge \cdots \wedge \varepsilon^{j_s}$$

(see Problem 6.12)

(ii) If $v \in V$ and $B^* \in \Lambda^3(V^*)$, then $i_v B^* = 3v^i B^*_{ijk} \varepsilon^j \wedge \varepsilon^k,\ j < k$.

PROBLEM 6.14. Prove Theorem 6.10.

PROBLEM 6.15. Prove Theorem 6.11.

PROBLEM 6.16. We noted above that once we choose B^* in $\Lambda^n(V^*)$ (an "orientation"), then $\rfloor B^*$ defines an isomorphism from $\Lambda^p(V)$ to $\Lambda^{n-p}(V^*)$. Further, with a nondegenerate b we have the linear mapping, $b^\flat$, from V to V^*, which has an inverse, $b^\# = (b^\flat)^{-1}$, according to the corollary of Theorem 4.1, and this induces a linear map from $\Lambda^{n-p}(V^*)$ to $\Lambda^{n-p}(V)$ according to eq. (6.4). Form the composition

$$\Lambda^p(V) \to \Lambda^{n-p}(V^*) \to \Lambda^{n-p}(V)$$

If $\{e_1, \ldots, e_n\}$ is an orthonormal basis of V, $B^* = \varepsilon^1 \wedge \cdots \wedge \varepsilon^n$, and b is positive definite, then this composition reduces to the $*$ map defined above multiplied by $(n - p)!/n!$.

7

THE TANGENT MAP OF REAL CARTESIAN SPACES

Up to now we have been discussing certain abstract vector spaces, namely, tensor product spaces. These will be important in the context of our objective of modeling physical phenomena by various mathematical spaces with mappings between them. Tensor product spaces will be part of the structure of the image spaces of our mappings. For now we will focus on their domains.

For vector spaces, in general, and $\mathbb{R}^n$, in particular, we have linear mappings, and, more generally, nonlinear mappings. For vector spaces we have an important device for dealing with the more general nonlinear mappings. We can approximate them by linear mappings—derivatives. We will recall some of the basic ideas of the differential calculus on $\mathbb{R}^n$ in the first section.

In our applications we will need to use for our domains spaces more general than $\mathbb{R}^n$, including spaces which are not vector spaces (surfaces in $\mathscr{E}^3_0$, for example). We will use spaces for which we can describe a device such as we have for vector spaces; a way of approximating mappings between such spaces (real-valued functions on surfaces in $\mathscr{E}^3_0$, for example) by linear ones—"derivatives." These spaces will be our differentiable manifolds.

The structures, tangent and cotangent spaces, that we need for the concept of a "derivative" of a mapping of differentiable manifolds can be described for the special cases of vector spaces, and, in particular, of cartesian spaces. It will turn out that we will be able to make a straightforward generalization of these concepts in the general setting of manifolds. Thus, we can temporarily avoid some of the abstract properties of manifolds, and see, in the special case of cartesian spaces, some of the important concepts that can be defined on them.

7.1 Maps of real cartesian spaces

Real cartesian n-space, $\mathbb{R}^n$, is the set of all ordered n-tuples of real numbers. A point in this space is an n-tuple $(a^1, a^2, \ldots, a^n)$, which we will frequently abbreviate by a. We will focus our attention on an arbitrary but fixed point $a \in \mathbb{R}^n$ and its neighborhoods $\mathscr{U}_a$. These are defined by the "usual" topology of $\mathbb{R}^n$; that is, the topology given by the distance $d(a, b) = \left[\sum_{i=1}^{n} (a^i - b^i)^2\right]^{1/2}$.

We can, on the one hand, consider *functions on* $\mathscr{U}_a$, $f: \mathscr{U}_a \to \mathbb{R}$, from a neighborhood of a to the reals, and, on the other hand, we can consider *curves in* $\mathbb{R}^n$ *through* a, $\gamma_a: I \to \mathbb{R}^n$, from an open interval of $\mathbb{R}$ containing 0, and such that $\gamma_a(0) = a$.

There are very important examples of functions where $\mathscr{U}_a = \mathbb{R}^n$, namely, the n *natural coordinate functions*, (or *natural projections*), $\pi^i: \mathbb{R}^n \to \mathbb{R}$ given by $a \mapsto a^i$. There are n corresponding special curves through a, $\vartheta_{ak}: \mathbb{R} \to \mathbb{R}^n$ given by $u \mapsto (a^1, a^2, \ldots, a^k + u, \ldots, a^n)$. They are the n *natural coordinate curves through* a, (or *natural injections*). Note that the π^i are linear, but the ϑ_{ak} are not.

With a given curve through a we can form its *component functions*, $\gamma_a^i = \pi^i \circ \gamma_a$. With a given function on $\mathscr{U}_a$ we can form the *partial functions* $f \circ \vartheta_{ak}$. The value of $f \circ \vartheta_{ak}$ at u is $f(a^1, a^2, \ldots, a^k + u, \ldots, a^n)$.

Since $\mathbb{R}$ and $\mathbb{R}^n$ are topological spaces, and we have the concept of continuity for mappings between any two topological spaces, we have the concept of continuity for functions and curves through a. (In particular, it is clear that π^i and ϑ_{ak} are continuous on their domains.)

Moreover, for any function, f, the $f \circ \vartheta_{ak}$ are functions from subsets of $\mathbb{R}$ to $\mathbb{R}$ so, for $k = 1, \ldots, n$, the ordinary derivatives, $D_0 f \circ \vartheta_{ak}$, at $0 \in \mathbb{R}$ are defined. We write these in the usual partial derivative notation, i.e., we define the notation $\partial f / \partial \pi^k |_a$ by

$$\left.\frac{\partial f}{\partial \pi^k}\right|_a = D_0 f \circ \vartheta_{ak}, \qquad k = 1, \ldots, n$$

Similarly, since for any curve, γ_a, through a, the $\gamma_a^i = \pi^i \circ \gamma_a$ are functions from subsets of $\mathbb{R}$ to $\mathbb{R}$ we have $D_0 \pi^i \circ \gamma_a = D_0 \gamma_a^i$, $i = 1, \ldots, n$. We read the notation on the right as "the components of the tangent of γ_a at a."

Finally, if $\phi: \mathscr{U}_a \to \mathbb{R}^p$ (where $\mathscr{U}_a \subset \mathbb{R}^n$), and if $\bar{\pi}^i$, $i = 1, \ldots, p$ are the coordinate functions on $\mathbb{R}^p$, then we can form the compositions $\phi^i = \bar{\pi}^i \circ \phi$, $i = 1, \ldots, p$, *the component functions of the map* ϕ, and we can form the compositions $\phi \circ \vartheta_{ak}$, $k = 1, \ldots, n$, *the partial maps of* ϕ (Fig. 7.1). Composing these we have

$$\phi^i \circ \vartheta_{ak} = \bar{\pi}^i \circ (\phi \circ \vartheta_{ak}), \qquad i = 1, \ldots, p, \qquad k = 1, \ldots, n$$

So, for each $i = 1, \ldots, p$ we have n functions from neighborhoods of $0 \in \mathbb{R}$ to $\mathbb{R}$, and hence the ordinary derivatives at $0 \in \mathbb{R}$,

$$D_0 \phi^i \circ \vartheta_{ak} = \left.\frac{\partial \phi^i}{\partial \pi^k}\right|_a \tag{7.1}$$

which as indicated by the notation on the right are the n partial derivatives of the p component functions ϕ^i of ϕ at a.

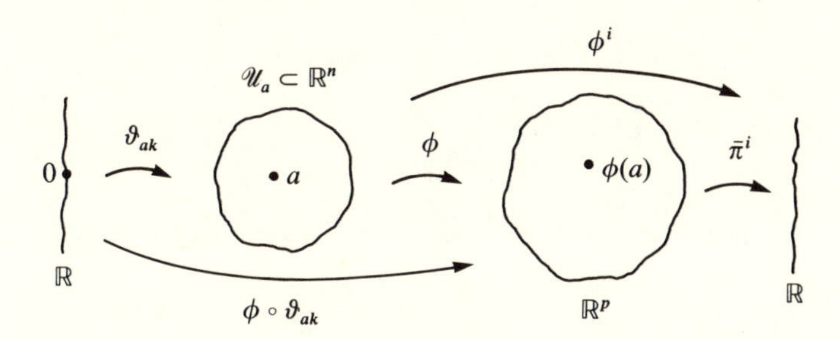

FIGURE 7.1

In summary, we point out that the ability to define partial derivatives depends on the existence in cartesian spaces of the natural projections, $\bar{\pi}^i$, and the coordinate curves ϑ_{ak}. Note that we have not yet invoked the natural vector space structure of cartesian spaces.

Now, since cartesian spaces are normed vector spaces, we can go further and make the following definitions for maps between $\mathbb{R}^n$ and $\mathbb{R}^p$ (and, more generally, for maps between any two normed vector spaces V and W.)

Definitions
Two maps ϕ and ψ (continuous at a) are *tangent* (*to each other*) *at* a if

$$\lim_{v \to a} \frac{\phi(v) - \psi(v)}{v - a} = 0$$

A map $\phi\colon \mathscr{U}_a \to \mathbb{R}^p$ ($\mathscr{U}_a \subset \mathbb{R}^n$) is *differentiable at* a if it is tangent at a to an affine map, A, that is, a map $A\colon \mathbb{R}^n \to \mathbb{R}^p$ given by $A(v) = \phi(a) + (D_a\phi)(v - a)$ where $D_a\phi$ is a linear map from $\mathbb{R}^n$ to $\mathbb{R}^p$. $D_a\phi$ is called *the derivative of* ϕ *at* a. If ϕ has a derivative at each $a \in \mathscr{U}_a \subset \mathbb{R}^n$ then $D\phi\colon \mathbb{R}^n \to \mathscr{L}(\mathbb{R}^n, \mathbb{R}^p)$ is given by $D\phi\colon a \mapsto D_a\phi$; i.e., $D\phi(a) = D_a\phi$.

From the calculus on cartesian spaces we have the results: (1) If $\phi\colon \mathscr{U}_a \to \mathbb{R}^p$ is differentiable at a, then $\partial\phi^i/\partial\pi^k|_a$ exist, and (2) If ϕ is a C^1 map at a (that is, $\partial\phi^i/\partial\pi^k|_a$ exist in a neighborhood of a and $\partial\phi^i/\partial\pi^k$ are continuous at a), then ϕ is differentiable at a. In both cases, $D_a\phi$ is the Jacobian matrix, $(\partial\phi^i/\partial\pi^k|_a)$.

In particular, if ϕ is a curve, γ, taking a to $\gamma(a)$, and if $\gamma^i = \bar{\pi}^i \circ \gamma$ are the component functions of γ, then $D_a\gamma = (D_a\gamma^i)$, a column matrix, and if ϕ is a function, f on $\mathscr{U}_a \subset \mathbb{R}^n$, then $D_a f = (\partial f/\partial\pi^i|_a)$, a row matrix.

(More generally, if vector spaces V and W are not cartesian spaces we get partial derivatives after we choose bases, and then for each choice, the derivative is *represented by* a matrix.)

Now we recall several results from calculus in $\mathbb{R}^n$; namely, a couple of "chain rules" and a "product rule."

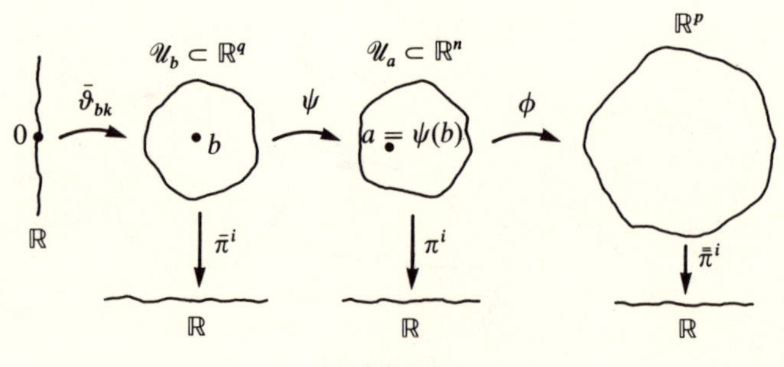

FIGURE 7.2

(i) If f is in C^1 at a, and γ_a is in C^1 at 0, then $f \circ \gamma_a$ is in C^1 at 0, and

$$D_0 f \circ \gamma_a = \left.\frac{\partial f}{\partial \pi^i}\right|_a D_0 \gamma_a^i \tag{7.2}$$

written as a sum of products of components, or

$$D_0 f \circ \gamma_a = \left(\left.\frac{\partial f}{\partial \pi^i}\right|_a\right) \circ (D_0 \gamma_a^i)$$

written as a composition of derivatives.

(ii) More generally, suppose $\mathscr{U}_b \subset \mathbb{R}^q$, $\mathscr{U}_a \subset \mathbb{R}^n$, ψ: $\mathscr{U}_b \to \mathbb{R}^n$, ϕ: $\mathscr{U}_a \to \mathbb{R}^p$ and $\psi(b) = a$ (Fig. 7.2). If ϕ is C^1 at a, and ψ is C^1 at b, then the functions $\bar{\bar{\pi}}^i \circ \phi \circ \psi \circ \bar{\vartheta}_{bk} = \phi^i \circ \psi \circ \bar{\vartheta}_{bk}$ are C^1 at b, and

$$\left.\frac{\partial \phi^i \circ \psi}{\partial \bar{\pi}^k}\right|_b = \left.\frac{\partial \phi^i}{\partial \pi^j}\right|_a \left.\frac{\partial \psi^j}{\partial \bar{\pi}^k}\right|_b \qquad \begin{cases} i = 1, \ldots, p \\ j = 1, \ldots, n \\ k = 1, \ldots, q \end{cases} \tag{7.3}$$

or, equivalently, $\phi \circ \psi$ is C^1 at b, and

$$\left(\left.\frac{\partial \phi^i \circ \psi}{\partial \bar{\pi}^k}\right|_b\right) = \left(\left.\frac{\partial \phi^i}{\partial \pi^j}\right|_a\right) \circ \left(\left.\frac{\partial \psi^j}{\partial \bar{\pi}^k}\right|_b\right)$$

(iii) For functions f, g,

$$\begin{aligned} \left.\frac{\partial fg}{\partial \pi^i}\right|_a &= D_0 fg \circ \vartheta_{ai} = D_0 (f \circ \vartheta_{ai})(g \circ \vartheta_{ai}) \\ &= f(a) \left.\frac{\partial g}{\partial \pi^i}\right|_a + g(a) \left.\frac{\partial f}{\partial \pi^i}\right|_a \end{aligned} \tag{7.4}$$

As an example of the use of this notation, consider the function described in the usual notation by $f(x, y) = \sin(xy^2)$. We write $f(x^1, x^2) = \sin(x^1(x^2)^2)$, and using the projections we can also write $f = \sin \circ (\pi^1(\pi^2)^2)$. Then

$$\left.\frac{\partial f}{\partial \pi^1}\right|_a = D_0 \sin \circ (\pi^1(\pi^2)^2) \circ \vartheta_1 = (D_{a^1(a^2)^2} \sin)(D_0\pi^1(\pi^2)^2 \circ \vartheta_1)$$

by the chain rule. Now the second factor

$$D_0\pi^1(\pi^2)^2 \circ \vartheta_1 = D_0(\pi^1 \circ \vartheta_1)(\pi^2)^2 \circ \vartheta_1) = \pi^1(a) \left.\frac{\partial(\pi^2)^2}{\partial \pi^1}\right|_a + (\pi^2(a))^2 \left.\frac{\partial \pi^1}{\partial \pi^1}\right|_a = (a^2)^2$$

So $\partial f/\partial \pi^1|_a = (a^2)^2 \cos(a^1(a^2)^2)$.

There are two other results from the differential calculus in $\mathbb{R}^n$ that we will need later; the inverse and implicit function theorems, and an existence and uniqueness theorem for ordinary differential equations. We will recall these when the need for them arises in Section 9.4 and Section 13.1, respectively.

PROBLEM 7.1. (a) Find $\partial f/\partial \pi^2|_a$ in the example above.

(b) If $f(x, y, z) = c_1x + c_2y + c_3z$ in the usual notation (c_i are constants), write f and compute $\partial f/\partial \pi^k|_a$ in our notation.

(c) If $f(x, y, z) = xyz$ in the usual notation, write f, and compute $(\partial f/\partial \pi^k)|_a$ in our notation.

PROBLEM 7.2. (a) If $f(x, y, z) = x^y$ write f in terms of the projections and compute $(\partial f/\partial \pi^i)|_a$.

(b) If $\phi(x, y, z) = (x^y, z)$ write an expression for ϕ and compute $(\partial \phi^i/\partial \pi^k)|_a$.

7.2 The tangent space and the cotangent space at a point of $\mathbb{R}^n$

In order to define a "derivative" for mappings between more general types of spaces, i.e., other than normed vector spaces, we introduce certain structures on them. Such structures can be introduced on the very special spaces $\mathbb{R}^n$, and when we eventually come to the description of our more general spaces—differentiable manifolds—it will be a straightforward matter to generalize these structures for our more general spaces.

For cartesian spaces we have available the concept of a C^1 map, i.e., a map whose domain is an open set on which it is continuously differentiable. So we can make the following definitions. (Since, for the time being, we will be working in neighborhoods of one fixed point, a, of $\mathbb{R}^n$, we will usually suppress the subscript a on the γ's, the partial derivatives, etc.)

Definitions

Two C^1 curves, γ_1 and γ_2, through a, are *equivalent* (at a) if $D_0 f \circ \gamma_1 = D_0 f \circ \gamma_2$ for all C^1 functions, f, on neighborhoods of a. Two C^1 functions, f^1 and f^2, on neighborhoods of a are *equivalent* (at a) if $D_0 f^1 \circ \gamma = D_0 f^2 \circ \gamma$ for all C^1 curves, γ, through a. These definitions define equivalence relations. The equivalence class containing γ is called *the tangent of* γ (*at* a) and will be denoted by $[\gamma]$. That is, $[\gamma] = \{\alpha: D_0 f \circ \gamma = D_0 f \circ \alpha$ for all C^1 functions on neighborhoods of $a\}$. The equivalence class containing f is called *the differential of* f (*at* a) and will be denoted by $[f]$. That is, $[f] = \{g: D_0 f \circ \gamma = D_0 g \circ \gamma$ for all C^1 curves through $a\}$.

We wish to make several observations about these definitions.

(i) Note that though, for simplicity, we are restricting ourselves to $\mathbb{R}^n$, these definitions make sense for any normed vector space, since we have the concepts of differentiable curves and functions for such spaces. With a little more effort we could carry out the following development in this more general context. However, we choose to simply go from $\mathbb{R}^n$ to the full generalization of differentiable manifolds in Chapter 9 and note there that normed vector spaces are special cases.

(ii) It is important to note that for each curve, γ, through a we have defined a tangent *at a only*. We have not defined tangents at other points of γ. This will be done later, in Section 7.3. In the present development, eventually the curve will not be important, but the point will be. Thus, "the tangent of γ" will evolve into "the tangent at a."

(iii) The curve γ_1 given by $\gamma_1(u) = (a^1 + u^2, a^2 + \sin u)$ is equivalent to the coordinate curve ϑ_2. The curve γ_2 given by $\gamma_2(u) = (a^1, a^2 + 2u)$ is not equivalent to ϑ_2. This shows, in particular, that two curves can have the same set of values and have different tangents. Note that we define the tangent of γ so that we have a unique tangent vector at each point in contrast to the common picture of lots of tangent vectors at a point having different lengths.

(iv) From the chain rule (7.2) (taking $f = \pi^i$) we see that $\gamma_1 \sim \gamma_2 \Leftrightarrow D_0\gamma_1^i = D_0\gamma_2^i$, and (taking $\gamma = \vartheta_i$), $f^1 \sim f^2 \Leftrightarrow \partial f^1/\partial\pi^i = \partial f^2/\partial\pi^i$. That is, we can say that two curves are equivalent if they have the same "tangent vector," in the usual calculus sense, and two functions are equivalent if their level surfaces have the same "normal vector" as in calculus.

Let T_a be the set of all tangents of curves at a, and let T^a be the set of all differentials of functions at a.

Theorem 7.1. *T^a is a real n-dimensional vector space.*

Proof. With the usual definitions of addition of functions and multiplications of functions by real numbers, the corresponding operations

$[f] + [g] = [(f + g)]$, and $c[f] = [cf]$ are well-defined. With these operations T^a is a vector space whose zero element is the equivalence class of functions constant on neighborhoods of a. Note that neither this theorem nor the definitions above depend on coordinate functions or coordinate curves. ∎

Theorem 7.2. *The set of differentials $\{[\pi^i]\}$ is a basis for T^a and for each $[f] \in T^a$ we have*

$$[f] = \frac{\partial f}{\partial \pi^i}[\pi^i] \tag{7.5}$$

Proof. (1) Suppose $c_i[\pi^i] = 0$. (We are using the "summation convention"). Then $[c_i\pi^i] = 0$ by the definition of operations in T^a. But $0 \in T^a$ is the equivalence class of functions which are constant on $\mathcal{U}_a$. For such a function, g, $\partial g/\partial \pi^k = 0$ and since $c_i\pi^i$ and g are equivalent $\partial c_i\pi^i/\partial\pi^k = 0$, which implies that $c_i = 0$ for all i, so $\{[\pi^i]\}$ is a linearly independent set.

(2) Let $[f]$ be an arbitrary element of T^a. For any f, $f \sim (\partial f/\partial\pi^i)\pi^i$ since the partial derivatives of both sides are equal. So $[f] = [(\partial f/\partial\pi^i)\pi^i] = \partial f/\partial\pi^i\,[\pi^i]$ and hence the $[\pi^i]$ span T^a. ∎

Definitions

T^a is called *the cotangent space at a* and $\{[\pi^i]\}$ is *the natural basis of* T^a.

Now we will also make T_a into a vector space. The procedure is not quite so direct as for T^a since $\gamma_1 + \gamma_2$ is not a curve through a. We first note that in each equivalence class there is precisely one differentiable curve, γ, of the form $\gamma^i(u) = a^i + r^i u$.*

We define $[\gamma_1] + [\gamma_2]$ to be the equivalence class containing the curve α given by $\alpha^i(u) = a^i + (r_1^i + r_2^i)u$ where γ_1 given by $\gamma_1^i(u) = a^i + r_1^i u$ is in $[\gamma_1]$ and γ_2 given by $\gamma_2^i(u) = a^i + r_2^i u$ is in$[\gamma_2]$. Similarly, $c[\gamma]$ is defined to be the equivalence class containing the curve given by $\beta^i(u) = a^i + cr^i u$, if $[\gamma]$ contains the curve given by $\gamma^i(u) = a^i + r^i u$.

Theorem 7.3. *With the given definitions of addition and scalar multiplication, T_a is a real vector space.*

Proof. Straightforward verification of vector space properties. ∎

It should be noted that T_a as defined here cannot be constructed when we generalize beyond $\mathbb{R}^n$, as addition and scalar multiplication of vectors

* This is the affine map in the class of curves tangent to each other at 0 (definition in Section 7.1). Thus, the class of curves equivalent at a is the same as the class of curves tangent to each other at 0.

cannot be defined the way we did it. However, in Problem 7.4 we make a definition of T_a which gives the same space and can be generalized to spaces where we have derivatives.

Theorem 7.4. *The set* $\{[\vartheta_k]\}$ *is a basis for* T_a, *and for each* $[\gamma] \in T_a$ *we have*

$$[\gamma] = (D_0\gamma^i)[\vartheta_i] \tag{7.6}$$

Proof. Problem 7.5.

Definitions
T_a is called *the tangent space at a*, and $\{[\vartheta_k]\}$ is *the natural basis* of T_a.

Theorem 7.5. T_a *and* T^a *are dual vector spaces and* $\{[\pi^i]\}$ *and* $\{[\vartheta_k]\}$ *are dual bases.*

Proof. We have to show that the function from $T_a \times T^a$ to $\mathbb{R}$ given by $([\gamma], [f]) \mapsto D_0 f \circ \gamma$ is nondegenerate bilinear. To show nondegeneracy, choose successively $f = \pi^1, \pi^2, \ldots$, and $\gamma = \vartheta_1, \vartheta_2, \ldots$. To show linearity in the first argument we have to show $D_0 f \circ \gamma = D_0 f \circ \gamma_1 + D_0 f \circ \gamma_2$ for γ in the equivalence class $[\gamma_1] + [\gamma_2]$ and also that $D_0 f \circ \gamma = cD_0 f \circ \gamma_1$ for γ in the equivalence class $c[\gamma_1]$. These two equalities are verified easily when we use the special representatives α and β (described above Theorem 7.3) in each of the corresponding equivalence classes. ∎

Since T_a and T^a are finite-dimensional they can be considered *the* duals of one another, i.e., $T^a = T_a^*$ and $T_a = T^{a*}$, by Theorem 2.15. Thus, in particular, the differential, $[f]$ of f can be considered to be a linear function from T_a to $\mathbb{R}$, and from eqs. (7.5), (7.6), (7.2) we have

$$\langle [f], [\gamma] \rangle = D_0 f \circ \gamma \tag{7.7}$$

In the above development we defined the tangent of a curve at a point, in terms of the numbers $D_0 f \circ \gamma$, in a way which is strongly reminiscent of the ordinary geometrical definition; i.e., a tangent is a class of curves all of which have the same "tangent vector." Now we change our point of view somewhat and notice that the numbers $D_0 f \circ \gamma$ define certain mappings. Namely, for each γ through a we have a mapping, $\mathfrak{F}_a \to \mathbb{R}$, from the set of C^1 functions defined on neighborhoods of a to the real numbers, given by $f \mapsto D_0 f \circ \gamma$. Furthermore, all the mappings corresponding to the curves in an equivalence class are the same, so with each $[\gamma]$ we have a mapping

$$L_{[\gamma]}\colon \mathfrak{F}_a \to \mathbb{R} \tag{7.8}$$

given by $f \mapsto D_0 f \circ \gamma$.

The fact that the functions in $\mathfrak{F}_a$ do not all have the same domain causes problems in trying to give $\mathfrak{F}_a$ a simple standard algebraic structure, but we can define addition and multiplication in $\mathfrak{F}_a$ (in particular, scalar multiplication) on intersections of domains in the usual way.

Theorem 7.6. *The function $L_{[\gamma]}$ is a φ-derivation of $\mathfrak{F}_a$ to $\mathbb{R}$, i.e., $L_{[\gamma]}$ is linear, φ is the evaluation map, and*

$$L_{[\gamma]} \cdot fg = f(a) L_{[\gamma]} \cdot g + g(a) L_{[\gamma]} \cdot f \tag{7.9}$$

Proof. Immediate from the properties of the derivative. ■

Theorem 7.7. *The set $\{L_{[\gamma]}: [\gamma] \in T_a\}$, with the usual definition of addition and scalar multiplication of maps, is a vector space isomorphic to T_a.*

Proof. To show that the map $[\gamma] \mapsto L_{[\gamma]}$ is linear we have to show $L_{[\gamma_1]} \cdot f + L_{[\gamma_2]} \cdot f = L_{[\gamma_1]+[\gamma_2]} \cdot f$ and $cL_{[\gamma]} \cdot f = L_{c[\gamma]} \cdot f$. That is, we must have, for all f, $D_0 f \circ \gamma_1 + D_0 f \circ \gamma_2 = D_0 f \circ \gamma$ for γ in the equivalence class of $[\gamma_1] + [\gamma_2]$ and $cD_0 f \circ \gamma = D_0 f \circ \gamma_1$ for γ_1 in the equivalence class of $c[\gamma]$. But these are the same two conditions we already verified in the proof of Theorem 7.5. ■

On the basis of Theorem 7.7 we can think of tangents of curves as derivations, $L_{[\gamma]}$, of $\mathfrak{F}_a$ to $\mathbb{R}$ instead of equivalence classes, $[\gamma]$, of curves. The concept of a tangent being a function from $\mathfrak{F}_a$ to $\mathbb{R}$ also has a motivation from ordinary calculus, where the directional derivative gives a function from $\mathfrak{F}_a$ to $\mathbb{R}$ determined by the "vector" of coefficients.

Before proceeding we will pause to bring our notation into closer conformity with more common usage. We introduced the notations $[\gamma]$ and $[f]$ to emphasize the duality of the roles of curves and functions. However, $[f]$, which we called the differential of f, does have the properties and/or behavior we attribute (usually in a vague sort of way) to the ordinary calculus differential. In particular, if in eq. (7.5) we replace $[f]$ by df and $[\pi^i]$ by $d\pi^i$ we get

$$df = \frac{\partial f}{\partial \pi^i} d\pi^i \tag{7.10}$$

a familiar notation. Defining df as a vector or, more precisely, a covector, clarifies the usual calculus explanations of the relation between differentials and derivatives, the difference between differentials of dependent and independent variables, etc.

For $[\gamma]$, the situation is somewhat different. We do not have a good standard notation from calculus. We will keep $[\gamma]$ for the tangent vector of γ, except for the coordinate curves ϑ_k. Corresponding to $[\vartheta_k]$ we have $L_{[\theta_k]}$, and

$$L_{[\theta_k]} \cdot f = D_0 f \circ \vartheta_k = \frac{\partial f}{\partial \pi^k} \tag{7.11}$$

Equation (7.11) suggests replacing $L_{[\theta_k]}$ (and $[\vartheta_k]$) by $\partial/\partial\pi^k$, and writing (7.11) as

$$\frac{\partial}{\partial \pi^k} \cdot f = \frac{\partial f}{\partial \pi^k} \tag{7.12}$$

Replacing $[\vartheta_k]$ by $\partial/\partial\pi^k$, eq. (7.6) becomes

$$[\gamma] = (D_0 \gamma^i) \frac{\partial}{\partial \pi^i} \tag{7.13}$$

or, using $L_{[\gamma]}\pi^i = D_0\gamma^i$,

$$L_{[\gamma]} = L_{[\gamma]} \cdot \pi^i \frac{\partial}{\partial \pi^i} \tag{7.14}$$

In three dimensions, the tangent vector $\partial/\partial\pi^i$ of the coordinate curves are the **i**, **j**, **k** of calculus, and (7.13) is usually written something like

$$t_\gamma = \frac{dx}{du}\mathbf{i} + \frac{dy}{du}\mathbf{j} + \frac{dz}{du}\mathbf{k}$$

Finally, using (7.10) and (7.14), (7.7) can be written in the form

$$\langle df, L_{[\gamma]} \rangle = L_{[\gamma]} \cdot f \tag{7.15}$$

Now that we have discussed the concepts of tangent space and cotangent space at a point of $\mathbb{R}^n$ we would like to have these structures available on more general spaces. Our discussion involved at times (e.g., in the definitions of addition and scalar multiplication in T_a) certain specific properties of $\mathbb{R}^n$. Any generalization would have to be independent of such properties. There are several approaches, most of which require a prior description of the spaces (differentiable manifolds), on which these concepts are to be defined.

However, while continuing to work in $\mathbb{R}^n$, a slight abstraction of our description of T_a as a certain set of derivations can be made which can be taken over to the more general spaces.

Instead of taking the specific functions $L_{[\gamma]}$ given by (7.8) for our tangents we could consider the functions from $\mathfrak{F}_a$ to $\mathbb{R}$ with one of the properties of $L_{[\gamma]}$, i.e., the set of all φ-derivations of $\mathfrak{F}_a$ to $\mathbb{R}$. This space is too large. We consider rather the set, $\{\Delta\}$, of derivations, Δ, of $\mathfrak{F}_a$ to $\mathbb{R}$ with the additional property that there exists an n-tuple $(c^1, \ldots, c^n)$ of numbers such that

$$\text{each derivation has the form } c^k \frac{\partial}{\partial \pi^k} \tag{7.16}$$

We now have a set of derivations corresponding to n-tuples $(c^1, \ldots, c^n)$, instead of corresponding to curves, γ. There are no curves involved in this definition. We call such a function simply *a tangent vector at a.*

Notice that this definition explicitly involves the structure of $\mathbb{R}^n$, in particular, we can interpret (7.16) to say that the tangent space at a point of $\mathbb{R}^n$ is just $\mathbb{R}^n$.

Theorem 7.8. *The space of tangents at a and the space of tangents of curves (given by (7.8)) are the same.*

Proof. If $L_{[\gamma]}$ is a tangent of a curve, then

$$L_{[\gamma]} \cdot f = D_0 f \circ \gamma = D_0 \gamma^i \frac{\partial}{\partial \pi^i} \cdot f$$

by (7.14) so $L_{[\gamma]}$ has the property (7.16). On the other hand, if Δ is a tangent vector, then $\Delta = c^k\, \partial/\partial \pi^k$, which is a tangent to a curve, namely, the equivalence class containing the curve with component functions $\gamma^k\colon u \mapsto a^k + c^k u$. ∎

Now we finally identify T_a with the space of derivations of C^∞ functions on neighborhoods of a to $\mathbb{R}$.

Lemma. *A derivation, v, has the property that if f is a constant function then* $v \cdot f = 0$.

Proof. Problem 7.6.

Theorem 7.9. *A derivation, v_a, on the set of C^∞ functions on neighborhoods of a satisfies property (7.16).*

Proof. For any C^1 function f we have the first-order Taylor expansion

$$f(x) = f(a) + \sum_i (\pi^i(x) - \pi^i(a)) \int_0^1 \left.\frac{\partial f}{\partial \pi^i}\right|_{a+t(x-a)} dt$$

or

$$f = f(a) + \sum_i (\pi^i - \pi^i(a)) f^i \tag{7.17}$$

where the f^i are the functions (of x) defined by the integrals. If f is in C^2 we can take the partial derivative on both sides of (7.17) and apply the product rule on the terms on the right. We get $\partial f/\partial \pi^k|_a = f^k(a)$. If v_a is a derivation and we apply it to both sides of (7.17), and use the product rule on the terms on the right, we get, using the lemma above,

$$v_a \cdot f = \sum_i v_a \cdot \pi^i (f^i(a))$$

This step can be justified only if $f \in C^\infty$, since if $f \in C^k$, $f^i \in C^{k-1}$. Thus, using $f^i(a) = \partial f/\partial \pi^i|_a$ we get

$$v_a \cdot f = \sum_i v_a \cdot \pi^i \left.\frac{\partial f}{\partial \pi^i}\right|_a$$

which is of the form (7.16). ■

Theorem 7.9 says $\{v\colon v \text{ is a derivation defined on } C^\infty \text{ functions}\} \subset \{c^k(\partial/\partial\pi^k) \text{ with domain } C^\infty \text{ functions}\}$. The inclusion the other way is clear. The space of tangent vectors were derivations of the form $c^k(\partial/\partial\pi^k)$ with domain C^1 functions; i.e.,

$$\left\{\boldsymbol{\Delta} = c^k \frac{\partial}{\partial \pi^k} \colon c^k \frac{\partial}{\partial \pi^k} \text{ has domain } C^1 \text{ functions}\right\}$$

Now there is a 1–1 correspondence between maps $c^k(\partial/\partial\pi^k)$ with domain C^∞ functions and maps $c^k(\partial/\partial\pi^k)$ with domain C^1 functions, so we can conclude that $\{v\} \cong \{\boldsymbol{\Delta}\}$.

To summarize, we have described T_a, the tangent space at a point of $\mathbb{R}^n$ in four slightly different ways: a set of equivalence classes of C^1 curves, $[\gamma]$; a set of derivations, $L_{[\gamma]}$, on $\mathfrak{F}_a$; the set of derivations, $\boldsymbol{\Delta}$ on $\mathfrak{F}_a$ with the property (7.16); and the set of all derivations on C^∞ functions.

There are at least two other ways of describing T_a. With the exception of one "generally unimportant" situation, these descriptions can all be carried over to the spaces more general than $\mathbb{R}^n$ in which we will be interested, and they will all be equivalent.

In the future we will use the notation $v_a, w_a, \ldots$ for generic elements of T_a, and $\sigma_a, \tau_a, \ldots$ for generic elements of T^a. Since, for the time being, we will be working in neighborhoods of one fixed point, a, of $\mathbb{R}^n$, we will usually suppress the subscript a.

PROBLEM 7.3. Show that $f = (\pi^1)^2/2 + (\pi^2)^2/2 + (1 - a^1)\pi^1 - a^2\pi^2$ is equivalent to π^1 at (a^1, a^2).

PROBLEM 7.4. (i) We can make T_a into a vector space without using the specific form of $\mathbb{R}^n$. Having made T^a into a vector space, we have its dual, T^{a*}. Now define a map $\rho\colon T_a \to T^{a*}$ by $\rho[\gamma]\cdot[f] = D_0 f \circ \gamma$. Show that ρ is 1–1 and define $[\gamma_1] + [\gamma_2] = \rho^{-1}(\rho[\gamma_1] + \rho[\gamma_2])$ and $c[\gamma] = \rho^{-1}(\rho c[\gamma])$.

(ii) If our space is $\mathbb{R}^n$, then this vector space is the same as the space T_a of Theorem 7.3.

PROBLEM 7.5. Prove Theorem 7.4.

PROBLEM 7.6. Prove the Lemma above Theorem 7.9.

PROBLEM 7.7. We abstracted certain properties of the functions $L_{[\gamma]}$, and called functions with these properties tangent vectors. Another set of properties of functions $\mathbf{v}\colon \mathfrak{F}_a \to \mathbb{R}$ giving the same set of functions is

(i) $\mathbf{v}$ is linear on $\mathfrak{F}_a$.
(ii) For f of the form $F \circ \psi$ where $\psi\colon \mathscr{U}_a \to \mathbb{R}^p$ and $F\colon \mathscr{U}_{\psi(a)} \subset \mathbb{R}^p \to \mathbb{R}$, and ψ and F are C^1,

$$\mathbf{v}(F \circ \psi) = \mathbf{v}(\psi^i(a)) \left.\frac{\partial F}{\partial \pi^i}\right|_{\psi(a)}$$

(cf., Willmore, 1959, p. 197 ff.)

PROBLEM 7.8. Show that the following functions are C^∞ and nonanalytic on $\mathbb{R}$ and sketch their graphs.

(a) $f(u) = \begin{cases} e^{-1/u} & u > 0 \\ 0 & u \le 0 \end{cases}$

(b) $g(u) = \begin{cases} e^{1/u(u-1)} & 0 < u < 1 \\ 0 & u \le 0 \text{ or } u \ge 1 \end{cases}$

(c) $h(u) = \dfrac{f(u)}{f(u) + f(1-u)}$ where $f(u)$ is defined in part (a)

7.3 The tangent map

Now we will study mappings from open sets of $\mathbb{R}^n$ to $\mathbb{R}^p$. For simplicity (to avoid having to deal simultaneously with different differentiability classes of functions) we will restrict ourselves to C^∞ mappings and C^∞ functions on $\mathbb{R}^n$ and $\mathbb{R}^p$. At each point of $\mathbb{R}^n$ we have a tangent space, and at each point of $\mathbb{R}^p$ there is a tangent space. In particular, if ϕ maps an open set containing a to $\mathbb{R}^p$, then we can look at the tangent space T_a at a and the tangent space $T_{\phi(a)}$ at $\phi(a)$.

If g is a C^∞ function in a neighborhood of $\phi(a)$, then $g \circ \phi$ is a C^∞ function in a neighborhood of a. Hence with each $v \in T_a$ we have a function w given by

$$w: g \mapsto v \cdot (g \circ \phi) \tag{7.18}$$

It is easy to check that this is a derivation so $w \in T_{\phi(a)}$ and we have a map from T_a to $T_{\phi(a)}$ given by $v \mapsto w$ (Fig. 7.3).

If γ is a curve through a, then $\phi \circ \gamma$ is a curve through $\phi(a)$. Hence with each $[\gamma]$ (or $L_{[\gamma]}$) in T_a we have a $[\phi \circ \gamma]$ (or $L_{[\phi \circ \gamma]}$) in $T_{\phi(a)}$. This mapping from T_a to $T_{\phi(a)}$ given by $L_{[\gamma]} \mapsto L_{[\phi \circ \gamma]}$ is the same as that given by $v \mapsto w$ above, since if $v = L_{[\gamma]}$, then

$$L_{[\phi \circ \gamma]} \cdot g = D_0 g \circ (\phi \circ \gamma) = D_0 (g \circ \phi) \circ \gamma = L_{[\gamma]} \cdot g \circ \phi = v \cdot (g \circ \phi) = w \cdot g$$

so $L_{[\phi \circ \gamma]} = w$.

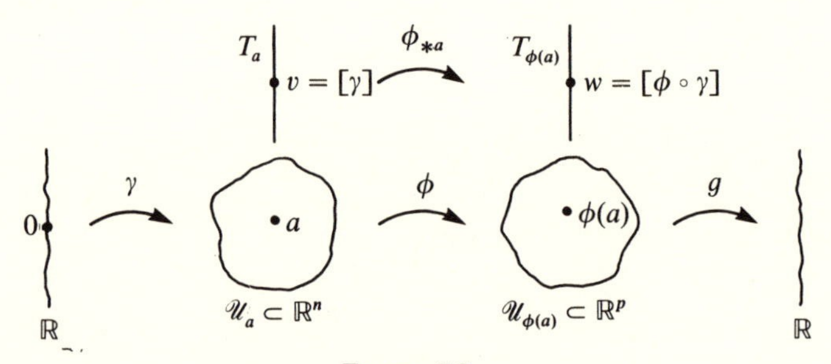

FIGURE 7.3

Definition

The mapping ϕ_{*a}: $T_a \to T_{\phi(a)}$ defined by $v \mapsto w$ where w is given by (7.18), or defined by $[\gamma] \mapsto [\phi \circ \gamma]$, is called *the tangent map* (or *the differential*, or *the derivative*) *of* ϕ *at* a.

Theorem 7.10. *ϕ_{*a} is linear.*

Proof. Problem 7.9.

Let us now evaluate ϕ_{*a} on the elements $\partial/\partial\pi^k$, $k = 1, \ldots, n$, of the natural basis of T_a. If $\partial/\partial\bar{\pi}^j$, $j = 1, \ldots, p$, are the elements of the natural basis of $T_{\phi(a)}$, then, as in eq. (1.6), for each k we should be able to write $\phi_{*a}\cdot\partial/\partial\pi^k$ as a linear combination of the $\partial/\partial\bar{\pi}^j$. If g is a C^∞ function on $\mathscr{U}_{\phi(a)}$, then evaluating $\phi_*\cdot\partial/\partial\pi^k$ on g,

$$\begin{aligned}
\left(\phi_*\cdot\frac{\partial}{\partial\pi^k}\right)\cdot g &= \frac{\partial}{\partial\pi^k}\cdot g\circ\phi && \text{by (7.18)} \\
&= \frac{\partial g\circ\phi}{\partial\pi^k} && \text{by the definition of } \frac{\partial}{\partial\pi^k} \\
&= \frac{\partial g}{\partial\bar{\pi}^j}\frac{\partial\phi^j}{\partial\pi^k} && \text{by the chain rule, (7.3)} \\
&= \frac{\partial\phi^j}{\partial\pi^k}\frac{\partial}{\partial\bar{\pi}^j}\cdot g && \text{by the definition of } \frac{\partial}{\partial\bar{\pi}^j}
\end{aligned}$$

so

$$\phi_*\cdot\frac{\partial}{\partial\pi^k} = \frac{\partial\phi^j}{\partial\pi^k}\frac{\partial}{\partial\bar{\pi}^j} \tag{7.19}$$

and the coefficients of the linear transformation are $\partial\phi^j/\partial\pi^k$. Notice that if we arrange these as a $p \times n$ matrix we get the Jacobian matrix, or derivative, of ϕ. Finally, if v^k are the components of v in the basis $\partial/\partial\pi^k$ of T_a, and w^j are the components of $\phi_{*a}\cdot v$ in the basis $\partial/\partial\bar{\pi}^j$ of $T_{\phi(a)}$, then

$$w^j = \frac{\partial\phi^j}{\partial\pi^k}v^k \tag{7.20}$$

We consider two special cases: $n = 1$ and $p = 1$.

(i) If $n = 1$ then ϕ maps an open set $\mathscr{U}$ in $\mathbb{R}$ to $\mathbb{R}^p$, that is, ϕ is a curve, γ. Note that up to now we have been focusing mainly on curves *through a given point* in order to define the concept of tangent vector at a point. In the process we defined the tangent of γ, but not the tangents to arbitrary points of γ. Now γ is more general—its domain need not contain 0, and even our restricted concept of tangent of γ does not apply.

To define a general concept of tangents to a curve, we first make explicit that eq. (7.19) is valid for each point, a, in the domain of ϕ, and in this case it reduces to

$$\gamma_{*a} \cdot \left.\frac{d}{d\pi}\right|_a = \left.\frac{d\gamma^i}{d\pi}\right|_a \left.\frac{\partial}{\partial \bar{\pi}^i}\right|_{\gamma(a)}$$

(Recall that $d/d\pi|_a = [\vartheta_a]$ is the tangent of the coordinate curve, or "translation," ϑ_a). But $d\gamma^i/d\pi|_a = D_a\gamma^i = D_0\gamma^i \circ \vartheta_a$ by the chain rule. So

$$\gamma_{*a}\colon [\vartheta_a] \mapsto D_0\gamma^i \circ \vartheta_a \left.\frac{\partial}{\partial \bar{\pi}^i}\right|_{\gamma(a)} = [\gamma \circ \vartheta_a]_{\gamma(a)}$$

by eq. (7.13). This result is also immediate from the definition of ϕ_{*a}. Further, we have a map $d/d\pi\colon a \mapsto [\vartheta_a]$ for all $a \in \mathbb{R}$. Thus, finally, for each a in the domain of γ we have a vector in $T_{\gamma(a)}$ given by the composition

$$\dot{\gamma} = \gamma_{*a} \circ \frac{d}{d\pi}\colon a \mapsto [\vartheta_a] \mapsto [\gamma \circ \vartheta_a]_{\gamma(a)} \tag{7.21}$$

Definitions
The map $\dot{\gamma}$ from the domain of γ to the set of tangent spaces of $\mathbb{R}^p$ is called *the canonical lift of* γ. $\dot{\gamma}(a) = [\gamma \circ \vartheta_a]_{\gamma(a)}$ is *the tangent* (or *velocity*) *of* γ *at* $\gamma(a)$.

Note that in the special case that γ is a curve through $\gamma(a)$, $\gamma(a) = \gamma(0)$, $a = 0$, and the tangent of γ at $\gamma(a)$ is the tangent of γ in our original sense; that is, $\dot{\gamma}(0) = [\gamma]$.

Since

$$[\gamma \circ \vartheta_a]_{\gamma(a)} = D_a\gamma^i \left.\frac{\partial}{\partial \bar{\pi}^i}\right|_{\gamma(a)}$$

we can also write

$$\dot{\gamma}(a) = D_a\gamma^i \left.\frac{\partial}{\partial \bar{\pi}^i}\right|_{\gamma(a)} \tag{7.22}$$

so that $\dot{\gamma}(a)$ is a derivation, Δ, on C^1 functions. Then,

$$\dot{\gamma}(a) \cdot f = D_a\gamma^i \left.\frac{\partial}{\partial \bar{\pi}^i}\right|_{\gamma(a)} \cdot f = D_a\gamma^i \left.\frac{\partial f}{\partial \bar{\pi}^i}\right|_{\gamma(a)} = D_a f \circ \gamma$$

The equation $\dot{\gamma}(a)\cdot f = D_a f \circ \gamma$ is sometimes used to define $\dot{\gamma}(a)$. With this definition $\dot{\gamma}(a)$ is clearly a derivation on C^∞ functions, and hence a tangent at $\gamma(a)$.

Finally, since ϕ_{*a} is a linear mapping on a one-dimensional space when ϕ is a curve, it is sometimes identified with its image $\phi_{*a}\cdot d/d\pi|_a$ at $d/d\pi$ according to Theorem 2.4; that is, if ϕ is a curve, the notation ϕ_{*a} is sometimes used for $\dot{\phi}(a)$.

(ii) If $p = 1$, then $\mathbb{R}^p = \mathbb{R}$, and the tangent spaces of $\mathbb{R}^p$ have a basis consisting of the single vector $d/d\bar{\pi}$. ϕ is a function $\mathcal{U}_a \subset \mathbb{R}^n \to \mathbb{R}$. $\phi_{*a}\cdot v = c(d/d\bar{\pi})$ evaluated at $\bar{\pi}$ gives $c = (\phi_{*a}\cdot v)\cdot\bar{\pi} = v\cdot(\bar{\pi}\cdot\phi) = v\cdot\phi$. Finally, putting $v = L_{[\gamma]}$ we get $c = \langle d\phi, v\rangle|_a$ by (7.15), so $\phi_{*a}\cdot v = \langle d\phi, v\rangle|_a(d/d\bar{\pi})$. This relationship accounts for the fact that ϕ_{*a} is often called the differential of ϕ at a. When ϕ is a function we get $\phi_{*a} = d\phi$ if we equate $\phi_{*a}\cdot v$ and its coefficient, $\langle d\phi, v\rangle = d\phi\cdot v$.

Theorem 7.11. (The chain rule) *If ϕ: $\mathcal{U}_a \to \mathbb{R}^p$, and ψ: $\mathcal{U}_{\phi(a)} \to \mathbb{R}^q$ where $\mathcal{U}_a \subset \mathbb{R}^n$ and $\mathcal{U}_{\phi(a)} \subset \mathbb{R}^p$, then $(\psi\circ\phi)_{*a} = \psi_{*\phi(a)}\circ\phi_{*a}$.*

Proof. Problem 7.11.

With ϕ_{*a} we have the transpose (or dual map) (see Section 2.4) $\phi^*_{\phi(a)}$: $T^*_{\phi(a)} \to T^*_a$ given by $\langle\phi^*_{\phi(a)}\cdot dg, v\rangle = \langle dg, \phi_{*a}\cdot v\rangle$ for all $v \in T_a$ and for all $dg \in T^*_{\phi(a)}$.

Theorem 7.12. $\phi^*_{\phi(a)}\cdot dg = d(g\circ\phi)$.

Proof.
$$\begin{aligned}\langle\phi^*_{\phi(a)}\cdot dg, v\rangle &= \langle dg, \phi_{*a}\cdot v\rangle && \text{by definition of transpose}\\ &= (\phi_{*a}\cdot v)\cdot g && \text{by (7.15)}\\ &= v\cdot(g\circ\phi) && \text{by (7.18)}\\ &= \langle d(g\circ\phi), v\rangle && \text{by (7.15)}\end{aligned}$$
■

The coefficients of $\phi^*_{\phi(a)}$ in the natural bases of T^*_a and $T^*_{\phi(a)}$ are given by

$$\phi^*_{\phi(a)}\cdot d\bar{\pi}^k = \frac{\partial\phi^k}{\partial\pi^j}\,d\pi^j \tag{7.23}$$

If τ_k are the components of τ in the basis $d\bar{\pi}^k$ of $T^*_{\phi(a)}$, and σ_j are the components of $\phi^*_{\phi(a)}\cdot\tau$ in the basis $d\pi^j$ of T^*_a, then

$$\sigma_j = \frac{\partial\phi^k}{\partial\pi^j}\,\tau_k \tag{7.24}$$

One can proceed now to build the various tensor spaces and their algebras which we described in Chapters 4–6 on the spaces T_a, $T_{\phi(a)}$, T_a^*, and $T_{\phi(a)}^*$. Thus, in particular, we have the spaces $T_a \otimes \cdots \otimes T_a = (T_a)_0^r$ and $T_a^* \otimes \cdots \otimes T_a^* = (T_a)_s^0$ and their symmetric and skew-symmetric subspaces $S^r(T_a)$, $S^s(T_a^*)$, $\Lambda^r(T_a)$, and $\Lambda^s(T_a^*)$.

Moreover, according to (4.7), and (4.8), ϕ_{*a} induces a linear map $(\phi_r)_a$: $(T_a)_0^r \to (T_{\phi(a)})_0^r$, and $\phi_{\phi(a)}^*$ induces a linear map $(\phi^s)_{\phi(a)}$: $(T_{\phi(a)})_s^0 \to (T_a)_s^0$. In particular, we have the restrictions of these maps to $\Lambda^r(T_a)$ and $\Lambda^s(T_{\phi(a)}^*)$ described in Section 6.3.

Now these maps can be extended in another direction. We have a space T_a for each $a \in \mathbb{R}^n$, so we also have the spaces $(T_a)_0^r$ for each $a \in \mathbb{R}^n$. Hence we can form $\bigcup_{a\in\mathbb{R}^n} (T_a)_0^r$ and define a map, ϕ_r, on $\bigcup_{a\in\mathscr{U}_a} (T_a)_0^r$ by its restrictions to the individual T_a's; i.e.,

$$\phi_r\colon \bigcup_{a\in\mathscr{U}_a} (T_a)_0^r \to \bigcup_{b\in\mathbb{R}^p} (T_b)_0^r \text{ by } \phi_r|_{(T_a)_0^r} = (\phi_r)_a.$$

Similarly, we have

$$\phi^s\colon \bigcup_{\phi(a)} (T_{\phi(a)})_s^0 \to \bigcup_{a\in\mathbb{R}^n} (T_a)_s^0 \qquad \text{given by} \qquad \phi^s|_{(T_{\phi(a)})_s^0} = (\phi^s)_{\phi(a)}$$

Notice, however, that while ϕ_r and ϕ^s are perfectly well-defined mappings we cannot say much about them until we know more about their domains and ranges. Thus, while ϕ itself is a C^∞ map, the statement that ϕ_r is a C^∞ map has no meaning until we have a meaning for differentiability of functions on $\bigcup_{a\in\mathscr{U}_a} (T_a)_0^r$. We will elucidate this point when we come to tensor fields in Chapter 11.

PROBLEM 7.9. Prove Theorem 7.10.

PROBLEM 7.10. Illustrate eq. (7.22) with $\gamma\colon u \mapsto (1 + u^2, 1 + \sin u)$ and $a = \pi/2$.

PROBLEM 7.11. Prove Theorem 7.11.

PROBLEM 7.12. Derive eq. (7.3) from the result of Theorem 7.11. (The notation will be a little different.)

PROBLEM 7.13. Here is another way to describe the tangent map, ϕ_*. We define a mapping, ψ, from the set of C^∞ functions at $\phi(a)$ to the set of C^∞ functions at a by $\psi\colon g \mapsto g \circ \phi$. ψ is a linear map (strictly speaking, ψ should be a map between vector spaces of "germs" of functions; see comment below eq. (7.8)). The transpose, ψ^*, of ψ will map the space of linear functions on the C^∞ functions at a into the space of linear functions on the C^∞ functions at $\phi(a)$. The restriction of ψ^* to the derivations on the C^∞ functions at a will be the tangent map at a.

PROBLEM 7.14. Show that $\psi^* \cdot v$, where ψ^* is the map defined in Problem 7.13, depends only on dg (that is, it is the same for equivalent functions) so that the transpose of ψ^* restricted to derivations is ϕ^*. In particular then, we can write Theorem 7.12 as $\psi \cdot dg = d\psi \cdot g$.

8

TOPOLOGICAL SPACES

Chapter 1 was supposed to serve to recall some linear algebra needed for the following exposition of tensor algebra. This chapter is supposed to serve a similar purpose, to recall some topology which we will need to discuss differentiable manifolds.

8.1 Definitions, properties, and examples

Definitions

If S is a set, and $\mathbf{T}$ is a set of subsets of S such that (i) every union of elements of $\mathbf{T}$ is an element of $\mathbf{T}$, (ii) the intersection of two elements of $\mathbf{T}$ is in $\mathbf{T}$, and finally, (iii) both S and the empty set are in $\mathbf{T}$, then $(S, \mathbf{T})$ is a *topological space* with *topology*, $\mathbf{T}$. The elements of $\mathbf{T}$ are *open sets*. A (open) neighborhood of a point is an open set containing that point. The complement (in S) of an open set is a *closed set*. If $A \subset S$ then $s \in S$ is *a limit point* of A if for all neighborhoods, $\mathscr{U}_s$ of s $(\mathscr{U}_s - s) \cap A \neq \varnothing$. The *closure*, $\bar{A}$, of $A \subset S$ is the union of A and all its limit points. $(S, \mathbf{T})$ is a *Hausdorff space* if distinct elements of S are contained in disjoint elements of $\mathbf{T}$.

There are many simple consequences of the axioms, and many simple relations among the concepts defined. Thus, for example, finite unions of closed sets are closed, any intersection of closed sets is closed, A is closed if and only if $A = \bar{A}$, and so on.

A given set S can have many topologies defined on it, some of which may be quite "pathological." For example, every set has the *trivial* topology in which the only members of $\mathbf{T}$ are the empty set and S itself, and every set has the *discrete* topology in which the elements of $\mathbf{T}$ are all the subsets of S. If $\mathbf{T}_1 \subset \mathbf{T}_2$ then $\mathbf{T}_1$ is *weaker* than $\mathbf{T}_2$, or $\mathbf{T}_2$ is *finer* than $\mathbf{T}_1$. (See Problem 8.10.)

If $(S, \mathbf{T})$ is a topological space, and $A \subset S$, then $\mathbf{T}' = \{A \cap \mathscr{U} : \mathscr{U} \in \mathbf{T}\}$ is called the *relative* or *induced* topology of A, and $(A, \mathbf{T}')$ is a *subspace* of $(S, \mathbf{T})$. (See Problem 8.2.)

The most important examples of topological spaces are the real cartesian spaces, $\mathbb{R}^n$, with the topology defined by means of the "Euclidean" distance $d(a, b) = [\sum_{i=1}^{n} (a^i - b^i)^2]^{\frac{1}{2}}$, the open sets being the unions of open "balls," $\{b \in \mathbb{R}^n : d(a, b) < r, a \in \mathbb{R}^n, r \in \mathbb{R}^+\}$. These spaces are called *Euclidean metric spaces* (cf., Section 5.4(i)).

Euclidean metric spaces can be generalized in several directions.

1. If V^n is an n-dimensional vector space, then, once a basis is chosen, its elements are in 1–1 correspondence with points of $\mathbb{R}^n$. A set in V^n will be open if its image in $\mathbb{R}^n$ is open. This definition is independent of the choice of basis since the components in the two bases are related by the change of basis matrix, a nonsingular continuous mapping of $\mathbb{R}^n$ onto $\mathbb{R}^n$ under which open sets are preserved (see Section 8.2). Thus, every finite-dimensional vector space can be considered to have a natural topology.

2. For any set on which we can define a distance function, or metric, we can define open sets just as we did for Euclidean metric spaces and get a topology.

Definitions
A *metric* or *distance function* on a set S is a function $d: S \times S \to \mathbb{R}$ such that (i) $d(x, y) > 0$ if $x \neq y$, (ii) $d(x, y) = 0$ if $x = y$, (iii) $d(x, y) = d(y, x)$, and (iv) $d(x, z) \leq d(x, y) + d(y, z)$ for all x, y, z in S. A set on which a metric is defined is *a metric space.*

Other important examples of metric spaces are (i) $\mathbb{R}^n$ with the distance function $d(a, b) = \max_{1 \leq i \leq n} |a^i - b^i|$; (ii) the set of infinite sequences $\{(a^1, a^2, \ldots, a^n, \ldots): a^i \in \mathbb{R}\}$ with $\sum_{i=1}^{\infty} (a^i)^2 < \infty$, so that we can define

$$d(a, b) = \left[\sum_{i=1}^{\infty} (a^i - b^i)^2\right]^{\frac{1}{2}};$$

and (iii) the set of all continuous functions defined on the closed interval $[a, b]$, $(a, b \in \mathbb{R})$ with $d(f, g) = \max_{a \leq t \leq b} |f(t) - g(t)|$.

3. The Euclidean distance of the cartesian product $\mathbb{R}^n$ was defined in terms of the metric $d(a, b) = |a - b|$ of $\mathbb{R}$. If $X_1, \ldots, X_n$ are metric spaces, we can define a distance, and hence a topology, on the cartesian product $X_1 \times \cdots \times X_n$ by $d = (\sum (d_i)^2)^{\frac{1}{2}}$ where d_i is the metric on X_i. Example (i) above, in which the balls are cartesian products of open intervals, suggests a further generalization for when the X_i are not necessarily metric spaces. If $X_1, \ldots, X_n$ are topological spaces we define a topology on the cartesian product $X_1 \times \cdots \times X_n$ whose open sets are the unions of subsets of the form $\mathscr{U}_1 \times \cdots \times \mathscr{U}_n$ where $\mathscr{U}_i$ are open in X_i. With this *product topology* $X_1 \times \cdots \times X_n$ is called a *product space.*

Theorem 8.1. *If (X, d_X) and (Y, d_Y) are metric spaces, the product topology on $X \times Y$ is the same as the metric topology induced by* $\max(d_X, d_Y)$.

Proof. Problem 8.7.

While vector spaces, metric spaces, and product spaces provide a vast generalization of real cartesian spaces, they do not include many topological spaces occurring in modern analysis, geometry, and applications. So we go back to general topological spaces.

Definitions
Suppose $(S, \mathbf{T})$ is a given topological space, and suppose $\mathbf{B}$ is a set of open sets such that every open set of X is a union of elements of $\mathbf{B}$. Then $\mathbf{B}$ *is a basis for* $\mathbf{T}$. Alternatively, $\mathbf{B}$ is a set of open sets such that for every open set $\mathscr{U} \in \mathbf{T}$ and $s \in \mathscr{U}$ there is a $B \in \mathbf{B}$ contained in $\mathscr{U}$ containing s. Since $\mathbf{B}$ determines $\mathbf{T}$ we can say that $\mathbf{T}$ is the topology *generated* by $\mathbf{B}$.

Theorem 8.2. *The set of open balls with rational centers and rational radii is a basis of Euclidean metric space.*

Proof. Given $\mathscr{U} \in \mathbf{T}$ and $s \in \mathscr{U}$, then s will be in some ball in $\mathscr{U}$. Inside of this ball there will be one with center s and rational radius, r. Finally, a ball, $\mathscr{O}$, with rational center a distance less than $r/3$ from s and radius $r/2$ will have the required property, $s \in \mathscr{O} \subset \mathscr{U}$. ∎

Corollary. *Euclidean metric space has a countable basis.*

Not every metric space has a countable basis, see Problem 8.5. But, clearly, metric spaces are contained in the class characterized by the weaker property: each point has a countable family of neighborhoods at least one of which is contained in every neighborhood of the point (called *1st countability*).

In the definition of basis we started with a given topological space. If we start with just a set S, and pick a set of subsets of S, in general, this family will not constitute a basis for a topology in S. To be a basis it is necessary and sufficient that the family includes S itself (or covers S) and also includes a set $\mathscr{U}$, for each s in an intersection $\mathscr{U}_1 \cap \mathscr{U}_2$ such that $s \in \mathscr{U} \subset \mathscr{U}_1 \cap \mathscr{U}_2$. Two different bases can generate different topologies.

An array of concepts related to the idea of a covering will be important for us.

Definitions
If $\{A_\alpha\}$ and $(B_\beta\}$ are coverings of S, then $\{B_\beta\}$ is called a *refinement* of $\{A_\alpha\}$ if each B_β is contained in some A_α. If $\{A_\alpha\}$ is a covering of S such that each $s \in S$ has a neighborhood which intersects only a finite number of members of $\{A_\alpha\}$, then $\{A_\alpha\}$ is called *locally finite.* A subset, A, of S is *compact* if every open covering of A has a finite subcovering. S is *locally compact* if each $s \in S$

has a neighborhood whose closure is compact. S is *paracompact* if every open covering of S has an open locally finite refinement.

Lots of examples of these definitions can be found in $\mathbb{R}^n$. Thus, the open disks of radius 1, with centers (m, n) where m, and n are integers, cover $\mathbb{R}^2$. So do the open disks of radius $\frac{1}{2}$ with centers $(m/2, n/2)$. The latter is not a refinement of the former. The set of all intersections $\{A_\alpha \cap B_\beta\}$, where A_α belongs to the first covering and B_β belongs to the second, is a refinement of both. Also, a subcovering is a refinement of a covering. Each of the examples given is a locally finite covering. Each example also shows that $\mathbb{R}^2$ is not compact. An important theorem in analysis, the Heine–Borel theorem, says that the compact subsets of $\mathbb{R}^n$ are the closed bounded sets. Thus, an open disk is not compact. We can cover it by concentric open disks of radius $r - 1/n$, for which there is no finite subcovering. Also, this is an example of a covering which is not locally finite.

$\mathbb{R}^n$ and open sets of $\mathbb{R}^n$ are locally compact, because each point is in a ball whose closure is compact. On the other hand, the set of rationals in $\mathbb{R}$ in the relative topology is not locally compact. The closure of every neighborhood of every point in this space can be covered by a set of open sets which has no finite subcovering.

Finally, a theorem we will need later, but whose proof is a little long, guarantees that a Hausdorff space which is locally compact and has a countable basis is paracompact, so, in particular, $\mathbb{R}^n$ and open sets in $\mathbb{R}^n$ are paracompact.

PROBLEM 8.1. Given an example of a topological space which is not a Hausdorff space.

PROBLEM 8.2. (i) Show that $\mathbf{T}'$ is a topology on A (in the definition of relative topology). (ii) Show that if $\mathcal{U}_i$ is open in X_i, $i = 1, \ldots, n$, then the subsets of the form

$$\mathcal{U}_1 \times \mathcal{U}_2 \times \cdots \times \mathcal{U}_n$$

are a topology on

$$X_1 \times X_2 \times \cdots \times X_n.$$

PROBLEM 8.3. Prove that a metric space is Hausdorff.

PROBLEM 8.4. Consider the functions $d_1(a, b) = \sum_{i=1}^n |a^i - b^i|$ and $d_\infty(a, b) = \max_{1 \leq i \leq n} \{|a^i - b^i|\}$. Show they are both metrics on $\mathbb{R}^n$ and determine the same topology in $\mathbb{R}^n$ as the Euclidean distance.

PROBLEM 8.5. Show that for any set

$$d(x, y) = \begin{cases} 0 & x = y \\ 1 & x \neq y \end{cases}$$

is a metric. What is the topology determined by this metric?

PROBLEM 8.6. Give an example of a nonmetrizable topological space.

PROBLEM 8.7. Prove Theorem 8.1.

PROBLEM 8.8. Show that the two given descriptions of a basis for **T** both describe the same thing.

PROBLEM 8.9. Prove the statement (p. 110) giving necessary and sufficient conditions for a basis of a topology.

8.2 Continuous mappings

Definitions
If X and Y are two topological spaces, a mapping $\phi: X \to Y$ is *continuous* if the inverse image of each open set in Y is open in X. ϕ is *continuous at* $s \in X$ if for every neighborhood $\mathscr{V}$ of $\phi(s)$ there is a neighborhood $\mathscr{U}$ of s such that $\phi(\mathscr{U}) \subset \mathscr{V}$. If ϕ is 1–1, and onto Y and ϕ and ϕ^{-1} are both continuous, then ϕ is *a homeomorphism.* If there exists a homeomorphism between two spaces they are *homeomorphic.* If (X, d) and (Y, D) are metric spaces, and $D(\phi(x), \phi(y)) = d(x, y)$, then ϕ is *an isometry* (cf. eq. (5.21)).

Theorem 8.3. *ϕ is continuous if and only if ϕ is continuous at s for each $s \in X$.*

Proof. Problem 8.12.

Theorem 8.4. (1) *If $\phi: (X, d) \to (Y, D)$ is an isometry, then X and $\phi(X)$ are homeomorphic.* (*ii*) *$i: (X, d) \to (X, D)$ is continuous at x if there is a positive number, c, such that for all y in some d-neighborhood of x, $D(x, y) \leq cd(x, y)$.*

Proof. Problem 8.13.

We will need a variety of basic topological results in the following chapters. For the most part, we will deal with these situations as they arise. However, we insert here one important example of the sort of fact we will need.

Theorem 8.5. *Suppose X and Y are topological spaces and $\phi: X \to Y$ is continuous,* 1–1 *and onto. If, moreover, X is compact and Y is Hausdorff, then ϕ is a homeomorphism.*

Proof. We have to show that ϕ^{-1} is continuous. We will use the criterion that a map is continuous if (and only if) the inverse image of every closed set is closed. That is, we will show that if $F \subset X$ is closed then $(\phi^{-1})^{-1}(F) = \phi(F)$ is closed. Now, F is compact because a closed subset of a compact space is compact. Then $\phi(F)$ is compact because the image of a compact set under a continuous mapping is compact. Finally, $\phi(F)$ is closed because a compact subset of Hausdorff space is closed. ■

PROBLEM 8.10. (i) If $Y \subset X$, then $(Y, \mathbf{T}_2)$ is a subspace of $(X, \mathbf{T}_1)$ iff the inclusion map $i: (Y, \mathbf{T}_2) \to (X, \mathbf{T}_1)$ is continuous. (ii) If $\mathbf{T}_1$ and $\mathbf{T}_2$ are two topologies on X, then the identity map $id: (X, \mathbf{T}_1) \to (X, \mathbf{T}_2)$ is continuous iff $\mathbf{T}_2 \subset \mathbf{T}_1$.

PROBLEM 8.11. With the natural topology on V^n, and the product topologies on $V^n \times V^n$ and $\mathbb{R} \times V^n$, show that the vector space operations on V^n are continuous and all linear functions on V^n are continuous.

PROBLEM 8.12. Prove Theorem 8.3.

PROBLEM 8.13. Prove Theorem 8.4.

PROBLEM 8.14. Our proof of Theorem 8.5 essentially consisted of stating four "lemmas" none of which we have proved. Prove any one or more of these.

9

DIFFERENTIABLE MANIFOLDS

At the beginning of Chapter 7 we mentioned that for the description of physical phenomena we have to generalize the concept of maps between cartesian spaces to those between more general domains and image spaces. We want to keep the domain as general as possible; i.e., with as little structure as possible. This is the perspective that motivates the study of sets on which local coordinates are defined. After general definitions, examples, and a brief discussion of mappings we impose the structures of Sections 7.2 and 7.3 on manifolds.

9.1 Definitions and examples

A manifold is a Hausdorff space, M, with certain additional structure on it.

Definitions
An n-dimensional chart (or *a local coordinate system*) *at* $p \in M$ is a pair $(\mathscr{U}, \mu)$ where $\mathscr{U}$ is an open set of M containing p, and μ is a homeomorphism of $\mathscr{U}$ onto an open set of $\mathbb{R}^n$ (with the Euclidean topology). $\mathscr{U}$ *is a coordinate neighborhood*, and μ is *a coordinate map*. If π^i are the coordinate functions on $\mu(\mathscr{U})$ then $\mu^i = \pi^i \circ \mu$ are *local coordinate functions* (or *local coordinates*) *on* $\mathscr{U}$. A set of n-dimensional charts $\{(\mathscr{U}_\alpha, \mu_\alpha)\}$ such that $\{\mathscr{U}_\alpha\}$ covers M is an *atlas*. A Hausdorff space with an atlas is *a locally Euclidean space* (or *a topological manifold*).

Definitions
A C^k *manifold* is a locally Euclidean space with a countable basis and with an atlas with the properties (1) if $(\mathscr{U}, \mu)$ and $(\mathscr{V}, \nu)$ are two charts such that $\mathscr{U} \cap \mathscr{V} \neq \varnothing$, then $\nu \circ \mu^{-1}$ is C^k on $\mu(\mathscr{U} \cap \mathscr{V})$ and $\mu \circ \nu^{-1}$ is C^k on $\nu(\mathscr{U} \cap \mathscr{V})$, and (2) if $(\mathscr{W}, \xi)$ has property (1) for every chart in the atlas then $(\mathscr{W}, \xi)$ is in the atlas. The mappings $\nu \cdot \mu^{-1}$ and $\mu \circ \nu^{-1}$ are called *overlap mappings*, or *coordinate transformations*, and a pair of charts having property (1) are said to be *C^k-related*, or, *C^k-compatible*. An atlas with property (1) is called a *C^k atlas*, and an atlas with property (2) is said to be *maximal* (see Fig. 9.1).

We do not usually concern ourselves with property (2) in practice because we can get a C^k manifold by first putting a C^k atlas on a Hausdorff space with a countable basis and then throwing in all C^k-related (or,

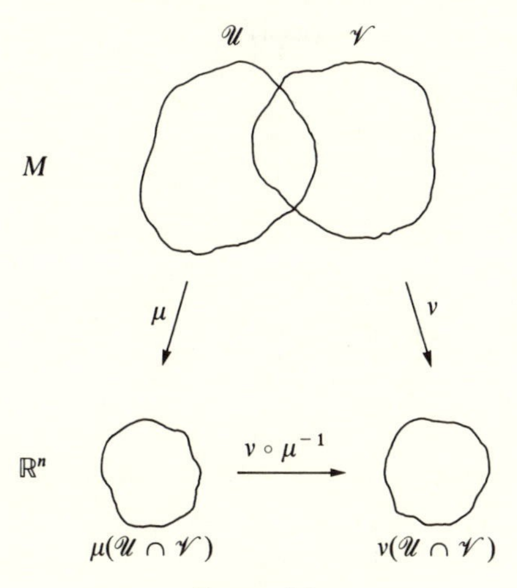

FIGURE 9.1

admissible) charts. An important result in differential topology says that when we do this, we can always find a subset of the set of these charts which cover M and are all C^∞-related; that is, M has a C^∞ atlas.

We note also that, according to the result mentioned in Section 8.2, since a manifold is locally compact, Hausdorff, and has a countable basis, it is paracompact.

Let us look at some examples of manifolds.

1. We can make $\mathbb{R}^n$ into a manifold by putting on it the atlas consisting of the single chart $(\mathbb{R}^n, id)$. The corresponding coordinate functions $\mu^i = \pi^i \circ id = \pi^i$ are called *the standard* (or *natural*) *coordinates* (see Section 7.1). If we add all possible C^ω-related charts we get a C^ω manifold. For example, on $\mathbb{R}^2$ we could add the chart $(\mathscr{U}, \mu)$ given by

$$\mu^1: (a^1, a^2) \mapsto b_1 = \tan^{-1} \frac{a^2}{a^1}$$

$$\mu^2: (a^1, a^2) \mapsto b_2 = \sqrt{(a^1)^2 + (a^2)^2}$$

defined on any open set, $\mathscr{U} \subset \mathbb{R}^2$ for which $a^1 > 0$. The manifold constructed on $\mathbb{R}^n$ in this way gives *the standard manifold structure* of $\mathbb{R}^n$. There are other manifold structures on $\mathbb{R}^n$. Thus, for $\mathbb{R}^1$, the single chart $(\mathbb{R}, \mu_0)$ where $\mu_0: a \mapsto a^3$ is an atlas. This determines a C^ω manifold different from the standard one, since $(\mathbb{R}, id)$ and $(\mathbb{R}, \mu_0)$ are not C^ω-related. In fact they are not C^k-related for any k.

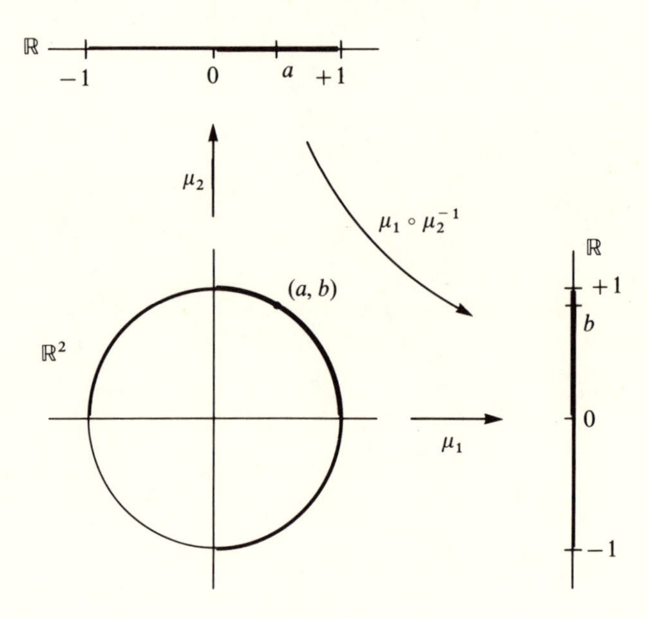

FIGURE 9.2

2. Any finite-dimensional vector space, V^n, with the standard, or natural, topology can be made into a manifold by choosing a basis, and then the single chart, (V^n, μ), will be an atlas, where $\mu: V^n \to \mathbb{R}^n$ maps v to its set of components. With another basis we get another chart (V^n, ν) and $\nu \circ \mu^{-1}$ is given by the change of basis matrix.

3. To a large extent the concept of a manifold is motivated by the desire to generalize and/or abstract the intuitive ideas of curves in $\mathbb{R}^2$ and $\mathbb{R}^3$, and surfaces in $\mathbb{R}^3$. For an example of this kind consider the unit circle $S^1 = \{(a, b): a^2 + b^2 = 1, a, b \in \mathbb{R}\}$ in $\mathbb{R}^2$ (see Fig. 9.2). S^1 is given the topology induced from $\mathbb{R}^2$. Let $\mathscr{U}_1 = \{(a, b) \in S^1: a > 0\}$ and let $\mu_1: \mathscr{U}_1 \to \mathbb{R}$ by $(a, b) \mapsto b$. Then $(\mathscr{U}_1, \mu_1)$ is a chart on S^1. Further, let $\mathscr{U}_2$, $\mathscr{U}_3$, $\mathscr{U}_4$ be the subsets of S^1 for which $b > 0$, $a < 0$, $b < 0$, respectively, and μ_2, μ_3, μ_4 take (a, b) to a, b, and a, respectively. Then the four charts $(\mathscr{U}_i, \mu_i)$, $i = 1, 2, 3, 4$, form an atlas for S^1. On $\mathscr{U}_1 \cap U_2$, since $\mu_1(a, b) = b$ and $\mu_2(a, b) = a$, $\mu_1 \circ \mu_2^{-1}: a \mapsto (a, b) \mapsto b = \sqrt{1 - a^2}$ for $0 < a < 1$, which is C^ω. We can check that all the other overlap maps are also C^ω so S^1 is a C^ω manifold.

4. Consider the set, E (figure-of-eight shape) of points, (a, b) in $\mathbb{R}^2$ such that $(a, b) = (\sin 2u, \sin u)$, $0 < u < 2\pi$. These formulas determine a bijective map, μ, from E to $(0, 2\pi)$. (E, μ) will be a chart on E, and hence E will be a C^ω manifold if μ is a homeomorphism. $\mu(0, 0) = \pi$, but no $\mathbb{R}^2$ neighborhood of (0, 0) maps into a neighborhood of π, so μ is not continuous if E has the induced, or relative, subset topology from $\mathbb{R}^2$, and (E, μ) is not a chart on E. Moreover, E, with the induced topology, cannot be made into a manifold since the only connected 1-dimensional manifolds are open intervals of $\mathbb{R}$ and S^1 (Bishop and Goldberg, 1968, p. 26). On the other hand, (E, μ) is a chart and E is a manifold if E is given the topology induced by μ from $\mathbb{R}$, the identification topology.

Examples 3 and 4 above illustrate two different ways of prescribing subsets of $\mathbb{R}^n$. Other examples appear in Problems 9.2, 9.3, 9.4, and 9.6. They prompt a generalization discussed in Section 9.4.

From the standpoint of the abstract definition of a manifold, Examples 3 and 4 (and their generalizations) are somewhat special in that we used an "ambient" space to construct them. Clearly, this is not always necessary, and when we do use an ambient space we have to be careful about thinking of the manifold as being "contained in" that space. This is suggested in Example 4, in which the topology of the manifold is not the subset topology of $\mathbb{R}^2$. Another example which illustrates this is the graph of $b = |a|$, $-1 < a < 1$. As a subset of $\mathbb{R}^2$ it has a corner, but as a 1-dimensional manifold it is "smooth." The concept of a manifold lying in a larger space will be elucidated in Chapter 10. It is important because (i) the basic familiar geometrical prototypes of manifolds, curves and surfaces, are *in* Euclidean affine space, $\mathscr{E}_0^3$, and (ii) certain important theorems (cf., Auslander and MacKenzie, 1977, p. 116) show that in some sense every manifold is contained in $\mathbb{R}^n$ for large enough n. However, for many important manifolds this relation is not intuitively clear, and, in any event, it is instructive to study a manifold on its own (i.e., instrinsic properties) without reference to any $\mathbb{R}^n$ in which it may or may not be contained. This is illustrated by the following examples.

5. Consider the set of lines ℓ through a point of Euclidean space, $\mathscr{E}_0^3$. We make this set into a topological space by giving it the topology determined by that of a 2-sphere, S^2, with center at the given point, and with opposite points identified. This space is Hausdorff and has a countable basis. (See Example 6.) We represent $\mathscr{E}_0^3$ by $\mathbb{R}^3$ with (0, 0, 0) corresponding to the given point.

Let $\mathscr{U}_1 = \{\ell : \ell \text{ is not in the plane } a^1 = 0\}$ and let $\mu_1 : \mathscr{U}_1 \to \mathbb{R}^2$ be given by

$$\mu_1^1 : \ell \mapsto \frac{a^2}{a^1} = \text{slope of the projection of } \ell \text{ on the } a^1a^2 \text{ plane}$$

$$\mu_1^2 : \ell \mapsto \frac{a^3}{a^1} = \text{slope of the projection of } \ell \text{ on the } a^1a^3 \text{ plane}$$

That is,

$$\mu_1(\ell) = \left(\frac{a^2}{a^1}, \frac{a^3}{a^1}\right) \qquad \text{and } \mu_1 \text{ is 1–1 onto } \mathbb{R}^2$$

We define two more charts analogously. That is, $\mathscr{U}_2$ is the set of all lines except those in $a^2 = 0$, and $\mu_2(\ell) = (a^3/a^2, a^1/a^2)$, and $\mathscr{U}_3$ is the set of all lines except those in $a^3 = 0$, and $\mu_3(\ell) = (a^1/a^3, a^2/a^3)$. Clearly, $\{\mathscr{U}_1, \mathscr{U}_2, \mathscr{U}_3\}$ covers the space. We will rely on the assumption that it is intuitively evident that the $\mathscr{U}$'s are open, and the μ's are homeomorphisms onto $\mathbb{R}^2$, and defer the proof of these facts to Problem 9.8. To complete the verification that we have a C^k manifold, we have to check the overlap maps of property (1).

Thus, consider, for example

$$\mu_2 \circ \mu_1^{-1} : \mu_1(\mathscr{U}_1 \cap \mathscr{U}_2) \to \mu_2(\mathscr{U}_1 \cap U_2).$$

$\mathscr{U}_1 \cap \mathscr{U}_2$ consists of all lines except those in the planes $a^1 = 0$ or $a^2 = 0$, and

$$\mu_2 \circ \mu_1^{-1} : \left(\frac{a^2}{a^1}, \frac{a^3}{a^1}\right) \mapsto \left(\frac{a^3}{a^2}, \frac{a^1}{a^2}\right) = \left(\frac{a^3/a^1}{a^2/a^1}, \frac{1}{a^2/a^1}\right)$$

or

$$\mu_2 \circ \mu_1^{-1} : (b^1, b^2) \mapsto \left(\frac{b^2}{b^1}, \frac{1}{b^1}\right) \qquad \text{for all } b^1 \neq 0$$

This mapping is C^ω and, similarly, checking the other overlap mappings we find they are all C^ω, so we have constructed a C^ω 2-manifold, *the analytic real projective plane*, $P^2(\mathbb{R})$.

6. Example 5, the projective plane, is a particular example of a class of manifolds called *Grassmann manifolds* of $\mathbb{R}^n$, or, more generally, of any vector space V^n.

Definitions

We denote by $G(k, \mathbb{R}^n)$ the set of all *k-planes through the origin of* $\mathbb{R}^n$; that is, with $\mathbb{R}^n$ considered a vector space, the set of all k-dimensional subspaces of $\mathbb{R}^n$. We denote by $F(k, \mathbb{R}^n)$ the set of all *k-frames of* $\mathbb{R}^n$; that is, the set of all linearly independent sets of k n-tuples,

$$\left\{ \begin{pmatrix} a_{11} \\ \vdots \\ a_{n1} \end{pmatrix}, \begin{pmatrix} a_{12} \\ \vdots \\ a_{n2} \end{pmatrix}, \ldots, \begin{pmatrix} a_{1k} \\ \vdots \\ a_{nk} \end{pmatrix} \right\}$$

Two k-frames are *equivalent* if they determine the same k-plane. Two k-frames will be equivalent iff their matrices A and B are related by $A = BC$ where C is a nonsingular $k \times k$ matrix.

Now corresponding to each k-plane, Π, we have an equivalence class of k-frames and hence an equivalence class of matrices. For each matrix some set of k rows must be linearly independent.

We can consider the set, $\mathscr{U}_1$, of all k-planes having a matrix representation with its first k rows linearly independent. Then each of these k-planes has a representation of the form

$$\begin{bmatrix} 1 & 0 & \cdots & 0 \\ 0 & 1 & \cdots & 0 \\ \vdots & \vdots & & \\ 0 & 0 & \cdots & 1 \\ a_{k+1\,1} & & \cdots & a_{k+1\,k} \\ \vdots & & & \vdots \\ a_{n1} & & \cdots & a_{nk} \end{bmatrix}$$

Similarly, consider the set $\mathscr{U}_i$ of all k-planes whose matrices have some other set of k rows linearly independent. This set of matrices will have a representative with the identity $k \times k$ matrix in these k rows.

In this way all k-planes lie in one or more of the sets $\mathscr{U}_i$, $i = 1, \ldots, \binom{n}{k}$. The $\mathscr{U}_i$ will be coordinate neighborhoods in $G(k. \mathbb{R}^n)$ and with each Π in each $\mathscr{U}_i$ we can associate a unique element of $\mathbb{R}^{(n-k)k}$ given by stringing out the elements of the complement of the unit submatrix in the canonical representation of Π. If Π is in both $\mathscr{U}_i$ and $\mathscr{U}_j$, then the canonical representation in $\mathscr{U}_j$ is a multiple by a nonsingular $k \times k$ matrix of the canonical representation in $\mathscr{U}_i$, and this relation gives the overlap map.

To illustrate, for $G(2, \mathbb{R}^3)$ we have the atlas $\{(\mathscr{U}_i, \mu_i): i = 1, 2, 3\}$ where $\mathscr{U}_i$ are those 2-planes whose canonical matrices are

$$\begin{pmatrix} a_{11} & a_{12} \\ 1 & 0 \\ 0 & 1 \end{pmatrix}, \quad \begin{pmatrix} 1 & 0 \\ a_{21} & a_{22} \\ 0 & 1 \end{pmatrix} \quad \text{and} \quad \begin{pmatrix} 1 & 0 \\ 0 & 1 \\ a_{31} & a_{32} \end{pmatrix}$$

respectively. Then $\mu_i: \Pi \mapsto (a_{i1}, a_{i2})$. Now if Π is in $\mathscr{U}_1$ and also in $\mathscr{U}_2$, and if Π is represented in $\mathscr{U}_1$ by

$$\begin{pmatrix} a_{11} & a_{12} \\ 1 & 0 \\ 0 & 1 \end{pmatrix}$$

and in $\mathscr{U}_2$ by

$$\begin{pmatrix} 1 & 0 \\ a_{21} & a_{22} \\ 0 & 1 \end{pmatrix}$$

then

$$\begin{pmatrix} 1 & 0 \\ a_{21} & a_{22} \\ 0 & 1 \end{pmatrix} C = \begin{pmatrix} a_{11} & a_{12} \\ 1 & 0 \\ 0 & 1 \end{pmatrix}$$

for some nonsingular matrix C. So

$$C = \begin{pmatrix} a_{11} & a_{12} \\ 0 & 1 \end{pmatrix}$$

where $a_{11} \neq 0$ and $a_{21} = 1/a_{11}$ and $a_{22} = -a_{12}/a_{11}$. Hence the overlap map $\mathbb{R}^2 - (0, r) \to \mathbb{R}^2$ is C^ω.

Now let us recall that, in this example, we started with a set $G(k, \mathbb{R}^n)$ of k-dimensional vector spaces. Out of this set we selected subsets, $\mathscr{U}_i$, which, together, cover it. With each $\mathscr{U}_i$ we have a map $\mu_i: \mathscr{U}_i \to \mathbb{R}^{(n-k)k}$ and hence we have an "atlas" on $G(k, \mathbb{R}^n)$. For $G(2, \mathbb{R}^3)$ we have shown that the overlap maps are C^ω. In the general case also, it can be shown we have a C^ω "atlas" on $G(k, \mathbb{R}^n)$. Thus, we have given $G(k, \mathbb{R}^n)$ a "differentiable structure", even though no topology has been defined on $G(k, \mathbb{R}^n)$. According to our definition, a manifold must be a topological space.

In this example, we can define a topology on the given set by recalling that its elements are in 1–1 correspondence with equivalence classes of $n \times k$ matrices, so we can give $G(k. \mathbb{R}^n)$ the quotient topology; i.e., a set in $G(k, \mathbb{R}^n)$ is open if its inverse image in $F(k, \mathbb{R}^n)$ under the natural projection is open. We can then show that the $\mathscr{U}_i$ defined above are open, the μ_i are homeomorphisms and $G(k. \mathbb{R}^n)$ has a countable basis and is Hausdorff (Brickell and Clark, 1970, pp. 41, p. 43), so $G(k, \mathbb{R}^n)$ has all the properties required in the definition of a C^ω manifold. In general, if one simply has a set with a differentiable structure, as we had in this example, we can define an induced topology on the set for which the set of all coordinate domains forms a basis. Thus, the topology will have a countable basis if the set of coordinate domains in countable, otherwise it may not. Neither will it necessarily be Hausdorff.

7. Finally, we note that many important manifolds are obtained by a method of constructing new manifolds from old ones. Thus, one can take any open set of a manifold M and make it into a manifold of the same dimension as M by using the induced topology and restricting the coordinate maps. Such a manifold is called *an open submanifold of M*. For an important example, we note first that the set of all $n \times p$ real matrices can be given a natural manifold structure. For the $n \times n$ matrices we have the determinant function, which is continuous. So, since $\mathbb{R} - 0$ is open, the nonsingular $n \times n$ matrices form an open submanifold of the manifold of $n \times n$ matrices.

Also, starting with two manifolds M and N, and using the product topology in $M \times N$, we can make it a manifold of dimension $\dim M + \dim N$. If $(\mathscr{U}, \mu)$ is a chart of M and $(\mathscr{V}, \nu)$ is a chart of N, then $(\mathscr{U} \times \mathscr{V}, \mu \times \nu)$ will have all the required properties for a chart of $M \times N$. Further, the set of all products $\mathscr{U} \times \mathscr{V}$ will cover $M \times N$, and on the intersection of $\mathscr{U}_1 \times \mathscr{V}_1$ and $\mathscr{U}_2 \times \mathscr{V}_2$, if $\mu_1 \times \nu_1(p, q) = (a, b)$ and $\mu_2 \times \nu_2(p, q) = (c, d)$, then the map $(a, b) \mapsto (\mu_1^{-1}(a), \nu_1^{-1}(b)) \mapsto (\mu_2 \circ \mu_1^{-1}(a), \nu_2 \circ \nu_1^{-1}(b))$ will be C^k if M and N are C^k manifolds. Thus, the set of charts so constructed is a C^k atlas and $M \times N$ is *the product manifold of M and N*. As an important example, we can construct the torus, $S^1 \times S^1$, from Example 3.

PROBLEM 9.1. (i) Check the remaining overlap maps of Example 3. (ii) For S^1 as given in Example 3, let $\mathscr{U}_5 = S^1 - (0, 1)$, $\mu_5: \mathscr{U}_5 \to \mathbb{R}$ by $(a, b) \mapsto a/(1 - b)$, $\mathscr{U}_6 = S^1 - (0, -1)$ and $\mu_6: \mathscr{U}_6 \to \mathbb{R}$ by $(a, b) \mapsto a/(1 + b)$. Then $\{(\mathscr{U}_5, \mu_5), (\mathscr{U}_6, \mu_6)\}$ is a C^ω atlas for S^1. (iii) This atlas gives the same manifold as the one constructed in Example 3.

PROBLEM 9.2. For $S^n = \{(a^1, \ldots, a^{n+1}): (a^1)^2 + \cdots + (a^n)^2 + (a^{n+1})^2 = 1, a_i \in \mathbb{R}\}$, we can choose $\mathscr{U}_1 = S^n - (0, \ldots, 1)$, $\mu_1: \mathscr{U}_1 \to \mathbb{R}^n$ by

$$\mu_1^i: (a^1, \ldots, a^{n+1}) \mapsto \frac{a^i}{1 - a^{n+1}},$$

$\mathscr{U}_2 = S^n - (0, \ldots, -1)$ and $\mu_2: \mathscr{U}_2 \to \mathbb{R}^n$ by

$$\mu_2^i: (a^1, \ldots, a^{n+1}) \mapsto \frac{a^i}{1 + a^{n+1}}$$

This is a C^ω atlas for S^n.

PROBLEM 9.3. Note that S^1 is also given by $S^1 = \{(a, b) \in \mathbb{R}^2: a = \cos 2\pi u, b = \sin 2\pi u, u \in \mathbb{R}\}$. Make S^1 into a manifold with an atlas having two charts.

PROBLEM 9.4. S^{n-1} is the set of n-tuples, $(a^1, a^2, \ldots, a^n)$ of $\mathbb{R}^n$ given by

$$
\begin{aligned}
a^1 &= \cos u^1 \\
a^2 &= \sin u^1 \cos u^2 \\
a^3 &= \sin u^1 \sin u^2 \cos u^3 \\
&\vdots \\
a^{n-1} &= \sin u^1 \quad \cdots \quad \sin u^{n-2} \cos u^{n-1} \\
a^n &= \sin u^1 \quad \cdots \quad \sin u^{n-2} \sin u^{n-1}, \qquad u^i \in \mathbb{R}
\end{aligned}
$$

Make S^{n-1} into a manifold with an atlas having two charts.

PROBLEM 9.5. (i) Check the remaining overlap maps of Example 5. (ii) Show that $P^2(\mathbb{R}) = G(1, \mathbb{R}^3)$. (iii) Generalize to projective n space, $P^n(\mathbb{R})$.

PROBLEM 9.6. The torus in $\mathbb{R}^3$ can be given by either

$$
\begin{aligned}
x &= (a + b \sin v) \cos u \\
y &= (a + b \sin v) \sin u \\
z &= b \cos v
\end{aligned}
$$

or $(\sqrt{x^2 + y^2} - a)^2 + z^2 = b^2$. Make it into a manifold with three charts and with six charts.

PROBLEM 9.7. Exhibit an atlas for $S^1 \times S^1$.

PROBLEM 9.8. As a set $P^2(\mathbb{R}) = \{[a]: a \in \mathbb{R}^3 - 0, \text{ and } a \sim b \text{ if } \exists\, t \neq 0 \rightarrow a^i = tb^i\}$. Let $\pi =$ the natural projection of $\mathbb{R}^3 - 0 \rightarrow P^2(\mathbb{R})$ given by $a \mapsto [a]$. Give $P^2(\mathbb{R})$ the quotient topology. Let

$$
\begin{aligned}
&\mathscr{U}_1 = \pi(\mathscr{V}_1) && \text{where} && \mathscr{V}_1 = \{a \in \mathbb{R}^3 - 0: a^1 \neq 0\} \\
&\mu_1: \mathscr{U}_1 \rightarrow \mathbb{R}^2 && \text{given by} && [a] \mapsto \left(\frac{a^2}{a^1}, \frac{a^3}{a^1}\right)
\end{aligned}
$$

Prove that $\mathscr{U}_1$ is open, and μ_1 is a homeomorphism of $\mathscr{U}_1$ onto $\mathbb{R}^2$.

9.2 Mappings of manifolds

We said at the beginning of Chapter 7 that we want to introduce a concept of differentiability and a definition of a derivative for a class of spaces greater than just the class of normed vector spaces. In the last section we defined a differentiable manifold or a topological space with a superimposed "differentiable structure." On such a space we can say what we mean by maps between two manifolds being differentiable. In particular, we can say what we mean by a curve in M or a function on M being differentiable, as these are special cases of maps between manifolds. Having these two concepts available on M, we can, and will, in the next section, build the same structures on M that we did on $\mathbb{R}^n$ in Chapter 7. To define differentiability, the basic idea is to refer maps of manifolds back to maps of cartesian spaces by means of charts and then show that differentiability is independent of coordinates so that we have a well-defined concept.

Theorem 9.1. *If ϕ is a continuous mapping from M to N, and $\nu \circ \phi \circ \mu^{-1}$ is C^k for one chart $(\mathscr{U}, \mu)$ at p and one chart $(\mathscr{V}, \nu)$ at $\phi(p)$, then $\nu_1 \circ \phi \circ \mu_1^{-1}$ is C^k for any other charts $(\mathscr{U}_1, \mu_1)$ and $(\mathscr{V}_1, \nu_1)$ at p and $\phi(p)$.*

Proof. $\nu_1 \circ \phi \circ \mu_1^{-1} = (\nu_1 \circ \nu^{-1}) \circ (\nu \circ \phi \circ \mu^{-1}) \circ (\mu \circ \mu_1^{-1})$ so, since $\nu \circ \phi \circ \mu^{-1}$ is C^k and $\nu_1 \circ \nu^{-1}$ and $\mu \circ \mu_1^{-1}$ are C^k by the definition of C^k manifold, then by the chain rule on cartesian spaces we have our result. ■

Definitions
If ϕ is a continuous mapping from a manifold M to a manifold N, and if $(\mathscr{U}, \mu)$ is a chart at p and $(\mathscr{V}, \nu)$ is a chart at $\phi(p)$, then $\hat{\phi} = \nu \circ \phi \circ \mu^{-1}$ is *a coordinate expression* (or *representation*) *of ϕ on U. ϕ is C^l at $p \in M$* if all coordinate expressions at p are C^l, or, by Theorem 9.1, if any one coordinate expression at p is C^l, $l \leq k$. In particular, *a curve, γ, through p is differentiable* at p if for any chart $(\mathscr{U}, \mu)$, $\hat{\gamma} = \mu \circ \gamma$ is differentiable at 0, and *a function, f, is differentiable* at p if $\hat{f} = f \circ \mu^{-1}$ is differentiable at $\mu^{-1}(p)$.

Theorem 9.2. *If μ is a coordinate map on a C^k manifold, then μ and μ^{-1} are C^k.*

Proof. Problem 9.9.

Theorem 9.3. *On a C^k manifold,* (i) *the injection $\mathbf{i}_{q_0}$: $M \to M \times N$ defined by $p \mapsto (p, q_0)$ where q_0 is fixed in N is C^k,* (ii) *the projection $\boldsymbol{\pi}^1$: $M \times N \to M$ defined by $(p, q) \mapsto p$ for all $q \in M$ is C^k.*

Proof. (i) For every chart $(\mathscr{U}, \mu)$ at p and $(\mathscr{V}, \nu)$ at q_0 the coordinate expression $\hat{\mathbf{i}}_{q_0}$: $\mu(p) \mapsto a \mapsto (a, b_0) \mapsto (\mu(p), \nu(q_0))$, or $\hat{\mathbf{i}}_{q_0}$: $(\mu^1(p), \ldots, \mu^m(p)) \mapsto$

$(\mu^1(p), \ldots, \mu^m(p), \nu^1(q_0), \ldots, \nu^n(q_0))$, or

$$\hat{\mathbf{i}}^j_{q_0}(\mu^1(p), \ldots, \mu^m(p)) = \mu^j(p), \qquad j = 1, \ldots, m$$
$$\hat{\mathbf{i}}^{m+j}_{q_0}(\mu^1(p), \ldots, \mu^m(p)) = \nu^j(q_0) \qquad j = 1, \ldots, n$$

So the component functions of $\hat{\mathbf{i}}_{q_0}$ are C^k functions.

(ii) *Problem 9.10.*

Theorem 9.4. *ϕ is a C^k map at p if and only if for all C^k functions, f, on neighborhoods of $\phi(p)$, the functions $f \circ \phi$ are C^k at p.*

Proof. Problem 9.11.

From the point of view of the desire to define a concept of differentiability for functions and maps for spaces other than vector spaces, there is a more direct approach to the idea of a differentiable manifold. We impose directly the structure of (i.e., postulate the existence of) a class of differentiable functions on open sets of a topological space (see Chevalley, 1946; Chern, 1959; Sternberg, 1983). From there we can go in the opposite direction and construct a differentiable atlas (see Sternberg). In this approach, Theorem 9.4 is a definition and we can prove that a map ϕ is differentiable iff its coordinate expressions are—our definition.

Theorem 9.5. *On a C^∞ manifold, if f is differentiable on a neighborhood of p, then there exists a neighborhood $\mathscr{V}$ of p, and a differentiable function g on M such that $g(q) = f(q)$ for all $q \in \mathscr{V}$.*

Proof. Choose a chart $(\mathscr{U}, \mu)$ at p such that $\mu(p) = (0, \ldots, 0)$, and choose r_2 such that $\{a: |a| < r_2\} \subset f(\mathscr{U})$. For $r_1 < r_2$ there is a C^∞ function, h, on $\mathbb{R}^n$ such that

$$h(a) = \begin{cases} 1 & \text{for } |a| \leq r_1 \\ 0 & \text{for } |a| \geq r_2 \end{cases}$$

Define g on $\mathscr{U}$ by $\hat{g}(a) = \hat{f}(h(a)a^1, \ldots, h(a)a^n)$ and $g(q) = g(p)$ for q outside of $\mathscr{U}$. ■

When we generalize from cartesian spaces to manifolds it is useful to introduce an additional concept to help classify manifolds, or to describe when two manifolds are essentially the same.

Definition

If M and N are C^k manifolds, then a 1–1 map, ϕ, from M onto N such that both ϕ and ϕ^{-1} are C^k is a *C^k-diffeomorphism.*

Let us consider the following *examples.*

1. Let $\mathbb{R}^2$ have the standard structure, and let $D^2 = \{(a^1, a^2) \in \mathbb{R}^2: (a^1)^2 + (a^2)^2 < 1\}$, the unit disk, have the structure given by the atlas $\{(D^2, id)\}$. Then $\phi: D^2 \to \mathbb{R}^2$ given by

$$\phi^1: (a^1, a^2) \mapsto \frac{a^1}{1 - (a^1)^2 - (a^2)^2}$$

$$\phi^2: (a^1, a^2) \mapsto \frac{a^2}{1 - (a^1)^2 - (a^2)^2}$$

for $(a^1)^2 + (a^2)^2 < 1$ is a diffeomorphism.

2. In Section 9.1, Example 1, we made $\mathbb{R}$ into a manifold M with atlas $\{(\mathbb{R}, id)\}$ and into a manifold N with atlas $\{(\mathbb{R}, \mu_0)\}$. Let $\phi: N \to M$ be given by $a \mapsto a^3$. Then ϕ is a diffeomorphism. This illustrates the fact that two differentiable manifolds with different differentiable structures can be diffeomorphic. As a matter of fact, all topologically equivalent manifolds of the same dimension less than or equal to 4 are diffeomorphic. There are examples of nondiffeomorphic manifolds of dimension greater than or equal to 7.

3. $S^2 = \{(a^1, a^2, a^3): (a^1)^2 + (a^2)^2 + (a^3)^2 = 1\}$ with the induced topology from $\mathbb{R}^3$ is made into a manifold determined by six charts $(\mathscr{V}_i, v_i)$ where $\mathscr{V}_1 = \{(a^1, a^2, a^3) \in S^2: a^1 > 0\}$, $v_1: (a^1, a^2, a^3) \mapsto (a^2, a^3)$, etc. Now we can map S^2 onto the real projective plane, $P^2(\mathbb{R})$, as follows.

Let ϕ be a map from D^2 onto $\mathbb{R}^2$ given by

$$\phi^1: (a^2, a^3) \mapsto \frac{a^2}{\sqrt{1 - (a^2)^2 - (a^3)^2}}$$

$$\phi^2: (a^2, a^3) \mapsto \frac{a^3}{\sqrt{1 - (a^2)^2 - (a^3)^2}}$$

for $(a^2)^2 + (a^3)^2 < 1$. For a point on S^2 in $\mathscr{V}_1$, the values on the right are a^2/a^1 and a^3/a^1, which are the values of the map μ_1 of Example 5 in Section 9.1. Thus, we have $\mathscr{V}_1 \xrightarrow{v_1} D^2 \xrightarrow{\phi} \mathbb{R}^2 \xrightarrow{\mu_1^{-1}} \mathscr{U}_1$, and each map is a 1–1 onto map, so we have a 1–1 map of a coordinate neighborhood of S^2 onto a coordinate neighborhood of $P^2(\mathbb{R})$. v_1, ϕ, and μ_1^{-1} have differentiable inverses, so we have a diffeomorphism between $\mathscr{V}_1$ and $\mathscr{U}_1$. However, if we try to map *all* of S^2 onto $P^2(\mathbb{R})$ by this method we get a differentiable map, but not 1–1, and hence not a diffeomorphism.

PROBLEM 9.9. Prove Theorem 9.2.

PROBLEM 9.10. Prove Theorem 9.3(ii).

PROBLEM 9.11. Prove Theorem 9.4.

9.3 The tangent and cotangent spaces at a point of M

Having the concepts of differentiability for curves and functions, we can now proceed exactly as in Section 7.2. Thus, for differentiable curves through p and differentiable functions on a neighborhood of p we can form $f \circ \gamma$ and define the tangent, $[\gamma]_p$, of a curve γ_p through p and the differential, $[f]_p$, of a function f as equivalence classes, exactly as in Section 7.2.

We can proceed to characterize the equivalence of curves and functions in terms of local coordinates, generalizing what we did in observation (iv) in Section 7.2. In order to do so we note that, just as for cartesian spaces, along with coordinate functions $\mu^i = \pi^i \circ \mu$ on neighborhoods of p, we also have *coordinate curves* ζ_{pi} *through* p given by

$$\zeta_{pi} = \mu^{-1} \circ \vartheta_{\mu(p)i}: u \mapsto \mu^{-1}(\mu^1(p), \ldots, \mu^i(p) + u, \ldots, \mu^n(p))$$

for some open interval containing 0 (Fig. 9.3). (Recall that the $\vartheta_{\mu(p)i}$ are the natural coordinate curves through $\mu(p)$ in $\mathbb{R}^n$.) Then along with the

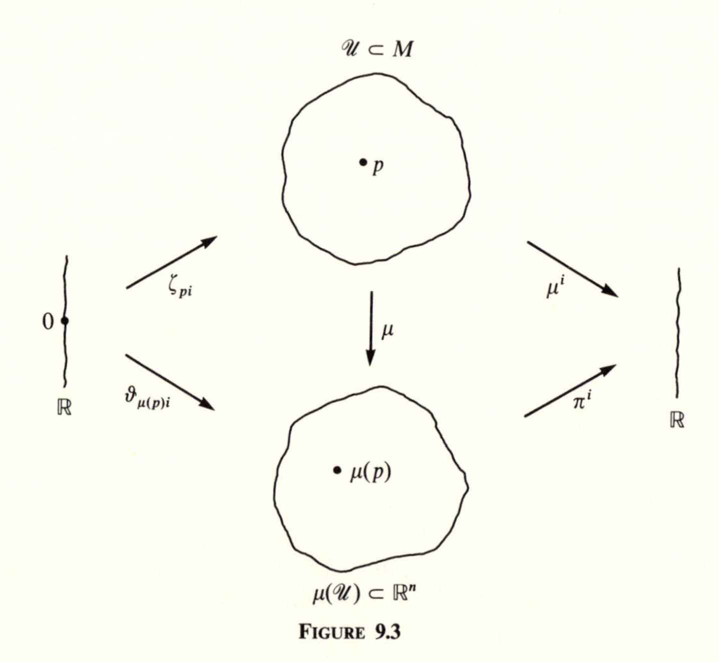

FIGURE 9.3

component functions, $\gamma^i = \mu^i \circ \gamma$, of γ, we have the partial functions, $f \circ \zeta_i$, of f, and we have the derivatives $D_0\gamma^i$, $D_0 f \circ \zeta_i = D_0 f \circ \mu^{-1} \circ \vartheta_i = D_0 \hat{f} \circ \vartheta_i = \partial \hat{f}/\partial \pi^i|_{\mu(p)}$, and $D_0 f \circ \gamma$. Introducing the notation

$$\left.\frac{\partial f}{\partial \mu^i}\right|_p = \left.\frac{\partial \hat{f}}{\partial \pi^i}\right|_{\mu(p)} \tag{9.1}$$

and using the chain rule, eq. (7.2), for cartesian spaces we get a chain rule

$$D_0 f \circ \gamma = \left.\frac{\partial f}{\partial \mu^i}\right|_p D_0\gamma^i \tag{9.2}$$

relating these derivatives. Finally, from eq. (9.2), as in observation (iv) in Section 7.2,

$$\gamma_1 \sim \gamma_2 \Leftrightarrow D_0\gamma_1^i = D_0\gamma_2^i \qquad \text{and} \qquad f^1 \sim f^2 \Leftrightarrow \left.\frac{\partial f^1}{\partial \mu^i}\right|_p = \left.\frac{\partial f^2}{\partial \mu^i}\right|_p$$

The set of differentials at p, T^p, is a vector space. The differentials of the coordinate functions $[\mu^1], [\mu^2], \ldots, [\mu^n]$ constitute a coordinate basis of T^p. The proofs are just as in Section 7.2. With the notation $[\mu^i] = d\mu^i$ and $[f] = df$ we have, corresponding to (7.10), for all coordinate systems,

$$df|_p = \left.\frac{\partial f}{\partial \mu^i}\right|_p d\mu^i|_p \tag{9.3}$$

Using a local coordinate system, the constructions used in Section 7.2 to make the set of tangents of curves through p a vector space, T_p, can be duplicated for general manifolds. An additional argument is now required to guarantee that our definitions are independent of coordinates. See eq. (9.7) and comments below eq. (9.8). Alternatively, we can use the method of Problem 7.4. The tangents of the coordinate curves $[\zeta_1], [\zeta_2], \ldots, [\zeta_n]$ constitute a coordinate basis of T_p. With the notation $[\zeta_i] = \partial/\partial\mu^i$ we have, corresponding to eq. (7.13), for all coordinate systems,

$$[\gamma] = D_0\gamma^i \left.\frac{\partial}{\partial \mu^i}\right|_p$$

As in Section 7.2, the elements, v_p, of the tangent space, T_p, at p may be described as derivations on functions. Thus, T_p is a set of derivations on

$\mathfrak{F}_p$, and we have, corresponding respectively to eqs. (7.12) and eq. (7.14),

$$\left.\frac{\partial}{\partial \mu^i}\right|_p \cdot f = \left.\frac{\partial f}{\partial \mu^i}\right|_p$$

and

$$v_p = (v_p \cdot \mu^i) \left.\frac{\partial}{\partial \mu^i}\right|_p \tag{9.4}$$

If $v = [\gamma] \in T_p$ and $f \in \mathfrak{F}_p$, then $b: T_p \times T^p \to \mathbb{R}$ given by $(v, df) \mapsto v \cdot f = D_0 f \circ \gamma$ (from (9.4) and (9.2)) is nondegenerate bilinear, so T_p and T^p are dual with respect to b, df can be considered to be a function from T_p to $\mathbb{R}$, and (since they are finite-dimensional) we can write $T^p = T_p^*$. Then

$$\langle df, v \rangle = v \cdot f = D_0 f \circ \gamma \tag{9.5}$$

In particular, $\langle d\mu^i, \partial/\partial\mu^j \rangle = \delta^i_j$; i.e., $\{\partial/\partial\mu^i\}$ and $\{d\mu^i\}$ are dual bases.

Finally, if $\sigma_p \in T^p$, then $\sigma_p = \sigma_i \, d\mu^i|_p$, and since $\sigma_i = \sigma_p \cdot \partial/\partial\mu^i|_p$,

$$\sigma_p = \left(\sigma_p \cdot \left.\frac{\partial}{\partial \mu^i}\right|_p\right) d\mu^i|_p \tag{9.6}$$

If M is a C^∞ manifold, T_p is the set of derivations on C^∞ functions at p. We get higher-order partial derivatives at p as follows. Having defined $\partial f/\partial\mu^j|_p$, eq. (9.1), we can define other functions on a neighborhood of p. Thus, if f is differentiable at each point of the coordinate neighborhood, $\mathscr{U}$, then we have the functions

$$\frac{\partial f}{\partial \mu^j}: p \mapsto \left.\frac{\partial f}{\partial \mu^j}\right|_p$$

on $\mathscr{U}$. Further, by definition, $\partial f/\partial\mu^j$ will be differentiable at p if $\partial f/\partial\mu^j \circ \mu^{-1}$ is differentiable at $\mu(\mathrm{p})$. But $\partial f/\partial\mu^j \circ \mu^{-1} = \partial \hat{f}/\partial\pi^j$. So if f is C^2 at p, $\partial f/\partial\mu^j$ will be C^1 functions at p, and we have

$$\left.\frac{\partial^2 f}{\partial \mu^i \, \partial \mu^j}\right|_p = \left.\frac{\partial}{\partial \mu^i}\left(\frac{\mathrm{d}f}{\partial \mu^j}\right)\right|_p = \frac{\partial}{\partial \pi^i}\left(\frac{\partial f}{\partial \mu^j} \circ \mu^{-1}\right)_{\mu(p)} = \frac{\partial}{\partial \pi^i}\left(\frac{\partial \hat{f}}{\partial \pi^j}\right)_{\mu(p)}$$

Continuing, we get the result that if f is C^k at p, then the $(k-1)$th-order partial derivative functions will be C^1 functions at p, and we can take their partial derivatives.

We have generalized the concepts and formulas of Section 7.2 for $\mathbb{R}^n$ to concepts and formulas valid for a manifold, M. A new feature arises for manifolds which did not exist for $\mathbb{R}^n$ in Section 7.2 (but is relevant for $\mathbb{R}^n$ in Section 9.1 when we made it into a manifold). A formula may be given in terms of coordinates like eq. (9.4), or without the use of coordinates like eq. (9.5). In the former case, since the concepts described in the formula are independent of coordinates, the formula must transform properly on the intersection of coordinate domains. If $v^i = v \cdot \mu^i$ on $\mathcal{U}$ and $\bar{v}^i = v \cdot \bar{\mu}^i$ on $\bar{\mathcal{U}}$, then on $\mathcal{U} \cap \bar{\mathcal{U}}$, by eq. (9.4), $v^i\, \partial/\partial\mu^i \cdot d\bar{\mu}^j = \bar{v}^j$. By eq. (9.5) $\partial/\partial\mu^i \cdot \mathrm{d}\bar{\mu}^j = \partial/\partial\mu^i \cdot \bar{\mu}^i$, so

$$\bar{v}^j = \frac{\partial \bar{\mu}^j}{\partial \mu^i} v^i \tag{9.7}$$

Similarly, putting $\sigma_i = \sigma \cdot \mathrm{d}/\mathrm{d}\mu^i$ and $\bar{\sigma}_i = \sigma \cdot \partial/\partial\bar{\mu}^i$ we get, from (9.6), and (9.5),

$$\bar{\sigma}_j = \frac{\partial \mu^i}{\partial \bar{\mu}^j} \sigma_i \tag{9.8}$$

Equations (9.7) and (9.8) for the transformations of the components of vectors and 1-forms can be generalized to formulas for tensors of any type. See eqs. (9.13) and (9.14). Also, compare these results with eq. (1.4), which relates the components v^i and $\bar{v}^i$ by the change of basis matrix. If a concept is defined or a relation is given in terms of components in a given basis, these formulas are used to show that the concept or relation is independent of coordinates.

Having now the concepts of tangent and cotangent spaces at a point $p \in M$ we can reproduce practically verbatim the discussion of Section 7.3 of the tangent map ϕ_{*p} at a point, and its transpose, or dual, $\phi^*_{\phi(p)}$.

In particular, if $\phi\colon M \to N$, $(\partial/\partial\mu^k)$ is a basis of T_p, $(\partial/\partial v^j)$ is a basis of $T_{\phi(p)}$, and $\phi^j = v^j \circ \phi$ are the component functions of ϕ, then eq. (7.19) becomes

$$\phi_* \cdot \frac{\partial}{\partial \mu^k} = \frac{\partial \phi^j}{\partial \mu^k} \frac{\partial}{\partial v^j} \tag{9.9}$$

and if $(d\mu^j)$ is a basis of T^*_p and dv^k is a basis of $T^*_{\phi(p)}$, eq. (7.23) becomes

$$\phi^* \cdot dv^k = \frac{\partial \phi^k}{\partial \mu^j} d\mu^j \tag{9.10}$$

Since $\partial\phi^j/\partial\mu^i = \partial\hat{\phi}^j/\partial\pi^i$, eq. (9.9) says that the (matrix) representation (with respect to the bases $\{\partial/\partial\mu^i\}$ and $\{\partial/\partial\nu^j\}$) of the tangent map of ϕ is the derivative of the representation $\hat{\phi} = \nu \circ \phi \circ \mu^{-1}$ (with respect to the coordinate maps μ and ν) of ϕ. Similarly, eq. (9.10) says that the (matrix) representation of the transpose (or dual) of the tangent map of ϕ is the transpose of the derivative of the representation of ϕ. Equations (7.20) and (7.24) become respectively,

$$w^j = \frac{\partial\phi^j}{\partial\mu^k} v^k \tag{9.11}$$

and

$$\sigma_j = \frac{\partial\phi^k}{\partial\mu^j} \tau_k \tag{9.12}$$

It is interesting to compare these equations for the transformation of components under a mapping, ϕ, with eqs. (9.7) and (9.8) for the transformation of components under a transformation of coordinates.

Just as for mappings, ϕ, between cartesian spaces in Section 7.3, we sort out two important cases.

(i) If $\phi = \gamma$, a curve defined on some open set in $\mathbb{R}$, we have

$$\dot{\gamma}\colon a \mapsto [\vartheta_a] = \left.\frac{d}{d\pi}\right|_a \mapsto [\gamma \circ \vartheta_a]_{\gamma(a)} = D_a\gamma^i \left.\frac{\partial}{\partial\mu^i}\right|_{\gamma(a)}$$

the canonical lift of γ, and

$$\dot{\gamma}(a) = D_a\gamma^i \left.\frac{\partial}{\partial\mu^i}\right|_{\gamma(a)} \tag{9.13}$$

the tangent (or, velocity) of γ at $\gamma(a)$. Since $\dot{\gamma}(a) = \gamma_{*a} \cdot d/d\pi|_a$, the tangent of γ at a is sometimes denoted by γ_{*a}.

(ii) If ϕ is a function, f, since in this case, as before in Section 7.3, $\phi_{*a} \cdot v = \langle d\phi, v\rangle\, d/d\bar{\pi}$, the tangent map is sometimes called the differential of f.

With T_p, $T_{\phi(p)}$, T^p, and $T^{\phi(p)}$ we have all the various tensor product spaces and subspaces built on these vector spaces at p and $\phi(p)$, just as in Section 7.3. In a coordinate neighborhood, $\mathscr{U}$, we have coordinate representations such as $A^{i_1 \cdots i_r}\, \partial/\partial\mu^{i_1} \otimes \cdots \otimes \partial/\partial\mu^{i_r}$ and $A_{j_1 \cdots j_s}\, d\mu^{j_1} \otimes \cdots \otimes d\mu^{j_s}$ from which we get, on $\mathscr{U} \cap \mathscr{V}$, respectively,

$$\bar{A}^{j_1 \cdots j_r} = A^{i_1 \cdots i_r} \frac{\partial\bar{\mu}^{j_1}}{\partial\mu^{i_1}} \cdots \frac{\partial\bar{\mu}^{j_r}}{\partial\mu^{i_r}} \tag{9.14}$$

and

$$\bar{A}_{j_1 \cdots j_s} = A_{i_1 \cdots i_s} \frac{\partial \mu^{i_1}}{\partial \bar{\mu}^{j_1}} \cdots \frac{\partial \mu^{i_s}}{\partial \bar{\mu}^{j_s}} \tag{9.15}$$

(Compare with eq. (4.5).)

Moreover, ϕ_{*p} and $\phi^*_{\phi(p)}$ are extended to the contravariant and covariant tensor algebras at p and $\phi(p)$, respectively, just as in Section 7.3. That is, we have $\phi_r|_p: (T_p)^r_0 \to (T_{\phi(p)})^r_0$ and $\phi^s|_{\phi(p)}: (T_{\phi(p)})^0_s \to T_p)^0_s$.

Finally, note that we have been working all this time at a fixed point $p \in M$. As in Section 7.3, we extend $\phi_r|_p$ and $\phi^s|_{\phi(p)}$ to maps

$$\phi_r: \bigcup_{p \in M} T_p)^r_0 \to \bigcup_{p \in N} T_p)^r_0 \qquad \text{and} \qquad \phi^s: \bigcup_{\phi(p) \in N} T_{\phi(p)})^0_s \to \bigcup_{p \in M} T_p)^0_s$$

In particular, we will write ϕ_* for ϕ_1, the extension of the tangent map at a point, and we will write ϕ^* for ϕ^1, the extension of the transpose (or dual) of the tangent map.

PROBLEM 9.12. Describe geometrically

(i) the set of values, $\zeta_i(u)$, of the coordinate curves
(ii) the hypersurfaces $\{p: \mu^i(p) = \text{constant}\}$

on the chart $(\mathscr{U}, \mu)$ on

(1) $\mathbb{R}^3$ given by "cylindrical coordinates"
(2) S^2 given by "stereographic coordinates" (see Problem 9.2).

PROBLEM 9.13. Derive eq. (9.2).

PROBLEM 9.14. (i) Define $[\gamma_1] + [\gamma_2]$ and $c[\gamma]$ in T_p as in Section 7.2 and show that these definitions are independent of coordinates. (ii) If M is a finite-dimensional vector space, V^n, then there are natural isomorphisms $V^n \cong T_p$ for each $p \in V^n$ given by $v \mapsto [\gamma]_p$ where $\gamma: u \mapsto p + uv$.

PROBLEM 9.15. Derive eq. (9.13).

PROBLEM 9.16. Derive the "transformation law" on intersections of coordinate domains for (r, s) tensors.

9.4 Some properties of mappings

The first thing we will show is that every differentiable mapping has a simple local representation; that is, in appropriate coordinates the equations of the mapping are simple. Then we will use these representations to describe certain class of manifolds generalizing examples in Section 9.1.

Our main tool will be the "Inverse Function Theorem" for cartesian spaces.

Theorem 9.6. *Suppose $\Phi: \mathscr{U} \subset \mathbb{R}^n \to \mathbb{R}^n$, Φ is C^k on $\mathscr{U}$, and* $\det D_{a_0}\Phi \neq 0$ *for some $a_0 \in \mathscr{U}$. Then there exist open neighborhoods $\mathscr{V}_{a_0}$ and $\bar{\mathscr{V}}_{\Phi(a_0)}$ of a_0 and $\Phi(a_0)$ respectively such that*

(i) *$\Phi|_{\mathscr{V}_{a_0}}$ is 1–1 and onto $\bar{\mathscr{V}}_{\Phi(a_0)}$*
(ii) *Φ^{-1} is C^k on $\bar{\mathscr{V}}_{\Phi(a_0)}$*

Proof. Woll (1966, pp. 20 ff).

Suppose $\phi: M \to N$ is differentiable and $\dim M = m$ and $\dim N = n$. Since the overlap maps are all differentiable, the rank of the Jacobian matrix (the derivative) is the same for all representations of ϕ, and so we can talk about *the rank of ϕ*.

Theorem 9.7. (The rank theorem, Boothby, 1975, p. 70.) *A necessary and sufficient condition that $\phi: M \to N$ has rank r in a neighborhood of a point, p, is that there is an admissible chart $(\mathscr{U}, \mu)$ at p and an admissible chart $(\mathscr{V}, \nu)$ at $\phi(p)$ such that*

$$\nu^i \circ \phi = \mu^i, \qquad i = 1, \ldots, r$$

$$\nu^j \circ \phi = 0, \qquad j = r + 1, \ldots, n$$

on $\phi^{-1}(\mathscr{V}) \cap \mathscr{U}$.

Another way of stating this condition is that if $q \in \phi^{-1}(\mathscr{V}) \cap \mathscr{U}$, and $\mu(q) = (a^1, \ldots, a^m)$, then $\hat{\phi} = \nu \circ \phi \circ \mu^{-1}: (a^1, \ldots, a^m) \mapsto (a^1, \ldots, a^r, 0, \ldots, 0)$. Or, finally, if $\nu \circ \phi(q) = (b^1, \ldots, b^n)$ then the equations of the representation, $\hat{\phi}$, in these coordinates, of ϕ are $b^1 = a^1, \ldots, b^r = a^r, b^{r+1} = 0, \ldots, b^n = 0$.

Proof. The proof is based on the following lemma.

Lemma. *If ϕ maps an open set $\mathscr{U} \subset \mathbb{R}^m$ to $\mathbb{R}^n$ and has rank r then for each $a_0 \notin \mathscr{U}$ there is a diffeomorphism $\Phi: \mathscr{V} \to \bar{\mathscr{V}}$ with $a_0 \in \mathscr{V}$, $\mathscr{V} \subset \mathscr{U}$, and $\bar{\mathscr{V}} \subset \mathbb{R}^m$, and a diffeomorphism $\Psi: \mathscr{W} \to \bar{\mathscr{W}}$ with $\phi(a_0) \in \mathscr{W}$, $\mathscr{W} \subset \phi(\mathscr{U})$, and $\bar{\mathscr{W}} \subset \mathbb{R}^n$ such that*

$$\Psi \circ \phi \circ \Phi^{-1}: (\bar{a}^1, \ldots, \bar{a}^m) \mapsto (\bar{a}^1, \ldots, \bar{a}^r, 0, \ldots, 0)$$

where $(\bar{a}^1, \ldots, \bar{a}^m) \in \bar{\mathscr{V}}$ (see Fig. 9.4).

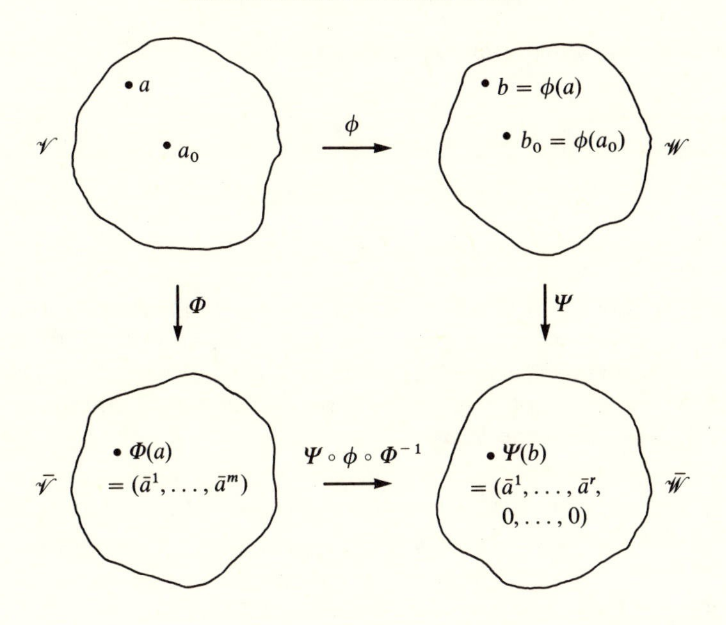

FIGURE 9.4

Proof (*of lemma*). We can assume, without loss of generality, that the upper left hand $r \times r$ submatrix of the Jacobian matrix of ϕ is nonsingular at a_0. Then we define Φ on $\mathscr{U}$ by

$$\begin{aligned} \Phi^i &= \phi^i, \qquad i = 1, \ldots, r \\ \Phi^j &= \pi^j, \qquad j = r+1, \ldots, m \end{aligned} \tag{9.16}$$

$\det D_{a_0}\Phi \neq 0$ so by the inverse function theorem, we have $\bar{\mathscr{V}}$, and $\mathscr{V}$, with Φ a diffeomorphism between them. Writing out (9.16) we have

$$\begin{aligned} \bar{a}^1 &= \Phi^1(a) = \phi^1(a) = b^1 \\ &\vdots \\ \bar{a}^r &= \Phi^r(a) = \phi^r(a) = b^r \\ \bar{a}^{r+1} &= \Phi^{r+1}(a) = \pi^{r+1}(a) = a^{r+1} \\ &\vdots \\ \bar{a}^m &= \Phi^m(a) = \pi^m(a) = a^m \end{aligned}$$

Now $\phi \circ \Phi^{-1}$ is a map from $\bar{\mathscr{V}}$ to $\mathbb{R}^n$ given by

$$\begin{aligned}\phi \circ \Phi^{-1}: (a^1, \ldots, a^m) &\overset{\Phi^{-1}}{\mapsto} (a^1, \ldots, a^m)\\ &\overset{\phi}{\mapsto} (\phi^1(a), \ldots, \phi^n(a))\\ &= (\bar{a}^1, \ldots, \bar{a}^r, \phi^{r+1} \circ \Phi^{-1}(\bar{a}), \ldots, \phi^n \circ \Phi^{-1}(\bar{a}))\end{aligned}$$

$\phi \circ \Phi^{-1}$ has rank r on $\bar{\mathscr{V}}$ since ϕ has rank r on $\mathscr{V}$. This implies that for all $j = r+1, \ldots, m$ and $k = r+1, \ldots, n$, $\partial\phi^k \circ \Phi^{-1}/\partial\pi^j = 0$ on $\mathscr{V}$. That is, $\phi^k \circ \Phi^{-1}$ are functions of $\bar{a}^1, \ldots, \bar{a}^r$ only. $\phi^k \circ \Phi^{-1}$ are defined in a neighborhood of $\bar{a}_0^1 = b_0^1, \ldots, \bar{a}_0^r = b_0^r$ so they are defined in a neighborhood of b_0.

Thus, we can define Ψ by

$$\begin{aligned}\Psi^i &= \pi^i, & i &= 1, \ldots, r\\ \Psi^k &= \pi^k - \phi^k \circ \Phi^{-1}, & k &= r+1, \ldots, n\end{aligned} \tag{9.17}$$

or, expressed alternatively,

$$\begin{aligned}\Psi: (b^1, \ldots, b^n) \mapsto (b^1, \ldots, b^r, b^{r+1} - \phi^{r+1} \circ \Phi^{-1}(b^1, \ldots, b^r), \ldots,\\ b^n - \phi^n \circ \Phi^{-1}(b^1, \ldots, b^n))\end{aligned}$$

det $D_{b_0}\Psi \neq 0$, so we have $\mathscr{W}$ and $\bar{\mathscr{W}}$, with Ψ a diffeomorphism between them.

Finally, putting $b^i = \bar{a}^i$, $i = 1, \ldots, r$, and noting that $\phi^k \circ \Phi^{-1}(\bar{a}^1, \ldots, \bar{a}^r) = \phi^k(a) = b^k$, $k = r+1, \ldots, n$, the mapping (9.17) becomes

$$\Psi: (b^1, \ldots, b^n) \mapsto (\bar{a}^1, \ldots, \bar{a}^r, 0, \ldots, 0)$$

and

$$\Psi \circ \phi \circ \Phi^{-1}: (\bar{a}^1, \ldots, \bar{a}^m) \mapsto (\bar{a}^1, \ldots, \bar{a}^r, 0, \ldots, 0) \qquad \blacksquare$$

There are two special cases for $\phi: M \to N$ of maximum rank; immersions and submersions.

Definitions
ϕ is *an immersion of M into N* if rank $\phi = \dim M$. ϕ is *a submersion of M into N* if rank $\phi = \dim N$.

For these two special cases, the proof of the theorem shows:

1. A necessary and sufficient condition that ϕ is an immersion is that if $(\mathscr{V}, \nu)$ is any chart at $\phi(p)$, then there is an admissible chart $(\mathscr{U}, \mu)$ at p with $\mu^i = \nu^i \circ \phi$, $i = 1, \ldots, m$. (To get $i = 1, \ldots, m$ it may be necessary to relabel the ν^i.)

2. A necessary and sufficient condition that ϕ is a submersion is that if $(\mathscr{V}, \nu)$ is any chart at $\phi(p)$ and $(\mathscr{W}, \xi)$ is any chart at p, then there is an admissible chart $(\mathscr{U}, \mu)$ at p with $\mu^i = \nu^i \circ \phi$ for $i = 1, \ldots, n$, and $\mu^j = \xi^j$ for $j = n + 1, \ldots, m$. (Again we may have to relabel the ν^i.)

We illustrate the theorem and the two special cases with the following examples.

(i) $m = 3$, $n = 3$, *and* $r = 2$. Suppose (a^1, a^2, a^3) are coordinate values in a neighborhood of p and (b^1, b^2, b^3) are coordinate values in a neighborhood of $\phi(p)$ and

$$\hat{\phi}: \begin{cases} b^1 = a^1 + a^2 + a^3 \\ b^2 = a^1 + a^2 + a^3 \\ b^3 = a^1 + 2a^2 + a^3 \end{cases}$$

If we choose coordinates $(\bar{a}^1, \bar{a}^2, \bar{a}^3) = (a^1 + a^2 + a^3, a^1 + 2a^2 + a^3, a^3)$ around p and $(\bar{b}^1, \bar{b}^2, \bar{b}^3) = (b^1, b^3, b^1 - b^2)$ around $\phi(p)$, then on the image of ϕ we have

$$\bar{b}^1 = \bar{a}^1, \qquad \bar{b}^2 = \bar{a}^2, \qquad \bar{b}^3 = 0$$

(ii) $m = 2$, $n = 3$, *and* $r = 2$. Suppose, in a neighborhood of p, ϕ is represented by

$$\hat{\phi}: \begin{cases} b^1 = a^1 + a^2 \\ b^2 = a^1 + a^2 \\ b^3 = a^1 + 2a^2 \end{cases}$$

If we choose coordinates $(\bar{a}^1, \bar{a}^2) = (a^1 + a^2, a^1 + 2a^2)$ around p, then on the image of ϕ we have

$$\bar{a}^1 = b^1, \qquad \bar{a}^2 = b^3$$

and if we relabel b^2 and b^3 we get the required result.

(iii) $m = 3$, $n = 2$, *and* $r = 2$. Suppose, in a neighborhood of p, ϕ is represented by

$$\hat{\phi}: \begin{cases} b^1 = a^1 + a^2 + a^3 \\ b^2 = a^1 + 2a^2 + a^3 \end{cases}$$

If we choose coordinates $(\bar{a}^1, \bar{a}^2, \bar{a}^3) = (a^1 + a^2 + a^3, a^1 + 2a^2 + a^3, a^3)$ around p, then on the image of ϕ we have

$$\bar{a}^1 = b^1, \qquad \bar{a}^2 = b^2, \qquad \bar{a}^3 = a^3$$

In Section 9.1 we had examples in which subsets of $\mathbb{R}^2$ and $\mathbb{R}^3$, obtained as images of subsets of $\mathbb{R}$ and $\mathbb{R}^2$, respectively, were made into manifolds. We can generalize these cases.

Theorem 9.8. *The image of a* 1–1 *immersion of M into N can be made into a manifold diffeomorphic with M.*

Proof. Give $\phi(M)$ the topology induced by ϕ; i.e., $\phi(\mathscr{U})$ is open if $\mathscr{U}$ is open. Then ϕ^{-1} is a homeomorphism. If $(\mathscr{V}, \mu)$ is a chart at p then $(\phi(\mathscr{V}), \mu \circ \phi^{-1})$ will be a chart at $\phi(p)$. Any two charts at $\phi(p)$ will be C^k-related (if the corresponding charts at p are C^k-related), and $(\mu \circ \phi^{-1}) \circ \phi \circ \mu^{-1}$ is a diffeomorphism. ∎

(In the literature, a 1–1 immersion is sometimes called an imbedding. We will save this term for a stronger concept in Section 10.2.)

Another characterization of immersion is given in terms of the relation between the classes of differentiable functions on M and N. Clearly, if $\phi: M \to N$ is differentiable, then corresponding to each differentiable function g on N, $g \circ \phi$ is a differentiable function on M. To go the other way we need a little more.

Theorem 9.9. *$\phi: M \to N$ is an immersion if and only if corresponding to each differentiable function, f, on M and each point $p \in M$ there is a differentiable function, g, on N such that $g \circ \phi = f$ on some neighborhood of p.* (Borrowing the terminology of the special case in Section 10.1 we can paraphrase this as every differentiable function on M is the "restriction" of some differentiable function on N.)

Proof. **If:** Let $\mathscr{U}$ be a coordinate neighborhood of p and $f^1, \ldots, f^m$ a set of coordinate functions. Then on $\mathscr{U}$ there are functions g^i such that $g^i \circ \phi = f^i, i = 1, \ldots, m$. Now let v be a vector in T_p and form $\langle df^i, v\rangle$. Since $\langle df^i, v\rangle = \langle dg^i \circ \phi, v\rangle = \langle \phi^* \circ dg^i, v\rangle = \langle dg^i, \phi^* \circ v\rangle$ it follows that $\phi^* \circ v = 0$ implies $v = 0$ and so ϕ^* has rank m.

Only if: At each point $p \in M$, there is a neighborhood, $\mathscr{W}$, of $\phi(p)$, and a differentiable map, ψ, on $\mathscr{W}$ such that $\psi \circ \phi = id$ on $\tau(\mathscr{W})$ (Auslander and MacKenzie, 1977, p. 47). Then $f \circ \psi \circ \phi = f$ on $\psi(\mathscr{W})$. Let $f \circ \psi = g$. Finally, extend g to a function on N by Theorem 9.5. ∎

When a differentiable manifold is defined as indicated below Theorem 9.4 by prescribing a class of differentiable functions, then Theorem 9.9 can be used as a definition to transfer the differentiable structure of M to the image of M in N. (See Auslander and MacKenzie, 1977, pp. 36 ff.)

In Section 9.1, we had examples in which subsets of $\mathbb{R}^2$ and $\mathbb{R}^3$, obtained as inverse images of points of $\mathbb{R}$, were made into manifolds. These are examples coming from submersions in the special cases where $N = \mathbb{R}^n$.

If $N = \mathbb{R}^n$, then $\phi: M \to N$ has the form $\phi = (f^1, \ldots, f^n)$ and the second special case of the rank theorem says for a submersion there is a chart $(\mathscr{U}, \mu)$ at p with $\mu^i = \pi^i \circ \phi = f^i$, $i = 1, \ldots, n$. (Using the identity chart on $\mathbb{R}^n$.) That is, a set of functions, $f^1, \ldots, f^n$, with $\operatorname{rank}(\partial f^i/\partial \xi^j) = n$ can be enlarged to a local coordinate system at p. Moreover, this chart has the property that if $b \in \phi(\mathscr{U})$, then on $\mathscr{U} \cap \phi^{-1}(b)$ the values of $f^1, \ldots, f^n$ are constant.

If we denote the level sets, or fibers, $\phi^{-1}(b)$, of a submersion $\phi: M \to \mathbb{R}^n$ by F_b, we can summarize these comments more formally as follows.

Theorem 9.10. *If $\phi: M \to \mathbb{R}^n$ is a submersion, then at each point of F_b there is a coordinate neighborhood $\mathscr{U}$ of M in which $q \in F_b \cap \mathscr{U}$ satisfy $f^1(q) = b^1, \ldots, f^n(q) = b^n$.*

Theorem 9.11. *The level sets of a submersion can be made into manifolds.*

Proof. We will assume that $m = \dim M > n$. At each point of F_b there is a chart $(\mathscr{U}, \mu)$ with $\mu(q) = (b^1, \ldots, b^n, a^{n+1}, \ldots, a^m)$ with $(b^1, \ldots, b^n)$ fixed for all $q \in F_b \cap \mathscr{U}$. The projection $\pi^a: (b^1, \ldots, b^n, a^{n+1}, \ldots, a^m) \mapsto (a^{n+1}, \ldots, a^m)$ is a diffeomorphism so $(F_b \cap \mathscr{U}, \pi^a \circ \mu)$ is a chart on F_b. Since the coordinate maps $\pi^a \circ \mu$ are diffeomorphisms, the overlap maps are differentiable. ∎

Definition

A level set of a submersion $\phi: M \to \mathbb{R}^n$ with the manifold structure of Theorem 9.11 will be called *a differentiable variety of ϕ*.

It is instructive to compare Theorem 9.11 and Theorem 9.8. In particular note that F_b has the topology of M in Theorem 9.11, but in Theorem 9.8 $\phi(M)$ need not have the topology of N. Also, F_b is closed in M, but $\phi(M)$ need not be closed in N.

PROBLEM 9.17. Prove Theorem 9.7 using the given lemma.

PROBLEM 9.18. Prove the results 1 and 2 for the special cases of Theorem 9.7.

PROBLEM 9.19. The subset S^{n-1} of $\mathbb{R}^n$ is a differentiable variety.

PROBLEM 9.20. Verify the properties of F_b and $\phi(M)$ as stated in the last paragraph above.

PROBLEM 9.21. At each point $p \in F_b$ the tangent space of F_b is a subspace of T_pM, and T_pF_b is the kernel of ϕ_{*p}.

10

SUBMANIFOLDS

The prototypical submanifold is a surface in ordinary space. There are various ways of describing surfaces in ordinary space. The two main methods are by parametric equations, $x = x(u, v)$, $y = y(u, v)$, $z = z(u, v)$, and the points (x, y, z) satisfying $F(x, y, z) = \text{constant}$. We will extend these two descriptions to general differentiable manifolds. They correspond to, and derive, roughly, from the concepts of immersion and submersion, respectively.

10.1 Parametrized submanifolds

Let M be a given manifold. In analogy with parametrized curves, we focus on a mapping, ϕ, from another given manifold to M.

Definition
(P, ϕ) is *an immersed submanifold of* M if P is a differentiable manifold and ϕ is a 1–1 immersion of P into M.

We will shorten "immersed submanifold" to simply "submanifold" in the following. A warning about terminology: in the literature sometimes the term "submanifold" is used as we are using it, and sometimes it stands for the stronger concept we have called an "imbedded submanifold" in Section 10.2. Compare the comment below Theorem 9.8.

It is usual (as in the case of curves) to think of a submanifold of M as a subset of M. Hence, under the conditions of the above definition, we frequently say that $\phi(P)$ is a (immersed) submanifold of M. That is, from this point of view, submanifolds of M are certain manifolds contained in M, specifically, the manifolds constructed in Theorem 9.8. Not every manifold in M is a submanifold of M.

Theorem 10.1. *If* $S \subset M$ *and* S *is a manifold, then* S *is a submanifold of* $M \Leftrightarrow$ *the inclusion map,* $i: S \to M$, *is an immersion.*

Proof. $\Rightarrow$: Let ψ be the diffeomorphism $\phi^{-1}: \phi(S) \to P$. Then $i = \phi \circ \psi$ is the inclusion map with the required rank.

$\Leftarrow$: Let $P = S$ and $\phi = i$. ∎

Along with the point of view that a submanifold is the image set of a 1–1 immersion, we think of the tangent space, T_pP, of P at p, as a subspace of $T_{\phi(p)}M$, the tangent space of M at $\phi(p)$. More precisely, we define the tangent space of the submanifold to be $\phi_*(T_pP)$ and then identify T_pP and $\phi_*(T_pP)$.

Consider the following examples, where $\mathbb{R}^n$ and open subsets of $\mathbb{R}^n$ have the standard structure.

1. P is the interval $(0, 1)$ in $\mathbb{R}$, and $\phi: (0, 1) \to \mathbb{R}^2$ according to $u \mapsto (\cos 2\pi u, \sin 2\pi u)$.
2. Same as 1 with $P = \mathbb{R}$.
3. $\phi: \mathbb{R} \to \mathbb{R}^2$ given by $u \mapsto (u^2, u^3)$.
4. $\phi: \mathbb{R} \to \mathbb{R}^2$ given by $u \mapsto (2u^2/1 + u^2, 2u^3/1 + u^2)$.
5. $\phi: \mathbb{R} \to \mathbb{R}^2$ given by $u \mapsto (0, \sin 2\pi u)$.
6. $\phi: \mathbb{R} \to S^1 \times S^1$ with the product structure on $S^1 \times S^1$, given by $u \mapsto (e^{2\pi i u}, e^{2\pi i \alpha u})$ where α is an irrational number.
7. P is an open submanifold of M, and $\phi = i$ the inclusion map of P into M.
8. $\phi: \mathbb{R}^2 \to \mathbb{R}^3$ given by $(u, v) \mapsto (u, v, \sqrt{u^2 + v^2})$.
9. $\phi: (0, \pi) \to \mathbb{R}^2$ given by $u \mapsto (1 + \cos u + \cos 2u, \sin u + \sin 2u)$.
10. P is the open rectangle $\mathcal{U} = \{(u^1, u^2, u^3): (0, 0, 0) < (u^1, u^2, u^3) < (\pi, \pi, 2\pi)\}$ in $\mathbb{R}^3$, and $\phi: \mathcal{U} \to \mathbb{R}^4$ according to $(u^1, u^2, u^3) \mapsto (\cos u^1, \sin u^1 \cos u^2, \sin u^1 \sin u^2 \cos u^3, \sin u^1 \sin u^2 \sin u^3)$.

Examples 1, 6, 7, 10 are submanifolds. The others are not. See Problem 10.1.

The results on immersions in Section 9.4 are valid, in particular, for submanifolds. Thus, if we represent a submanifold according to Theorem 10.1 by (S, i) where $S \subset M$ and i is the inclusion map, then Theorem 9.7 says that at each point of S there are charts with $\mu^i = v^i$, $i = 1, \ldots, p$ (p is the dimension of S), and $v^j = 0$, $j = p + 1, \ldots, m$.

Just as in the case of parametrized curves and surfaces in ordinary Euclidean space, more than one submanifold can have the same image set in M. Moreover, we can get more than one differentiable manifold with the same underlying set as the image of different submanifolds. For example, $P_1 = (0, 2\pi)$, $\phi_1: u \mapsto (\sin 2u, \sin u)$ (Example 4, Section 9.1) and $P_2 = (-\pi, \pi)$, $\phi_2: u \mapsto (\sin 2u, \sin u)$ have the same image, E. E is diffeomorphic with P_1 and also with P_2. However, as manifolds, E with atlas $\{(E, \phi_1^{-1})\}$ and E with atlas $\{(E, \phi_2^{-1})\}$ are different, since $\phi_2^{-1} \circ \phi_1$ is not even continuous at π.

Theorem 10.2. *If M is a manifold, and S is a subset with the induced topology, and S is a submanifold of M, then S has a unique differentiable structure.*

Proof. *Problem 10.2.*

Theorem 10.3. *If P is compact, and (P, ϕ) is a submanifold of M, then $\phi(P)$ has the induced topology from M.*

Proof. This is a corollary of Theorem 8.5. ■

Finally, Theorem 9.9 says that if $S \subset N$ and S is a submanifold of N, then every differentiable function on S is the restriction of a differentiable function on N.

PROBLEM 10.1. Explain why each of the examples above is or is not a submanifold.

PROBLEM 10.2. Prove Theorem 10.2.

10.2 Differentiable varieties as submanifolds

We saw in Section 10.1 that if (P, ϕ) is a submanifold of M, then each point of $\phi(P)$ has a $(\phi(P))$ neighborhood in a coordinate neighborhood of M whose points, q, satisfy $v^{p+1}(q) = 0, \ldots, v^m(q) = 0$ where v^i are local coordinates on M. We saw also, in Theorem 9.10, that if $F_b = \phi^{-1}(b)$ is a differentiable variety in M, then at each point of F_b there is a coordinate neighborhood $\mathcal{U}$ of M in which all $q \in F_b \cap \mathcal{U}$ satisfy $f^1(q) = b^1, f^2(q) = b^2, \ldots, f^n(q) = b^n$ where f^i are local coordinate functions on M. This suggests that we draw our attention to subsets of coordinate neighborhoods of the types arising in these cases, and formulate our results in terms of them.

Definition
Sets, U_k, in a coordinate neighborhood, $\mathcal{U}$, of M for which $m - k$ of the coordinate values are constant ($m = \dim M$) are called *k-dimensional coordinate slices of $\mathcal{U}$.*

Coordinate slices are, clearly, generalizations of coordinate curves, and, according to the paragraph above, they are special cases of differentiable varieties. Comparing the two situations described in that paragraph in the terminology of coordinate slices, we can say that for submanifolds at every point there is a chart, $(\mathcal{U}, \mu)$, of M having a coordinate slice, U_k such that $U_k \subset \phi(P) \cap \mathcal{U}$. That is, every point has a coordinate slice neighborhood. For differentiable varieties we have the stronger property that there is a chart at each point such that all points of F_b in $\mathcal{U}$ are in some coordinate slice of $\mathcal{U}$. That is, there is a coordinate slice, U_k of $\mathcal{U}$ such that $U_k = \phi(P) \cap \mathcal{U}$ (Fig. 10.1). We formalize this stronger property in the following definition.

Definitions
A subset $S \subset M$ has the *k-submanifold property* if at each point of S, M has a chart, $(\mathcal{U}, \mu)$, such that $S \cap \mathcal{U}$ is a k-dimensional coordinate slice of $\mathcal{U}$. Such charts are *adapted to S*.

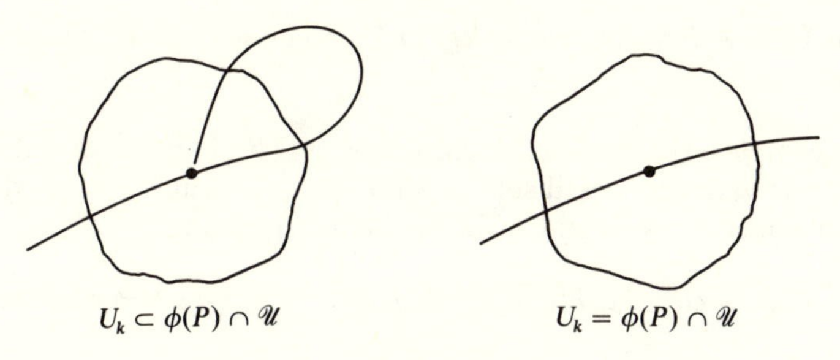

FIGURE 10.1

Theorem 10.4. *A differentiable variety has the k-submanifold property.*

Now, via the k-submanifold property, we will show that differentiable varieties are submanifolds of a special type; namely, with the relative topology of M, and closed in M.

Definitions
$\phi: M \to N$ is *an imbedding* if ϕ is a 1–1 immersion and, with $\phi(M)$ having the relative topology of N, ϕ is a homeomorphism of M onto $\phi(M)$. If ϕ is an imbedding $\phi(M)$ is *an imbedded submanifold* of N (cf., comment below Theorem 9.8).

Theorem 10.5. *Locally, an immersion is an imbedding.*

Proof. Problem 10.3.

Theorem 10.6. *If $S \subset M$ has the k-submanifold property then it is an imbedded submanifold of M.*

Proof. We will show that S is a manifold and the inclusion map is an imbedding. If $(\mathcal{U}, \mu)$ is a chart of M adapted to S, then using the relative topology for S, $\tilde{\mathcal{U}} = \mathcal{U} \cap S$ is a coordinate neighborhood of S, and $\tilde{\mu} = \pi \circ \mu | \tilde{\mathcal{U}}$ (where π is the projection from $\mathbb{R}^m$ to $\mathbb{R}^k$) is a coordinate map, so $(\tilde{\mathcal{U}}, \tilde{\mu})$ is a chart on S. Since $\tilde{\nu}^{-1} = \nu^{-1} \circ \mathbf{i}$, where $\mathbf{i}$ is an injection of $\mathbb{R}^k$ to $\mathbb{R}^m$, the overlap maps, $\tilde{\mu} \circ \tilde{\nu}^{-1}$, are differentiable, and S is a k-dimensional manifold. Finally, in the coordinate system $(\tilde{\mathcal{U}}, \tilde{\mu})$ on S and the coordinate system $(\mathcal{U}, \mu)$ on M, the inclusion map is represented by $\mu \circ \tilde{\mu}^{-1}: (a^1, \ldots, a^k) \mapsto (a^1, \ldots, a^k, a_0^{k-1}, \ldots, a_0^m)$ where $a_0^{k+1}, \ldots, a_0^m$ are fixed. So this mapping from S to M is 1–1 differentiable and a homeomorphism between S and its image. ∎

Combining Theorem 10.4 and Theorem 10.6 and the fact that a differentiable variety is closed, Problem 9.20, we have the following.

Theorem 10.7. *A differentiable variety in M is a closed imbedded submanifold of M.*

We complete the circle of implications establishing the equivalence of differentiable varieties, closed subsets with the k-submanifold property, and closed imbedded submanifolds with the following result.

Theorem 10.8. *A closed imbedded submanifold of M is a differentiable variety in M.*

Proof. We noted in Section 10.1 that at each point, q, of $\phi(P)$ there is a chart $(\mathscr{V}, v)$ of M with $v^j = 0, j = p + 1, \ldots, m$, on a $\phi(P)$ neighborhood of q. For any covering of M containing such charts we can use a partition of unity subordinate to the covering to construct a submersion, $\phi: M \rightarrow \mathbb{R}^{m-p}$, under which the submanifold goes to $(0, \ldots, 0)$. ■

We also have the converse of Theorem 10.6. This result is part of another triad of equivalent concepts.

Theorem 10.9. *$S \subset M$ is an imbedded submanifold of M if and only if each point of S has an M-neighborhood, $\mathscr{U}$, such that $S \cap \mathscr{U}$ is the homeomorphic image, $\phi(\mathscr{V})$, under a 1–1 differentiable map, ϕ, of an open set, $\mathscr{V} \subset \mathbb{R}^p$, and ϕ has rank p at each point of V.*

Proof. The "only if" part is immediately evident, and the "if" part may be based on the following lemma.

Lemma. *If S has the property described in Theorem 10.9, then S has the k-submanifold property.*

Proof. Problem 10.5.

Theorem 10.9 gives a local description of an imbedded submanifold. Thus, we have three equivalent ways of describing a set $S \subset M$; namely, S is the image of an imbedding, S has the property of Theorem 10.9, or S has the k-submanifold property.

PROBLEM 10.3. Prove Theorem 10.5.

PROBLEM 10.4. (i) Which of the examples in Section 10.1 are imbedded submanifolds? (ii) Which of the examples in Section 10.1 are differentiable varieties?

PROBLEM 10.5. Prove the lemma for Theorem 10.9.

PROBLEM 10.6. The dimension of a differentiable variety in M is $\dim M - n$.

11

VECTOR FIELDS, 1-FORMS, AND OTHER TENSOR FIELDS

Our discussion of vector fields will include their interpretation as derivations, construction of the tangent manifold, C^∞ vector fields, and the Lie bracket of two vector fields. Comments on 1-forms will be focused on analogies with vector fields. Finally, under mappings of tensor fields, we discuss interior multiplication of forms, the pull-back of $(0, s)$ tensor fields, and ϕ-related $(r, 0)$ tensor fields.

11.1 Vector fields

In Section 9.3 we defined, for each $p \in M$, a tangent space, T_p. If $S \subset M$, then we write $TS = \bigcup_{p \in S} T_p$, and π: $TS \to S$ for the mapping which projects each T_p to p.

Definition
A vector field on S is a map $X: S \to TS$ such that for $p \in S$, $X(p) \in T_p$, or equivalently, $\pi \circ X = id$.

Theorem 11.1. *If* $\mathbb{R}^S$ *is the ring of functions on S, then with the definitions*

$$(X + Y)(p) = X(p) + Y(p)$$
$$(fX)(p) = f(p)X(p)$$

the set of vector fields on S is a module over $\mathbb{R}^S$.

Proof. Problem 11.1.

In Section 9.3 we described a tangent vector, $v_p \in T_p$ at a point $p \in M$ as a derivation on the C^1 functions, $\mathfrak{F}_p$, to the reals. So, in particular, given a vector field, X, we have $X(p) = v_p$ and

$$X(p)\colon \mathfrak{F}_p \to \mathbb{R}$$

according to

$$f \mapsto v_p \cdot f = X(p) \cdot f \tag{11.1}$$

Thus, we can say

> *With a vector field X on an open set $\mathscr{U}$, if we pick a point, $p \in \mathscr{U}$, then with each function, $f \in \mathfrak{F}_{\mathscr{U}}$ (the set of functions differentiable on $\mathscr{U}$) we can form $X_p \cdot f = X(p) \cdot f$.*

The fact that X depends on two arguments, $p \in \mathscr{U}$ and $f \in \mathfrak{F}_{\mathscr{U}}$, suggests that we may commute the choosing of p and f in this statement. Thus, we can also say

> *With a vector field X on an open set $\mathscr{U}$, if we first pick a function, $f \in \mathfrak{F}_{\mathscr{U}}$, we get a function given by*

$$p \to X_p \cdot f \tag{11.2}$$

Denoting this function by Xf we have

$$Xf(p) = X_p \cdot f \tag{11.3}$$

The map $f \mapsto Xf$ is frequently thought of as just another interpretation of a vector field. Denoting this interpretation of X by L_X, we have

$$L_X \colon \mathfrak{F}_{\mathscr{U}} \to \mathbb{R}^{\mathscr{U}}$$

according to

$$f \mapsto Xf$$

(cf., notation introduced in eq. (7.8)).

Theorem 11.2. *L_X is an R-linear derivation from $\mathfrak{F}_{\mathscr{U}}$ to $\mathbb{R}^{\mathscr{U}}$.*

Proof. Problem 11.2.

Theorem 11.3. *There is a 1–1 correspondence between vector fields on $\mathscr{U}$ and derivations in $\mathfrak{F}_{\mathscr{U}}$.*

Proof. (i) If Δ is a derivation on $\mathfrak{F}_{\mathscr{U}}$, define X_Δ by

$$X_\Delta(p) \colon f \mapsto (\Delta \cdot f)(p)$$

X_Δ is a vector field $\mathscr{U}$. Now $L_{X_\Delta} \cdot f(p) = X_\Delta f(p) = X_\Delta(p) \cdot f = (\Delta \cdot f)(p)$. So the mapping from vector fields to derivations is onto.

(ii) If $L_X = L_Y$, then $L_X \cdot f(p) = L_Y \cdot f(p)$ for all f and p. So $X(p) \cdot f = Y(p) \cdot f$ for all f and p. Hence $X = Y$ and the correspondence is 1–1. ∎

Finally, instead of fixing X and letting f vary as we have done, we can fix $f \in \mathfrak{F}_{\mathscr{U}}$ and consider the set of vector fields on $\mathscr{U}$. For each f we have a mapping from the module of vector fields on $\mathscr{U}$ to the module of functions on $\mathscr{U}$ given by

$$X \mapsto Xf \tag{11.4}$$

Theorem 11.4. *The map given by eq. (11.4) is a module homomorphism.*

Proof. The proof follows immediately from the following Lemma.

Lemma. *Given $f \in \mathfrak{F}_{\mathscr{U}}$. For any vector fields X and Y on $\mathscr{U}$ and any $g \in \mathfrak{F}_{\mathscr{U}}$,*

$$(X + Y)f = Xf + Yf \qquad \textit{and} \qquad (gX)f = g(Xf)$$

Proof. Problem 11.3.

In a local coordinate system $(\mathscr{U}, \mu)$ we have *the coordinate vector fields*

$$\frac{\partial}{\partial \mu^i}: p \mapsto \left.\frac{\partial}{\partial u^i}\right|_p \tag{11.5}$$

and the functions

$$\frac{\partial}{\partial \mu^i} f: p \mapsto \left.\frac{\partial}{\partial \mu^i}\right|_p \cdot f \tag{11.6}$$

But in Section 9.3 we defined $\partial/\partial\mu^i|_p$ to be $[\alpha_i]_p$ so that $\partial/\partial\mu^i|_p \cdot f = D_0 f \circ \alpha_i = \partial f/\partial\mu^i|_p$, and we defined $\partial f/\partial\mu^i$ by $\partial f/\partial\mu^i$: $p \mapsto \partial f/\partial\mu^i|_p$, so

$$\frac{\partial f}{\partial \mu^i}(p) = \left.\frac{\partial}{\partial \mu^i}\right|_p \cdot f \tag{11.7}$$

From (11.6) and (11.7) we have $(\partial/\partial\mu^i)f(p) = (\partial f/\partial\mu^i)(p)$, or

$$\frac{\partial}{\partial \mu^i} f = \frac{\partial f}{\partial \mu^i} \tag{11.8}$$

In particular, since $L_{\partial/\partial\mu^i}: f \mapsto (\partial/\partial\mu^i)f$, we have $L_{\partial/\partial\mu^i}: f \mapsto \partial f/\partial\mu^i$; i.e., the derivations $L_{\partial/\partial\mu^i}$ map C^1 functions to their partial derivatives.

If X^i are functions on a coordinate neighborhood, $\mathscr{U}$, by Theorem 11.1 we have vector fields $X^i\, \partial/\partial\mu^i$. Conversely, given a vector field X on a neighborhood $\mathscr{U}$ we can define *component functions*, X^i, of X, with respect to a basis of coordinate vector fields by $X^i: p \mapsto X^i(p)$ where the $X^i(p)$ are the v_p^i given by $X(p) = v_p^i\, \partial/\partial\mu^i|_p$ according to eq. (9.4).

Theorem 11.5. $X^i = X\mu^i$.

Proof.

$$\begin{aligned} X\mu^i(p) &= X(p)\cdot\mu^i \qquad \text{by (11.1)} \\ &= v_p\cdot\mu^i = v_p^i \\ &= X^i(p) \qquad \text{by the definition above} \end{aligned}$$ ■

Theorem 11.6. *The set of vector fields on a coordinate domain, $\mathscr{U}$, form an $\mathfrak{F}_{\mathscr{U}}$-module with basis $\{\partial/\partial\mu^i: 1 = 1, \ldots, n\}$. That is, for every vector field X,*

$$X = X^i \frac{\partial}{\partial\mu^i} \tag{11.9}$$

or, in terms of functions,

$$Xf = X^i \frac{\partial f}{\partial\mu^i} \qquad \textit{for all } f \in \mathfrak{F}_{\mathscr{U}} \tag{11.10}$$

Proof. These results follow immediately from the fact that $\partial/\partial\mu^i|_p$ form a basis at each point $p \in \mathscr{U}$. □

We have seen that, given a vector field, X, on $\mathscr{U}$, and a function $f \in \mathfrak{F}_{\mathscr{U}}$, we get a function Xf on $\mathscr{U}$. What can we say about Xf? The properties of Xf can be investigated by means of its coordinate representation,

$$\widehat{Xf}: b = \mu(p) \mapsto Xf(p) = X\mu^i(p)\frac{\partial f}{\partial\mu^i}(p) = \hat{X}^i(b)\frac{\partial \hat{f}}{\partial\pi^i}(b) \tag{11.11}$$

Alternatively, we have the representation (11.10) for Xf.

We will assume from now on (unless otherwise explicitly stated) that the manifold, M, is C^∞, so we can talk about C^∞ functions. Recall, from Section 9.3 that if f is C^∞ then $\partial f/\partial\mu^i$ are C^∞. Then from either (11.10) or (11.11) we have the following result.

Theorem 11.7. *For a given vector field, X,*

$$Xf \in C^{\infty} \qquad \text{for all } f \in C^{\infty} \Leftrightarrow X^i \in C^{\infty}$$

This gives us information about Xf in terms of a local representation of X. We could use this to define $X \in C^{\infty}$. To give this an intrinsic definition (i.e., coordinate-free) we need some structure on the image space, TM, of X. Thus, we will make TM into a manifold.

Let $\{(\mathscr{U}, \mu)\}$ be an atlas on M. We get a naturally induced atlas on TM in which each coordinate domain $\mathscr{V}$ is the union $\bigcup_{p \in \mathscr{U}} T_p$ over the coordinate neighborhood $\mathscr{U}$ of M, and each coordinate map $v: \mathscr{V} \to \mu(\mathscr{U}) \times \mathbb{R}^n$ is given by $v_p \mapsto (\mu^1(p), \ldots, \mu^n(p), v_p^1, \ldots, v_p^n)$ where the v_p^i are given by $v_p = v_p^i\, \partial/\partial\mu^i|_p$. That is, for $p \in \mathscr{U}$, if we write $v^i: p \mapsto v_p^i$, then for $v \in \pi^{-1}(p)$, $v^i = \mu^i \circ \pi$ for $i = 1, \ldots, n$ and $v^{i+n} = v^i \circ \pi$ for $i = 1, \ldots, n$.

Theorem 11.8. *With the given differentiable structure and the induced coordinate domain topology TM is a manifold, the tangent manifold of M.*

Proof. *Problem 11.5.*

Theorem 11.9. *X is C^{∞} at $p \Leftrightarrow$ the component functions, X^i, are C^{∞} at p.*

Proof. We have to show that a representation $v \circ X \circ \mu^{-1}$ of X on pairs of charts is C^{∞} iff the representations of X^i are C^{∞}. Now

$$v \circ X \circ \mu^{-1}: (\mu^1(p), \ldots, \mu^n(p)) \mapsto p \mapsto$$
$$X(p) \mapsto (\mu^1(p), \ldots, \mu^n(p), X^1(p), \ldots, X^n(p)).$$

The first n components are projections, and these are C^{∞} by Theorem 9.2. The second n components are $v^{i+n} \circ X \circ \mu^{-1} = X^i \circ \mu^{-1}$, which are the representations of X^i and these are C^{∞} iff the X^i are C^{∞}. ■

Corollary.

(i) *$\partial/\partial\mu^i$ are C^{∞} vector fields on a coordinate domain, $\mathscr{U}$.*
(ii) *$X \in C^{\infty}$ on an open set, $\mathscr{U} \Leftrightarrow Xf \in C^{\infty}$ for all $f \in \mathfrak{F}^{\infty}_{\mathscr{U}}$.*
(iii) *$X \in C^{\infty}$ on an open set, $\mathscr{U} \Leftrightarrow L_X: \mathfrak{F}^{\infty}_{\mathscr{U}} \to \mathfrak{F}^{\infty}_{\mathscr{U}}$.*

Theorem 11.10. *The set, $\mathfrak{X}(M)$, of C^{∞} vector fields on a manifold, M, form an $\mathfrak{F}^{\infty}_M$-module, and on each chart $\{\partial/\partial\mu^i\}$ form a local basis.*

Proof. We have to show that $X + Y$ and fX are C^{∞} if X, Y, and f are, then we can use Theorem 11.6. By Corollary (ii) above, $X, Y \in C^\infty \Rightarrow Xf + Yf \in C^\infty$. Then by the Lemma for Theorem 11.4 $(X + Y)f \in C^\infty$ and so by Corollary (ii) again $X + Y \in C^\infty$. Similarly for fX. ∎

Notice that in Theorem 11.10 we referred to vector fields defined on an entire given manifold. While it does tell us something about a class of such vector fields, it leaves much to be desired. In particular, while, on the one hand, the module of vector fields described in Theorem 11.1 contains any vector field defined pointwise in an arbitrary manner on all of M, and on the other hand, we have, on a coordinate domain, the C^∞ vector fields $\partial/\partial u^i$, and hence any linear combination with C^∞ functions, we have not yet established the very existence of nontrivial C^∞ vector fields on a given manifold M. The way we get C^∞ vector fields on a manifold is by extending a locally given C^∞ vector field. This can be done in any way such that the component functions X^i and $\bar{X}^i$ in intersecting coordinate domains $\mathscr{U}$ and $\bar{\mathscr{U}}$ satisfy

$$\bar{X}^i = X^j \frac{\partial \bar{\mu}^i}{\partial \mu^j} \qquad \text{on } \bar{\mathscr{U}} \cap \mathscr{U} \tag{11.12}$$

cf., eq. (9.7).

For example, let $\{(\mathscr{U}, \mu), (\bar{\mathscr{U}}, \bar{\mu})\}$ be the stereographic atlas on S^2. That is,

$$\mathscr{U} = S^2 - (0, 0, 1), \qquad \mu\colon (a^1, a^2, a^3) \mapsto \left(\frac{a^1}{1 - a^3}, \frac{a^2}{1 - a^3}\right)$$

$$\bar{\mathscr{U}} = S^2 - (0, 0, -1) \qquad \text{and } \bar{\mu}\colon (a^1, a^2, a^3) \mapsto \left(\frac{a^1}{1 + a^3}, \frac{a^2}{1 + a^3}\right)$$

Then

$$\bar{\mu}^i = \frac{\mu^i}{(\mu^1)^2 + (\mu^2)^2}, \, i = 1, 2, \qquad \text{and} \qquad (X^1, X^2) = (\mu^1 - \mu^2, \mu^1 + \mu^2)$$

and $(\bar{X}^1, \bar{X}^2) = (-\bar{\mu}^1 - \bar{\mu}^2, \bar{\mu}^1 - \bar{\mu}^2)$ satisfy (11.12), so the local vector fields

$$(\mu^1 - \mu^2)\frac{\partial}{\partial \mu^1} + (\mu^1 + \mu^2)\frac{\partial}{\partial \mu^2} \qquad \text{on } \mathscr{U}$$

$$(-\bar{\mu}^1 - \bar{\mu}^2)\frac{\partial}{\partial \mu^1} + (\bar{\mu}^1 - \bar{\mu}^2)\frac{\partial}{\partial \bar{\mu}^2} \qquad \text{on } \bar{\mathscr{U}}$$

define a C^∞ vector field on S^2. (Note that at the points $(0, 0, \pm 1)$, $X = 0$.)

Finally, we want to define a certain "multiplication" for vector fields.

Given vector fields X, Y and a function f, we can form the functions Xf, Yf, YXf, and XYf.

Theorem 11.11. *The mapping given by $f \to XYf - YXf$ is a derivation.*

Proof. Problem 11.7.

We denote this derivation by $L_{[X,Y]}$. That is,

$$L_{[X,Y]} \cdot f = XYf - YXf \tag{11.13}$$

We saw, Theorem 11.3, that with each derivation, Δ, we have a corresponding vector field. In this case, $\Delta = L_{[X,Y]}$: $f \mapsto XYf - YXf$ and the corresponding vector field is given by $Z_\Delta(p)$: $f \mapsto (XYf - YXf)(p)$.

Definition
The vector field, Z_Δ, corresponding to the derivation, $L_{[X,Y]}$, of Theorem 11.11 is denoted by $[X, Y]$, and called *the Lie bracket of X and Y*.

That is,

$$\begin{aligned}[X, Y](p)\colon f \mapsto L_{[X,Y]} \cdot f(p) &= (XYf - YXf)(p) \\ &= X(p)Yf - Y(p)Xf\end{aligned}$$

or

$$[X, Y]f = XYf - YXf \tag{11.14}$$

In coordinates, using (11.10),

$$\begin{aligned}[X, Y]f &= X^i \frac{\partial}{\partial \mu^i} Yf - Y^i \frac{\partial}{\partial \mu^i} Xf \\ &= X^i \frac{\partial Y^j}{\partial \mu^i} \frac{\partial f}{\partial \mu^j} - Y^i \frac{\partial X^j}{\partial \mu^i} \frac{\partial f}{\partial \mu^j}\end{aligned}$$

and

$$[X, Y] = X^i \frac{\partial Y^j}{\partial \mu^i} \frac{\partial}{\partial \mu^j} - Y^i \frac{\partial X^j}{\partial \mu^i} \frac{\partial}{\partial \mu^j} \tag{11.15}$$

Theorem 11.12. *The map $\mathfrak{X}(M) \times \mathfrak{X}(M) \to \mathfrak{X}(M)$ given by $(X, Y) \mapsto [X, Y]$ has the following properties.*

(i) *It is $\mathbb{R}$-bilinear.*
(ii) *It is skew-symmetric.*

(iii) $[[X, Y]Z] + [[YZ], X] + [[ZX], Y] = 0$. (The Jacobi identity)
(iv) $[fX, gY] = fg[X, Y] + f(Xg)Y - g(Yf)X$.

Proof. Problem 11.8.

By property (i), the set of C^∞ vector fields over $\mathbb{R}$ forms an algebra. Notice that the multiplication defined on $\mathfrak{X}$ in Theorem 11.12 is not associative.

Definition
A vector space with a multiplication with the properties of Theorem 11.12 is called a Lie algebra.

PROBLEM 11.1. Prove Theorem 11.1.

PROBLEM 11.2. Prove Theorem 11.2.

PROBLEM 11.3. Prove the Lemma for Theorem 11.4.

PROBLEM 11.4. For points in the intersection, $\mathscr{U} \cap \bar{\mathscr{U}}$, of the two coordinate neighborhoods $\partial/\partial\bar{\mu}^i = (\partial\mu^j/\partial\bar{\mu}^i)\,\partial/\partial\mu^j$ and $\bar{X}^i = X^j\,\partial\bar{\mu}^i/\partial\mu^j$ (eq. (11.12)).

PROBLEM 11.5. Prove Theorem 11.8.

PROBLEM 11.6. Using the atlas for $P^2(\mathbb{R})$ given in Section 91, show that the local vector fields

$$\mu_1^1 \frac{\partial}{\partial\mu_1^1} - \mu_1^2 \frac{\partial}{\partial\mu_1^2}$$

$$-2\mu_2^1 \frac{\partial}{\partial\mu_2^1} - \mu_2^2 \frac{\partial}{\partial\mu_2^2}$$

$$\mu_3^1 \frac{\partial}{\partial\mu_3^1} + 2\mu_3^2 \frac{\partial}{\partial\mu_3^2}$$

define a C^∞ vector field on $P^2(\mathbb{R})$.

PROBLEM 11.7. Prove Theorem 11.11.

PROBLEM 11.8. Prove Theorem 11.12.

11.2 1-Form fields

There are many analogies between vector fields and 1-form fields. In the following discussion, more or less "parallel" to that of vector fields, we will focus on these analogies.

Definition
If $S \subset M$ and $T^*S = \bigcup_{p \in S} T_p^*$, then a *1-form field on S* is a map $\sigma: S \to T^*S$ such that $\sigma(p) \in T_p^*$.

Theorem 11.13. *The set of* 1*-forms on S is an* R^S*-module.*

Recall that an element of $T_p^* = T^p$ has the form $df|_p$ for some $f \in \mathfrak{F}_p$, and we saw in eq. (9.5) that $df|_p$ is a function $T_p \to R$ given by $df|_p: v_p \mapsto \langle df|_p, v_p \rangle = v_p \cdot f$. So, in particular, if X is a given vector field, and σ is a given 1-form field, then $\sigma(p) = df|_p$ for some $f \in \mathfrak{F}_p$ and

$$\sigma(p): T_p \to \mathbb{R}$$

according to

$$X(p) \mapsto \langle df|_p, X(p) \rangle = \langle \sigma(p), X(p) \rangle$$

Now, as in Section 11.1, let $S = \mathscr{U}$ be an open set in M. Given σ on $\mathscr{U}$, for each vector field, X, on $\mathscr{U}$ we have a function $\langle \sigma, \mathrm{X} \rangle: \mathscr{U} \to \mathbb{R}$ given by $p \mapsto \langle \sigma(p), X(p) \rangle$. That is,

$$\langle \sigma, X \rangle(p) = \langle \sigma(p), X(p) \rangle \tag{11.16}$$

Equation (11.16) corresponds to eq. (11.3). Finally, for each σ we have, in analogy with the map L_X in Section 11.1, a map, $\rfloor\sigma$, from the vector fields on $\mathscr{U}$ to $\mathbb{R}^{\mathscr{U}}$ given by $X \mapsto \langle \sigma, X \rangle$. (See definition of $\rfloor B^*$ in Section 6.3 with $s = r = 1$.)

Theorem 11.14. $\rfloor\sigma$ *is a module homomorphism from the module of vector fields on* $\mathscr{U}$ *to* $\mathbb{R}^{\mathscr{U}}$.

The interpretation of a 1-form field, σ, as a module homomorphism, $\rfloor\sigma$, generalizes the definition of a 1-form as a linear function from a vector space to the reals.

Continuing the analogy with Section 1.1, instead of fixing σ and letting X vary, we can fix X and let σ vary so that we have a map

$$\sigma \mapsto \langle \sigma, X \rangle \tag{11.17}$$

Lemma. *Given a vector field X, if σ and τ are 1-form fields, and $g \in \mathfrak{F}_{\mathscr{U}}$*

$$\langle \sigma + \tau, X \rangle = \langle \sigma, X \rangle + \langle \tau, X \rangle \qquad \text{and} \qquad \langle g\sigma, X \rangle = g\langle \sigma, X \rangle$$

Theorem 11.15. *The map given by eq.* (11.17) *is a module homomorphism.*

Definition
If $f \in \mathfrak{F}_{\mathscr{U}}$, the 1-form field $df\colon \mathscr{U} \to T^*\mathscr{U}$ given by $p \mapsto df|_p \in T_p^*$ is *the differential of* f.

In a local coordinate system $(\mathscr{U}, \mu)$ we have *the coordinate 1-forms*

$$d\mu^i\colon p \mapsto d\mu^i|_p \tag{11.18}$$

Then, using eq. (9.5), $\langle d\mu^i, X \rangle\colon p \mapsto X\mu^i(p)$, and hence $\lrcorner d\mu^i\colon X \mapsto X\mu^i$.

If σ_i are functions on $\mathscr{U}$, we can form $\sigma_i \, d\mu^i$, and conversely, given a 1-form σ on $\mathscr{U}$ we can define *component functions*, σ_i, of σ by $\sigma_i\colon p \mapsto \sigma_i(p)$ where the $\sigma_i(p)$ are the coefficients of σ_p according to eq. (9.6). With these component functions we have the following results corresponding to Theorems 11.5 and 11.6.

Theorem 11.16. $\sigma_i = \langle \sigma, \partial/\partial\mu^i \rangle$.

Theorem 11.17. *The set of* 1*-form fields on a coordinate domain form an* $\mathfrak{F}_{\mathscr{U}}$*-module with basis* $\{d\mu^i\colon i, \ldots, n\}$. *That is, for every* 1*-form* σ

$$\sigma = \sigma_i \, d\mu^i \tag{11.19}$$

or, in terms of functions,

$$\langle \sigma, X \rangle = \sigma_i X \mu^i \tag{11.20}$$

for all vector fields X on $\mathscr{U}$.

At this point we might note a certain breakdown in the "parallel" treatments of vector fields and 1-forms. On the one hand, we have a special class of 1-form fields for which we have no corresponding class of vector fields, namely, the 1-form fields *df* coming from functions $f \in \mathfrak{F}_{\mathscr{U}}$. On the other hand, while eqs. (9.3) and (9.6) are simply different notations for the same thing, we can extend the coefficients of (9.6) to all of $\mathscr{U}$ as we did in Theorems 11.16 and 11.17, but, in general we cannot extend (9.3) to the familiar equation $df = (\partial f/\partial\mu^i)\, d\mu^i$ by writing Theorem 11.16 in the simpler form $\sigma_i = \partial f/\partial\mu^i$. We get these results only if σ comes from a function, that is, $\sigma = df$. However, though to every $f \in \mathfrak{F}_{\mathscr{U}}$ we have a 1-form $\sigma = df$, we cannot say that every 1-form on $\mathscr{U}$ can be written df. In order for the latter to hold, the component functions σ_i of the given σ must satisfy "integrability conditions." A rough way of describing this distinction between vector fields and 1-form fields is that we can generally expect to be able to integrate vector fields to get curves, but more stringent conditions are required to get functions from 1-form fields. We will have more to say about this in Chapters 12 and 13.

Restricting ourselves to C^∞ manifolds again, we have from (11.20) that

$$\langle \sigma, X \rangle \in C^\infty \text{ for all } X \in C^\infty \Leftrightarrow \sigma_i \in C^\infty \tag{11.21}$$

This corresponds to Theorem 11.7.

Corresponding to our introduction of C^∞ vector fields in Section 11.1, we now have C^∞ 1-form fields. Specifically, we can make $T^*M = \bigcup_{p \in M} T_p^*$ into a manifold, *the cotangent manifold of M*, in a manner analogous to what we did for TM, and then get results corresponding to Theorems 11.8–11.10. In particular, the set, $\mathfrak{X}^*(M)$, of C^∞ 1-form fields on M forms an $\mathfrak{F}_M^\infty$ module.

Finally, the manifold T^*M has a special property other manifolds do not have, and which will be important when we study mechanics.

Theorem 11.18. *T^*M has a natural (canonical) 1-form field, θ_M.*

Proof. If $\sigma \in T^*M$, $v \in T_\sigma T^*M$, and π is the natural projection, π: $T^*M \to M$ given by $\sigma \mapsto p$, then a 1-form θ_M: $T^*M \to T^*T^*M$ is defined by $\langle \theta_M(\sigma), v \rangle = \langle \sigma, \pi_* \cdot v \rangle$. That is, on each vector space T_p^*M, θ_M is the transpose of π_* (see Fig. 11.1). ■

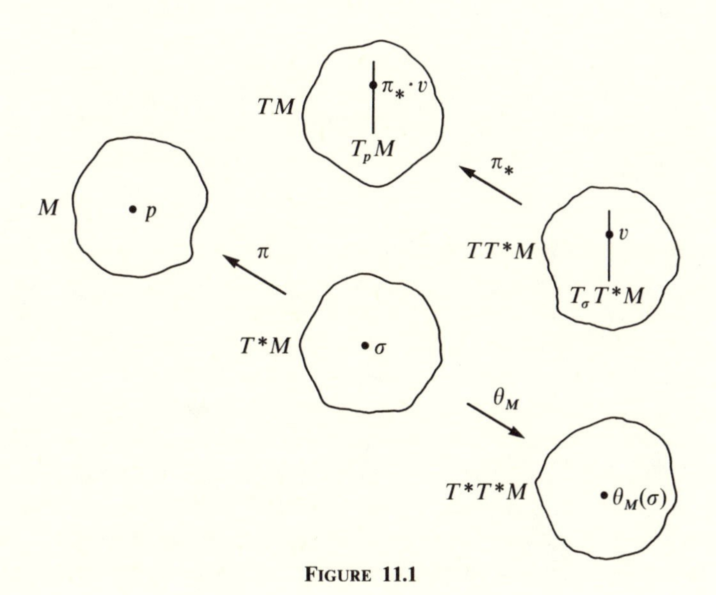

FIGURE 11.1

PROBLEM 11.9. Make T^*M into a manifold.

PROBLEM 11.10. State and prove any one of the results corresponding to Theorems 11.8–11.10 for T^*M and 1-form fields.

PROBLEM 11.11. For points in the intersection, $\mathscr{U} \cap \bar{\mathscr{U}}$, of two coordinate neighborhoods (i) $\bar{\sigma}_i = \sigma_j \, \partial\mu^j/\partial\bar{\mu}^i$; (ii) $d\bar{\mu}^i = (\partial\bar{\mu}^i/\partial\mu^j)\, d\mu^j$; and (iii) $d\bar{\mu}^1 \wedge \cdots \wedge d\bar{\mu}^n = \det(\partial\bar{\mu}^i/\partial\mu^j)\, d\mu^1 \wedge \cdots \wedge d\mu^n$ (cf., Section 6.3).

11.3 Tensor fields

We will generalize the two previous sections (mainly introducing a lot more notation).

Recall that at the end of Section 9.3 we pointed out that having T_p and T_p^* at each point we can construct at each point the tensor product spaces $(T_p)_s^r$ as we did in Chapter 4. We also have the symmetric and skew-symmetric subspaces,

$$\left.\begin{matrix} \mathsf{S}^r(T_p) \\ \Lambda^r(T_p) \end{matrix}\right\} \subset (T_p)_0^r \qquad \text{the contravariant tensors}$$

and

$$\left.\begin{matrix} \mathsf{S}^s(T_p^*) \\ \Lambda^s(T_p^*) \end{matrix}\right\} \subset (T_p)_s^0 \qquad \text{the covariant tensors.}$$

We can take the set, for all $p \in M$, of vector spaces of each type, and form differentiable manifolds. Thus, for example, we have

$$T_s^r M = \bigcup_{p \in M} (T_p)_s^r, \qquad \Lambda^s(T^*M) = \bigcup_{p \in M} \Lambda^s(T_p^*), \qquad \text{etc.}$$

Definitions

An (r, s) *tensor field* K, *on* M is a map $K\colon M \to T_s^r M$ given by $p \mapsto A \in (T_p)_s^r$. In particular, an *s-form field on* M is a map $\omega\colon M \to \Lambda^s(T^*M)$ given by $p \mapsto \omega(p) \in \Lambda^s(T_p^*)$, and an *r-vector field on* M is a map $F\colon M \to \Lambda^r(TM)$ given by $p \mapsto F(p) \in \Lambda^r(T_p)$.

That is, the values of an s-form field are skew-symmetric covariant tensors and the values of an r-vector field are skew-symmetric contravariant tensors. (Frequently, s-form fields are called differential forms, and skew-symmetric covariant tensors are called exterior forms. We will follow the common practice, and simply use the word *form* for either concept, expecting the meaning to be clear from the context.)

Given an (r, s) tensor field, K, 1-form fields $\sigma^1, \ldots, \sigma^r$, and vector fields $X_1, \ldots, X_s$ we can define a function $M \to R$ by

$$K(\sigma^1, \ldots, \sigma^r, X_1, \ldots, X_s)\colon p \mapsto K(p)(\sigma^1(p), \ldots, X_s(p))$$

This generalizes the functions Xf and $\langle \sigma, X \rangle$ in Sections 11.1 and 11.2. Corresponding to Theorem 11.7 and eq. (11.21) we have that $K(\sigma^1, \ldots, \sigma^r, X_1, \ldots, X_s)$ is a C^∞ function if $\sigma^1, \ldots, \sigma^r, X_1, \ldots, X_s$, and K are C^∞.

Finally, given an (r, s) tensor field, K, the mapping

$$\mathfrak{X}^*(M) \times \cdots \times \mathfrak{X}^*(M) \times \mathfrak{X}(M) \times \cdots \times \mathfrak{X}(M) \to \mathfrak{F}_M^\infty$$

from r 1-form fields and s vector fields to $\mathfrak{F}_M^\infty$ given by

$$(\sigma^1, \ldots, \sigma^r, X_1, \ldots, X_s) \mapsto K(\sigma^1, \ldots, \sigma^r, X_1, \ldots, X_s) \tag{11.22}$$

generalizes the maps L_X and $\lrcorner\sigma$ defined in Sections 11.1, and 11.2. In particular for s-forms we get $\omega(X_1, \ldots, X_s)$ and for r-vector fields we get $F(\sigma^1, \ldots, \sigma^r)$. (Sometimes an (r, s) tensor field is *defined* as a multilinear map of the form (11.22), and then a tensor field as we have defined it is derived; cf. O'Neill, 1983, pp. 35–37).

As before, we can describe differentiability of K, ω, F, etc., in terms of the functions $K(\sigma^1, \ldots, X_s)$, $\omega(X_1, \ldots, X_s)$, etc., or in terms of *the component functions* of K, ω, F, etc. in a local coordinate system. For K we have, in a local coordinate system

$$K^{i_1 \cdots i_r}_{j_1 \cdots j_s} = K\left(d\mu^{i_1}, \ldots, d\mu^{i_r}, \frac{\partial}{\partial \mu^{j_1}}, \ldots, \frac{\partial}{\partial \mu^{j_s}}\right) \tag{11.23}$$

and

$$K = K^{i_1 \cdots i_r}_{j_1 \cdots j_s} \frac{\partial}{\partial \mu^{i_1}} \otimes \cdots \otimes d\mu^{j_s} \tag{11.24}$$

For an s-form field, ω, we have

$$\omega_{j_1 \cdots j_s} = \omega\left(\frac{\partial}{\partial \mu^{j_1}}, \ldots, \frac{\partial}{\partial \mu^{j_s}}\right) \tag{11.25}$$

and

$$\omega = \omega_{j_1 \cdots j_s} \, d\mu^{j_1} \wedge \cdots \wedge d\mu^{j_s}, \qquad j_1 < \cdots < j_s \tag{11.26}$$

Of course, as before, for eqs. (11.24) and (11.26) to make sense we have to make the set of (r, s) tensor fields and the set of s-form fields into $\mathfrak{F}_M^\infty$-modules.

Finally, recall that in Section 6.1 we constructed the tensor algebras $\bigoplus V^r_0$, $\bigoplus V^0_s$, ΛV^*, etc. At each point $p \in M$ the value of each of the various tensor (form) fields we have been discussing is a member of one of these algebras. Thus, we can extend these algebras to algebras of tensor (form) fields defining addition, multiplication, and scalar multiplication pointwise.

PROBLEM 11.12. Make $\mathrm{S}^2(T^*M) = \bigcup_{p \in M} \mathrm{S}^2(T^*_p)$ into a manifold.

PROBLEM 11.13. State and prove any one of the results corresponding to Theorems 11.8–11.10 for $\mathrm{S}^2(T^*M)$ and symmetric (0, 2) tensor fields.

11.4 Mappings of tensor fields

We will now describe two important mappings of tensor fields.

1. There is an operator on the algebra of differential forms on M which takes s-forms on M to $s - 1$ forms on M.

Definition
Given an r-vector field F we define the map $F\rfloor$, called *interior multiplication by* F on the exterior algebra of forms taking s-forms to $s - r$ forms ($s \geq r$) by $F\rfloor: \omega \to F\rfloor\omega$ where $(F\rfloor\omega)(p) = F(p)\rfloor\omega(p)$, the interior product defined in Section 6.3.

Theorem 11.19. *If* ω *is an s-form field,* $s > 1$, *and* $F = X$ *is a vector field then* $X\rfloor\omega(X_1, \ldots, X_{s-1}) = \omega(X, X_1, \ldots, X_{s-1})$.

Proof. By definition (see Section 6.3) $\langle X\rfloor\omega, Y\rangle = \langle\omega, X \wedge Y\rangle$. For $Y = X_1 \wedge \cdots \wedge X_{s-1}$, then

$$\langle X\rfloor\omega, X_1 \wedge \cdots \wedge X_{s-1}\rangle = \langle\omega, X \wedge X_1 \wedge \cdots \wedge X_{s-1}\rangle \qquad (11.27)$$

But, for any s-form field ω, and s vector fields $Y_1, \ldots, Y_s$, extending eq. (6.8) pointwise,

$$\langle\omega, Y_1 \wedge \cdots \wedge Y_s\rangle = \omega(Y_1, \ldots, Y_s) \qquad (11.28)$$

Applying (11.28) to (11.27) we get the result. ■

Definition
If ω is an s-form field, we write

$$\begin{aligned} i_X\omega &= sX \rfloor \omega, \qquad s \geq 1 \\ i_X\omega &= 0, \qquad s = 0 \end{aligned} \tag{11.29}$$

The operation i_X is also called *interior multiplication by X*.

Theorem 11.20. (i) *i_X is a skew-derivation in the algebra of differential forms on M; i.e., i_X is linear, and if σ is a p-form field, and τ is a q-form field, then*

$$i_X(\sigma \wedge \tau) = (i_X\sigma) \wedge \tau + (-1)^p\sigma \wedge (i_X\tau) \tag{11.30}$$

(ii) $i_{X+Y}\omega = i_X\omega + i_Y\omega$ *and* $i_{fX}\omega = fi_X\omega$.
(iii) $i_X\,df = L_Xf = Xf$.

Proof. We show these results are all valid at an arbitrary point $p \in M$. They all come immediately from the definition of $X \rfloor \omega$, except eq. (11.30) which we prove by using linearity. Then (11.30) is verified for basis elements using Problem 6.13(i). ■

Corollary. *If $\omega = \sigma^1 \wedge \cdots \wedge \sigma^s$ where the σ^i are 1-forms, then*

$$i_X\omega = \sum_{j=1}^{s} (-1)^{j+1}(\sigma^j \cdot X)\sigma^1 \wedge \cdots \wedge \hat{\sigma}^j \wedge \cdots \wedge \sigma^s \tag{11.31}$$

2. Mappings of tensor fields may be induced by a differentiable map $\phi: M \to N$. Near the end of Section 9.3 we mentioned that with ϕ we have $\phi_r|_p: (T_p)^r_0 \to (T_{\phi(p)})^r_0$ for contravariant tensor spaces at each $p \in M$, and we have $\phi^s|_{\phi(p)}: (T_{\phi(p)})^0_s \to (T_p)^0_s$ for covariant tensor spaces at each $p \in M$.

In particular, if we are given a covariant tensor field, K, on N, then $\phi^s|_{\phi(p)}$ maps $K(\phi(p))$ "backward" into an element of $(T_p)^0_s$.

Definitions
The covariant tensor field on M, $\phi^*K: M \to T^0_sM$ given by $\phi^*K: p \mapsto \phi^s|_{\phi(p)} \cdot K(\phi(p))$ is called *the pull-back of K by ϕ*, and we define an operator, ϕ^*, on covariant tensor-fields by $\phi^*: K \mapsto \phi^*K$. (The reader will recall that the notation ϕ^* was introduced in Section 9.3 to denote the transpose of the tangent map and then extended to denote maps of cotangent manifolds. See comments on notation at the end of this section.)

Theorem 11.21. $\phi^*df = d\phi^*f$.

Proof. Extend Theorem 7.12 to 1-form fields on manifolds. ■

Theorem 11.22. *If K and ϕ are in C^∞, then $\phi^*K \in C^\infty$.*

Proof. From eq. (11.24) $\phi^*K(p) = K_{j_1\cdots j_s}|_{\phi(p)}\phi^s_{\phi(p)}\cdot(d\bar{\mu}^{j_1} \otimes \cdots \otimes d\bar{\mu}^{j_s})|_{\phi(p)}$ where $\{d\bar{\mu}^{j_1}, \ldots, d\bar{\mu}^{j_s}\}$ is a coordinate basis of $T^*_{\phi(p)}$. So,

$$\begin{aligned}\phi^*K(p) &= (K_{j_1\cdots j_s} \circ \phi)(p)(\phi^* \cdot d\bar{\mu}^{j_1} \otimes \cdots \otimes \phi^* \cdot d\bar{\mu}^{j_s})_{\phi(p)} \\ &= (K_{j_1\cdots j_s} \circ \phi)(p)\left(\frac{\partial \phi^{j_1}}{\partial \mu^{i_1}} d\mu^{i_1} \otimes \cdots \otimes \frac{\partial \phi^{j_s}}{\partial \mu^{i_n}} d\mu^{i_n}\right)(p)\end{aligned}$$

by eq. (9.10). Our result now follows from the generalizations of Theorem 11.9 and its corollaries. ■

We have seen above that $\phi: M \to N$ induces a "backward" mapping of covariant tensor fields. Now suppose we start with a contravariant tensor field K on M and we want to define a contravariant vector field on N by means of the maps $\phi_r|_p$. Then $\phi_r|_p$ maps $K(p)$ "forward" into an element of $(T_{\phi(p)})^r_0$. A problem arises now that did not occur in the covariant case, for if $p \neq q$, then, in general, $\phi_r|_p K(p) \neq \phi_r|_q K(q)$. So, if $\phi(p) = \phi(q)$, then K will not have a well-defined image.

Definition
If $\phi: M \to N$, a contravariant tensor field, K_M, on M is *ϕ-related* to a contravariant tensor field, K_N, on N if $\phi_r|_p \cdot K_M(p) = K_N(\phi(p))$ for all $p \in M$.

Theorem 11.23. *A vector field X on M is ϕ-related to a vector field Y on N iff $X(g \circ \phi) = (Yg) \circ \phi$ for all $g \in \mathfrak{F}_N$.*

Proof. $X(g \circ \phi)(p) = (Yg) \circ \phi(p)$ iff $X(p)\cdot(g \circ \phi) = Y(\phi(p))\cdot g$ iff $\phi_{*p}\cdot X(p)\cdot g = Y(\phi(p)\cdot g$ iff $\phi_{*p}\cdot X(p) = Y(\phi(p))$. ■

Theorem 11.24. *If vector fields X_1 and X_2 are ϕ-related respectively to vector fields Y_1 and Y_2, then $[X_1, X_2]$ is ϕ-related to $[Y_1, Y_2]$.*

Proof. Problem 11.16.

Theorem 11.25. *If ϕ is regular on M (ϕ_{*p} is 1–1 for all $p \in M$) and K is a contravariant tensor field on N, then there is a unique contravariant tensor field, ϕ^*K, on M, called* the pull-back of K *by ϕ, which is ϕ-related to K.*

Proof. Define a linear map, ϕ^{-1r}, from $\bigcup_{\phi(p)} (T_{\phi(p)})^r_0$ to $\bigcup_p (T_p)^r_0$ by

$$\phi^{-1r}|_{(T_{\phi(p)})^r_0}: w_1 \otimes \cdots \otimes w_r \mapsto \phi^{-1}_{*\phi(p)} \cdot w_1 \otimes \cdots \otimes \phi^{-1}_{*\phi(p)} \cdot w_r$$

where $w_i \in T_{\phi(p)}$ (see Problem 4.18). Then define ϕ^*K by $\phi^*K(p) = \phi^{-1r}|_{(T_{\phi(p)})^r_0} \cdot K(\phi(p))$. ϕ^*K will have the required property. ■

Theorem 11.26. *If ω is an s-form on N and X is a vector field on N and ϕ satisfies the hypotheses of Theorem 11.23, then*

$$\phi^* i_X \omega = i_{\phi^* X} \phi^* \omega$$

That is, the operators i_X and ϕ^* commute.

Proof. Problem 11.19.

We must, finally, make another comment about notation. In order to construct pull-backs of covariant tensor fields we made use of the "backward" maps, ϕ^s. If ϕ is regular, ϕ^{-1r} give us pull-backs of contravariant tensor fields and the $\phi^{r,s}$ of Problem 4.18 give us pull-backs of mixed tensor fields. The operator on tensor fields in all three cases will be denoted by ϕ^*. In order to simplify our notation we will frequently also write ϕ^* in place of ϕ^s, ϕ^{-1r}, or $\phi^{r,s}$. Which of the two slightly distinct meanings is intended will, hopefully, be clear in context.

PROBLEM 11.14. The values of the pull-back by ϕ of a covariant tensor field, K, are $\phi^*K(X_1, \ldots, X_s) = K(\phi_* \cdot X_1, \ldots, \phi_* \cdot X_s)$.

PROBLEM 11.15. Let $M = \mathbb{R}^2$, $N = \mathbb{R}^3$, and ϕ: $(a, b) \mapsto (\cos a, \sin a \cos b, \sin a \sin b)$. Then, if $\omega = x\, dy \wedge dz + y\, dz \wedge dx + z\, dx \wedge dy$ where x, y, z are the natural coordinate functions on $\mathbb{R}^3$, write $\phi^*\omega$ in terms of the natural coordinate functions u, v on $\mathbb{R}^2$.

PROBLEM 11.16. Prove Theorem 11.24.

PROBLEM 11.17. If K_M is ϕ-related to K_N, its values are $K_M(\sigma^1, \ldots, \sigma^r) = K_N(\phi^{*-1} \cdot \sigma^1, \ldots, \phi^{*-1} \cdot \sigma^r)$.

PROBLEM 11.18. If π^1: $M \times N \to M$ and π^2: $M \times N \to N$ are the projections and X is a vector field on M, then there is a unique vector field, $\tilde{X}$, on $M \times N$ such that $\tilde{X}$ is π^1-related to X and $\tilde{X}$ is π^2-related to the zero vector field on N. (Similarly, for Y a vector field on N.)

PROBLEM 11.19. Prove Theorem 11.26.

PROBLEM 11.20. The natural 1-form, θ_M, on T^*M (Theorem 11.18) can be characterized by its behavior under pull-backs to M; θ_M is the natural 1-form on T^*M iff $\phi^*\theta_M = \theta_M$ for all maps ϕ: $M \to T^*M$.

12

EXTERIOR DIFFERENTIATION AND INTEGRATION OF DIFFERENTIAL FORMS

In Section 11.2 we mapped functions $f \in \mathfrak{F}_M$ into 1-forms, df. In Section 12.1 we will extend this mapping to an operation on the algebra of differential forms, and obtain some properties of this operation. In Section 12.2 we will describe a concept of integration on manifolds which reduces to the usual integrals on $\mathbb{R}^n$, and we prove a generalization of Stokes' theorem.

12.1 Exterior differentiation of differential forms

In Section 11.4, we defined, for each vector field, X, on M, an operator, i_X, on the algebra of differential form fields on M which took s-form fields on M into $s-1$-form fields on M. We defined i_X by giving its values on ω, and it turned out that i_X was a skew-derivation, and $i_X df = Xf$ and $i_X f = 0$. Now, we define an operator, d, on the algebra of differential form fields on M taking s-form fields to $s+1$-form fields. We define d, *the exterior derivative*, by listing properties it is to have. It will turn out that there is precisely one such operator.

Theorem 12.1. *There exists one and only one operator, d, on the algebra of differential forms with the following properties.*

(1) *d is a skew-derivation; i.e., d is $\mathbb{R}$-linear and if σ is a p-form and τ is a q-form then*

$$d(\sigma \wedge \tau) = (d\sigma) \wedge \tau + (-1)^p \sigma \wedge d\tau \tag{12.1}$$

(2) *For $f \in \mathfrak{F}_M$, $d: f \mapsto df$ (recall that df was defined in Section 11.2).*
(3) $d(df) = 0$.

Proof. (i) (*Local existence*) If

$$\omega = \omega_{i_1 \cdots i_s} d\mu^{i_1} \wedge \cdots \wedge d\mu^{i_s}, \qquad i_1 < \cdots < i_s,$$

on (U, μ) let

$$d\omega = d\omega_{i_1 \cdots i_s} \wedge d\mu^{i_1} \wedge \cdots \wedge d\mu^{i_s}, \qquad i_1 < \cdots < i_s \tag{12.2}$$

Property (2) is just a matter of definition, and $\mathbb{R}$-linearity comes from $\mathbb{R}$-linearity of d on functions. For eq. (12.1) let $\sigma = f\, d\mu^{i_1} \wedge \cdots \wedge d\mu^{i_p}$ and $\tau = g\, d\mu^{j_1} \wedge \cdots \wedge d\mu^{j_q}$. Then

$$\begin{aligned} d(\sigma \wedge \tau) &= d(fg \wedge d\mu^{i_1} \wedge \cdots \wedge d\mu^{j_q}) \\ &= (df)g \wedge d\mu^{i_1} \wedge \cdots \wedge d\mu^{j_q} + f\, dg \wedge d\mu^{i_1} \wedge \cdots \wedge d\mu^{j_q} \\ &= (d\sigma) \wedge \tau + (-1)^p \sigma \wedge d\tau, \end{aligned}$$

since d is a derivation on functions. We then get (12.1) for any σ and τ by the $\mathbb{R}$-linearity of d. Finally, $df = \partial f/\partial \mu^i\, d\mu^i$, so

$$d(df) = (d\, \partial f/\partial \mu^i) \wedge d\mu^i = (\partial^2 f/\partial \mu^i\, \partial \mu^j) d\mu^j \wedge d\mu^i = 0.$$

(ii) (*Local uniqueness*) If $(\mathscr{U}, \mu)$ is a local coordinate system, then for a 1-form $\omega = \omega_i\, d\mu^i$ property (1) gives $d\omega = d\omega_i \wedge d\mu^i + \omega_i \wedge d\, d\mu^j$, which is determined by properties (2) and (3). Similarly, if ω is a higher-order form, writing $\omega = \omega_{i_1 \cdots i_s}\, d\mu^{i_1} \wedge \cdots \wedge d\mu^{i_s}$, then properties (1) and (3) give

$$d\omega = d\omega_{i_1 \cdots i_s} \wedge d\mu^{i_1} \wedge \cdots \wedge d\mu^{i_s},$$

which is determined by property (2).

(iii) Finally, if ω is defined on M, then by local existence and uniqueness we have a unique $d\omega$ on the coordinate neighborhoods $\mathscr{U}$ and $\mathscr{V}$ and hence on $\mathscr{U} \cap \mathscr{V}$, so $d\omega$ is unambiguously defined on all of M. ■

Corollary. *For any form, ω, $d\, d\omega = 0$.*

EXAMPLE. Let $M = \mathbb{R}^3$ with the standard structure. Denote the standard coordinate functions, π^i, by x, y, z. Then $(\partial/\partial x, \partial/\partial y, \partial/\partial z)$ form a basis of the module of vector fields, $\mathfrak{X}$, and (dx, dy, dz) form a basis of the module 1-forms, $\mathfrak{X}^*$.

(i) For a 0-form, f,

$$df = \frac{\partial f}{\partial x} dx + \frac{\partial f}{\partial y} dy + \frac{\partial f}{\partial z} dz \tag{12.3}$$

(ii) For a 1-form, $\sigma = f\,dx + g\,dy + h\,dz$,

$$d\sigma = \left(\frac{\partial h}{\partial y} - \frac{\partial g}{\partial z}\right) dy \wedge dz + \left(\frac{\partial f}{\partial z} - \frac{\partial h}{\partial x}\right) dz \wedge dx + \left(\frac{\partial g}{\partial x} - \frac{\partial f}{\partial y}\right) dx \wedge dy \tag{12.4}$$

(iii) For a 2-form, $\omega = f\,dy \wedge dz + g\,dz \wedge dx + h\,dx \wedge dy$,

$$d\omega = \left(\frac{\partial f}{\partial x} + \frac{\partial g}{\partial y} + \frac{\partial h}{\partial z}\right) dx \wedge dy \wedge dz \tag{12.5}$$

(iv) For higher-order forms, ω, $d\omega = 0$.

Except for the fact that d operates on form fields instead of vector fields, it appears that the operator d unifies and generalizes the differential operators, grad, curl, and div of vector analysis.

To obtain the relation between d and grad we define a nondegenerate symmetric $(2, 0)$ tensor field, b, on $\mathbb{R}^3$ by $b(dx, dx) = 1$, $b(dx, dy) = 0$, etc. Corresponding to b we have the linear mapping, $b^\sharp$, from $\mathfrak{X}^*$ to $\mathfrak{X}$ given by $b^\sharp\colon \sigma \mapsto X$ where $X\colon \tau \mapsto b(\tau, \sigma)$. Now we define $\operatorname{grad} f$ by $\operatorname{grad} f = b^\sharp \cdot df$. By eq. (12.3)

$$b^\sharp \cdot df = \frac{\partial f}{\partial x} b^\sharp \cdot dx + \frac{\partial f}{\partial y} b^\sharp \cdot dy + \frac{\partial f}{\partial z} b^\sharp \cdot dz = \frac{\partial f}{\partial x}\frac{\partial}{\partial x} + \frac{\partial f}{\partial y}\frac{\partial}{\partial y} + \frac{\partial f}{\partial z}\frac{\partial}{\partial z}$$

so

$$\operatorname{grad} f = \frac{\partial f}{\partial x}\frac{\partial}{\partial x} + \frac{\partial f}{\partial y}\frac{\partial}{\partial y} + \frac{\partial f}{\partial z}\frac{\partial}{\partial z}$$

(In vector analysis notation $\partial/\partial x = \mathbf{i}$, $\partial/\partial y = \mathbf{j}$, and $\partial/\partial z = \mathbf{k}$.)

To obtain the relation between d and curl and the relation between d and div, we fix the order of the basis as (dx, dy, dz) and introduce the Hodge star operator (see Section 6.3). Then from (12.4) we get

$$*d\sigma = \left(\frac{\partial h}{\partial y} - \frac{\partial g}{\partial z}\right) dx + \left(\frac{\partial f}{\partial z} - \frac{\partial h}{\partial x}\right) dy + \left(\frac{\partial g}{\partial x} - \frac{\partial f}{\partial y}\right) dz$$

i.e., $*d$ maps 1-forms to 1-forms. From (12.5) we get

$$*d * \sigma = \frac{\partial f}{\partial x} + \frac{\partial g}{\partial y} + \frac{\partial h}{\partial z}$$

i.e., $*d*$ maps 1-forms to 0-forms. Finally, using $b^{\#}$ we obtain curl and div from $*d$ and $*d*$, respectively.

To indicate briefly a generalization, we note that $*d*$ can be applied to forms on $\mathbb{R}^n$ and takes s-forms to $s-1$ forms. $*d*$ can be applied to forms on certain manifolds ("orientable", Section 12.2) having a little more structure ("a metric tensor").

Definitions
$\delta = (-1)^{n(p+1)+1} *d*$ is called *the codifferential operator* and $\Delta = \delta d + d\delta$ is called *the Laplace–Beltrami operator* ($n = \dim M$ and p is the order of the form). Clearly, on functions on $\mathbb{R}^3$, Δ is the ordinary Laplacian. A form, ω, for which $\Delta\omega = 0$ is called a *harmonic form.*

Recall that we defined $i_X\omega$ by giving its values on vector fields in terms of the values of ω (i.e., $i_X\omega(X_1, \ldots, X_{s-1}) = s\omega(X, X_1, \ldots, X_{s-1})$). We can describe $d\omega$ by its values on vector fields in terms of the values of ω on the vector fields and their Lie brackets.

Theorem 12.2. (i) *If ω is a 0-form,*

$$d\omega(X) = X\omega$$

(ii) *If ω is a 1-form,*

$$d\omega(X, Y) = \tfrac{1}{2}(X\omega \cdot Y - Y\omega \cdot X - \omega \cdot [X, Y])$$

(iii) *If ω is an s-form,*

$$d\omega(X_1, \ldots, X_{s+1}) = \frac{1}{s+1}\left(\sum_{1}^{s+1} (-1)^{i+1} X_i\omega(X_1, \ldots, \hat{X}_i, \ldots, X_{s+1}\right)$$

$$+ \sum_{i<j}^{s+1} (-1)^{i+j}\omega([X_i, X_j], X_1, \ldots, \hat{X}_i, \ldots, \hat{X}_j, \ldots, X_{s+1}))$$

($\hat{X}_i$ *means that the argument is deleted.*)

Proof. (i) Directly from the definitions of df and Xf.

(ii) Using a coordinate system and the $\mathbb{R}$-linearity of d we can restrict ourselves to $\omega = f\,d\mu$. For these,

$$d\omega(X, Y) = (df \wedge d\mu)(X, Y) = \tfrac{1}{2}(df \otimes d\mu - d\mu \otimes df)(X, Y)$$
$$= \tfrac{1}{2}(df \cdot X\, d\mu \cdot Y - d\mu \cdot X\, df \cdot Y)$$
$$= \tfrac{1}{2}(Xf\, Y\mu - X\mu\, Yf) \tag{12.6}$$

Further,

$$X\langle \omega, Y\rangle = X\langle f\, d\mu, Y\rangle = X(fY\mu) = Xf\, Y\mu + fX\, Y\mu \tag{12.7}$$

Similarly,

$$-Y\langle \omega, X\rangle = -X\mu\, Yf - fY\, X\mu \tag{12.8}$$

Now form $2d\omega(X, Y) - X\langle\omega, Y\rangle + Y\langle\omega, X\rangle$. Replacing the terms in this expression by their expansions in (12.6), (12.7) and (12.8) we get $-fX\, Y\mu + fY\, X\mu = -f[X, Y]\mu = -f\, d\mu[X, Y] = -\omega \cdot [X, Y]$.

(iii) *Problem 12.4.* ■

Definitions
An s-form, θ, is (i) *closed* if $d\theta = 0$; (ii) *exact* if $\theta = d\omega$ for some ω; (iii) *locally exact* if $\theta = d\omega$ for some neighborhood of each point in the domain of θ.

The corollary of Theorem 12.1 says that if θ is exact then θ is closed. (This is sometimes called the Poincaré lemma, and sometimes its converse is so designated, see Theorem 12.3.) In terms of the differential operator, d, θ is closed if $\theta \in \ker d$ and θ is exact if θ is in the image of d; cf., *Problem 2.29.*

In the example above, eq. (12.3) gives $d\, df = 0$ by direct computation. We can express this by saying that, if a form is the "gradient" of a function, then its "curl" vanishes. Similarly, from eq. (12.4), $d\, d\sigma = 0$ says that if a form is the "curl" of another form then its "divergence" vanishes. These are familiar results in vector analysis.

Theorem 12.3. *If θ is closed, then it is locally exact. More specifically, if θ is a closed s-form, then at each point there exists a cubical coordinate neighborhood, $\mathscr{U} = \{p: a^i < \mu^i(p) < b^i\}$, on which there exists an $s - 1$ form ω such that $d\omega = \theta$.*

Outline of proof. Exhibit a map, H, on forms on $\mathscr{U}$ taking s forms to $s - 1$ forms and such that $d\, H\theta = \theta$. This is done indirectly by obtaining H on forms on $\mathscr{U}$ such that $H\, d\theta + d\, H\theta = \theta$. Such maps can be constructed from the maps i_X. (See Singer and Thorpe, 1967, pp. 116 ff.)* (Also note that

* H is called a homotopy operator. A homotopy operator comes from a homotopy from the identity on $\mathscr{U}$ to the constant map $\mathscr{U} \to p$. (Woll, 1966, pp. 193 ff.).

this method is simply an abstraction and generalization of the classical method; see Kaplan, 1959, pp. 281–283.) ■

In Section 11.4 we saw that a differentiable mapping $\phi: M \to N$ can induce mappings of tensor fields. In particular, for a covariant tensor field there is a pull-back, and for a differential form ω we have $\phi^*\omega$. Now we can generalize Theorem 11.21.

Theorem 12.4. *$\phi^* \, d\omega = d\phi^* \, \omega$. That is, exterior differentiation commutes with ϕ^*, or d is invariant under differentiable maps, ϕ.*

Proof. Note that ϕ^* is linear, so we need only consider monomials for ω. Proceeding by induction we have

(i) for $\omega = f \, d\mu$,

$$\begin{aligned} \phi^* \, d\omega &= \phi^*(df \wedge d\mu) = \phi^* \, df \wedge \phi^* \, d\mu \\ &= d\phi^* f \wedge d\phi^*\mu = d(\phi^* f \wedge d\phi^*\mu) = d(\phi^* f \wedge \phi^* d\mu) \\ &= d\phi^*(f \, d\mu) = d\phi^*\omega \end{aligned}$$

(ii) Let $\omega = \omega_1 \wedge \omega_2$ where ω_1 is of order 1 and ω_2 is of order s. Then

$$\begin{aligned} \phi^* \, d\omega &= \phi^*(d\omega_1 \wedge \omega_2 - \omega_1 \wedge d\omega_2) \\ &= \phi^*(d\omega_1) \wedge \phi^*\omega_2 - \phi^*\omega_1 \wedge \phi^*(d\omega_2) \\ &= d\phi^*\omega_1 \wedge \phi^*\omega_2 - \phi^*\omega_1 \wedge d\phi^* \, \omega_2 = d(\phi^*\omega_1 \wedge \phi^*\omega_2) \\ &= d\phi^*(\omega_1 \wedge \omega_2) = d\phi^*\omega \end{aligned}$$

■

PROBLEM 12.1. Show that the components of $d\omega$ are

$$\frac{1}{s!} \sum_{\pi} (\operatorname{sgn} \pi) \frac{\partial \omega_{i_{\pi(2)} \cdots i_{\pi(s+1)}}}{\partial \mu^{i_{\pi(1)}}}$$

and that they transform properly under coordinate transformations.

PROBLEM 12.2. Let f and g be functions on $\mathbb{R}^3$, and σ and τ be 1-form fields on $\mathbb{R}^3$. Define dot and cross products by $\sigma \cdot \tau = b(\sigma, \tau)$ $(= *(\sigma \wedge *\tau) = *(\tau \wedge *\sigma))$, see Section 6.3) and $\sigma \times \tau = *(\sigma \wedge \tau)$ respectively, and write grad, curl, and div for d, $*d$, and $*d*$ respectively. Prove the vector analysis formulas

(i) $\operatorname{grad} fg = g \operatorname{grad} f + f \operatorname{grad} g$
(ii) $\operatorname{curl} f\sigma = \operatorname{grad} f \times \sigma + f \operatorname{curl} \sigma$
(iii) $\operatorname{div} f\sigma = \operatorname{grad} f \cdot \sigma + f \operatorname{div} \sigma$
(iv) $\operatorname{div} \sigma \times \tau = \operatorname{curl} \sigma \cdot \tau - \sigma \cdot \operatorname{curl} \tau$

PROBLEM 12.3. (i) If σ is a 1-form on $\mathbb{R}^3$, and ω is a 2-form on $\mathbb{R}^3$ and Δ is the Laplace–Beltrami operator, express σ and ω in terms of the natural basis of $\mathfrak{X}^*$ and compute $\Delta\sigma$ and $\Delta\omega$. (ii) Show that for a 1-form (or, a vector field) using the terminology in Problem 12.2, $\Delta\sigma = \operatorname{grad} \operatorname{div} \sigma - \operatorname{curl} \operatorname{curl} \sigma$.

PROBLEM 12.4. Prove part (iii) of Theorem 12.2.

12.2 Integration of differential forms

There are two apparently different types of integration situations that occur in calculus. In one case we want to integrate a given function over a given "3-dimensional" or "solid" subset of $\mathbb{R}^3$, and in the other case we form surface, or line integrals over surfaces, or curves, respectively, in $\mathbb{R}^3$.

While the first case generalizes "automatically" to $\mathbb{R}^n$, the generalization of the second case is not easy and is rarely presented in the classical textbooks (cf., Coburn, 1955). A presentation that simultaneously generalizes the second case to $\mathbb{R}^n$ and unifies the two situations would seem to be desirable. We can actually go beyond $\mathbb{R}^n$ to arbitrary differentiable manifolds. However, on manifolds it turns out that the second case, line and surface integrals, generalizes more naturally than the first, apparently simpler, case. Let us see what difficulties arise when we try to generalize the latter.

First, recall that in $\mathbb{R}^n$ for the latter case, given a certain class of subsets, S, (e.g., bounded sets whose boundary has zero content), and a certain class of functions, f (e.g., bounded continuous functions), we can define an integral as a function $\int : (S, f) \mapsto \int_S f$. This function has certain properties (e.g., linear in f, additive in S).

We want to define a concept on manifolds which reduces to this in the case $M = \mathbb{R}^n$. If $\mathbb{R}^n$ is to be a special case of M, we have to think of it as a manifold, and not just a set of n-tuples, and any concept defined on $\mathbb{R}^n$ as a manifold has to be independent of coordinates.

We have to define $\int_S f$ on $\mathbb{R}^n$ as a manifold, so that if $(\mathscr{U}, \mu)$ is the natural coordinate system, and hence any coordinate system, on $\mathbb{R}^n$, and $S \subset \mathscr{U}$, then $\int_S f = \int_{\mu(S)} f_\mu$, where f_μ is the representative of f on $\mu(S)$. That is, we require that $\int_{\mu(S)} f_\mu = \int_{\bar{\mu}(S)} f_{\bar{\mu}}$ for any two coordinate systems with $S \subset \mathscr{U} \cap \bar{\mathscr{U}}$. But by the "change of variables" theorem in the integral calculus on $\mathbb{R}^n$,

$$\int_{\bar{\mu}(S)} f_{\bar{\mu}} = \int_{\mu(S)} f_{\bar{\mu}} \circ (\bar{\mu} \circ \mu^{-1}) |\det J_{\bar{\mu} \circ \mu^{-1}}| \tag{12.9}$$

which, in general, is not $\int_{\mu(S)} f_\mu$. Clearly, the same situation occurs for any manifold, M, if we try to define $\int_S f$ in terms of representations via coordinate systems. So "the integral of a function" is not coordinate-invariant, and hence we cannot expect to be able to integrate *functions* on M.

Equation (12.9) suggests the proper way to proceed, if we write it in terms of iterated integrals, namely,

$$\underset{\mu(S)}{\int \cdots \int} f(y^1, \ldots, y^n)\, dy^1 \cdots dy^n$$

$$= \underset{\mu(S)}{\int \cdots \int} f(y^1(x^1 \cdots x^n), \ldots, y^n(x^1, \ldots, x^n)) \left| \det\left(\frac{\partial x^i}{\partial y^j}\right) \right| dx^1 \cdots dx^n \tag{12.10}$$

Recalling the transformation formula

$$d\bar{\mu}^1 \wedge \cdots \wedge d\bar{\mu}^n = \det\left(\frac{\partial \bar{\mu}^i}{\partial \mu^j}\right) d\mu^1 \wedge \cdots \wedge d\mu^n$$

(Problem 11.11) for n-forms; or slightly more generally,

$$g\, d\bar{\mu}^1 \wedge \cdots \wedge d\bar{\mu}^n = g \det\left(\frac{\partial \bar{\mu}^i}{\partial \mu^j}\right) d\mu \wedge \cdots \wedge d\mu^n \tag{12.11}$$

we see that the integrands in eq. (12.10) transform something like n-forms. This suggests trying to define integrals of n-forms rather than of functions. This will lead, in particular, to a generalization of line and surface integrals of calculus.

We first make a definition for $\mathbb{R}^n$ for functions (equivalently, forms) of compact support. Recall that the support, $\operatorname{supp} f$, of a function, f, is the closure of the set on which $f \neq 0$.

Definition
If ω is an n-form of compact support on $\mathbb{R}^n$, then

$$\int_S \omega|_S = \int_{x_1^n}^{x_2^n} \cdots \int_{x_1^1}^{x_2^1} f\, dx^1 \cdots dx^n \tag{12.12}$$

where $x^1, \ldots, x^n$ are the natural coordinate functions on $\mathbb{R}^n$,

$$\omega = f\, dx^1 \wedge \cdots \wedge dx^n,$$

and $\{x: x_1^i \leq x^i \leq x_2^i\}$ is any box containing $\operatorname{supp} \omega$ and S is any set containing $\operatorname{supp} \omega$.

Now we proceed to define the integral of an n-form, θ, over an n-dimensional manifold, M, in two steps: first, when $\operatorname{supp} \theta$ is contained in a coordinate neighborhood, $\mathscr{U}$, and then for any θ of compact support.

As usual, we try to define $\int_M \theta$ by means of coordinates in terms of a representation in $\mathbb{R}^n$. Thus, if $\operatorname{supp} \theta \subset \mathscr{U}$, where $(\mathscr{U}, \mu)$ is a chart, we put

$$\int_M \theta = \int_{\mathbb{R}^n} \omega \tag{12.13}$$

where

$$\omega = \begin{cases} (\mu^{-1})^*\theta|_{\mathscr{U}} & \text{on } \mu(\mathscr{U}) \\ 0 & \text{outside of } \mu(\mathscr{U}) \end{cases}$$

so the right side of eq. (12.13) is defined by eq. (12.12).

Somewhat more explicitly, writing $\theta|_{\mathscr{U}} = f\, d\mu^1 \wedge \cdots \wedge d\mu^n$, then

$$\begin{aligned}(\mu^{-1})^*\theta|_{\mathscr{U}} &= (\mu^{-1})^* f\, d\mu^1 \wedge \cdots \wedge d\mu^n = f_\mu (\mu^{-1})^* d\mu^1 \wedge \cdots \wedge d\mu^n \\ &= f_\mu \frac{\partial(\mu^{-1})^1}{\partial x^{i_1}} dx^{i_1} \wedge \cdots \wedge \frac{\partial(\mu^{-1})^n}{\partial x^{i_n}} dx^{i_n}\end{aligned}$$

by eq. (9.10), where x^i are the natural coordinate functions on $\mathbb{R}^n$. But $(\mu^{-1})^i = \mu^i \circ \mu^{-1} = x^i$, so $(\mu^{-1})^*\theta|_{\mathscr{U}} = f_\mu\, dx^1 \wedge \cdots \wedge dx^n$ and eq. (12.13) can be written

$$\int_M \theta = \int_{\mu(\mathscr{U})} f_\mu\, dx^1 \wedge \cdots \wedge dx^n \tag{12.14}$$

Since the definition, eq. (12.13), and the description, eq. (12.14), are given in terms of a coordinate system, it is necessary to show they are actually independent of coordinates.

Theorem 12.5. *If $(\mathscr{U}, \mu)$ and $(\bar{\mathscr{U}}, \bar{\mu})$ are two coordinate systems, and*

$$\operatorname{supp} \theta \subset \mathscr{U} \cap \bar{\mathscr{U}},$$

then $\int_M \theta$ has the same value for both coordinate systems if and only if $\det(\partial\bar{\mu}^i/\partial\mu^j) > 0$.

Proof. Using eq. (12.14), we have for the $(\bar{\mathscr{U}}, \bar{\mu})$ coordinates,

$$\int_M \theta = \int_{\bar{\mu}(\bar{\mathscr{U}})} g_{\bar{\mu}}\, dy^1 \cdots dy^n = \int_{\bar{\mu}(\mathscr{U} \cap \bar{\mathscr{U}})} g_{\bar{\mu}}\, dy^1 \cdots dy^n$$

$$= \int_{\mu(\mathscr{U} \cap \bar{\mathscr{U}})} g_{\bar{\mu}} \det\left(\frac{\partial y^i}{\partial x^j}\right) dx^1 \cdots dx^n$$

by the "change of variables" theorem if and only if $\det(\partial y^i/\partial x^j) > 0$. But by eq. (12.11), $g_{\bar{\mu}} \det(\partial y^i/\partial x^j) = f_\mu$, so

$$\int_{\mu(\mathscr{U} \cap \bar{\mathscr{U}})} g_{\bar{\mu}} \det\left(\frac{\partial y^i}{\partial x^j}\right) dx^1 \cdots dx^n = \int_{\mu(\mathscr{U} \cap \bar{\mathscr{U}})} f_\mu\, dx^1 \cdots dx^n$$

$$= \int_{\mu(\bar{\mathscr{U}})} f_\mu\, dx^1 \cdots dx^n \qquad \blacksquare$$

Since the condition $\det(\partial \bar{\mu}^i/\partial \mu^j) > 0$ cannot be satisfied for all coordinate systems whose coordinate neighborhoods contain supp θ, we need to impose further conditions in order to define the concept of an integral over M. It will turn out that we can define an "oriented integral" over a manifold which has an atlas with the property $\det(\partial v^i/\partial \mu^j) > 0$ for all coordinate systems in the atlas.

Definitions
If there exists on M an atlas such that for all coordinate systems $\det(\partial \mu^i/\partial v^j) > 0$ then M is called *orientable* and the atlas is *an oriented atlas.* An n-form on M which never vanishes is called *a volume form.*

Theorem 12.6. *M has an oriented atlas if and only if M has a volume form.*

Proof. Problem 12.5.

Clearly, if M has a volume form it has others. Also, for every coordinate system, (μ^i), there is a coordinate system (v^i) such that $\det(\partial \mu^i/\partial v^j) < 0$. Thus, if M is orientable we have two classes of atlases for which $\det(\partial \mu^i/\partial v^j) > 0$. If we choose one, we say that we *orient M*, and that M is then *an oriented manifold.*

Another way of describing this property of a manifold is that it is possible to order the bases of the tangent spaces the same way at all points. That is, for all points and for all charts, the bases, $\{\mu_* \partial/\partial \mu^1, \ldots, \mu_* \partial/\partial \mu^n\}$, of the tangent spaces of $\mathbb{R}^n$ all have the same order. That this property is equivalent to the definition is proved by using the relation $d\mu^1 \wedge \cdots \wedge d\mu^n = \det(\partial \mu^i/\partial v^j)\, dv^1 \wedge \cdots \wedge dv^n$. With this more "geometric" way of describing orientability, we can convince ourselves of the non-orientability of the standard example, the Möbius Strip.

Definition
If M is an oriented n-dimensional manifold, $(\mathscr{U}, \mu)$ is any chart in the orientation of M, and θ is an n-form with $\operatorname{supp} \theta \subset \mathscr{U}$, then $\int_M \theta$ is defined by eq. (12.13).

Now that we have a clearly defined concept of the integral over an oriented manifold, M, of an n-form whose support is contained in a coordinate neighborhood, we can extend the definition to encompass the more general case of n-forms with compact support.

Theorem 12.7. *If θ has compact support in M, then there exist functions $f_1, \ldots, f_k$ on M such that*

$$\theta = f_1\theta + \cdots + f_k\theta \tag{12.15}$$

where each $f_i\theta$ has support in a coordinate neighborhood of M.

Proof. Since $\operatorname{supp} \theta$ is compact, we can cover M by $M - \operatorname{supp} \theta$ and coordinate neighborhoods $\mathscr{U}_1, \ldots, \mathscr{U}_k$. Since M is paracompact it has a partition of unity $f_1, \ldots, f_n$ subordinate to this cover. (The need to extend functions or forms defined locally to ones defined globally is the main reason for the assumption of paracompactness for manifolds.) ∎

Theorem 12.8. *The representation (12.15) of θ is not unique, but*

$$\sum_i \int_M f_i\theta = \sum_i \int_M g_i\theta$$

so we can define the integral of an n-form, θ, of compact support over an oriented n-dimensional manifold M by

$$\int_M \theta = \sum_i \int_M f_i\theta \tag{12.16}$$

for any covering and any partition of unity.

Proof. If $\{g_i\}$ is a partition of unity with $\operatorname{supp} g_j \subset \mathscr{V}_j$ for $j = 1, \ldots, l$ and $g_j = 0$ for $j > l$ then

$$\theta = g_i\theta + \cdots + g_l\theta$$

and

$$g_j\theta = f_1 g_j\theta + \cdots + f_k g_j\theta \tag{12.17}$$

for each j and supp $f_i g_j \subset \mathscr{V}_j$. Integrating (12.17), and summing on j we get

$$\sum_{j=1}^{l} \int g_j\theta = \sum_{j=1}^{l} \sum_{i=1}^{k} \int f_i g_j \theta = \sum_{i=1}^{k} \int f_i\theta \qquad \blacksquare$$

We must emphasize at this point that, from the point of view of generalizing integration on $\mathbb{R}^n$, our development so far has been very limited. Specifically, we usually want to integrate over given subsets of $\mathbb{R}^n$ in ordinary calculus—not all of $\mathbb{R}^n$—while so far all we have defined is an integral over all of M, and not over subsets of M. Of course, for subsets of M that are manifolds, the above development applies.

Whether or not we can integrate over a given subset of $\mathbb{R}^n$ depends to a large extent, as we noted at the beginning of this section, on the nature of its boundary. As in ordinary calculus, we have to describe sufficient conditions for integrability on subsets of M.

When we generalize to manifolds we can take either of two essentially equivalent approaches (cf., Boothby, 1975, pp. 251–252); we can start with a given open submanifold of M with a smooth enough boundary or we can generalize the concept of a manifold to that of *a manifold with boundary* and deal with subsets of M intrinsically. We will take the former approach. The latter is indicated in Problem 12.7.

Definitions
A regular domain in M is a subset $D \subset M$ whose boundary points have the property: at each boundary point, p, there is a coordinate system $(\mathscr{U}, \mu)$ such that $\mathscr{U} \cap \bar{D} = \{q \in \mathscr{U} : \mu^n(q) \geq \mu^n(p)\}$. If M is oriented, θ has compact support on M, D is a regular domain in M, and c_D is the characteristic function of D, then

$$\int_D \theta = \int_M c_D \theta \tag{12.18}$$

is *the integral of θ over D*. (The integral on the right exists because on each chart $\mu(\mathscr{U} \cap \partial D)$ has content zero.)

Theorem 12.9. *If D is a regular domain in M, then the boundary, ∂D, of D is an $n-1$ dimensional submanifold of M.*

Proof. First of all, in the coordinate system $(\mathscr{U}, \mu)$ the boundary points of D are points q such that $\mu^n(q) = \mu^n(p)$, or, equivalently the interior points of D are points q such that $\mu^n(q) > \mu^n(p)$. This follows from the fact that if q is such that $\mu^n(q) > \mu^n(p)$, then there exists a neighborhood of q such that $\mu^n(\tilde{q}) > \mu^n(p)$, by continuity. Then, the inclusion map $i: \partial D \to M$ is an immersion, since $(\mathscr{V}, v)$, where $\mathscr{V} = \partial D \cap \mathscr{U}$ and $v^i = \mu^j \circ i, j = 1, \ldots, n-1$, is an admissible chart on ∂D.

Theorem 12.10. *If M is orientable, and D is a regular domain in M, then ∂D is orientable and the orientation of ∂D can be described in terms of that of D.*

Proof. On the intersection of the coordinate neighborhoods $\mathscr{U}$ and $\bar{\mathscr{U}}$ the Jacobian matrix of $\bar{\mu} \circ \mu^{-1}$ is $(\partial \bar{\mu}^i / \partial \mu^j)$. But $\bar{\mu}^n$ is constant along the boundary, given by $\mu^n = \text{constant}$, so along that boundary $\partial \bar{\mu}^n / \partial \mu^i = 0$ for $i = 1, \ldots, n-1$. Hence,

$$\left(\frac{\partial \bar{\mu}^i}{\partial \mu^j}\right) = \begin{bmatrix} \dfrac{\partial \bar{\mu}^1}{\partial \mu^1} & \dfrac{\partial \bar{\mu}^2}{\partial \mu^1} & \cdots & 0 \\ \dfrac{\partial \bar{\mu}^1}{\partial \mu^2} & \cdots & \cdots & 0 \\ \vdots & \cdots & \cdots & \vdots \\ 0 & \cdots & \cdots & 0 \\ \dfrac{\partial \bar{\mu}^1}{\partial \mu^n} & \cdots & \cdots & \dfrac{\partial \bar{\mu}^n}{\partial \mu^n} \end{bmatrix}$$

and using the coordinates of Theorem 12.9 on ∂D we get

$$\det\left(\frac{\partial \bar{\mu}^i}{\partial \mu^j}\right) = \det\left(\frac{\partial \bar{v}^i}{\partial v^j}\right) \frac{\partial \bar{\mu}^n}{\partial \mu^n}$$

Since the interior of $\mu(\mathscr{U} \cap \bar{D})$ maps into the interior of $\bar{\mu}(\mathscr{U} \cap \bar{D})$ and both nth coordinates increase as we go into the interior from the boundary, $\partial \bar{\mu}^n / \partial \mu^n > 0$, and so ∂D is orientable if M is and with the coordinates of Theorem 12.9 ∂D has the induced orientation. ∎

Now that we have defined the concept of a regular domain and the integral over such a subset of M, and we know that the boundary of a regular domain is an orientable manifold if M is, we are able to state Stokes' theorem.

Theorem 12.11. (Stokes') *Suppose D is a regular domain of an n-dimensional, oriented manifold, M, and θ is an $n-1$ form on $\bar{D}$ with compact support. Then*

$$\int_D d\theta = (-1)^n \int_{\partial D} i^*\theta \tag{12.19}$$

In eq. (12.19) i is the inclusion map, $i: \partial D \to D$, and the orientation of ∂D is the induced orientation.

Proof. (Sketch) Just as in the theorem on local exactness, Theorem 12.3, the proof is based on the classical proof generalized to n dimensions and extended to manifolds. By compactness, we can reduce the proof to the consideration of one of the forms $f_i\theta$ with compact support in a coordinate neighborhood in Theorem 12.7. Then we pull it back to $\mu(\mathscr{U})$ and put it in a box whose edges are coordinate hypersurfaces in $\mathbb{R}^n$. If the box does not intersect the boundary, then both sides of eq. (12.19) are zero. If the box does intersect the boundary, the $n - 1$ form ω on $\mathbb{R}^n$ can be written

$$\omega = \sum_{j=1}^{n} (-1)^{j-1}\omega_j \, dx^1 \wedge \cdots \wedge dx^{j-1} \wedge dx^{j+1} \wedge \cdots \wedge dx^n$$

so that

$$d\omega = \sum_{j=1}^{n} \frac{\partial \omega_j}{\partial x^j} dx^1 \wedge \cdots \wedge dx^n$$

We integrate these forms using the definition given in eq. (12.12). Now in the integrations over the box of $d\omega$ and over the boundary points in the box of ω, all terms except the one with coefficient ω_n evaluated on $x^n = 0$ drop out, and the terms remaining are the same except for the factor $(-1)^n$. ■

We can eliminate the factor $(-1)^n$ in Stokes' theorem if ∂D has *the compatible orientation*, that is, the induced orientation if n is even and the opposite orientation if n is odd.

The subject of integration on manifolds can be pursued further in at least two different directions. We will only indicate these.

1. We want to be able to integrate over a class of subsets of M larger than the set of regular domains, since, in particular, regular domains have smooth boundaries. This is done by constructing "p-chains" with their algebra (cf. Spivak, 1965; Bishop and Goldberg, 1968; Warner, 1971). Then we extend the class of subsets over which we can integrate to the class of "parametrizable oriented subsets" of M (Bishop and Goldberg, 1968, p. 195).
2. With Stokes' theorem we have integration by parts for s-forms and then with the Laplace–Beltrami operator (Section 12.1) we get generalizations of Green's identities which carry us finally into a generalization of classical potential theory (cf. Woll, 1966, pp. 111 ff.).

PROBLEM 12.5. Prove Theorem 12.6.

PROBLEM 12.6. The 2-sphere, $x^2 + y^2 + z^2 = 1$, in $\mathbb{R}^3$, can be parametrized by the 2-cube $c:(u, v) \mapsto (\cos u, \sin u \cos v, \sin u \sin v)$ on $C = [0, \pi] \times [0, 2\pi]$ so $\int_{S^2} \theta = \int_C c^*\theta$. Evaluate $\int_{S^2} \theta$ if

$$\theta = (x\, dy \wedge dz + y\, dz \wedge dx + z\, dx \wedge dy)/r^3$$

and $r = \sqrt{x^2 + y^2 + z^2}$.

PROBLEM 12.7. A *manifold with boundary* is defined precisely the same as a manifold except that $\mathbb{R}^n$ is replaced by H^n where $H^n = \{a \in \mathbb{R}^n : a^n \geq 0\}$ with the relative topology of $\mathbb{R}^n$. Let $\partial H^n = \{a \in \mathbb{R}^n : a^n = 0\}$ and $\partial M = \{p \in M : \mu(p) \in \partial H^n\}$ (note that this is independent of μ). Show that ∂M is an $n - 1$-dimensional manifold imbedded in M.

PROBLEM 12.8. Fill in the details of the proof of Stokes' theorem.

13

THE FLOW AND THE LIE DERIVATIVE OF A VECTOR FIELD

The concept of a vector field on a manifold, M (or, on an open subset, $\mathscr{U}$, of M), has a simple intuitive definition, but also has broad and deep ramifications. In particular, we want to look at vector fields from two rather different and important points of view. Roughly, (1) a vector field can be thought of as determining, or being determined by, a family of curves, and (2) a vector field can be thought of as determining, or being determined by, a certain derivation on tensor fields on M (or $\mathscr{U}$). The first correspondence well known for cartesian spaces, $\mathbb{R}^n$, as the subject matter of ordinary differential equations will be examined in Sections 13.1–13.3. The second correspondence, introduced in restricted form in Section (11.1), will be described in Section 13.4.

13.1 Integral curves and the flow of a vector field

In Chapters 7 and 9 a major theme was the relation between tangents at $p \in M$ and curves through p. Now we wish, in a sense, to extend this relation to vector (i.e., tangent) fields and families of curves. Thus, suppose we start with a given field, X.

Definition

A curve, γ, whose values lie in the domain of a vector field, X, and such that

$$\dot{\gamma}(u) = X(\gamma(u)) \tag{13.1}$$

for all u in the domain of γ is *an integral curve of* X. (Recall, from Section 9.3, that $\dot{\gamma}(u)$ is the tangent of γ at $\gamma(u)$.)

Alternatively, γ is an integral curve of X if $\gamma_* \circ d/d\pi = X \circ \gamma$, or if the following diagram commutes:

$$\begin{array}{ccc} TR & \xrightarrow{\gamma_*} & TM \\ \Big\uparrow{\scriptstyle \frac{d}{d\pi}} & & \Big\uparrow{\scriptstyle X} \\ \mathbb{R} & \xrightarrow{\gamma} & M \end{array}$$

$\gamma_* \circ d/d\pi$ is the canonical lift of γ, and $X \circ \gamma$ is a vector field over γ.

This definition simply describes in a coordinate-free manner a condition on curves more commonly described by a system of differential equations. Thus, if we introduce a local coordinate system $(\mathscr{U}, \mu)$, at a point, $\gamma(u)$, in the domain of X, then

$$\dot{\gamma}(u) = D_u\gamma^i \left.\frac{\partial}{\partial \mu^i}\right|_{\gamma(u)} \qquad \text{where} \qquad \gamma^i = \mu^i \circ \gamma \text{ by Problem 9.15}$$

and

$$X(\gamma(u)) = \left(X^i \frac{\partial}{\partial \mu^i}\right)(\gamma(u)) = X^i(\gamma(u)) \left.\frac{\partial}{\partial \mu^i}\right|_{\gamma(u)}$$

So, in terms of local coordinates, (13.1) becomes

$$D_u\gamma^i = X^i(\gamma(u)), \qquad i = 1, \ldots, n$$

Finally, in terms of the coordinate representations, $\hat{X}^i$, of the functions $X^i = \hat{X}^i \circ \mu$, this becomes the system

$$D_u\gamma^i = \hat{X}^i(\gamma^1(u), \ldots, \gamma^n(u)), \qquad i = 1, \ldots, n \tag{13.2}$$

of ordinary differential equations. Thus, (13.2) is the local expression of the fact that γ is an integral curve of X.

Now having imposed condition (13.1) (or (13.2)) with a given vector field, X, we ask about the existence and properties of curves satisfying it.

Theorem 13.1. *Given $u_0 \in \mathbb{R}$, and $(a_0^1, \ldots, a_0^n)$ in the domains of all the $\hat{X}^i$. Then there exists an interval containing u_0, and a unique set of functions γ^i (on that interval) such that $\gamma^i(u_0) = a_0^i$ and γ^i satisfy (13.2) on that interval.**

Proof. See any book on ordinary differential equations.

One can get more on the nature of the γ^i in terms of whatever properties of the $\hat{X}^i$ may be available, and, of course we get much more in special cases such as in the case of linearity.

From the results of Theorem 13.1 for $\mathbb{R}^n$ we get corresponding results for curves on M.

* Note that we have omitted a hypothesis in the statement of Theorem 13.1; i.e., conditions on the $\hat{X}^i$. $\hat{X}^i \in C^1$ will suffice. In general, when such conditions are omitted it is to be understood that whatever differentiability is required is assumed.

We can also describe these results in a different way. First of all we notice that (13.2) is an autonomous system, i.e., the functions $\hat{X}^i$ do not have the argument u. A consequence of this is the following.

Theorem 13.2. (Admissible change of parameter for integral curves) *If γ_1 is an integral curve of X and γ_2 has the same range as γ_1, then γ_2 is an integral curve of X iff their parameters are related by a translation.*

Proof. The proof is based on the following lemma.

Lemma. *Suppose $\gamma_2 = \gamma_1 \circ \rho$, $S \subset \mathbb{R}$ and $\rho(S) \subset$ domain of γ_1. Then $\dot{\gamma}_2 = \dot{\gamma}_1 \circ \rho \Leftrightarrow \rho = \chi$ where $\chi(t) = t + t_0$.*

Proof of lemma. $\gamma_2 = \gamma_1 \circ \rho \Rightarrow \dot{\gamma}_2(t) = \dot{\gamma}_1(\rho(t))D_t\rho$. Then $\dot{\gamma}_2(t) = \dot{\gamma}_1(\rho(t)) \Leftrightarrow D_t\rho = 1 \Leftrightarrow \rho = \chi$. ∎

Proof of theorem. **If:** If $\rho = \chi$, then $\gamma_1(u) = \gamma_2(t) \Rightarrow \dot{\gamma}_1(u) = \dot{\gamma}_2(t)$ by the lemma. Then $\dot{\gamma}_1(u) = X(\gamma_1(u)) \Rightarrow \dot{\gamma}_2(t) = X(\gamma_2(t))$.

Only if: If $\dot{\gamma}_1(u) = X(\gamma_1(u))$ and $\dot{\gamma}_2(t) = X(\gamma_2(t))$ and $\gamma_2(t) = \gamma_1(\rho(t))$, then $\dot{\gamma}_1(\rho(t)) = \dot{\gamma}_2(t) \Rightarrow \rho = \chi$ by the lemma. ∎

From this theorem we see that there is no important loss of generality if we always take $u_0 = 0$; that is, consider only curves, γ_p, through p (where $p \in M$ is in the domain, E, of X and corresponds to $a_0 \in \mathbb{R}^n$ in Theorem 13.1). Now if I_1 is one interval with γ^i given by Theorem 13.1 and I_2 is another interval with α^i given by Theorem 13.1, then on $I_1 \cap I_2$, $\alpha^i = \gamma^i$ and we can define unique functions on $I_1 \cup I_2$ which satisfy the conditions of Theorem 13.1.* Hence, there is a largest interval, I_p (containing 0), on which γ_p, the integral curve through p, is defined, and we have, for each $p \in E$,

$$\gamma_p: I_p \to E \tag{13.3}$$

with

$$u \mapsto \gamma_p(u)$$

Let E_u be the subset of E for each of whose points, p, $\gamma_p(u)$ is defined. That is, in E_u, the integral curves through p given by the existence theorem can be extended at least as far as u. Thus, for each $u \in \mathbb{R}$ we have a map

$$\Theta_u: E_u \to E \tag{13.4}$$

* This is not as obvious as it may seem. In fact, this is one of the points where the Hausdorff property of M is required.

given by

$$p \mapsto \gamma_p(u)$$

Now we can rephrase the results of our existence theorem for integral curves in the form (1) a vector field X defined on E determines a family, $\{\gamma_p\}_{p \in E}$ of curves through points of E; or (2) a vector field X defined on E determines a family $\{\Theta_u\}_{u \in R}$ of maps of subsets of E; or, finally (3) a vector field X defined on E determines a map

$$\Theta\colon W \to E \tag{13.5}$$

given by

$$(u, p) \mapsto \Theta(u, p) = \Theta_u(p) = \gamma_p(u) = q$$

where W is some subset of $\mathbb{R} \times E$, Θ_u is given by (13.4), and γ_p is the integral curve of X through p.

Definitions
The map, Θ, defined in (13.5) is called *the flow of the given vector field, X*. (We sometimes call the maps Θ_u in (13.4) *translations*, and say the flow is the family, $\{\Theta_u\}$, of translations.)

EXAMPLE. $M = \mathbb{R}^2$ with the standard atlas $\{(\mathscr{U}, \mu)\} = \{(\mathbb{R}^2, id)\}$ with the given vector field $X = (\pi^1)^2\, \partial/\partial\pi^1$. The integral curves of X are represented by the solutions (γ^1, γ^2) of

$$\frac{d\gamma^1}{du} = (\gamma^1)^2, \qquad \frac{d\gamma^2}{du} = 0$$

The solutions through (a_0, b_0) are given by

$$\gamma_{(a_0, b_0)}(u) = (\gamma^1_{(a_0, b_0)}(u), \gamma^2_{(a_0, b_0)}(u)) = \left(\frac{a_0}{1 - ua_0}, b_0\right).$$

The maps Θ_u take (a_0, b_0) to $(a_0/(1 - ua_0), b_0)$ and the flow is

$$\Theta\colon (u, (a_0, b_0)) \mapsto \left(\frac{a_0}{1 - ua_0}, b_0\right)$$

The domain W of Θ is the set of pairs $(u, (a_0, b_0))$ where (u, a_0) is in the region of Fig. 13.1 between the two branches of the cylinder $ua_0 = 1$.

For example, if $(a_0, b_0) = (1, 1)$, then the solution curve $\gamma_{(1,1)}$ is defined for $u \in (-\infty, 1)$. If $u = \frac{3}{4}$, then the domain, $E_{3/4}$, of $\Theta_{3/4}$ is the region of the $(a_0 b_0)$ plane with $a_0 < \frac{4}{3}$, and $\Theta_{3/4}$ takes $E_{3/4}$ into the piece of the plane in which $a_0 > -\frac{4}{3}$.

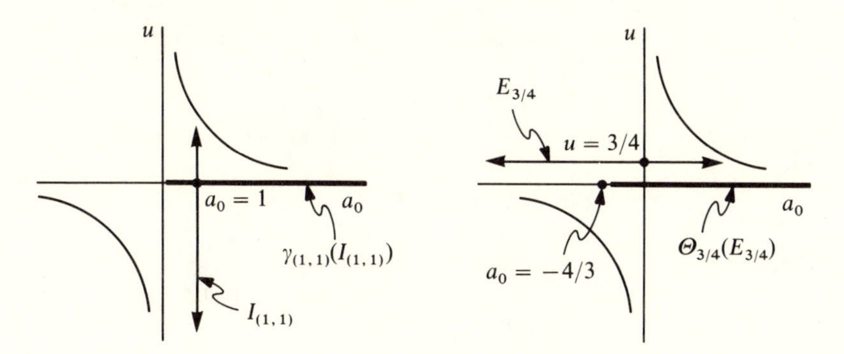

The section b_0 = constant

FIGURE 13.1

PROBLEM 13.1. Let $M = \mathbb{R}^2$ with the standard manifold structure, and define X in terms of the natural coordinates $(\mathscr{U}, \mu) = (\mathbb{R}^2, id)$ by $X = \pi^1\, \partial/\partial\pi^1 + \pi^2\, \partial/\partial\pi^2$. Find the maximal integral curve through (a, b).

PROBLEM 13.2. Same as Problem 13.1 for $X = (\exp \circ \mu^1)^{-1}\, \partial/\partial\pi^1$.

PROBLEM 13.3. Find the representation in the given stereographic atlas of the maximal integral curve through a point of the vector field on S^2 given in Section 11.1.

13.2 Flow boxes (local flows) and complete vector fields

The awkward situation in which the domain of Θ is not very well defined comes from the corresponding lack of information in Theorem 13.1 on the domain of the maximal integral curve through p. While we cannot do any better for individual integral curves, we can say a bit more when we consider families of curves.

Let S be any set in the common domain of the $\hat{X}^i$, and let $\mathscr{U} \subset S$ be the points of S which are centers of balls of radius r contained in S. Then a standard proof of Theorem 13.1 shows that if $\mathbf{K}$ is a bound for the $\hat{X}^i$ on their common domain, then for all $a_0 \in \mathscr{U}$ there is an integral curve through a_0 defined on the interval $(-r/\mathbf{K}, r/\mathbf{K})$.

Definition
If X is a vector field on $E \subset M$, a differentiable map $F\colon I \times \mathscr{V} \to M$ where $I = (-u_0, u_0) \subset \mathbb{R}$ and $\mathscr{V}$ is an open set in M is a *flow box* (or, *local flow*) *of* X if (1) for all $p \in \mathscr{V}$, $F_p\colon I \to M$ defined by $u \mapsto F(u, p)$ is an integral curve of X at p, and (2) for all $u \in I$, $F_u\colon \mathscr{V} \to M$ defined by $p \mapsto F(u, p)$ is a diffeomorphism of $\mathscr{V}$ onto $F_u(\mathscr{V})$. Briefly, a flow box is a flow for which W is a product and the maps Θ_u are diffeomorphisms.

For the example in Section 13.1 we have a flow box with $I = (-2, 2)$, $\mathscr{V} = \{(a_0 b_0): 0 < a_0 < \frac{1}{2}, b_1 < b < b_2\}$ and $F: (u, (a_0, b_0)) \mapsto (a_0/(1 - ua_0), b_0)$.

Theorem 13.3. *For every p in the domain of X there is a flow box of X containing p.*

Proof. Given p, choose a local coordinate system $(\mathscr{U}, \mu)$. The comments above the definition imply that there is a flow box, $\hat{F}$, of $\hat{X}$ on $\mu(\mathscr{U})$.* If $\hat{F}: I \times \hat{\mathscr{V}} \to \mu(\mathscr{U})$, then $F: I \times \mu^{-1}(\hat{\mathscr{V}}) \to \mathscr{U}$ given by $(u, p) \mapsto \mu^{-1}(\hat{F}(u, \mu(p)))$ will be a flow box of X containing p. ■

Theorem 13.4. *If the values of the maximal integral curve, γ_p, through p of X are contained in a compact subset, C, of M, then γ_p is defined for all of $\mathbb{R}$.*

Proof. Suppose the domain of γ_p is (a, b). Let (b_n) be a sequence converging to b. Then since C is compact, there is a subsequence (b_{n_k}) such that $\gamma_p(b_{n_k}) \to q \in C$. Let $F: I \times \mathscr{V} \to M$ be a flow box at q with $I = (-c, c)$. All $\gamma_p(b_{n_k})$ will be in $\mathscr{V}$ for large enough k. We can find a point $x = \gamma_p(b_{n_k})$ on γ_p such that $|b_{n_k} - b| < c$. If F_x is the integral curve of the flow box at x, and $\alpha_x(t) \equiv \gamma_p(t + b_{n_k})$, then $F_x = \alpha_x$ where they are both defined. Since F_x is defined on a larger interval than α_x (F_x is defined on $(-c, c)$, and α_x is defined for $t + b_{n_k} < b$, or $t < b - b_{b_k} < c$), α_x is not the maximal integral curve through x, and hence γ_p is not the maximal integral curve through p. ■

Definition
A vector field, X, is *complete* if the domains of all its integral curves can be extended to all of $\mathbb{R}$.

The following is an immediate consequence of Theorem 13.4.

Corollary. *A vector field on a compact manifold is complete.*

Note that if $M = \mathbb{R}^2$ (with the standard structure) and X is given by $X = X^1\, \partial/\partial\mu^1 + X^2\, \partial/\partial\mu^2$ where X^1 and X^2 are bounded functions on $\mathbb{R}^2$, then X is complete. However, if $M = \mathbb{R}^2 - (0, 0)$ and $X = \partial/\partial\mu^1 + \partial/\partial\mu^2$, then X is not complete. Also, in the example of Problem 13.1 X is complete, and in the example of Problem 13.2 X is not complete.

Theorem 13.5. *X is complete iff the domain, W, of the flow, Θ, of X is $\mathbb{R} \times E$.*

Proof. Problem 13.6.

* From the comments above the definition, property 1 is clear, but property 2 is not obvious. We need continuity and differentiability of solutions of differential equations *with respect to initial conditions.* The hypothesis $X \in C^1$ will suffice. See Abraham and Marsden, 1978, pp. 63 ff.)

Theorem 13.6. *If X is complete, then the flow, Θ, of X has the properties*

(1) $\Theta_0 = id.$
(2) $\Theta_u \circ \Theta_v = \Theta_{u+v}$ *for all* $u, u \in \mathbb{R}$.
(3) *Either*:
  (i) $u \neq v \Rightarrow \Theta_u \neq \Theta_v$ *for all* $u, v \in \mathbb{R}$, *or*
  (ii) *there exists a* $u_0 > 0$ *such that* $\Theta_{u_0} = \Theta_0 = id$, *and* $u \neq v \Rightarrow \Theta_u \neq \Theta_v$ *for all* $u, v \in [0, u_0)$.

Proof. (1) Immediate from Theorem 13.1.

(2) If γ_p is the integral curve of X through p, then for each fixed v, by Theorem 13.2, $u \mapsto \gamma_p(u + v)$ is the integral curve γ_q of X through $q = \gamma_p(v)$. Then $\gamma_q(u) = \Theta_u(q) = \Theta_u(\Theta_v(p))$ and $\gamma_q(u) = \gamma_p(u + v) = \Theta_{u+v}(p)$. So $\Theta_u \circ \Theta_v = \Theta_{u+v}$.

(3) We can find a point $p \in E$ and a neighborhood of p in which $X(q) \neq 0$ (assuming X is not identically zero). If $X^1(q) > 0$ in this neighborhood, and γ_p is an integral curve through p, then γ_p^1 is strictly increasing in some interval containing 0. Hence, if u, v are in such an interval, $u \neq v \Rightarrow \gamma_p^1(u) \neq \gamma_p^1(v) \Rightarrow \gamma_p(u) \neq \gamma_p(v) \Rightarrow \Theta_u(p) \neq \Theta_v(p) \Rightarrow \Theta_u \neq \Theta_v$. That is, there is some interval containing 0 for which $u \neq v \Rightarrow \Theta_u \neq \Theta_v$. Now suppose there exist $u, v > u$ such that $\Theta_u = \Theta_v$. Then $\Theta_{-u} \circ \Theta_u = \Theta_{v-u} \Rightarrow \Theta_{v-u} = \Theta_0$. So $w = v - u > 0$ is such that $\Theta_w = \Theta_0$. Let $u_0 > 0$ be the minimum w such that $\Theta_w = \Theta_0$. u_0 exists by the continuity of Θ with respect to u. Then $\Theta_u = \Theta_v$ for any $u, v > u$ in $[0, u_0)$ leads to a contradiction. ■

The prototypical example of a flow of type 3(ii) is that of the vector field $X = -\mu_2\, \partial/\partial\mu^1 + \mu_1\, \partial/\partial\mu^2$ on $\mathbb{R}^2$.

Theorem 13.6 says that if we start with a complete vector field we can construct a mapping $\Theta: \mathbb{R} \times E \to E$ with certain properties. We can reverse our point of view. Instead of starting with a vector field, if we start with a mapping $\mathscr{A}: \mathbb{R} \times M \to M$ with certain properties, we can then construct a certain complete vector field.

Definition
A mapping $\mathscr{A}: \mathbb{R} \times M \to M$ with the properties

(1) $\mathscr{A}(0, p) = p$ for all $p \in M$
(2) $\mathscr{A}(v, \mathscr{A}(u, p)) = \mathscr{A}(v + u, p) = \mathscr{A}(u, \mathscr{A}(v, p))$ for all $u \in \mathbb{R}$ and $p \in M$
is called *a 1-parameter group action.*

The first two properties of Theorem 13.6 are precisely those of the definition of $\mathscr{A}$, so the flow of a complete vector field is a 1-parameter group action. On the other hand, the concept of a 1-parameter group action is a special case of that of the action of a Lie group (i.e., a 1-parameter group action is an action of a 1-dimensional Lie group) and it can be shown that the only one-dimensional Lie groups are $\mathbb{R}$ and S^1. These two cases correspond to the two cases of Theorem 13.6(3), so that every 1-parameter group action has all the properties of Theorem 13.6.

Definition
Given an action, $\mathscr{A}$, for each $p \in M$ we have the curve $\mathscr{A}_p$: $\mathbb{R} \to M$ through p given by $\mathscr{A}_p$: $u \mapsto \mathscr{A}(u, p)$. *The infinitesimal generator of* $\mathscr{A}$ is the vector field, $X_{\mathscr{A}}$, defined by $X_{\mathscr{A}}$: $p \mapsto [\mathscr{A}_p]$ where $[\mathscr{A}_p]$ is the tangent vector of $\mathscr{A}_p$ at p.

We can complete the circle for the concepts and relations described above by noting that since a flow is an action and since an infinitesimal generator is a vector field, we can start with a vector field X, construct its flow, Θ_X and construct the infinitesimal generator, Y_{Θ_X}, of Θ_X. Or, starting with an action, $\mathscr{A}$, we can construct its infinitesimal generator, $X_{\mathscr{A}}$, and then construct the flow, $\Theta_{X_{\mathscr{A}}}$ of $X_{\mathscr{A}}$.

Theorem 13.7. (i) *The flow* $\Theta_{X_{\mathscr{A}}}$ *of the infinitesimal generator,* $X_{\mathscr{A}}$, *of a* 1-*parameter group action,* $\mathscr{A}$, is $\mathscr{A}$. (ii) *The infinitesimal generator* Y_{Θ_X} *of the flow,* Θ_X, *of a complete vector field,* X, *is* X.

Proof. (i) See Bishop and Goldberg (1968, p. 127). (ii) See Boothby (1975, p. 135).

Corollary. (i) *Every* 1-*parameter group action is the flow of a complete vector field.* (ii) *Every complete vector field is the infinitesimal generator of a* 1-*parameter group action.*

While the concept of a complete vector field is very neat, it is not adequate from the point of view of applications. Clearly, the terminology; i.e., the flow, comes from applications in fluid mechanics and in such applications "the flows" are rarely those of complete vector fields. In particular, the two general criteria we mentioned (bounded functions on $\mathbb{R}^n$ and compact manifolds) do not occur in those applications. In those cases "the flow" is only a *local* action; that is, a mapping defined only on a subset of $\mathbb{R} \times M$ and with (ii) and (iii) in the definition of $\mathscr{A}$ required to be valid only when the maps are defined (see Section 21.4). A local action has an infinitesimal generator just as before, but now it will not necessarily be a complete vector field. (See Bishop and Goldberg, 1968, pp. 126–127; Boothby, 1975, pp. 125–128.)

PROBLEM 13.4. Verify the statements for the four examples given in the paragraph above Theorem 13.5.

PROBLEM 13.5. Prove that X is complete iff there is an $r > 0$ such that for every $p \in E$ the integral curve of X through p is defined on the interval $(-r, r)$.

PROBLEM 13.6. Prove Theorem 13.5.

PROBLEM 13.7. Show that if Θ is the flow of a complete vector field with the property 3(ii) then Θ defines a map $\mathscr{A}: S^1 \times M \to M$. (Use $S^1 \cong \mathbb{R}/\langle u_0 \rangle$ where $\mathbb{R}/\langle u_0 \rangle$ is the quotient group of $\mathbb{R}$ mod u_0; i.e., $\mathbb{R}/\langle u_0 \rangle = \{[u]: u_1 \sim u_2$ if $u_1 - u_2$ is an integral multiple of $u_0\}$).

PROBLEM 13.8. Show that $\mathscr{A}: \mathbb{R} \times \mathbb{R}^2 \to \mathbb{R}^2$ given by $(u, (a_0, b_0)) \mapsto (u + a_0, b_0 e^u)$ is a 1-parameter group action, and its infinitesimal generator is $1\, \partial/\partial\pi^1 + \pi^2\, \partial/\partial\pi^2$.

PROBLEM 13.9. Prove Theorem 13.7.

13.3 Coordinate vector fields

In a coordinate neighborhood we have n vector fields $\partial/\partial\mu^i$, $i = 1, \ldots, n$, *the coordinate vector fields* (Section 11.1). For the vector field $\partial/\partial\mu^1$, equation (13.2) is $D_0\gamma^1 = 1$ and $D_0\gamma^i = 0$, $i = 2, \ldots, n$, so the integral curve through p is given by $\gamma_p^1(u) = u + a^1$, $\gamma_p^i(u) = a^i$, $i = 2, \ldots, n$, where $\mu(p) = (a^1, \ldots, a^n)$. Hence, $\Theta_u: p \mapsto \Theta_u(p) = \gamma_p(u)$ takes p to a point $\Theta_u(p)$ with coordinates $(u + a^1, a^2, \ldots, a^n)$. If $\partial/\partial\mu^2$ is another coordinate vector field with flow H, then H_v takes $\Theta_u(p)$ to a point $H_v \circ \Theta_u(p)$ with coordinates $(u + a^1, v + a^2, a^3, \ldots, a^n)$. Clearly, we can reverse the order of these operations and get $\Theta_u \circ H_v = H_v \circ \Theta_u$. That is, the flows of two coordinate vector fields commute.

Suppose, on the other hand, we start with an arbitrary vector field.

Theorem 13.8. *Any vector field, X, can be chosen to be a coordinate vector field in a neighborhood of a point, p, of a manifold at which $X(p) \neq 0$.*

Proof. Let $(\mathscr{U}, \mu)$ be a coordinate system at p such that $\mu(p) = (0, \ldots, 0)$ and $X(p) = \partial/\partial\mu^1|_p$. Define a map, ψ, from a neighborhood of $(0, \ldots, 0) \in \mathbb{R}^n$ to a neighborhood of p by $\psi: (u, a^2, \ldots, a^n) \mapsto \Theta_u \circ \mu^{-1}(0, a^2, \ldots, a^n)$ where Θ is the flow of X. The curves $u \mapsto \Theta_u \circ \mu^{-1}(0, a^2, \ldots, a^n)$ are integral curves of X through $\mu^{-1}(0, a^2, \ldots, a^n)$. Now, at $(0, \ldots, 0)$, $\psi_* \cdot \partial/\partial\pi^1 = X(p)$ by assumption, and $\psi_* \cdot \partial/\partial\pi^i|_{(0,0)} = \partial/\partial\mu^i|_p$, $i = 2, \ldots, n$, since on $u = 0$, $\psi = \mu^{-1}$. So ψ_* is nonsingular at $(0, \ldots, 0)$, and ψ has a local inverse. ψ^{-1} is a homeomorphism on $\mathscr{V} \subset \mathscr{U}$ so $(\mathscr{V}, \psi^{-1})$ is a local coordinate system with coordinate functions v^i and coordinate curves ζ_i. The integral curves of X are now the coordinate curves ζ_1, so $X = \partial/\partial v^1$. ■

Above Theorem 13.8 we saw that commutativity of their flows is a necessary condition for two vector fields to be coordinate vector fields. We also have the converse.

Theorem 13.9. *Two vector fields are coordinate vector fields of a local coordinate system iff their flows commute locally.*

Proof. **Only if:** given above.

If: Suppose $\mu(p) = 0$, and map a neighborhood of $(0, \ldots, 0) \in \mathbb{R}^n$ back to a neighborhood of p by $(u, v, a^3, \ldots, a^n) \mapsto H_v \circ \Theta_u \circ \mu^{-1}(0, 0, a^3, \ldots, a^n)$ where Θ is the flow of X and H is the flow of Y. As in the proof of Theorem 13.8 we can show that this map has a local inverse which is a local coordinate map. The coordinate curves corresponding to $\{u = \text{const}, a^3 = \text{const}, \ldots, a^n = \text{const}\}$ are $v \mapsto H_v(\Theta_u(\mu^{-1}(0, 0, a^3, \ldots, a^n))) = H_v(q) = \gamma_q(v)$, which are integral curves of Y. Now since $H_v \circ \Theta_u = \Theta_u \circ H_v$ (locally) $(u, v, a^3, \ldots, a^n) \mapsto \Theta_u \circ H_v \circ \mu^{-1}(0, 0, a^3, \ldots, a^n)$ is the inverse of the same coordinate map and so the coordinate curves $u \mapsto \Theta_u(H_v(\mu^{-1}(0, 0, \ldots, a^n)))$ which correspond to $\{v = \text{const}, a^3 = \text{const}, \ldots, a^n = \text{const}\}$ are the integral curves of X. ∎

We make several observations in connection with Theorem 13.9.

1. Theorem 13.9 can be generalized to: A set of vector fields are coordinate vector fields of a coordinate system if and only if all their flows commute pairwise.
2. The commutativity of two flows is precisely equivalent to the vanishing of the Lie bracket of their vector fields, so Theorem 13.9 and its generalization can be expressed in terms of the vanishing of Lie brackets.
3. If we go from p to $\Theta_u(p)$, then to $H_v(\Theta_u(p))$, then to $\Theta_{-u}(H_v(\Theta_u(p)))$, and finally to $H_{-v}(\Theta_{-u}(H_v(\Theta_u(p))))$ we traverse a "parallelogram," and $uv[X, Y](p)$ is a second-order measure of the amount by which the "parallelogram" fails to close.

PROBLEM 13.10. Prove any one or more of the statements above (cf. Bishop and Goldberg, 1968, pp. 134–138).

13.4 The Lie derivative

At the beginning of this chapter we said that a vector field can be considered from two different points of view. Up to now we have taken one point of view. Now we take the other.

We saw in Section 11.1 that there is a 1–1 correspondence between vector fields on M and derivations on functions on M. In particular, given X we have an operation, L_X, on $\mathfrak{F}_M$. Now compare this with what we did in Section 12.1. There, by defining an operation d, on $\mathfrak{F}_M$ and on 1-form fields we were able to define a unique extension to the entire algebra of differential forms. In our present case we can, in a similar manner, construct an extension of L_X to act on larger domains. Thus, with out definition of L_X on $\mathfrak{F}_M$, if we make the further definition

$$L_X: Y \mapsto [X, Y]$$

for $Y \in \mathfrak{X}$, then a unique derivation is determined on the algebra of contravariant tensor fields. Finally, we can go all the way to the algebra of all tensor fields on M if we postulate that our operation commutes with contractions. That is, for a vector field, Y, and a 1-form, ω,

$$L_X\langle \omega, Y \rangle = \langle \omega, L_X Y \rangle + \langle L_X \omega, Y \rangle$$

Thus, corresponding to each X we have, for each tensor field, K, $L_X K$, the Lie derivative of K with respect to X.

The proof of the existence and uniqueness of the Lie derivative proceeds by induction from tensor fields of one order to those of one order higher by means of the derivation property. Rather than formalize this description we will formalize a more "geometrical" description of the Lie derivative and show that the object defined has the properties described above, so that, by uniqueness, they are the same thing.

Suppose we are given a vector field, X, with a flow Θ, and a tensor field, K, of type (r, s). Pick $p \in M$. Then we have the integral curve γ_p through p defined on some interval, I, containing 0 (see Fig. 13.2). Fix $u \in I$. For each such u we get an element $K(\gamma_p(u)) \in (T_{\gamma_p(u)})^r_s$. Finally, recall (see page 159)

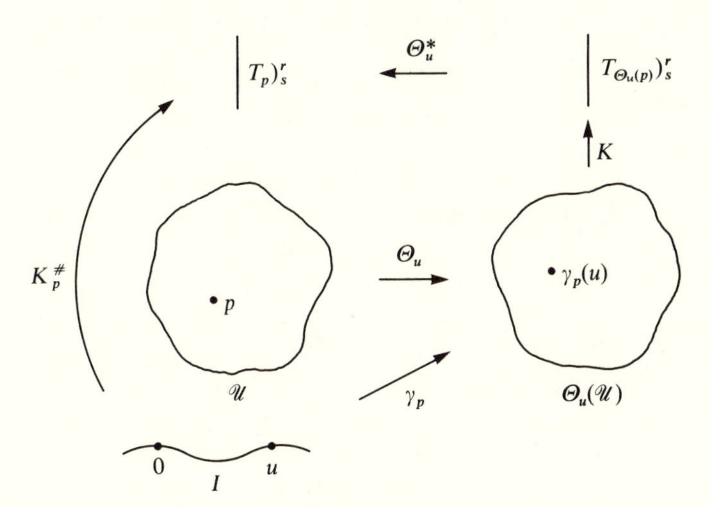

FIGURE 13.2

that Θ_u induces a linear map, Θ_u^*, from $(T_{\gamma_p(u)})_s^r$ back to $(T_p)_s^r$. Thus, for each fixed p, we have a curve $K_p^\#: I \to (T_p)_s^r$ given by $u \mapsto \Theta_u^* \circ K(\gamma_p(u)) = (\Theta_u^* K)(p)$ where $\Theta_u^* K$ is the pull-back of K by Θ_u (Section 11.4).

Definitions
The Lie derivative at p of K with respect to X, denoted by $L_X K(p)$, is the derivative, $D_0 K_p^\#$ where $K_p^\#$ is the curve described above. *The Lie derivative of K with respect to X* is the map $L_X K: p \mapsto L_X K(p)$.

Note that since $K_p^\#$ is a map from one vector space to another, the concept of derivative is defined. (See Section 7.1.) Its existence can be inferred from that of the corresponding tangent maps which in turn comes from that of the maps which compose it. Note also that, as we have done on previous occasions, we identify the linear map $D_0 K_p^\#$ with its value at 1, which is in $(T_p)_s^r$ (see Theorem 2.4), so that the Lie derivative, L_X, maps (r, s) tensor fields to (r, s) tensor fields.

In order to obtain general properties of the Lie derivative, as well as to obtain useful descriptions for particular important tensor fields, K, it will be convenient to describe $L_X K$ in terms of its component functions in a basis of coordinate tensor fields. Thus, in a local coordinate system at p, we have

$$L_X K = (L_X K)_{j_1 \cdots j_s}^{i_1 \cdots i_r} \frac{\partial}{\partial \mu^{i_1}} \otimes \cdots \otimes d\mu^{j_s}$$

$L_X K$ will be described by expressing the component functions $(L_X K)_{j_1 \cdots j_s}^{i_1 \cdots i_r}$ of $L_X K$ in terms of the component functions $K_{j_1 \cdots j_s}^{i_1 \cdots i_r}$ of K.

If $(\mathscr{U}, \mu)$ is a local coordinate system at p, then under the map Θ_u, $\partial/\partial\mu^i|_p$ go to $\Theta_{u*} \cdot \partial/\partial\mu^i|_p$ and $d\mu^j|_p$ to $(\Theta_u^{-1})^* \cdot d\mu^j|_p$. The tensor products of the $\Theta_{u*} \cdot \partial/\partial\mu^i|_p$ and the $(\Theta_u^{-1})^* \cdot d\mu^j|_p$ form a basis of $(T_{\gamma_p(u)})_s^r$ so

$$K(\gamma_p(u)) = \bar{K}_{j_1 \cdots j_s}^{i_1 \cdots i_r}(\gamma_p(u)) \Theta_{u*} \cdot \frac{\partial}{\partial \mu^{i_1}}\bigg|_p \otimes \cdots \otimes (\Theta_u^{-1})^* \cdot d\mu^{j_s}|_p$$

and

$$\Theta_u^* \cdot K(\gamma_p(u)) = \bar{K}_{j_1 \cdots j_s}^{i_1 \cdots i_r}(\gamma_p(u)) \frac{\partial}{\partial \mu^{i_1}}\bigg|_p \otimes \cdots \otimes d\mu^{j_s}|_p \tag{13.6}$$

Equation (13.6) gives the curve $K_p^\#$ in terms of its component functions in the coordinate basis of $(T_p)_s^r$, so the representation of $D_0 K_p^\#$ in this basis is an ordered n^{r+s}-tuple of numbers formed from the $D_0 \bar{K}_{j_1 \cdots j_s}^{i_1 \cdots i_r} \circ \gamma_p$, or as an element of $(T_p)_s^r$,

$$D_0 K_p^\# = D_0 \bar{K}_{j_1 \cdots j_s}^{i_1 \cdots i_r} \circ \gamma_p \frac{\partial}{\partial \mu^{i_1}}\bigg|_p \otimes \cdots \otimes d\mu^{j_s}|_p \tag{13.7}$$

Thus,

$$(L_X K)^{i_1 \cdots i_r}_{j_1 \cdots j_s}(p) = D_0 \bar{K}^{i_1 \cdots i_r}_{j_1 \cdots j_s} \circ \gamma_p \tag{13.8}$$

Equation (13.8) gives the component description of the Lie derivative in terms of coordinate component functions of K.

Here are two examples of the calculation of the components of the Lie derivative by differentiating $\bar{K}^{i_1 \cdots i_r}_{j_1 \cdots j_s} \circ \gamma_p$ (according to eq. (13.8)). Let $M = \mathbb{R}^2$ with the standard structure given by the atlas $\{(\mathbb{R}^2, id)\}$. Let $X = \partial/\partial\mu^1 + 2\mu^1\partial/\partial\mu^2 = \partial/\partial x + 2x\partial/\partial y$. The integral curves and the flow of X are obtained by integrating the system $(dx/du) = 1$, $(dy/du) = 2x$. We get $\Theta: \mathbb{R} \times \mathbb{R}^2 \to \mathbb{R}^2$ given by $\Theta: (u, (x_0, y_0)) \mapsto (u + x_0, u^2 + 2ux_0 + y_0)$.

1. Let

$$K = Y = (\mu^1 + 1)^2 \frac{\partial}{\partial\mu^1} + \mu^2 \frac{\partial}{\partial\mu^2} = (x + 1)^2 \frac{\partial}{\partial x} + y \frac{\partial}{\partial y}$$

To use eq. (13.8) we need $\bar{Y}^1(\gamma_p(u))$ and $\bar{Y}^2(\gamma_p(u))$. These are the components of Y with respect to the basis

$$E_1 = \Theta_{u*} \cdot \frac{\partial}{\partial x}\bigg|_p, \qquad E_2 = \Theta_{u*} \cdot \frac{\partial}{\partial y}\bigg|_p$$

of $T_{\gamma_p(u)}$.

Since $\Theta_u: (x_0, y_0) \mapsto (u + x_0, u^2 + 2ux_0 + y_0)$,

$$\begin{aligned}
E_1 = \Theta_{u*} \cdot \frac{\partial}{\partial x}\bigg|_p &= \frac{\partial \Theta_u^1}{\partial x_0} \frac{\partial}{\partial x}\bigg|_{\gamma_p(u)} + \frac{\partial \Theta_u^2}{\partial x_0} \frac{\partial}{\partial y}\bigg|_{\gamma_p(u)} \\
&= 1 \frac{\partial}{\partial x}\bigg|_{\gamma_p(u)} + 2u \frac{\partial}{\partial y}\bigg|_{\gamma_p(u)} \\
E_2 = \Theta_{u*} \cdot \frac{\partial}{\partial y}\bigg|_p &= \frac{\partial \Theta_u^1}{\partial y_0} \frac{\partial}{\partial x}\bigg|_{\gamma_p(u)} + \frac{\partial \Theta_u^2}{\partial y_0} \frac{\partial}{\partial y}\bigg|_{\gamma_p(u)} \\
&= 0 \frac{\partial}{\partial x}\bigg|_{\gamma_p(u)} + 1 \frac{\partial}{\partial y}\bigg|_{\gamma_p(u)}
\end{aligned}$$

Now the components of Y are $(x + 1)^2$ and y *in the* $\partial/\partial x$, $\partial/\partial y$ *basis*, so in the E_1, E_2 basis,

$$\begin{pmatrix} \bar{Y}^1 \\ \bar{Y}^2 \end{pmatrix} = \begin{pmatrix} 1 & 0 \\ -2u & 1 \end{pmatrix} \begin{pmatrix} (x + 1)^2 \\ y \end{pmatrix}$$

and, hence, on the integral curve through (x_0, y_0),

$$\bar{Y}^1(\gamma_p(u)) = (u + x_0 + 1)^2$$
$$\bar{Y}^2(\gamma_p(u)) = -2u(u + x_0 + 1)^2 + (u^2 + 2ux_0 + y_0)$$

Finally,

$$D_0 \bar{Y}^1 \circ \gamma_p = 2(x_0 + 1)$$
$$D_0 \bar{Y}^2 \circ \gamma_p = -2(x_0^2 + x_0 + 1)$$

That is,

$$L_X Y = 2(x + 1)\frac{\partial}{\partial x} - 2(x^2 + x + 1)\frac{\partial}{\partial y}$$

We can check this by noting the fact that $L_X Y$ is the Lie bracket $[X, Y]$ (Theorem 13.11(3)) whose components are $X^i(\partial Y^j/\partial \mu^i) - Y^i(\partial X^j/\partial \mu^i)$.

2. Let

$$K = \omega = (\mu^1 + 1)\, d\mu^1 + \mu^2\, d\mu^2 = (x + 1)\, dx + y\, dy$$

To use eq. (13.8) we need $\bar{\omega}_1(\gamma_p(u))$ and $\bar{\omega}_2(\gamma_p(u))$. These are the components of ω with respect to the basis

$$\Xi^1 = (\Theta_u^{-1})^* \cdot dx|_p, \qquad \Xi^2 = (\Theta_u^{-1})^* \cdot dy|_p$$

of $T^*_{\gamma_p(u)}$.

Since Θ_u is given by

$$x = u + x_0, \qquad y = u^2 + 2ux_0 + y_0$$

Θ_u^{-1} is given by

$$x_0 = x - u, \qquad y_0 = u^2 - 2ux + y$$

(note that $\Theta_u^{-1} = \Theta_{-u}$) and

$$\Xi^1 = (\Theta_u^{-1})^* \cdot dx|_p = \frac{\partial(\Theta_u^{-1})^1}{\partial x} dx|_{\gamma_p(u)} + \frac{\partial(\Theta_u^{-1})^1}{\partial y} dy|_{\gamma_p(u)}$$
$$= 1\, dx|_{\gamma_p(u)} + 0\, dy|_{\gamma_p(u)}$$
$$\Xi^2 = (\Theta_u^{-1})^* \cdot dy|_p = \frac{\partial(\Theta_u^{-1})^2}{\partial x} dx|_{\gamma_p(u)} + \frac{\partial(\Theta_u^{-1})^2}{\partial y} dy|_{\gamma_p(u)}$$
$$= -2u\, dx|_{\gamma_p(u)} + 1\, dy|_{\gamma_p(u)}$$

Now, the components of ω are $x + 1$ and y *in the* dx, dy *basis*, so in the Ξ^1, Ξ^2 basis,

$$\begin{pmatrix} \bar{\omega}_1 \\ \bar{\omega}_2 \end{pmatrix} = \begin{pmatrix} 1 & 2u \\ 0 & 1 \end{pmatrix} \begin{pmatrix} x + 1 \\ y \end{pmatrix}$$

and, hence, on the integral curve through (x_0, y_0),

$$\bar{\omega}_1(\gamma_p(u)) = u + x_0 + 1 + 2u(u^2 + 2ux_0 + y_0)$$

$$\bar{\omega}_2(\gamma_p(u)) = u^2 + 2ux_0 + y_0$$

Finally,

$$D_0\bar{\omega}_1 \circ \gamma_p = 1 + 2y_0, \qquad D_0\bar{\omega}_2 \circ \gamma_p = 2x_0$$

That is, $L_X\omega = (1 + 2y)\,dx + 2x\,dy$. We can check this by deriving the components

$$X\omega_1 + \omega_1 \frac{\partial X^1}{\partial x} + \omega_2 \frac{\partial X^2}{\partial x}, \qquad X\omega_2 + \omega_1 \frac{\partial X^1}{\partial y} + \omega_2 \frac{\partial X^2}{\partial y}$$

of $L_X\omega$ from Theorem 13.11(2).

We saw in Theorem 13.8 that (where $X \neq 0$) we can always choose local coordinates so that X is a coordinate vector field. With such a choice eq. (13.8) simplifies.

Theorem 13.10. *In any coordinate system in which* $X = \partial/\partial\mu^1$ *the component functions of* $L_X K$ *are*

$$(L_X K)^{i_1 \cdots i_r}_{j_1 \cdots j_s} = \frac{\partial K^{i_1 \cdots i_r}_{j_1 \cdots j_s}}{\partial \mu^1}$$

Proof. If $X = \partial/\partial\mu^1$, then the flow is just a translation, so $\bar{K}^{i_1 \cdots i_r}_{j_1 \cdots j_s} = K^{i_1 \cdots i_r}_{j_1 \cdots j_s}$. γ_p is the coordinate curve, α_1, through p with tangent $\partial/\partial\mu^1$, and in Section 9.3 we had, by definition of $(\partial f/\partial\mu^i)|_p$, for any function, f, $D_0 f \circ \alpha_i = (\partial f/\partial\mu^i)|_p$. So

$$D_0\bar{K}^{i_1 \cdots i_r}_{j_1 \cdots j_s} \circ \gamma_p = D_0 K^{i_1 \cdots i_r}_{j_1 \cdots j_s} \circ \gamma_p = \left.\frac{\partial K^{i_1 \cdots i_r}_{j_1 \cdots j_s}}{\partial \mu^1}\right|_p. \qquad \blacksquare$$

Using the components in Theorem 13.10 it is easy to confirm three basic properties of L_X, namely, (1) L_X is a derivation on the algebra of tensor fields on M, (2) $L_X K$ has the same symmetry and skew-symmetry properties as K does, and (3) L_X commutes with contractions.

Now we note explicitly coordinate-free descriptions of $L_X K$ in special cases.

Theorem 13.11.

(1) *For a function, f, $L_X f = Xf$.*
(2) *For an exact 1-form, df, $L_X\, df = dXf$.*
(3) *For a vector field Y, $L_X Y = [X, Y]$.*

Proof. (1) is immediate fromTheorem 13.10. (Note that in this case our notation agrees with that already used in Section 11.1.) For (2) observe that in the coordinate system of Theorem 13.10 the components on the left side are $\partial/\partial\mu^i\,(\partial f/\partial\mu^1)$. For (3) in the special coordinates we have

$$L_X Yf = (L_X Y)^i \frac{\partial f}{\partial \mu^i} = \frac{\partial Y^i}{\partial \mu^1}\frac{\partial f}{\partial \mu^i}$$

and

$$XYf - YXf = \frac{\partial}{\partial \mu^1}\left(Y^i \frac{\partial f}{\partial \mu^i}\right) - Y^i \frac{\partial Xf}{\partial \mu^i}$$

$$= \frac{\partial Y^i}{\partial \mu^1}\frac{\partial f}{\partial \mu^i} \qquad \blacksquare$$

From (1) and (2) we have $L_X\, df = dL_X f$, i.e., in this case L_X and d commute. We will soon see that this generalizes.

Theorem 13.12. *We have the following formula for the values of $L_X K$:*

$$\begin{aligned}(L_X K)(\omega^1, \ldots, \omega^r, X_1, \ldots, X_s) &= L_X(K(\omega^1, \ldots, \omega^r, X_1, \ldots, X_s)) \\ &\quad - \sum_{i=1}^{r} K(\omega^1, \ldots, L_X\omega^i, \ldots, \omega^r, X_1, \ldots, X_s) \\ &\quad - \sum_{i=1}^{s} K(\omega^1, \ldots, \omega^r, X_1, \ldots, L_X X_i, \ldots, X_s)\end{aligned}$$

Proof. Problem 13.14.

Theorem 13.13. *In a local coordinate system $(\mathscr{U}, \mu)$ the component functions of $L_X K$ are*

$$\begin{aligned}(L_X K)^{i_1\cdots i_r}_{j_1\cdots j_s} &= XK^{i_1\cdots i_r}_{j_1\cdots j_s} - \sum_{\alpha} K^{i_1\cdots i_{\alpha-1} h i_{\alpha+1}\cdots i_r}_{j_1\cdots j_s} \frac{\partial X^{i_\alpha}}{\partial \mu^h} \\ &\quad + \sum_{\alpha} K^{i_1\cdots i_r}_{j_1\cdots j_{\alpha-1} h j_{\alpha+1}\cdots j_s} \frac{\partial X^h}{\partial \mu^{j_\alpha}}\end{aligned}$$

Proof. In the formula in Theorem 13.12, replace ω^i by $d\mu^i$ and X_i by $\partial/\partial\mu^i$. $\blacksquare$

Using the components in Theorem 13.13 we see immediately that the Lie derivative is $\mathbb{R}$-linear in its subscript argument, i.e., $L_{aX+bY}K = aL_X K + bL_Y K$.

Recall that in Section 11.4 we described the skew-derivation, i_X, interior multiplication by X, on the algebra of differential forms, and in Section 12.1 we introduced the operation, d, of exterior differentiation, also a skew-derivation on differential forms. We have the following important relation among operations on the algebra of differential forms.

Theorem 13.14. *On differential forms,*

$$L_X = i_X \circ d + d \circ i_X \tag{13.9}$$

Proof. We will (1) show $i_X \circ d + d \circ i_X$ is a derivation; (2) confirm the formula for functions; (3) confirm the formula for 1-form fields.

$$\begin{aligned}
(1)\ (i_X \circ d + d \circ i_X)\sigma \wedge \tau &= i_X(d\sigma \wedge \tau + (-1)^p \sigma \wedge d\tau) \\
&\quad + d(i_X\sigma \wedge \tau + (-1)^p \sigma \wedge i_X\tau) \\
&= i_X\, d\sigma \wedge \tau + (-1)^{p+1}\, d\sigma \wedge i_X\tau + (-1)^p i_X\sigma \wedge d\tau \\
&\qquad + (-1)^{2p}\sigma \wedge i_X\, d\tau + di_X\sigma \wedge \tau + (-1)^{p-1} i_X\sigma \wedge d\tau \\
&\qquad + (-1)^p\, d\sigma \wedge di_X\tau + (-1)^{2p}\sigma \wedge di_X\tau \\
&= (i_X\, d\sigma + di_X\sigma) \wedge \tau + \sigma \wedge (i_X\, d\tau + di_X\tau)
\end{aligned}$$

(2) $L_X f = Xf \qquad$ and $\qquad (i_X \circ d + d \circ i_X) f = i_X\, df + di_X f = Xf$

(3) $L_X\, df = dXf \qquad$ and $\qquad (i_X \circ d + d \circ i_X)\, df = 0 + di_X\, df = dXf$ ∎

Corollary. *On differential forms* $d \circ L_X = L_X \circ d$.

The following formula giving $D_u K_p^\#$ in terms of the Lie derivative will be useful.

Theorem 13.15. $D_u K_p^\# = \Theta_u^* \cdot L_X K(\Theta_u(p))$.

Proof. From

$$K_p^\#\colon u \mapsto \Theta_u^* \cdot K(\gamma_p(u)) \qquad \text{and} \qquad K_{\Theta_u(p)}^\#\colon v \mapsto \Theta_v^* \cdot K(\gamma_{\Theta_u(p)}(v))$$

we have

$$K_p^\#\colon u + v \mapsto \Theta_u^* \cdot \Theta_v^* \cdot K(\gamma_p(u+v)) = \Theta_u^* \cdot K_{\Theta_u(p)}^\#(v)$$

So

$$D_u K_p^\# = \Theta_u^* \cdot D_0 K_{\Theta_u(p)}^\# = \Theta_u^* \cdot L_X K(\Theta_u(p))$$

from the definition of the Lie derivative. ∎

PROBLEM 13.11. Using Theorem 13.11(1) we can write eq. (11.13) as $L_{[X,Y]}f = L_X L_Y f - L_Y L_X f = [L_X, L_Y]f$. Generalize by showing that this is valid if f is replaced by any tensor field K. (That is, $L_{[X,Y]}K$ is a measure of the noncommutativity of the Lie derivative of K.)

PROBLEM 13.12. Prove that for vector fields, X and Y, and a 1-form, ω,

$$L_X\langle\omega, Y\rangle = \langle\omega, L_X Y\rangle + \langle L_X\omega, Y\rangle$$

PROBLEM 13.13. (i) Write the formula of Theorem 13.12 for tensor fields of type (1, 0), (0, 1), (1, 1) and (0, 2). (ii) Write your formulas in (i) in terms of contractions.

PROBLEM 13.14. Prove Theorem 13.12. (Hint: Problem 13.13 shows that Theorem 13.12 says that Lie differentiation commutes with successive contractions.)

PROBLEM 13.15. Prove that

$$i_{[X,Y]} = L_X \circ i_Y - i_Y \circ L_X$$

on differential forms and, hence, in particular, $L_X \circ i_X = i_X \circ L_X$.

14

INTEGRABILITY CONDITIONS FOR DISTRIBUTIONS AND FOR PFAFFIAN SYSTEMS

In Section 13.1 we met the equation $\dot{\gamma} = X \circ \gamma$ and in Section 12.1 we met the equation $\theta = d\omega$. We saw that the former is an intrinsic (i.e., coordinate-free) description of a system of ordinary differential equations, and the latter is an intrinsic description of various systems of partial differential equations, depending on the degree of θ. We saw that differentiability conditions on X suffice to give local solutions of $\dot{\gamma} = X \circ \gamma$, but for $\theta = d\omega$ to have local solutions the additional condition $d\theta = 0$ is necessary. We will now generalize the situation $\theta = d\omega$ for θ a 1-form to a system of equations given by a system of 1-forms.

14.1 Completely integrable distributions

A system of k linear homogeneous partial differential equations on an n-dimensional manifold, M, $k < n$, can be written in the form $X_i f = 0$, $i = 1, \ldots, k$, where the X_i are local vector fields on M. We saw in Section 13.3 that $[X_1, X_2] = 0$ if and only if there is a coordinate system $(\mathscr{U}, \mu)$ in which $X_1 = \partial/\partial\mu^1$ and $X_2 = \partial/\partial\mu^2$. Thus, if $[X_1, X_2] = 0$, then $X_1\mu^3 = 0, \ldots, X_1\mu^n = 0$ and $X_2\mu^3 = 0, \ldots, X_2\mu^n = 0$. That is $\mu^3, \ldots, \mu^n$ are solutions of the system $X_1 f = 0$, $X_2 f = 0$. Roughly, $\mu^3, \ldots, \mu^n$ are functions on $\mathscr{U}$ which are constant when we move in the X_1 and X_2 directions. More precisely, if $[X_1, X_2] = 0$, then at each point $p \in \mathscr{U}$, $X_1 = \partial/\partial\mu^1$ is tangent to the coordinate curve $\bigcap\{q: \mu^i(q) = \mu^i(p)\}$, $i = 2, \ldots, n$, and $X_2 = \partial/\partial\mu^2$ is tangent to the coordinate curve $\bigcap\{q: \mu^i(q) = \mu^i(p)\}$, $i = 1, 3, 4, \ldots, n$, so X_1 and X_2 are both in the 2-dimensional coordinate slice consisting of the intersection of the hypersurfaces $\mu^i(q) = \mu^i(p)$, $i = 3, \ldots, n$. That is, we have a 2-dimensional submanifold on which μ_1 and μ_2 are coordinates, X_1 and X_2 are tangent vectors, and $\mu^3, \ldots, \mu^n$ are constant.

For example, in $\mathbb{R}^4$, given X_1 and X_2, with $[X_1, X_2] = 0$ at each point in $\mathscr{U}$ the hypersurfaces $\mu^3 = \text{const.}$ and $\mu^4 = \text{const.}$ Intersect in a 2-dimensional manifold whose tangent space contains X_1 and X_2.

We can generalize the discussion above by means of the observations following Theorem 13.9.

Theorem 14.1. *If* $[X_i, X_j] = 0$ $i, j = 1, \ldots, k$ *on* $\mathscr{U}$, *then (1) there are*

k-dimensional submanifolds whose tangent spaces at each point contain $X_1, \ldots, X_k$, and, moreover, (2) there are functions $\mu^1, \ldots, \mu^n$ such that $\mu^1, \ldots, \mu^k$ are coordinates on the submanifolds, $\mu^{k+1}, \ldots, \mu^n$ are constant on the submanifolds, and $X_i \mu^j = 0$, $i = 1, \ldots, k$, $j = k + 1, \ldots, n$.

Proof. Problem 14.1.

The example $[y\partial/\partial x, \partial/\partial y] = -\partial/\partial x$ in $\mathbb{R}^3$ shows that the hypothesis of Theorem 14.1 is not necessary for its conclusion. We will find a weaker condition which is both necessary and sufficient. We first introduce some standard terminology which involves the relationship between systems of vector fields and sections of bundles of Grassmann manifolds.

Definitions

Let $G(k. T_p)$ be the Grassmann manifold of k-planes of T_p (Section 9.1). A mapping $\mathscr{D}_k: M \to \bigcup_{p \in M} G(k, T_p)$ such that $\mathscr{D}_k(p) \in G(k, T_p)$ is called *a k-dimensional distribution on M*. A set of local vector fields $X_1, \ldots, X_k$, on $\mathscr{U} \subset M$ such that for all $p \in \mathscr{U}$, $X_1(p), \ldots, X_k(p)$ is a basis of $\mathscr{D}_k(p)$ is *a local basis of $\mathscr{D}_k$*. Given a distribution, $\mathscr{D}_k$, a submanifold, N, of M is called *an integral submanifold of $\mathscr{D}_k$* if $T_p(N) \subset \mathscr{D}_k(p)$ for all p. A given distribution, $\mathscr{D}_k$, is called *completely integrable* if it has a k-dimensional integral submanifold at each point.

In these terms, Theorem 14.1 gives sufficient conditions for a distribution to be completely integrable.

Theorem 14.2. *If the system $X_i f = 0$, $i = 1, \ldots, k$ has $n - k$ independent solutions, then the distribution, $\mathscr{D}_k$, of $\{X_i\}$ is completely integrable.*

Proof. Let $\mu_{k+1}, \ldots, \mu_n$ be solutions of the system. Let $\mu_1, \ldots, \mu_k$ be functions such that $\{\mu_i\}$ is a coordinate system. Let N be the k-dimensional coordinate slice $\bigcap \{q: \mu^j(q) = \mu^j(p)\}$, $j = k + 1, \ldots, n$. Then $X(p) \in T_p(N)$ implies that $X(p)$ is in the tangent space of each of the hypersurfaces $\mu^j(q) = \mu^j(p)$, $j = k + 1, \ldots, n$. Now $\langle d\mu_j(p), X(p) \rangle = 0$ since, if i is the inclusion map $i: N \to M$, then $i^* d\mu_j = di^* \mu_j = 0$ which says that $d\mu_j$ vanishes on the tangent space of N. Since $\langle d\mu_j, X_i \rangle = 0$ for $i = 1, \ldots, k$, $j = k + 1, \ldots, n$, $X(p) \in \mathscr{D}_k(p)$. ∎

Definition

A distribution, $\mathscr{D}_k$, *is involutive* if for all X and Y in $\mathscr{D}_k$, $[X, Y]$ is in $\mathscr{D}_k$; i.e., $[X, Y] = f^i X_i$ where f^i are functions. (See Problem 14.3.)

Theorem 14.3. *If $\mathscr{D}_k$ is completely integrable, then $\mathscr{D}_k$ is involutive.*

Proof. Let N be an integral submanifold of $\mathscr{D}_k$. Then at each point

$p \in N$,

$$X(p), Y(p) \in \mathscr{D}_k(p) \Rightarrow X(p), Y(p) \in T_p(N) \Rightarrow [X(p), Y(p)] \in T_p(N)$$
$$\Rightarrow [X, Y](p) \in \mathscr{D}_k(p). \qquad \blacksquare$$

The converse of Theorem 14.3 follows from the following basic lemma.

Lemma. *If $\mathscr{D}_k$ is involutive, then at each point, $p_0 \in M$, there is a coordinate system $(\mathscr{U}, \mu)$ such that $\partial/\partial\mu^1, \ldots, \partial/\partial\mu^k$ span $\mathscr{D}_k$ at each point of $\mathscr{U}$.*

Proof. Proof by induction on k. For $k = 1$ the theorem is true by Theorem 13.8. Now suppose $X_1, \ldots, X_k$ is a local basis of $\mathscr{D}_k$. (The existence of such a basis is included in a careful statement of the hypothesis.). Let $(\mathscr{V}, v)$ be a coordinate system at p_0 with $v(p_0) = 0$, and $X_1 = \partial/\partial v^1$. Define $k - 1$ vector fields by $Y_i = X_i - (X_i v^1)X_1$, $i = 2, \ldots, k$. These form a $k - 1$-dimensional involutive distribution, $\mathscr{D}_{k-1}$, on $\mathscr{V}$. In particular, they form a $k - 1$-dimensional involutive distribution on the coordinate slice $v^1(p) = 0$ (see Fig. 14.1). Now the induction hypothesis gives us a coordinate system $\xi^2, \ldots, \xi^n$ on $v^1(p) = 0$ such that $\partial/\partial\xi^2, \ldots, \partial/\partial\xi^k$ span $\mathscr{D}_{k-1}$ on a neighborhood, $\mathscr{W}$, of p_0 on $v^1(p) = 0$.

Let π be the projection $\pi\colon \mathscr{V} \to v^1(p) = 0$ given by

$$q \mapsto v(q) \mapsto (0, v^2(q), \ldots, v^n(q)) \mapsto v^{-1}(0, v^2(q), \ldots, v^n(q)$$

Then on $\pi^{-1}(\mathscr{W})$ we have the coordinate functions

$$\mu^1 = v^1, \mu^2 = \xi^2 \circ \pi, \ldots, \mu^n = \xi^n \circ \pi.$$

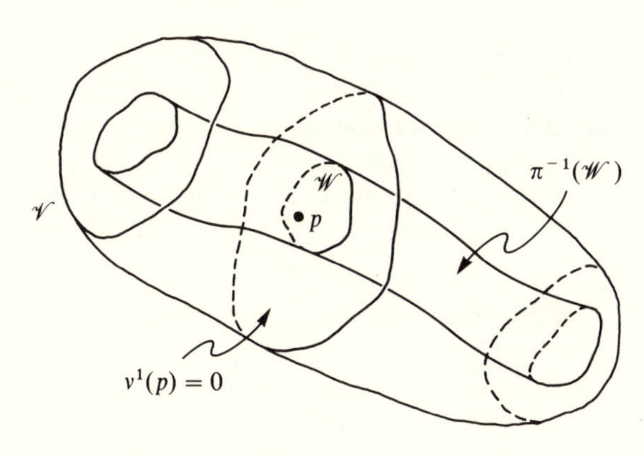

FIGURE 14.1

To show that $\partial/\partial\mu^1, \ldots, \partial/\partial\mu^k$ span $\mathscr{D}_k$ we will show that they span the same subspaces as $X_1, Y_2, \ldots, Y_k$. That is, putting $X_1 = Y_1$ we will show $Y_i\mu^j = 0$ for $i = 1, \ldots, k$, and $j = k + 1, \ldots, n$.

Now, $Y_1, \ldots, Y_k$ is the basis of an involutive distribution, so $[Y_i, Y_j] = f^l_{ij} Y_l$. Hence,

$$Y_1(Y_i\mu^j) = [Y_1, Y_i]\mu^j = f^l_{1i} Y_l \mu^j, \qquad i = 2, \ldots, k, \qquad j = k + 1, \ldots, n$$

That is, $Y_i\mu^j$ satisfy a system of linear homogeneous ordinary differential equations along μ^1 curves. On $v^1(p) = 0$, $\mu^i = \xi^i$ for $i \geq 2$ and $Y_i\xi^j = 0$ for $j > k$. So $Y_i\mu^j = 0$ for $i \geq 2$ and $j > k$ on $v^1(p) = 0$. Thus, by the uniqueness theorem for such systems of ordinary differential equations, $Y_i\mu^j \equiv 0$ for $i = 2, \ldots, k$ and $j > k$. Also $Y_1\mu^j = 0$ for $j > 1$, so $Y_i\mu^j = 0$ for $i = 1, \ldots, k$ and $j > k$. ■

Theorem 14.4. *If $\mathscr{D}_k$ is involutive, then (i) $\mathscr{D}_k$ is completely integrable, and (ii) the system $X_i f = 0$, $i = 1, \ldots, k$ has $n - k$ solutions.*

Proof. (i) On the coordinate slices

$$\{q \in \mathscr{U} : \mu^{k+1}(q) = c^{k+1}, \ldots, \mu^n(q) = c^n\}$$

of the coordinate system of the lemma, the tangent spaces are precisely those spanned by $\partial/\partial\mu^1, \ldots, \partial/\partial\mu^k$, so they are the k-planes, $\mathscr{D}_k(p)$, at each point.

(ii) Since $X_i = f^j_i\, \partial/\partial\mu^j$, $\mu^{k+1}, \ldots, \mu^n$ are solutions. ■

Finally, combining Theorem 14.3 and Theorem 14.4 on the one hand and from Theorem 14.2 on the other hand we have the following.

Theorem 14.5. *$\mathscr{D}_k$ is completely integrable iff $X_i f = 0$, $i = 1, \ldots, k$ has $n - k$ independent solutions.*

At this point we should put the results above on distributions on manifolds into perspective. These abstract results have deep and extensive historical roots in important problems in differential geometry (cf. Sections 17.4 and 24.2) and continuum mechanics, where we are confronted with the problem of solving systems of nonlinear partial differential equations in $\mathbb{R}^n$ of the form

$$\frac{\partial y^j}{\partial x^i} = \psi^j_i(x^1, \ldots, x^k, y^{k+1}, \ldots, y^n), \qquad i = 1, \ldots, k, \qquad j = k + 1, \ldots, n \tag{14.1}$$

A system of nonlinear partial differential equations of this special form has a couple of nice properties. (1) In matrix form the left side of (14.1) is the derivative of a map from $\mathbb{R}^k$ to $\mathbb{R}^{n-k}$ (Section 7.1), so (14.1) can be thought of as a generalization of an ordinary differential equation, and is sometimes called *a total differential equation* (Dieudonné, 1969, Vol. I, pp. 307 ff). (2) With eq. (14.1) we have *the associated linear system*

$$\frac{\partial f}{\partial x^i} + \psi_i^j \frac{\partial f}{\partial y^j} = 0, \qquad i = 1, \dots, k, \qquad j = k+1, \dots, n \tag{14.2}$$

Clearly, this is a special case of the system with which we started this section. By the implicit function theorem, a set of $n - k$ solutions, f^j, of this system, with $f^j(a_0^1, \dots, a_0^k, b_0^{k+1}, \dots, b_0^n) = 0$ corresponds to a solution $(y^{k+1}, \dots, y^n)$ of (14.1) with $y^j(a_0^1, \dots, a_0^k) = b_0^j$ and

$$f^j(a^1, \dots, a^k, y^{k+1}(a^1, \dots, a^k), \dots, y^n(a^1, \dots, a^k)) = 0$$

for $(a^1, \dots, a^k)$ in a neighborhood of $(a_0^1, \dots, a_0^k)$. A solution of (14.2) is called *an integral of* (14.1).

Theorem 14.6. *The system* (14.1) *has a solution through* (a, b) *if and only if*

$$\frac{\partial \psi_i^j}{\partial x^l} + \psi_l^h \frac{\partial \psi_i^j}{\partial u^h} = \frac{\partial \psi_l^j}{\partial x^i} + \psi_i^h \frac{\partial \psi_l^j}{\partial y^h}, \qquad i, l = 1, \dots, k, \qquad j = k+1, \dots, n \tag{14.3}$$

Proof. Problem 14.5.

PROBLEM 14.1. Prove Theorem 14.1.

PROBLEM 14.2. The vector fields

$$X_1 = \frac{\partial}{\partial \mu^1}, \qquad X_2 = \frac{\partial}{\partial \mu^2} + c\frac{\partial}{\partial \mu^3} \qquad \text{and} \qquad X_3 = -\frac{1}{(\mu^1)^2}\frac{\partial}{\partial \mu^2} + \frac{1}{c}\frac{\partial}{\partial \mu^3}$$

where c is a constant, determine three one-dimensional, three two-dimensional, and one three-dimensional distributions in the chart $(\mathscr{U}, \mu)$ on a three-dimensional manifold. Which are completely integrable and which are not? If $\mu^1 = r$, $\mu^2 = \vartheta$, and $\mu^3 = z$ are cylindrical coordinates on $\mathbb{R}^3$, interpret your results geometrically.

PROBLEM 14.3. Suppose $X_1, \dots, X_k$ is a local basis of $\mathscr{D}_k$ and $[X_i, X_j]$ is in $\mathscr{D}_k$ for $i, j = 1, \dots, k$. If $Y_i = f_i^k X_k$ where f_i^k are functions on $\mathscr{U} \subset M$ and $\det(f_i^k) \neq 0$, then $[Y_i, Y_j]$ is in $\mathscr{D}_k$.

PROBLEM 14.4. If a system of k first-order linear partial differential equations, $X_i f = 0$, is solved for k of the partial derivatives we can write it in the form (14.2). A system of the form (14.2) is called *a Jacobi system*. Prove that for a Jacobi system the integrability conditions reduce to $[X_i, X_j] = 0$.

PROBLEM 14.5. Prove Theorem 14.6.

14.2 Completely integrable Pfaffian systems

We have referred to the fact that the problem of solving certain systems of partial differential equations is a venerable one appearing in many branches of mathematics. It has been treated by many mathematicians over many years and has taken many different forms. In particular, our results above can be stated in other equivalent ways. In recent years the name of Frobenius has been assigned to each of our several equivalent theorems by one or more authors.

Definitions
A set of 1-forms $\{\sigma^i\}$ is called *a Pfaffian system* and equations $\sigma^i = 0$ are called *Pfaffian equations*. A Pfaffian system of $n - k$ linearly independent 1-forms defines *a codistribution* $\mathscr{D}^*_{n-k}: M \to \bigcup_{p \in M} G(k, T^*_p)$. *An integral submanifold of a Pfaffian system*, or of its *codistribution* $\mathscr{D}^*_{n-k}$ is a submanifold, N, such that at each point $\sigma^i(X) = 0$ for all i and for all X such that $X(p) \in T_p(N)$. A Pfaffian system or its codistribution, $\mathscr{D}^*_{n-k}$ is *completely integrable* if it has a k-dimensional integral submanifold at every point.

Note that with each codistribution $\mathscr{D}^*_{n-k}$ we have the associated distribution $\mathscr{D}_k$, given by $\langle \sigma^i, X_j \rangle = 0$. That is, at each point the corresponding subspaces of T_p and T^*_p are the orthogonal complements of one another (or annihilate one another) as described in Section 2.4.

We have the following characterizations of properties defined above.

Theorem 14.7. (1) *N is an integral submanifold of $\mathscr{D}^*_{n-k}$ iff N is an integral submanifold of its associated distribution.*

(2) (i) *A codistribution is completely integrable iff its associated distribution is completely integrable.* (ii) *A codistribution is completely integrable iff there exist $n - k$ functions f^j such that df^j span $\mathscr{D}^*_{n-k}$ at each point.*

Proof. (1) **If:** $X(p) \in T_p(N) \Rightarrow X_p \in \mathscr{D}_k(p) \Rightarrow \sigma^i(p)(X(p)) = 0$.
Only if: $X(p) \in T_p(N) \Rightarrow \sigma^i(p)(X(p)) \Rightarrow X_p \in \mathscr{D}_k(p)$.

(2) (i) Immediate from (1). (ii) Let df^j span the associated distribution, $\mathscr{D}_k$. Then $\langle df^j, X_i \rangle = 0 \Leftrightarrow X_i f^j = 0 \Leftrightarrow \mathscr{D}_k$ is completely integrable (Theorem 14.5) $\Leftrightarrow \mathscr{D}^*_{n-k}$ is completely integrable (by part (2)(i).) ∎

We should note a couple of interesting things about the condition that the df^j span $\mathscr{D}^*_{n-k}$ in Theorem 14.7 (2)(ii). First of all we see that it is the dual of the property $\partial/\partial\mu^i$ span $\mathscr{D}_k$ in the basic lemma above, and secondly we can think of it as a generalization of the property of exactness for a single 1-form in Section 12.1.

Finally, we have a necessary and sufficient condition for a codistribution to be completely integrable in terms of the codistribution itself. This corresponds to the condition that $\mathscr{D}_k$ is involutive for a distribution.

Lemma. *Let the vector fields $X_1, \ldots, X_n$ and the differential forms $\sigma^1, \ldots, \sigma^n$ be local bases of TM and T^*M, respectively, such that $\langle X_i, \sigma^j \rangle = \delta^j_i$. Let $[X_i, X_j] = f^l_{ij}X_l$ and $d\sigma^i = g^i_{jl}\sigma^j \wedge \sigma^l$. Then $g^i_{jl} = -\frac{1}{2}f^i_{jl}$.*

Proof. Problem 14.6.

Theorem 14.8. *The system $\sigma^{k+1}, \ldots, \sigma^n$ of $n-k$ 1-forms is completely integrable if and only if $d\sigma^i \wedge \sigma^{k+1} \wedge \cdots \wedge \sigma^n = 0$ for $i = k+1, \ldots, n$.*

Proof. From the lemma, if $f^l_{ij} = 0$ for $i, j = 1, \ldots, k$ and $l = k+1, \ldots, n$, then, for $i > k$, $d\sigma^i = g^i_{jl}\sigma^j \wedge \sigma^l$ with $g^i_{jl} = 0$ for $j, l \leq k$. ∎

From the proof of Theorem 14.8 we see that we can give the condition for complete integrability in another form.

Theorem 14.9. *The system $\sigma^{k+1}, \ldots, \sigma^n$ of $n-k$ 1-forms is completely integrable if and only if $d\sigma^j$, $j = k+1, \ldots, n$, is in the ideal generated by $\{\sigma^j\}$.*

Proof. Problem 14.8.

PROBLEM 14.6. Prove the lemma above.

PROBLEM 14.7. For a single 1-form σ, the integrability condition is $d\sigma \wedge \sigma = 0$. Compare this with the integrability condition $d\sigma = 0$ in Section 12.1. Write these conditions in terms of the natural coordinates on $\mathbb{R}^3$, and then interpret them geometrically.

PROBLEM 14.8. Prove Theorem 14.9.

PROBLEM 14.9. What are the integrability conditions for the system of 1-forms $dy^j - \psi^j_i\, dx^i, j = k+1, \ldots, n$. How do they relate to the integrability conditions for a Jacobi system (cf., Problem 14.4).

14.3 The characteristic distribution of a differential system

Whether or not a given distribution is completely integrable, i.e., has k-dimensional integral submanifolds, it may have lower-dimensional integral submanifolds. We have seen that the complete integrability of a distribution corresponds to the existence of solutions of systems of linear homogeneous partial differential equations. Now the problem of finding solutions of a more general system of partial differential equations can be formulated as a problem of finding lower-dimensional integral submanifolds of a distribution.

Thus, the problem of solving a given partial differential equation of the general form $f(x^1, \ldots, x^n, \partial z/\partial x^1, \ldots, \partial z/\partial x^n, z) = 0$ means we are given a function, f, on a region in $\mathbb{R}^{2n+1}$ and we want to find an n-dimensional submanifold, (P, ϕ), of $\mathbb{R}^{2n+1}$ with ϕ given locally by

$$(x^1, \ldots, x^n) \mapsto (x^1, \ldots, x^n, p_1(x^1, \ldots, x^n), \ldots, p_n(x^1, \ldots, x^n), z(x^1, \ldots, x^n))$$

such that on P

$$f(x^1, \ldots, x^n, p_1, \ldots, p_n, z) = 0$$

and

$$p_i = \frac{\partial z}{\partial x^i}$$

These conditions on P are respectively equivalent to the conditions that at each $p \in P$, and all $X_p \in T_pP$ for all p, the 1-forms

$$\sigma^1 = \frac{\partial f}{\partial x^1} dx^1 + \cdots + \frac{\partial f}{\partial x^n} dx^n + \frac{\partial f}{\partial p_1} dp_1 + \cdots + \frac{\partial f}{\partial p_n} dp_n + \frac{\partial f}{\partial z} dz$$

$$\sigma^2 = p_1\, dx^1 + \cdots + p_n\, dx_n - dz$$

vanish. In particular, we are looking for an n-dimensional integral submanifold of a $2n-1$-dimensional distribution on $\mathbb{R}^{2n+1}$ not a $2n-1$-dimensional integral submanifold.

The way we proceed is by noting that the pair of 1-forms, σ^1, σ^2, determine a certain 1-dimensional distribution, the characteristic distribution of σ^1, σ^2 on P, and then with initial conditions we construct P.

We will first briefly discuss the general concept of the characteristic distribution of a set of forms. This class of distributions has interesting properties; in particular such distributions are completely integrable. We will then go back and obtain the explicit form of the characteristics for $\{\sigma^1, \sigma^2\}$.

Definitions
If ω is an s-form, the space of vectors, X_p, satisfying

$$i_{X_p}\omega(p) = 0 \tag{14.4}$$

is *the associated space,* $A(\omega(p))$, *of* $\omega(p)$. The orthogonal complement, $A^*(\omega(p))$, of $A(\omega(p))$ in T_p^* is called *the enveloping space of* $\omega(p)$.

The condition (14.4) on X_p means that at $p \in M$ $i_X\omega(X_1, \ldots, X_{s-1}) = 0$, or $\omega(X, X_1, \ldots, X_{s-1}) = 0$ for all $X_i \in T_p$. By eq. (11.28) and the definition of $\rfloor$ in eq. (6.10), this condition is equivalent to

$$\langle F \rfloor \omega, X \rangle = 0 \tag{14.5}$$

for all decomposable $(s-1)$-vectors in $\Lambda^{s-1}(T_p)$. The terminology "enveloping space of $\omega(p)$" comes from the fact (usually the definition) that this subspace of T_p^* is the smallest one such that $\omega \in \Lambda^s(A^*(\omega))$. This characterization of $A^*(\omega)$ comes from (14.5) using a basis of T_p^* containing one of $A^*(\omega)$.

Definitions
If ω is an s-form, then a vector field which satisfies

$$i_X\omega = 0 \qquad \text{and} \qquad i_X\, d\omega = 0 \tag{14.6}$$

is called *a characteristic vector field of* ω. That is, at each point, p, *the characteristic vectors* are those in the intersection of $A(\omega(p))$ and $A(d\omega(p))$. The distribution of vector spaces spanned by the characteristic vectors, $X(p)$ where X is a characteristic vector field of ω is called *the characteristic distribution of* ω.

Theorem 14.10. $i_X\omega = 0$ *and* $i_X\, d\omega = 0 \Leftrightarrow i_X\omega = 0$ *and* $L_X\omega = 0$.

Proof. Problem 14.10.

Theorem 14.11. *The characteristic distribution of* ω *is completely integrable.*

Proof. From the identities

$$i_{[X,Y]}\omega = L_X i_Y\omega - i_Y L_X\omega \qquad \text{(Problem 13.15)}$$

$$L_{[X,Y]}\omega = L_X L_Y\omega - L_Y L_X\omega \qquad \text{(Problem 13.11)}$$

and from Theorem 14.10 we see that $[X, Y]$ is in the distribution if X and Y are. ■

EXAMPLE. If ω is represented locally by $\omega = \mu^3\, d\mu^1 \wedge d\mu^2$ on a 4-dimensional manifold, then the associated space of ω is the space spanned by $\partial/\partial\mu^3$ and $\partial/\partial\mu^4$, and $\partial/\partial\mu^4$ is a characteristic vector field of ω.

The conditions (14.6) describing a characteristic vector field relate a vector field, X, and an s-form, ω. We can reverse our point of view and start with a given vector field, X, instead of starting with a given s-form, ω.

Definition

If X is a vector field, then an s-form, ω, which satisfies (14.6) is called *an absolute integral invariant of* X.

Clearly, by Theorem 14.10 an absolute integral integral invariant of X is an invariant of X. This concept, due to Poincaré and Cartan, is useful in mechanics, where we will see its significance in an important example.

We can extend our definition of characteristic vector field of a single form to the concept of a characteristic vector field of a set of forms. Again we have to make some preliminary definitions.

Definitions

A set of forms on M is called *a differential system* (recall that a set of 1-forms is called a Pfaffian system). Let $\mathscr{I}_p$ be the ideal in the algebra ΛT_p^* generated by set $\{\omega^i(p)\}$ of forms. That is, $\mathscr{I}_p = \langle\{\omega^i(p)\}\rangle = \sum \theta^i(p) \wedge \omega^i(p)$ where $\theta^i(p) \in \Lambda T_p^*$. *The associated space*, $A(\mathscr{I})$, *of* $\mathscr{I}$ is the space of vectors X_p satisfying

$$i_{X_p}\omega(p) \in \mathscr{I} \tag{14.7}$$

for all $\omega(p) \in \mathscr{I}$ (or all $\omega^i(p)$). The orthogonal complement, $A^*(\mathscr{I})$, of $A(\mathscr{I})$ in T_p^* is called *the enveloping space of* $\mathscr{I}$.

Again one can show that $A^*(\mathscr{I})$ is the smallest subspace of T_p^* such that $\mathscr{I} \subset \Lambda A^*(\mathscr{I})$ (cf., Choquet-Bruhat and DeWitt-Morette, 1982, p. 234).

EXAMPLE. Let $\omega^1 = \varepsilon^1 + \varepsilon^2$ and $\omega^2 = \varepsilon^2 \wedge \varepsilon^3 + \varepsilon^1 \wedge \varepsilon^3 + \varepsilon^1 \wedge \varepsilon^2$ where $(\varepsilon^1, \varepsilon^2, \varepsilon^3)$ is a basis of T_p^*. Then $i_X\omega^1 \in \mathscr{I} \Rightarrow i_X\omega^1 = 0$ and $i_X\omega^2 \in \mathscr{I} \Rightarrow i_X\omega^2 = a\omega^1$ where $a \in \mathbb{R}$. If (e_1, e_2, e_3) is the dual basis in T_p, then the associated space of $\mathscr{I}$ is spanned by e_3, and the enveloping space of $\mathscr{I}$ is spanned by ε^1 and ε^2. Note that even though all three basis elements appear in the differential system, the associated space of $\mathscr{I}$ is nontrivial.

Definitions

If $\{\omega^i\}$ is a differential system, let $\mathscr{I} = \langle\{\omega^i, d\omega^i\}\rangle$, the ideal generated by $\{\omega^i\}$ and $\{d\omega^i\}$. A vector field which satisfies

$$i_X\omega^i \in \mathscr{I} \quad \text{and} \quad i_X\, d\omega^i \in \mathscr{I} \tag{14.8}$$

is called *a characteristic vector field of the differential system*, $\{\omega^i\}$. The distribution of the vector spaces spanned by *the characteristic vectors*, $X(p)$, *where* X is a characteristic vector field of $\{\omega^i\}$, is called *the characteristic distribution of the differential system* $\{\omega^i\}$.

Theorem 14.12. *The characteristic distribution of a differential system is completely integrable.*

Proof. Into $L_Z\theta = di_Z\theta + i_Z\, d\theta$ put, successively, $\theta = i_Y\omega$ and $Z = X$ and then $\theta = i_X\omega$ and $Z = Y$, and then substitute into

$$i_{[X,Y]}\omega = L_X i_Y \omega + L_Y i_X \omega.$$

We get

$$i_{[X,Y]}\omega = di_X i_Y \omega + i_X\, di_Y\omega - di_Y i_X \omega - i_Y\, di_X\omega \tag{14.9}$$

Then, if X and Y satisfy (14.8), since $\omega \in \mathscr{I} \Rightarrow d\omega \in \mathscr{I}$, ($\{\omega^i, d\omega^i\}$ is a *closed* differential system), each term on the right side of (14.9) is in $\mathscr{I}$ for all ω^i and $d\omega^i$ so $[X, Y]$ is in the distribution. ■

Consider the special case where ω^i are all 1-forms; i.e., $\{\omega^i\}$ is a Pfaffian system and determines a codistribution, $\mathscr{D}^*$. $\mathscr{I} = \sum \theta^i \wedge \omega^i$ has only the identically vanishing zero form, so the conditions on the characteristic vectors are

$$i_X\omega^i = 0 \qquad \text{and} \qquad i_X\, d\omega^i \in \mathscr{D}^*$$

In the case of the 1-forms σ^1 and σ^2 of the partial differential equation $f(x^1, \ldots, x^n, \partial z/\partial x^1, \ldots, \partial z/\partial x^n, z) = 0$ with which we started this section, the conditions on the characteristic vector fields are

$$\sigma^1 \cdot X = 0, \qquad \sigma^2 \cdot X = 0$$
$$i_X\, d\sigma^1 = c_1^1\sigma^1 + c_2^1\sigma^2, \qquad i_X\, d\sigma^2 = c_1^2\sigma^1 + c_2^2\sigma^2$$

Putting $X = X^i(\partial/\partial x^i) + P^i(\partial/\partial p_i) + Z(\partial/\partial z)$, these conditions become

$$\frac{\partial f}{\partial x^i} X^i + \frac{\partial f}{\partial p_i} P^i + \frac{\partial f}{\partial z} Z = 0 \tag{14.10}$$

$$Z - p_i X^i = 0 \tag{14.11}$$

$$P^i = c_1^2 \frac{\partial f}{\partial x^i} + c_2^2 p_i \tag{14.12}$$

$$X^i = -c_1^2 \frac{\partial f}{\partial p_i} \tag{14.13}$$

$$0 = -c_1^2 \frac{\partial f}{\partial z} + c_2^2 \tag{14.14}$$

The last three equations come from equating coefficients in the fourth equation above. The third equation above is satisfied with $c_1^1 = c_2^1 = 0$, since σ^1 is closed. If we substitute, c_2^2 from (14.14) into (14.12) we get

$$P^i = c_1^2 \left(\frac{\partial f}{\partial x^i} + p_i \frac{\partial f}{\partial z} \right) \tag{14.15}$$

and substituting X^i from (14.13) into (14.11) we get

$$Z = -c_1^2 \, p_i \frac{\partial f}{\partial p_i} \tag{14.16}$$

Thus, dividing through by the common factor $-c_1^2$ in (14.13), (14.15), and (14.16), we get a characteristic vector field

$$\frac{\partial f}{\partial p_i} \frac{\partial}{\partial x^i} - \left(\frac{\partial f}{\partial x^i} + p_i \frac{\partial f}{\partial z} \right) \frac{\partial}{\partial p_i} + p_i \frac{\partial f}{\partial p_i} \frac{\partial}{\partial z}$$

The integral curves of this vector field are the solutions of the system of ordinary differential equations

$$\begin{aligned} \frac{dx^i}{du} &= \frac{\partial f}{\partial p_i} \\ \frac{dp_i}{du} &= -\left(\frac{\partial f}{\partial x^i} + p_i \frac{\partial f}{\partial z} \right) \\ \frac{dz}{du} &= p_i \frac{\partial f}{\partial p_i} \end{aligned} \tag{14.17}$$

The n-dimensional integral submanifold, (P, ϕ), is constructed from families of these curves through an initial manifold. In brief, choosing initial conditions on an $n-1$-dimensional initial manifold, we get an $n-1$-parameter family of solutions of (14.17),

$$\begin{aligned} x^i &= x^i(u, t^\alpha), \qquad \alpha = 1, \dots, n-1 \\ p_i &= p_i(u, t^\alpha) \\ z &= z(u, t^\alpha) \end{aligned} \tag{14.18}$$

If we eliminate u and t^α between the first and third equations, the resulting function, z, of the x^i will be a solution of the partial differential equation. The set (14.18) describes a family of "strips" on the submanifold. For details, see Courant and Hilbert (1966, Vol. II, Chap. II).

PROBLEM 14.10. Prove Theorem 14.10.

PROBLEM 14.11. Apply the integrability conditions of Theorem 14.8 to the system $\{\omega^1\omega^2\}$ at the beginning of this section for $n = 1$ and $n = 2$. Draw the appropriate conclusions.

15

PSEUDO-RIEMANNIAN GEOMETRY

In order to have "geometrical" concepts in a manifold, M, such as lengths of curves, surface area, or curvature of curves or surfaces, we have to impose an additional "structure" on M. Here we assign a metric tensor, g, to M in terms of which we define the length of a curve and the distance between two points. We then examine the important special case when M has an atlas on each coordinate neighborhood of which the components of g are constant.

15.1 Pseudo-Riemannian manifolds

Definitions
A pseudo-Riemannian (*or semi-Riemannian*) *manifold*, (M, g), is a manifold, M, on which there is defined a nondegenerate, symmetric tensor field, g, of type $(0, 2)$; i.e., $g: M \to S^2(T^*M)$. g is called *a metric tensor field** on M. If g is positive definite (the index of g is 0) at each point of M, then (M, g) is a *Riemannian manifold*. If g has index 1, or index $\dim M - 1$ at each point of M, then (M, g) is *a Lorentzian manifold* (cf., Section 5.4).

Note that if M is connected, then since g is nondegenerate everywhere, the index of g must be the same at each point.

Clearly, it is easy to devise local tensor fields with the required properties. Thus, for example, we can define g_{ii} to be either plus or minus one for each i, and $g_{ij} = 0$ for $i \neq j$ on a coordinate patch. It is another matter whether one can construct a differentiable tensor field on all of a given manifold, M. Since, by our definition, manifolds are paracompact, and hence have partitions of unity, we have the following.

Theorem 15.1. *Every differentiable manifold can be made into a Riemannian manifold.*

Proof. Auslander and MacKenzie (1977, p. 105).

* Frequently, g is simply called a metric. Because the term "metric" occurs frequently with different meanings—we have already used it in connection with topological spaces, and see, for example, Laugwitz (1965, pp. 176 ff)—one must be aware of the context in which it appears.

The non-Riemannian case depends on the topology of M. Thus, for example, a 2-sphere cannot be made into a Lorentzian manifold, but a 3-sphere can. When we are only interested in local properties, the question of existence does not arise. This is the case for the larger part of "classical" differential geometry.

We note, first of all, certain structures, induced by g, on M. Recall that because of the symmetry, the two linear maps from T_p to T_p^* corresponding to a point of $S^2(T^*M)$ are the same. So g induces a linear mapping, $g^\flat$, from vector fields to 1-form fields; that is, $(g^\flat \cdot Y) \cdot Y = g(X, Y)$, Section 5.4.

In local coordinates, if $\{\partial/\partial\mu^i\}$ is a local coordinate basis of vector fields, then

$$g^\flat: \frac{\partial}{\partial\mu^i} \mapsto g\left(\frac{\partial}{\partial\mu^i}, -\right) = g_{ik}\, d\mu^k$$

and if $X = X^i\, \partial/\partial\mu^i$ then

$$g^\flat\left(X^i \frac{\partial}{\partial\mu^i}\right) = X^i g_{ik}\, d\mu^k$$

The components, $X^i g_{ik}$, of the 1-form field $g^\flat \cdot X$ *geometrically equivalent* to the vector field X are written X_k. In particular, for Euclidean space ($M = \mathbb{R}^n$ with the standard differential structure and $g = \delta_{ij}\, d\pi^i \otimes d\pi^j$) $g^\flat$: $\partial/\partial\pi^i \mapsto d\pi^i$ and $X_i = X^i$.

Since g is nondegenerate, $g^\flat$ has an inverse, $g^\sharp$, taking 1-form fields to vector fields. If $g^\sharp \cdot d\mu^i = g^{ij}\, \partial/\partial\mu^j$, then

$$g^{ij} g_{jk} = \delta^i_k \tag{15.1}$$

Moreover, we have the following.

Theorem 15.2. *g^{ij} are the components of a nondegenerate symmetric tensor field of type* (2, 0).

Proof. Applying the general result at the end of Section 2.3 to this case, the components of the bilinear function $\mathfrak{X}^* \times \mathfrak{X}^* \to \mathbb{R}$ corresponding to $g^\sharp$ are g^{ij}. ■

If ω_j are the components of a 1-form field ω then $g^{ij}\omega_j$ are the components of a vector field $g^\sharp \cdot \omega$ *geometrically equivalent* to the 1-form field ω, and are written ω^i. For higher orders there are many geometrically equivalent tensor fields, for example, $K_k{}^j = g^{jl} K_{kl}$ and $K^{ij} = g^{ik} g^{jl} K_{kl}$.

Definitions
If f is a function, grad $f = g^\# \cdot df$, a vector field. In local coordinates, grad $f = g^{ij}(\partial f/\partial \mu^i)\partial/\partial \mu^j$. If X is a vector field, curl $X = dg^\flat \cdot X$, a 2-form. In coordinates, the components of curl X are $\partial X_j/\partial \mu^i - \partial X_i/\partial \mu^j$.

At each point, p, of (M, g), the tangent space T_p will contain spacelike, timelike, and null vectors if g is not definite (see Section 5.4). If (M, g) is non-Riemannian, it is convenient to sort out three classes of curves: *spacelike, timelike* and *nonspacelike*, or *causal*, characterized, respectively, by their tangent vectors at all points.

We saw, in Section 12.2, that if M is orientable, then M has a volume form, i.e., an n-form field which never vanishes. Equivalently, M has an atlas such that for all coordinate systems $\det(\partial \mu^i/\partial v^j) > 0$.

Theorem 15.3. *On an orientable pseudo-Riemannian manifold a volume form is defined by* $\sqrt{|\det(g_{ij})|}\, d\mu^1 \wedge \cdots \wedge d\mu^n$.

Proof. Choose an oriented atlas for M. We have to show that the given expression is the same in all coordinate systems. Under a change of coordinates, we have a transformation of basis 1-form fields given by $d\bar{\mu}^i = \partial\bar{\mu}^i/\partial\mu^i\, d\mu^j$. This induces a transformation of $\Lambda^n(T^*M)$ (Problem 11.11) $d\bar{\mu}^1 \wedge \cdots \wedge d\bar{\mu}^n = \det(d\bar{\mu}^i/\partial\mu^j)\, d\mu^1 \wedge \cdots \wedge d\mu^n$. But, if $\bar{g}_{ij}$ are the components in $\bar{\mu}^i$ coordinates, and g_{ij} are the components in μ^i coordinates, then $g_{ij} = \partial\bar{\mu}^p/\partial\mu^i\, \partial\bar{\mu}^q/\partial\mu^j\, \bar{g}_{pq}$. Taking the determinant of both sides $|\det(g_{ij})| = (\det(\partial\bar{\mu}^i/\partial\mu^j))^2|\det(\bar{g}_{ij})|$, and then taking the square root, we get $\sqrt{|\det(g_{ij})|} = \det(\partial\bar{\mu}^i/\partial\mu^j)\sqrt{|\det(\bar{g}_{ij})|}$ since M is orientable. ■

Definitions
$\Omega = \sqrt{|\det(g_{ij})|}\, d\mu^1 \wedge \cdots \wedge d\mu^n$ is called *the volume form of an orientable pseudo-Riemannian manifold*, (M, g). If (M, g, Ω) is an oriented pseudo-Riemannian manifold with volume form Ω, and if D is a regular domain in M, f is a differentiable function of compact support on M, and X is a vector field on M, then *the volume of* D is $\int_M c_D\Omega$, *the integral of* f *over* D is $\int_D f = \int_D f\Omega$, and *the divergence of* X, div X, is defined by $L_X\Omega = (\text{div } X)\Omega$. (Note that these three definitions involve Ω but not g, so they have obvious generalizations.)

Theorem 15.4. *In a coordinate neighborhood,*

$$\text{div } X = \frac{1}{\sqrt{|\det(g_{ij})|}} \frac{\partial}{\partial \mu^k}\left(\sqrt{|\det(g_{ij})|}X^k\right)$$

Proof. Problem 15.2

Theorem 15.5. $\int_D \operatorname{div} X\Omega = \int_{\partial D} i^* i_X \Omega$ using the compatible orientation on ∂D. (Here i^* is the pullback of the inclusion map i: $\partial D \to D$ and i_X is interior multiplication by X.)

Proof. Problem 15.4.

Theorem 15.5 is a generalization of the divergence theorem in ordinary vector analysis. We can make it look more like that theorem if on the left side we use the definitions above and Theorem 15.4. The integrand on the right side is $i^*\Sigma_*(\operatorname{sgn}\pi)\sqrt{|\det(g_{ij})|}X^{\pi(1)}\, d\mu^{\pi(2)} \wedge \cdots \wedge d\mu^{\pi(n)}$ where Σ_* is the sum over $1, n-1$ shuffles (Section 6.2) by eq. (11.31). In a coordinate patch in which M and ∂D have the coordinates of Theorem 9.7, this reduces to $\Sigma_*(\operatorname{sgn}\pi)(\sqrt{|\det(g_{ij})|}X^{\pi(1)} \circ i)\, d\mu^{\pi(2)} \wedge \cdots \wedge d\mu^{\pi(n)}$.

Now we consider differentiable mappings between two given pseudo-Riemannian manifolds (M, g) and (N, h), or, in particular, mappings of a pseudo-Riemannian manifold to itself.

Definitions

If ϕ: $(M, g) \to (N, h)$ and $\phi^*h = g$, then ϕ is *an isometry.* We say it "preserves," or "leaves invariant" the pseudo-Riemannian structures, g and h (Fig. 15.1). More generally, if c is a nonzero constant, and $\phi^*h = cg$, then ϕ is called *a homothety.*

To clarify the meaning of the condition $\phi^*h = g$, we write it in more explicit detail. Recall the definition of the pullback, ϕ^*h, Section 11.4; if $p \in M$, then $\phi^*h(p) = \phi^* \cdot h(\phi(p))$. Now evaluating the right side on v_1 and v_2 in T_pM using the formula given by eq. (4.10) for $\phi^s \cdot A$, we have

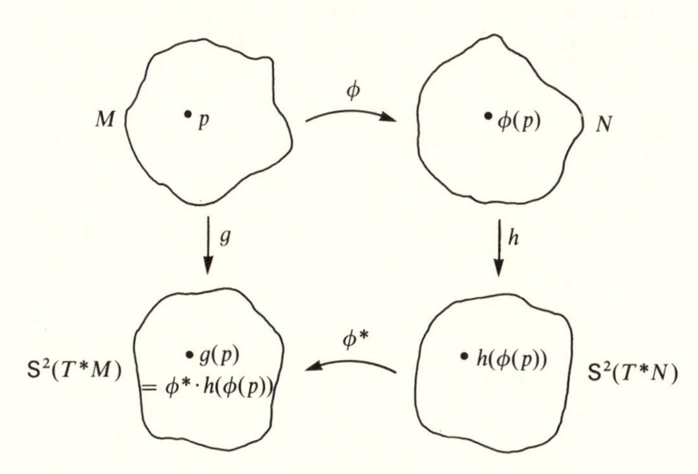

FIGURE 15.1

$\phi^* \cdot h(\phi(p))(v_1, v_2) = h(\phi(p))(\phi_* \cdot v_1, \phi_* \cdot v_2)$, so $\phi^* h = g$ can be written

$$h(\phi(p))(\phi_* \cdot v_1, \phi_* \cdot v_2) = g(p)(v_1, v_2) \tag{15.2}$$

If (μ^i) are local coordinates at p and (v^i) are local coordinates at $\phi(p)$, then in component notation (15.2) becomes

$$h_{ij}(\phi(p)) \frac{\partial \phi^i}{\partial \mu^k} \frac{\partial \phi^j}{\partial \mu^l} = g_{kl}(p) \tag{15.3}$$

Recall that in Section 5.4 we already used the term "isometry" for certain linear mappings of inner product vector spaces. Our present concept is a generalization of that earlier one. This is made evident by comparing eq. (15.2) with eq. (5.21) (and using the natural isomorphisms of Problem 9.14(ii)). In Section 8.2 we used the term "isometry" for maps of metric spaces. We will see, in the next section, that this concept can, in turn, be considered to be a generalization of our present one.

For a pseudo-Riemannian manifold (M, g) we can consider the maps, Θ_u, of a flow of a given vector field, X.

Definition
If g is preserved under all the maps, Θ_u, of a given vector field, X, then X is a *Killing field.*

Theorem 15.6. *X is a killing field iff $L_X g = 0$.*

Proof. By Theorem 13.15, if $\{\Theta_u\}$ is the flow of X, and $g_p^\#: u \mapsto \Theta_u^* g(p)$ then $D_u g_p^\# = \Theta_u^* L_X g(p) = \Theta_u^* \cdot L_X g(\Theta_u(p)) = \Theta_u^* \cdot L_X g(\gamma_p(u))$. Now, $L_X g = 0$ in some neighborhood of $p \Rightarrow D_u g_p^\# = 0$ in some neighborhood of $u = 0 \Rightarrow \Theta_u^* g(p) = \Theta_0^* g(p) = g(p)$, so $\Theta_u^* g = g$. Conversely, for a p, if $\Theta_u^* g = g$ for u in some neighborhood of $u = 0$, then $D_0 g_p^\# = L_X g(p) = 0$. ∎

In a local coordinate system, by Theorem 13.13, we can write

$$(L_X g)_{jk} = X^i \frac{\partial g_{jk}}{\partial \mu^i} + g_{ji} \frac{\partial X^i}{\partial \mu^k} + g_{ki} \frac{\partial X^i}{\partial \mu^j}$$

and we have the following result.

Corollary. *X is Killing field iff*

$$X^i \frac{\partial g_{jk}}{\partial \mu^i} + g_{ji} \frac{\partial X^i}{\partial \mu^k} + g_{ki} \frac{\partial X^i}{\partial \mu^j} = 0 \tag{15.4}$$

Equations (15.4) are called *the equation of Killing* (Eisenhart, 1949, p. 234).

We obtain important illustrations of the concepts introduced above if we start with the manifold $\mathbb{R}^n$ with the standard structure (Section 9.1). Using the standard coordinates (x^i) on $\mathbb{R}^n$ we put $g = dx^1 \otimes dx^1 + \cdots + dx^n \otimes dx^n$. Then $(M, g) = (\mathbb{R}^n, g_\delta) = \mathscr{R}_0^n$ is called *n-dimensional Euclidean space.* Again, with the standard coordinates on $\mathbb{R}^n$ if we put instead $g = -dx^0 \otimes dx^0 + dx^1 \otimes dx^1 + \cdots + dx^{n-1} \otimes dx^{n-1}$ then $(M, g) = (\mathbb{R}^n, g_\eta) = \mathscr{R}_1^n$ is called *n-dimensional Minkowski space.* (Compare with the terminology in Section 5.4.) The isometries of n-dimensional Euclidean space are *the Euclidean transformations*, and the Killing fields are

$$\left\langle \left\{ \frac{\partial}{\partial x^i}, x^j \frac{\partial}{\partial x^i} - x^i \frac{\partial}{\partial x^j}; i, j = 1, \ldots, n \right\} \right\rangle$$

The isometries of Minkowski space are *the Lorentz transformations* and the Killing fields are

$$\left\langle \left\{ \frac{\partial}{\partial x^\alpha}, x^0 \frac{\partial}{\partial x^i} + x^i \frac{\partial}{\partial x^0}, x^i \frac{\partial}{\partial x^j} - x^j \frac{\partial}{\partial x^i}; i, j = 1, \ldots, n-1, \alpha = 0, \ldots, n \right\} \right\rangle$$

Both of these are $n(n + 1)/2$ dimensional vector spaces over $\mathbb{R}$.

The existence of isometries of a pseudo-Riemannian manifold depends on the existence of "symmetries" of the space, so, in general, a pseudo-Riemannian space has no isometries (cf., Problem 20.12).

Finally, rather than studying maps between two given pseudo-Riemannian manifolds, we can alter our point of view, and start with a differentiable map, ϕ, between a manifold, P, and a pseudo-Riemannian manifold, (M, g). Then $\phi^* g$ is an induced structure on P. In particular, if (P, ϕ) is a submanifold of (M, g), and $\phi^* g$ is nondegenerate, then $\phi^* g$ is *the induced pseudo-Riemannian structure* on (P, ϕ).

PROBLEM 15.1. Show that the formula for the definition of div X is simply another form of the equation $dJ/dt = (\nabla \cdot \bar{v})J$ which appears in classical fluid mechanics (cf., Coburn, 1955, p. 101).

PROBLEM 15.2. Prove Theorem 15.4

PROBLEM 15.3. One can define div X, in terms of the Hodge star operator, by div $X = *d*\omega$ where $\omega = g^\flat \cdot X$. Show that the two definitions give the same thing.

PROBLEM 15.4. Prove Theorem 15.5.

PROBLEM 15.5. Show that the Killing fields of Euclidean space are as given above and find their integral curves.

PROBLEM 15.6. Same as Problem 15.5 for Minkowski space.

PROBLEM 15.7. Generalize Theorem 15.6; i.e., show that g can be replaced by any tensor field, K.

PROBLEM 15.8. S^3 is a submanifold of $\mathbb{R}^4$ (see Problem 9.19). Using the local coordinates of Example 10, Section 10.1, find the components $(\phi^*g)_{\alpha\beta}$ of the induced structure.

15.2 Length and distance

The first thing we do with our Riemannian or pseudo-Riemannian structure, g, on M is to use it to define the concept of length of a curve.

Recall, from Section 5.4, that if $v_p \in T_pM$ then the length of v_p is $\|v_p\| = |g_p(v_p, v_p)|^{1/2}$. Now if $\gamma: I \to M$ is a curve and $\dot{\gamma}(u)$ is its tangent (or, velocity) vector at u, then $g(\dot{\gamma}(u), \dot{\gamma}(u)) = g_{ij}D_u\gamma^i D_u\gamma^j$ (by eq. (9.13)), and we make the following definition.

Definition
The length of γ between $\gamma(u_1)$ and $\gamma(u_2)$ is

$$\mathbf{L}(\gamma) = \int_{u_1}^{u_2} \|\dot{\gamma}(u)\|\, du$$

Theorem 15.7. *The length of a curve is independent of its parametrization.*

Proof. Just as in elementary calculus, on each segment with $\beta(\rho(u)) = \gamma(u)$ such that $d\rho/du \neq 0$,

$$\int_{u_1}^{u_2} \|\dot{\beta}(\rho(u))\|\, |d\rho/du|\, du = \int_{u_1}^{u_2} \|\dot{\gamma}(u)\|\, du$$

so that $\int_{t_1}^{t_2} \|\dot{\beta}(t)\|\, dt = \int_{u_1}^{u_2} \|\dot{\gamma}(u)\|\, du$ if $d\rho/du > 0$, and $\int_{t_2}^{t_1} \|\dot{\beta}(t)\|\, dt = \int_{u_1}^{u_2} \|\dot{\gamma}(u)\|\, du$ if $d\rho/du < 0$. ■

A particularly important parametrization, *arc-length*, is obtained when $\dot{\gamma}(u)$ never vanishes along γ. In that case, s, defined by $s(u) = \int_a^u \|\dot{\gamma}(t)\|\, dt$ has an inverse on the domain γ, so γ can be reparametrized on the range of s.

Having the concept of length of a curve in (M, g), one can proceed to try to define a distance between two points of M. (Recall that a definition of distance has already been introduced in Section 8.1.) Let p and q be two points of M and consider the set of all curves joining them. This set is not empty if p and q are in the same component of M.

Definition
The distance, $d_g(p, q)$, between two points, p and q, on a Riemannian manifold, (M, g), is the greatest lower bound, $\inf_\gamma \mathbf{L}(\gamma)$, where $\inf_\gamma$ is taken over all curves joining p and q.

Note that on a non-Riemannian manifold, according to this definition, the distance between any two points joined by a timelike curve would be zero, because it is possible to find timelike curves joining the points having arbitrarily small lengths. Hence, this concept must be altered for non-Riemannian manifolds.

Theorem 15.8. *On a Riemannian manifold, d_g is a distance function.*

Proof. $d_g(p, p) = 0$ and symmetry are immediate from the definition. For any given positive number ε, there are curves γ_1 joining p and q and γ_2 joining q and r such that $\mathbf{L}(\gamma_1) < d_g(p, q) + \varepsilon$ and $\mathbf{L}(\gamma_2) < d_g(q, r) + \varepsilon$ respectively. So $d_g(p, r) \leq \mathbf{L}(\gamma_1 + \gamma_2) = \mathbf{L}(\gamma_1) + \mathbf{L}(\gamma_2) < d_g(p, q) + d_g(q, r) + 2\varepsilon$, from which we get the triangle inequality. The property if $p \neq q$ then $d_g(p, q) > 0$ comes out as a consequence of Theorem 15.9. Since the topology generated by d_g is the same as the original manifold topology, and the original manifold topology in Hausdorff, the topology generated by d_g is Hausdorff, from which the required property easily follows. ■

Definition
A curve joining p and q (on a Riemannian manifold) whose length is $d_g(p, q)$ is *a shortest curve joining p and q.*

For two given points p and q we need not have either the existence or uniqueness of shortest curves. Examples of the failure of existence and uniqueness are, respectively: (i) M is the Euclidean plane with the origin removed and with $p = (-1, 0)$ and $q = (1, 0)$; and (ii) M is a sphere in $\mathbb{R}^3$, and p and q are opposite poles.

We saw, in Section 8.1, that a distance function, d, on a set defines a Hausdorff topology. If we drop the property $d(p, q) > 0$ if $p \neq q$, then d, called a pseudodistance function, still defines a topology on M. Hence it is important to observe the following result.

Theorem 15.9. *The topology defined by d_g on a Riemannian manifold is the same as the original manifold topology.*

Proof. Define a distance function $d(p, q) = [\sum (\mu^i(p) - \mu^i(q))^2]^{1/2}$ on a coordinate domain $\mathscr{U}$. Then there are positive numbers, c_1 and c_2, such that $d_g(p, q) \leq c_1 d(p, q)$ and $d(p, q) \leq c_2 d_g(p, q)$ for all $p, q \in \mathscr{U}$. (See Hicks, 1965, pp. 70–71.) Thus, by Theorem 8.4(ii) $(\mathscr{U}, d_g)$ and $(\mathscr{U}, d)$ are homeomorphic, so that each coordinate domain of M is a coordinate domain with the topology of d_g, and these coordinate domains generate the same topology as those with the original topology.

Theorem 15.10. *For Riemannian manifolds ϕ is an isometry as defined above (i.e., ϕ preserves the Riemannian structures) iff ϕ is an isometry as defined in Section 8.2 (i.e., ϕ preserves distance functions).*

Proof. (i) Since the set of curves joining $\phi(p)$ and $\phi(q)$ is precisely the set of images of the curves, γ, joining p and q we can write $d(\phi(p), \phi(q)) = d(p, q)$ as

$$\inf_{\gamma} \int_a^b [h(\phi_* \dot{\gamma}(t), \phi_* \dot{\gamma}(t))]^{1/2}\, dt = \inf_{\gamma} \int_a^b (g(\dot{\gamma}(t), \dot{\gamma}(t))]^{1/2}\, dt$$

Thus, clearly (15.1) implies that d is preserved.

(ii) The converse seems to require a technically rather involved proof; cf., Kobayashi and Nomizu (1963, Vol. I, p. 169). ∎

As we have noted above, on a non-Riemannian manifold, where we sort out different classes of curves, a concept of distance between points p and q will have to depend on the kinds of curves that can join p and q. On a Lorentzian manifold (index $= 1$ or $n - 1$), one can show that for certain pairs of points, p and q, that can be joined by a timelike curve, $\sup_\gamma \mathbf{L}(\gamma)$ where $\sup_\gamma$ is taken over all timelike curves joining p and q, has a finite value, and there is a curve (a geodesic) whose length has this value (Hawking and Ellis, 1973, pp. 213–217; O'Neill, 183, p. 411). Such a curve is *a longest curve* joining the two points. These results, obviously more restrictive than for Riemannian manifolds since they do not apply to any arbitrary pair of points, are nevertheless important in the global study of spacetime. Pursuing this subject in the general case requires a careful examination of the causal structure of Lorentzian manifolds. We will look at certain special cases in the next section, and when we come to relativity.

Finally, there are slightly weaker concepts of "shortest curve" and "longest curve" which correspond to a relative rather than an absolute maximum or minimum in ordinary calculus, and which occur in more general form in the calculus of variations. Thus, given γ, consider all curves in a neighborhood of γ in the following sense. Let $I = [a, b]$, be in the domain of γ, and let $[a, b] \times [-\varepsilon, \varepsilon] \to M$ be a map which restricted to $[(a, 0), (b, 0)]$ is γ. Given a curve γ, a map, Q_γ, which this property is called *a variation of* γ.

If $(u, t) \in [a, b] \times [-\varepsilon, \varepsilon]$ then for each fixed u we get a "vertical" curve γ_u, and for each fixed t we get a "horizontal" curve γ_t with $\gamma_0 = \gamma$ and a length $\mathbf{L}(\gamma_t)$. Let $J_{Q_\gamma}\colon t \mapsto \mathbf{L}(\gamma_t)$.

Definitions
If for all variations of γ, Q_γ, such that $Q_\gamma(a, t) = p$ and $Q_\gamma(b, t) = q$ for all $t \in [-\varepsilon, \varepsilon]$, we have $D_0 J_{Q_\gamma} = 0$, then γ is called *an extremal of* $\mathbf{L}$. If for all such variations J_{Q_γ} has a maximum (minimum) at $t = 0$, then γ is *length-maximizing (minimizing)*.

Theorem 15.11. *A shortest (longest) curve is length-minimizing (maximizing).*

Proof. Problem 15.12.

Sometimes, instead of using the length of a curve, it is more convenient to use *the energy of a curve* defined by $\mathbf{E}(\gamma) = \int_a^b g(\dot{\gamma}(t), \dot{\gamma}(t))\, dt$. $\mathbf{E}(\gamma)$ is simply related to $\mathbf{L}(\gamma)$. It has the advantage of doing away with the square root, but the disadvantage of being parameter-dependent.

Theorem 15.12. $\mathbf{L}(\gamma)^2 \leq (b - a)|\mathbf{E}(\gamma)|$, *and* $\mathbf{L}(\gamma)^2 = (b - a)|\mathbf{E}(\gamma)|$ *iff the parameter of* γ *is proportional to arc-length.*

Proof. In the Cauchy–Schwarz inequality

$$\left(\int_a^b f(t)h(t)\, dt\right)^2 \leq \int_a^b (f(t))^2\, dt \int_a^b (h(t))^2\, dt$$

put $f = 1$ and $h = \|\dot{\gamma}\|$ and get the first result. For $f = 1$, Cauchy–Schwarz becomes an equality iff h is a constant. But by the definition of arc-length, s, $\|\dot{\gamma}(u)\| =$ constant iff the parameter of γ is proportional to s. ∎

PROBLEM 15.9. If $\beta = \gamma \circ s^{-1}$, then $\|\dot{\beta}(s)\| \equiv 1$.

PROBLEM 15.10. Fill in the details of the proof of Theorem 15.9.

PROBLEM 15.11. If $\gamma(u) = (u, u^2)$, construct a variation of γ, Q_γ: $[0, 1] \times [-\varepsilon, \varepsilon] \to \mathbb{R}^2$ with $Q_\gamma(0, t) = (0, 0)$, $Q_\gamma(1, t) = (1, 1)$, and with vertical curves $\{(c, t) \in \mathbb{R}^2: c = \text{const.}\}$. Is γ an extremal of $\mathbf{L}$?

PROBLEM 15.12. (i) Prove Theorem 15.11. (ii) Give an example of a length-minimizing curve which is not a shortest curve.

15.3 Flat spaces

We will consider a special class of pseudo-Riemannian manifolds, and for these, pursue some of the ideas of the previous section a bit further.

Definition
A pseudo-Riemannian manifold, (M, g), which has a coordinate covering on each domain of which the components g_{ij} are constant is *a flat space.*

Clearly, this class of pseudo-Riemannian manifolds includes the Euclidean and Minkowski spaces mentioned in Section 15.1. Cylinders and cones in 3-dimensional Euclidean space are examples of flat spaces. See Problem 15.13.

If there are coordinates, (x^i), for which g_{ij} are constant, then there are coordinates (y^i) given by $x^i = a^i_j y^j$ for which $g_{ij} = \delta_{ij}$. That is, g_{ij} can be "diagonalized" by the coordinate transformation equations $g_{ij}\,\partial x^i/\partial y^p\,\partial x^j/\partial y^p = \delta_{pq}$, so there is no loss of generality to write g locally in the form $dx^1 \otimes dx^1 + \cdots + dx^r \otimes dx^r - dx^{r+1} \otimes dx^{r+1} - \cdots - dx^n \otimes dx^n$ (cf., eq. (5.20)). In short, a flat space is locally isometric to a pseudo-Euclidean manifold.

Definitions

An affine atlas is such that on the intersection of the coordinate neighborhoods of two overlapping charts the coordinates (x^i) and $(\bar{x}^i)$ are affinely related (or, $\partial x^i/\partial \bar{x}^j$ are constants). *An affine manifold* is one which has an affine atlas.

Theorem 15.13. *A flat space is an affine manifold.*

Proof. Let g_{ij} be the components of g on the chart with coordinates (x^i) and let $\bar{g}_{ij}$ be the components of g on the chart with coordinates $(\bar{x}^i)$. Then

$$\bar{g}_{ij} = \frac{\partial x^p}{\partial \bar{x}^i}\frac{\partial x^q}{\partial \bar{x}^j} g_{pq}, \qquad \text{and} \qquad \frac{\partial^2 x^p}{\partial \bar{x}^k\,\partial \bar{x}^i}\frac{\partial x^q}{\partial \bar{x}^j} g_{pq} + \frac{\partial x^p}{\partial \bar{x}^i}\frac{\partial^2 x^q}{\partial \bar{x}^k\,\partial \bar{x}^j} g_{pq} = 0$$

This says that

$$\frac{\partial^2 x^p}{\partial \bar{x}^k\,\partial \bar{x}^i}\frac{\partial x^q}{\partial \bar{x}^j} g_{pq} = g\left(\frac{\partial^2 x^p}{\partial \bar{x}^k\,\partial \bar{x}^i}\frac{\partial}{\partial x^p}, \frac{\partial}{\partial \bar{x}^j}\right)$$

is skew-symmetric in i and j, and since it is clearly symmetric in i and k it follows, from Problem 5.2, that

$$g\left(\frac{\partial^2 x^p}{\partial \bar{x}^k\,\partial \bar{x}^i}\frac{\partial}{\partial x^p}, \frac{\partial}{\partial \bar{x}^j}\right) = 0 \qquad \text{for all } i, j, k$$

which implies $\partial^2 x^p/\partial \bar{x}^k\,\partial \bar{x}^i = 0$ for all i, k, p. ■

Definitions

In a flat space coordinate neighborhood, a curve, α, given by $\alpha^i(u) = v^i u + a^i$ is *a straight line segment*. Two straight line segments, α and β with $\beta^i(t) = w^i t + b^i$ are *parallel* if $v^i = cw^i$ where c is not zero.

Theorem 15.14. (i) *A straight line segment joining p and q on a Riemannian* *flat space is a shortest curve joining p and q.* (ii) *A straight line segment on a Lorentzian flat space which joins two points p and q, which can be joined by a timelike curve, is a longest curve joining p and q.*

Proof. (i) Let $\alpha^i(u) = v^i u + a^i$ be a straight line segment in a coordinate neighborhood, $\mathscr{U}$. Let γ be a curve in $\mathscr{U}$ joining p and q. At each point of γ we have a vector, v with components v^i. Then

$$\mathbf{L}(\gamma) = \int_{t_1}^{t_2} \|\dot{\gamma}\|\, dt \geq \int_{t_1}^{t_2} \frac{g(v, \dot{\gamma})}{\|v\|}\, dt \qquad \text{by the Cauchy–Schwarz inequality}$$

$$= \int_{t_1}^{t_2} \frac{\sum_i v^i \frac{d\gamma^i}{dt}}{\|v\|}\, dt = \sum_i \frac{v^i}{\|v\|} (\gamma^i(t_2) - \gamma^i(t_1))$$

$$= \sum \frac{v^i}{\|v\|} (\alpha^i(u_2) - \alpha^i(u_1)) = \mathbf{L}(\alpha)$$

since $\alpha^i(u_2) - \alpha^i(u_1) = v^i(u_2 - u_1)$. A curve, γ, which does not lie entirely within $\mathscr{U}$ can be broken into pieces each of which lies in a flat coordinate neighborhood and can be compared with a straight line segment in that neighborhood.

(ii) Exactly the same as part (i), using instead the backward Cauchy–Schwarz inequality (see Theorem 5.21). ■

Theorem 15.15. *A shortest curve on a flat Riemannian manifold, and a longest curve on a flat non-Riemannian manifold for which $\|\dot{\gamma}(t)\|$ never vanishes is a straight line.*

Proof. By Theorem 15.11 a shortest or longest curve is an extremal of $\mathbf{L}$; i.e., $D_0 J_{Q_\gamma} = 0$ for all variations, Q_γ. Further by Theorem 15.12 since $\|\dot{\gamma}(u)\| \neq 0$, such a curve can be reparametrized by arc-length, and hence, will be an extremal of $\mathbf{E}$. Now, consider the case where the end points, p and q, are contained in a coordinate neighborhood, and write the energy of the curves γ_t of a variation Q_γ in the neighborhood in component notation. Thus, for each γ_t we have

$$\mathbf{E}(\gamma_t) = \int_p^q g_{ij} D_u \gamma_t^i D_u \gamma_t^j \, du \tag{15.5}$$

Now, if we let t vary, and write $\gamma_t^i(u) = \gamma^i(u, t)$, then (15.5) can be written

$$f(t) = \int_p^q g_{ij} \frac{\partial \gamma^i}{\partial u}(u, t) \frac{\partial \gamma^j}{\partial u}(u, t)\, du$$

Finally, differentiating with respect to t, integrating by parts and setting $t = 0$ we get

$$f'(0) = 2g_{ij}\left[\frac{\partial\gamma^i}{\partial u}(q,0)\frac{\partial\gamma^j}{\partial t}(q,0) - \frac{\partial\gamma^i}{\partial u}(p,0)\frac{\partial\gamma^j}{\partial t}(p,0)\right] - 2\int_p^q g_{ij}\frac{\partial^2\gamma^i}{\partial u^2}(u,0)\frac{\partial\gamma^j}{\partial t}(u,0)\,du$$

Since

$$\frac{\partial\gamma^i}{\partial t}(q,0) = \frac{\partial\gamma^i}{\partial t}(p,0) = 0$$

$f'(0) = 0$ implies that

$$\int_p^q g_{ij}\frac{\partial^2\gamma^i}{\partial u^2}(u,0)\frac{\partial\gamma^j}{\partial t}(u,0)\,du = 0 \tag{15.6}$$

Suppose $\partial^2\gamma^i/\partial u^2 \neq 0$ at some point on γ_0. Then $g_{ij}\,\partial^2\gamma^i/\partial u^2\,\partial\gamma^j/\partial t \neq 0$ at that point for some $\partial\gamma^i/\partial t$ since g is nondegenerate. But $\partial\gamma^i/\partial t$ can be chosen arbitrarily since Q_γ can be chosen arbitrarily, and then $\partial\gamma^i/\partial t$ can be extended along γ_0 in such a way to contradict (15.6). This gives $(\partial^2\gamma^i/\partial u^2)(u,0)$, so that

$$\gamma^i(u) = a^i u + b^i$$

i.e., γ is a straight line segment. ■

PROBLEM 15.13. Take a nappe of a circular cone with angle ϑ in 3-dimensional Euclidean space, slice it along one of its generators and flatten it onto the u–v plane perpendicular to its axis, the z axis.

(i) Show that it has the equations

$$x = \sqrt{u^2+v^2}\cos\left(\frac{\tan^{-1}(v/u)}{\sin\vartheta}\right)\sin\vartheta$$

$$y = \sqrt{u^2+v^2}\sin\left(\frac{\tan^{-1}(v/u)}{\sin\vartheta}\right)\sin\vartheta$$

$$x = \sqrt{u^2+v^2}\cos\vartheta$$

(ii) Show that the cone is flat.

PROBLEM 15.14. Prove the statement in Theorem 15.15 that for an arbitrary choice of $\partial\gamma^i/\partial t$ along γ_0 a variation Q_{γ_0} can be constructed which has those values of $\partial\gamma^i/\partial t$ on γ_0.

PROBLEM 15.15. Extend Theorem 15.15 to the general case in which p and q are not in the same coordinate neighborhood.

16

CONNECTION 1-FORMS

We will describe the Levi–Civita connection and its covariant derivative on a pseudo-Riemannian manifold. Then we abstract the idea of a connection to define a more general class of manifolds. After briefly describing their main features, we go back to pseudo-Riemannian manifolds and discuss their possible connections.

16.1 The Levi–Civita connection and its covariant derivative

We have now, as in Chapter 15, a symmetric, nondegenerate tensor field, g, of type (0, 2) on M. We will construct a new structure on M by first working locally with g. Thus, we work in a chart $(\mathscr{U}, \mu)$ with coordinate neighborhood $\mathscr{U}$ and coordinate functions μ^i. In $\mathscr{U}$, g has components g_{ij} with respect to the coordinate basis $\{d\mu^i \otimes d\mu^j\}$.

Definitions
$[ij, k] = \frac{1}{2}(\partial g_{jk}/\partial\mu^i + \partial g_{ki}/\partial\mu^j - \partial g_{ij}/\partial\mu^k)$ are called *the Christoffel symbols of the first kind.* $\left\{ {l \atop i\ j} \right\} = g^{lk}[ij, k]$, where g^{lk} are given by eq. (15.1), are called *the Christoffel symbols of the second kind.*

We will now derive the relation between the Christoffel symbols in two coordinate systems (μ^i) and $(\bar{\mu}^i)$. From the basic relation

$$g_{ij} = \bar{g}_{rs} \frac{\partial\bar{\mu}^r}{\partial\mu^i} \frac{\partial\bar{\mu}^s}{\partial\mu^j},$$

putting $p_i^r = \partial\bar{\mu}^r/\partial\mu^i$ and differentiating with respect to μ^k, we get

$$\frac{\partial g_{ij}}{\partial\mu^k} = \frac{\partial\bar{g}_{rs}}{\partial\bar{\mu}^t} p_i^r p_j^s p_k^t + \bar{g}_{rs}\left(p_i^r \frac{\partial^2\bar{\mu}^s}{\partial\mu^i\,\partial\mu^k} + p_j^s \frac{\partial^2\bar{\mu}^r}{\partial\mu^i\,\partial\mu^k} \right)$$

By cyclically permuting the indices i, j, k and simultaneously permuting r, s, t, we get similar expansions for $\partial g_{jk}/\partial\mu^i$ and $\partial g_{ki}/\partial\mu^j$. Adding these last two and subtracting the one above, we get

$$\frac{\partial g_{jk}}{\partial \mu^i} + \frac{\partial g_{ki}}{\partial \mu^j} - \frac{\partial g_{ij}}{\partial \mu^k} = \left(\frac{\partial \bar{g}_{st}}{\partial \bar{\mu}^r} + \frac{\partial \bar{g}_{tr}}{\partial \bar{\mu}^s} - \frac{\partial \bar{g}_{rs}}{\partial \bar{\mu}^t}\right) p_i^r p_j^s p_k^t + 2\bar{g}_{rs} p_k^r \frac{\partial^2 \bar{\mu}^s}{\partial \mu^i \, \partial \mu^j}$$

or

$$[ij, k] = \overline{[rs, t]}\, p_i^r p_j^s p_k^t + \bar{g}_{rs} p_k^r \frac{\partial^2 \bar{\mu}^s}{\partial \mu^i \, \partial \mu^j} \tag{16.1}$$

Multiplying (16.1) on the left by $g^{lk}(\partial \bar{\mu}^m/\partial \mu^l)$ and on the right by $\bar{g}^{mn}(\partial \mu^k/\partial \bar{\mu}^n)$ (these two things are equal) we get

$$\begin{Bmatrix} l \\ i\ j \end{Bmatrix} p_l^m = \overline{\begin{Bmatrix} m \\ r\ s \end{Bmatrix}}\, p_i^r p_j^s + \frac{\partial^2 \bar{\mu}^m}{\partial \mu^i \, \partial \mu^j} \tag{16.2}$$

Equations (16.1) and (16.2) are the transformation formulas for the Christoffel symbols in the intersection, $\mathscr{U} \cap \mathscr{V}$, of the coordinate neighborhoods $\mathscr{U}$ and $\mathscr{V}$. Note that because of the appearance of the second derivative terms, the Christoffel symbols are not tensors.

With the Christoffel symbols $\begin{Bmatrix} l \\ i\ j \end{Bmatrix}$ we can form a set of $(\dim M)^2$ 1-forms on each coordinate domain, namely,

$$\lambda_i^l = \begin{Bmatrix} l \\ i\ j \end{Bmatrix} d\mu^j$$

Then, putting $\bar{\lambda}_r^m = \overline{\begin{Bmatrix} m \\ r\ s \end{Bmatrix}}\, d\bar{\mu}^s$, the transformation formula (16.2) yields the transformation formula

$$p_l^m \lambda_i^l = p_i^r \bar{\lambda}_r^m + dp_i^m \tag{16.3}$$

for these 1-forms on the intersection $\mathscr{U} \cap \mathscr{V}$.

Definition

The $(\dim M)^2$ 1-forms $\lambda_i^l = \begin{Bmatrix} l \\ i\ j \end{Bmatrix} d\mu^j$ defined on coordinate neighborhoods of M are called *the Levi–Civita connection 1-forms* of the pseudo-Riemannian manifold (M, g).

Now suppose $Y \in \mathfrak{X}$ is a given vector field on M. In the intersecting coordinate neighborhoods $\mathscr{U}$ and $\mathscr{V}$, Y has coordinate components Y^i and $\bar{Y}^i$ related by $\bar{Y}^i = p^i_j Y^j$ on $\mathscr{U} \cap \mathscr{V}$, and

$$\begin{aligned} d\bar{Y}^i &= p^i_j\, dY^j + (dp^i_j)Y^j \\ &= p^i_j\, dY^j + (p^i_r \lambda^r_j - p^s_j \bar{\lambda}^i_s)Y^j \qquad \text{by (16.3)} \\ &= p^i_r(dY^r + Y^j \lambda^r_j) - \bar{Y}^s \bar{\lambda}^i_s \end{aligned}$$

Hence, on $\mathscr{U} \cap \mathscr{V}$,

$$d\bar{Y}^i + \bar{Y}^s \bar{\lambda}^i_s = p^i_r(dY^r + Y^j \lambda^r_j) \tag{16.4}$$

Equation (16.4) is an interesting result.

On the one hand, if we evaluate the n 1-forms on both sides of (16.4) on a vector X at points in $\mathscr{U} \cap \mathscr{V}$ we get two sets of n functions which are related like the components of a vector. Thus, for each Y we have a mapping $dY: \mathfrak{X} \to \mathfrak{X}$; i.e., a vector-valued 1-form given by

$$dY: X \mapsto \langle dY^i + Y^j \lambda^i_j, X \rangle \frac{\partial}{\partial \mu^i} \tag{16.5}$$

On the other hand, if we take the tensor product of both sides of eq. (16.4) with $\partial/\partial\bar{\mu}^i$ we get

$$\frac{\partial}{\partial \mu^i} \otimes (dY^i + Y^j \lambda^i_j) = \frac{\partial}{\partial \bar{\mu}^i} \otimes (d\bar{Y}^i + \bar{Y}^j \bar{\lambda}^i_j)$$

That is, the (1, 1) tensor field defined on $\mathscr{U}$ by

$$\frac{\partial}{\partial \mu^i} \otimes (dY^i + Y^j \lambda^i_j)$$

and the (1, 1) tensor field defined on $\mathscr{V}$ by

$$\frac{\partial}{\partial \bar{\mu}^i} \otimes (d\bar{Y}^i + \bar{Y}^j \bar{\lambda}^i_j)$$

are the same on $\mathscr{U} \cap \mathscr{V}$.

We can summarize our results as follows.

Theorem 16.1. *On a pseudo-Riemannian manifold (M, g) for each vector field, Y there is a vector-valued 1-form, dY, given by (16.5) and there is a mapping*

$\nabla: \mathfrak{X} \to$ the module of (1, 1) *tensor fields on M given by*

$$\nabla: Y \mapsto \frac{\partial}{\partial \mu^i} \otimes (dY^i + Y^j \lambda_j^i) \tag{16.6}$$

Definitions
dY is called *the covariant differential* and $\nabla Y = \partial/\partial\mu^i \otimes (dY^i + Y^j\lambda_j^i)$ is called *the covariant derivative* of the vector field Y.

Note that at each point dY and ∇Y are corresponding elements in the isomorphism $\mathscr{L}(T_p, T_p) \cong T_p \otimes T_p^*$ of Theorem 4.1(ii) since for all vector fields, X, and 1-form fields, θ, $(dY \cdot X) \cdot \theta = \nabla Y(\theta, X)$.

Theorem 16.2. *∇ is R-linear on $\mathfrak{X}$, and has the property, for a function, f, and a vector field, Y,*

$$\nabla f Y = Y \otimes df + f\,\nabla Y \tag{16.7}$$

Proof. These properties come immediately from the definition (16.6). ∎

If $\{e_i\}$ is a basis of vector fields on an open set of M, and $\{\varepsilon^i\}$ is the dual basis, then $Y = Y^i e_i$ and from (16.7), $\nabla Y = e_i \otimes dY^i + Y^i \nabla e_i$, and the components of ∇Y are

$$Y^j_{,k} \equiv \nabla Y(\varepsilon^j, e_k) = dY^j(e_k) + Y^i \nabla e_i(\varepsilon^j, e_k) \tag{16.8}$$

In particular, if $e_i = \partial/\partial\mu^i$ are coordinate basis vectors, then

$$\nabla e_i(\varepsilon^j, e_k) = \begin{Bmatrix} j \\ i\ k \end{Bmatrix}$$

(see Problem 16.3) and

$$Y^j_{,k} = \nabla Y\left(d\mu^j, \frac{\partial}{\partial\mu^k}\right) = \frac{\partial Y^j}{\partial \mu^k} + \begin{Bmatrix} j \\ i\ k \end{Bmatrix} Y^i \tag{16.9}$$

Much of the preceding can be generalized, starting with a generalization of (16.4). Thus, if instead of Y, a tensor field is given we get a generalization of (16.4). For example, if K is a given (1, 1) tensor field, then differentiating

$$\bar{K}^i_j = \frac{\partial \mu^q}{\partial \bar{\mu}^j} \frac{\partial \bar{\mu}^i}{\partial \mu^p} K^p_q$$

leads to

$$d\bar{K}^i_j + \bar{K}^r_j \bar{\lambda}^i_r - \bar{K}^i_s \bar{\lambda}^s_j = q^l_j p^i_k (dK^k_l + K^r_l \lambda^k_r - K^k_s \lambda^s_l) \tag{16.10}$$

where $q^l_j = \partial \mu^l / \partial \bar{\mu}^j$, which leads to the definition of a (1, 2) tensor field, ∇K, on (M, g), or a (1, 1) tensor-valued 1-form, dK. Similarly, starting with an (r, s) tensor field, we get an $(r, s + 1)$ tensor field. Thus, we can extend the domain of ∇ to the entire mixed tensor field algebra. If K is an (r, s) tensor field its covariant derivative ∇K is an $(r, s + 1)$ tensor field.

PROBLEM 16.1. Let (g_{ij}) be a matrix of the linear mapping $g^\flat : T_p \to T^*_p$ corresponding to g. Then

$$\sum_i \begin{Bmatrix} i \\ i \; k \end{Bmatrix} = \frac{\partial}{\partial \mu^k} \ln \sqrt{\pm \det (g_{ij})}.$$

PROBLEM 16.2. The relation (16.3) between the λ's and the $\bar{\lambda}$'s in terms of $q^j_k = \partial \mu^j / \partial \bar{\mu}^k$ instead of the p's is

$$q^j_k \lambda^r_j = q^r_i \bar{\lambda}^i_k - dq^r_k \tag{16.11}$$

PROBLEM 16.3. Show that if $\{e_i\}$ are coordinate basis vectors and $\{\varepsilon^j\}$ is the dual basis then $\nabla e_i(\varepsilon^j, e_k) = \begin{Bmatrix} j \\ i \; k \end{Bmatrix}$.

PROBLEM 16.4. If $\{e_i\}$ is a local basis of vector fields and $\{\varepsilon^i\}$ is the dual basis, then define $A^i_{jk} = \nabla e_j(\varepsilon^i, e_k)$, where ∇ has the properties of Theorem (16.3). If $(\bar{e}_i)$ is another local basis, and $e_i = p^j_i \bar{e}_j$, then

$$A^i_{jk} p^m_i = \bar{A}^m_{rs} p^r_j p^s_k + \langle dp^m_j, e_k \rangle \tag{16.12}$$

and multiplying by ε^k,

$$A^i_{jk} \varepsilon^k p^m_i = \bar{A}^m_{rs} \bar{\varepsilon}^s p^r_j + dp^m_j \tag{16.13}$$

so $A^i_{jk} \varepsilon^k$ satisfy a relation of the same form as (16.3).

PROBLEM 16.5. If θ is a 1-form on M, then $\nabla\theta = d\mu^i \otimes (d\theta_i - \theta_j \lambda_i^j)$ defines a (0, 2) tensor field on M, and the components of $\nabla\theta$ are

$$\theta_{i,j} = \frac{\partial \theta_i}{\partial \mu^j} - \begin{Bmatrix} k \\ i\ j \end{Bmatrix} \theta_k \tag{16.14}$$

PROBLEM 16.6. If K is a (1, 1) tensor field on M, then the components of ∇K are

$$K^i_{j,k} = \frac{\partial K^i_j}{\partial \mu^k} + \begin{Bmatrix} i \\ p\ k \end{Bmatrix} K^p_j - \begin{Bmatrix} q \\ j\ k \end{Bmatrix} K^i_q \tag{16.15}$$

PROBLEM 16.7. If A is an (r, s) tensor field, and B is a (p, q) tensor field, then $\nabla AB = A\ \nabla B + (\nabla A)B$.

PROBLEM 16.8. X is a Killing field, Section 15.1, iff $X_{i,j} + X_{j,i} = 0$.

16.2 Geodesics of the Levi–Civita connection

Definitions

Suppose we are given a vector field, X and a curve γ. Then we can form ∇X, $\nabla X \circ \gamma$, a (1, 1) tensor field on γ, and $\dot{\gamma}$, a vector field on γ. Finally, the contraction $C \cdot \dot{\gamma}(\nabla X \circ \gamma)$, a vector field on γ, is called *the covariant derivative of X along γ*. If X is a vector field which satisfies $C \cdot \dot{\gamma}(\nabla X \circ \gamma) = 0$ on γ then X is *propagated* (*transported, translated,* or *displaced*) *parallelly along* γ.

The local expression for $C \cdot \dot{\gamma}(\nabla X \circ \gamma)$ is

$$D_u \gamma^k \left[\left(\frac{\partial X^j}{\partial \mu^k} + \begin{Bmatrix} j \\ i\ k \end{Bmatrix} X^i \right) \circ \gamma \right] = D_u(X^j \circ \gamma) + \left(\begin{Bmatrix} j \\ i\ k \end{Bmatrix} \circ \gamma \right) (D_u \gamma^k)(X^i \circ \gamma) \tag{16.16}$$

The right hand side of (16.16) shows that we can extend the concept of a vector field being propagated parallelly along a curve, to that of a vector field *on a curve* being so propagated. That is, we ask for vector fields, $\tilde{X}$, defined on γ, such that

$$D_u \tilde{X}^i + \left(\begin{Bmatrix} j \\ i\ k \end{Bmatrix} \circ \gamma)(D_u \gamma^k) \tilde{X}^i = 0 \tag{16.17}$$

on γ.

Equation (16.17) has two important properties. (i) The condition is independent of the parameter of the curve. (ii) Since the system is linear, there exists a unique solution with a given initial vector defined for all u. That is, a vector can be propagated arbitrarily far along a curve.

Definitions
If the tangent vector, $\dot{\gamma}$, of γ satisfies (16.17), that is, if

$$D_u^2\gamma^j + \left(\begin{Bmatrix} j \\ i\ k \end{Bmatrix} \circ \gamma\right) D_u\gamma^k D_u\gamma^i = 0 \tag{16.18}$$

then γ is *a geodesic*. For any curve, γ, the left side of (16.18) is called *the acceleration of* γ, or *the curvature vector of* γ.

Now we no longer have properties (i) and (ii) of eq. (16.17).

Theorem 16.3. *If γ^i satisfy (16.18), and $\bar{\gamma}^i(\bar{u}) = \gamma^i(u)$, then $\bar{\gamma}^i$ satisfy*

$$D_{\bar{u}}^2\bar{\gamma}^j + \left(\overline{\begin{Bmatrix} j \\ i\ k \end{Bmatrix}} \circ \bar{\gamma}\right) D_{\bar{u}}\bar{\gamma}^i D_{\bar{u}}\bar{\gamma}^k = 0$$

if and only if $\bar{u} = au + b$ and $a\ D_{\bar{u}}\bar{\gamma}^i = D_u\gamma^i$ where a and b are numbers. (Compare Theorem 13.2.)

Proof. Problem 16.10.

Thus, we see that if we have two parametric curves with the same values, one could be a geodesic and the other not. However, if a parametric curve with parameter u is a geodesic, then $\|\dot{\gamma}(u)\|$ is constant (as we will see in Section 16.5 (Theorem 16.10)), which implies that $s = au + b$. That is, on a geodesic (on a pseudo-Riemannian manifold), as long as $\|\dot{\gamma}(u)\| \neq 0$, we can always use arc-length as a parameter.

Theorem 16.4. *For each $u_0 \in \mathbb{R}$, $p \in M$, and $v \in T_p$ there is an interval containing u_0 and a unique curve γ defined on that interval such that $\gamma(u_0) = p$, $\dot{\gamma}(u_0) = v$, and γ^i satisfy (16.18).*

Proof. Follows immediately from Theorem 13.1 upon putting $D_u\gamma^i = v^i$. ■

Finally, we have the following generalization of Theorem 15.15.

Theorem 16.5. *Shortest and longest curves in a pseudo-Riemannian space are geodesics.*

Proof. The proof of Theorem 15.15 can be generalized. Thus, just as before a shortest or longest curve is an extremal of **E**, and we have eq. (15.5). Again we differentiate with respect to t, but now the g_{ij} are not constant. In the terms involving the derivatives of the g_{ij} we use

$$\frac{\partial g_{ij}}{\partial \mu^k} = g_{ih}\begin{Bmatrix} h \\ j\ k \end{Bmatrix} + g_{jh}\begin{Bmatrix} h \\ i\ k \end{Bmatrix} \tag{16.19}$$

(see eq. (16.35)) and in the other term we use

$$\frac{\partial}{\partial u}\left(g_{ij}\frac{\partial \gamma^i}{\partial t}\frac{\partial \gamma^j}{\partial u}\right) = g_{ij}\frac{\partial}{\partial u}\left(\frac{\partial \gamma^i}{\partial t}\right)\frac{\partial \gamma^j}{\partial u} + g_{ij}\frac{\partial \gamma^i}{\partial t}\frac{\partial^2 \gamma^j}{\partial u^2} + \frac{\partial g_{ij}}{\partial \mu^k}\frac{\partial \gamma^k}{\partial u}\frac{\partial \gamma^i}{\partial t}\frac{\partial \gamma^j}{\partial u}$$

The term on the left drops out after integration. The term on the right is expanded in terms of Christoffel symbols and some of these cancel with those above. We end up with

$$\int_p^q \left[g_{ij}\frac{\partial \gamma^i}{\partial t}\frac{\partial^2 \gamma^j}{\partial u^2} + g_{ih}\begin{Bmatrix} h \\ j\ k \end{Bmatrix}\frac{\partial \gamma^i}{\partial t}\frac{\partial \gamma^j}{\partial u}\frac{\partial \gamma^k}{\partial u}\right] du = 0$$

The integrand is

$$g_{ij}\frac{\partial \gamma^i}{\partial t}\left(\frac{\partial^2 \gamma^j}{\partial u^2} + \begin{Bmatrix} j \\ h\ k \end{Bmatrix}\frac{\partial \gamma^h}{\partial u}\frac{\partial \gamma^k}{\partial u}\right)$$

and now exactly as in the proof of Theorem 15.15 we conclude that the last factor must vanish. ■

PROBLEM 16.9. In the following examples you are given a chart on a 2-dimensional manifold such that the representation of the g_{ij} are as shown. Find the representation of the Christoffel symbols and the representations of the geodesics. (c is a real constant, and $(a, b) \in \mathbb{R}^2$)

	$\hat{g}_{11}(a, b)$	$\hat{g}_{12}(a, b)$	$\hat{g}_{22}(a, b)$
1	c^2	0	$c^2 \sin^2 a$
2	$1 + c^2$	0	a^2
3	$1 + 4a^2$	0	a^2

PROBLEM 16.10. Prove Theorem 16.3.

16.3 The torsion and curvature of a (linear or affine) connection

In Section 16.1 we saw that having a tensor field, g, on M we were able to define a set of 1-forms, λ^i_j, on each coordinate patch in terms of which we subsequently constructed $(r, s + 1)$ tensor fields on M. (With a given $(1, 0)$ field we constructed a $(1, 1)$ field, etc.)

We will now generalize somewhat as follows. Suppose we start with a manifold, M, not necessarily pseudo-Riemannian, and suppose there are given on each coordinate patch a set of 1-forms which satisfy eq. (16.3). That is, suppose we are given 1-forms, ω^l_i, such that

$$p^m_l \omega^l_i = p^r_i \bar{\omega}^m_r + dp^m_i \tag{16.20}$$

where $p^i_j = \partial\bar{\mu}^i/\partial\mu^j$ as before.

Definitions
A manifold with a collection of local 1-forms, ω^l_i, satisfying condition (16.20) is *a manifold with a (linear or affine) connection*. The 1-forms are called *connection 1-forms*.

Notice now that this structure is all that is needed to proceed to construct covariant derivatives. In particular, we have eqs. (16.4)–(16.8) and (16.10) and (16.11) with ω^l_i replacing λ^l_i. We can now define *the coefficients*, L^k_{ij}, *of a connection* (with respect to a coordinate basis) by

$$\omega^k_i = L^k_{ij}\, d\mu^j \tag{16.21}$$

Then eq. (16.2) and eqs. (16.9), (16.14), and (16.15) are valid with $\begin{Bmatrix} k \\ i\ j \end{Bmatrix}$ replaced by L^k_{ij}.

We can continue just as in Section 16.2 to define parallel propagation and geodesics with L^k_{ij} replacing $\begin{Bmatrix} k \\ i\ j \end{Bmatrix}$. The properties described in Section 16.2 of these concepts will still be valid, except that arc-length is no longer available since we do not have the length of a vector.

It is, however, important to notice that while $\begin{Bmatrix} k \\ i\ j \end{Bmatrix}$ are symmetric in i and j, this is not necessarily true of L^k_{ij}. From eq. (16.2) with $\begin{Bmatrix} k \\ i\ j \end{Bmatrix}$ replaced by L^k_{ij} we see that if L^k_{ij} is symmetric then $\bar{L}^k_{ij}$ is symmetric; i.e., the property of symmetry is independent of coordinates.

Theorem 16.6. *L^k_{ij} is symmetric (in its lower indices) on a chart $(\mathscr{U}, \mu)$ iff for* *each point $p \in \mathscr{U}$ there is a chart $(\bar{\mathscr{U}}, \bar{\mu})$ with $p \in \bar{\mathscr{U}}$ in which $\bar{L}^k_{ij}(p) = 0$.*

Proof. **If:** If $\bar{L}^k_{ij}(p) = 0$ in $(\bar{\mathscr{U}}, \bar{\mu})$ then at p, *eq.* (16.2) *becomes*

$$L^l_{ij} p^m_l = \frac{\partial^2 \bar{\mu}^m}{\partial \mu^i \, \partial \mu^j},$$

so that at p, L^l_{ij} is symmetric, and we can get this for each $p \in \mathscr{U}$.

Only if: Pick $p \in \mathscr{U}$, and suppose $(\mathscr{U}, \mu)$ is a chart with $\mu(p) = 0$. Define new coordinates on $\mathscr{U}$ of the form

$$\bar{\mu}^i = \mu^i + \tfrac{1}{2} f^i_{\ jk} \mu^j \mu^k$$

Then at p, $p^m_l = \delta^m_l$ and $\partial^2 \bar{\mu}^m / \partial \mu^i \, \partial \mu^j = \frac{1}{2}(f^m_{ij} + f^m_{ji})$. So from eq. (16.2) the condition $\bar{L}^m_{rs}(p) = 0$ is the same as the condition $L^m_{ij}(p) = \frac{1}{2}[f^m_{ij}(p) + f^m_{ji}(p)]$. Since $L^m_{ij}(p)$ is symmetric, this condition can be satisfied by $f^m_{ij}(p) = L^m_{ij}(p)$. ∎

Definition

Coordinates at p for which $L^k_{ij}(p) = 0$ are called *geodesic coordinates.*

In the special case when M is pseudo-Riemannian and L^k_{ij} are the coefficients of the Levi–Civita connection, the vanishing of the L^k_{ij} at p is equivalent to the vanishing of all the partial derivatives of the components of g.

In view of eq. (16.17), one sometimes says that the symmetry of L^k_{ij} corresponds to the existence of coordinate systems at a point of which the components, X^r, of a vector are constant under "infinitesimal parallel displacement."

There is another interesting geometrical interpretation of the symmetry condition. (Compare observation (3) Section 13.3.) We use the first-order approximation of the condition of parallel propagation, $\Delta X^r = -X^i L^r_{is} \, \Delta\mu^s$. If γ_1 and γ_2 intersect at p, let $\Delta_1 \mu^s$ and $\Delta_2 \mu^s$ be "small displacements" along γ_1 and γ_2 respectively (see Fig. 16.1). Now propagate $\Delta_2 \mu^s$ parallelly along γ_1, and propagate $\Delta_1 \mu^s$ parallelly along γ_2. We get, respectively,

$$\Delta_2 \mu^s + [-\Delta_2 \mu^i L^r_{is} \Delta_1 \mu^s]$$

and $\Delta_1 \mu^s + [-\Delta_1 \mu^i L^r_{is} \Delta_2 \mu^s]$ so that the small vector between the end points of the new vectors is, to the second-order approximation,

$$(L^r_{is} - L^r_{si}) \Delta_1 \mu^s \, \Delta_2 \mu^i.$$

Hence, in particular, if (and only if) the connection is symmetric we get an "infinitesimal parallelogram."

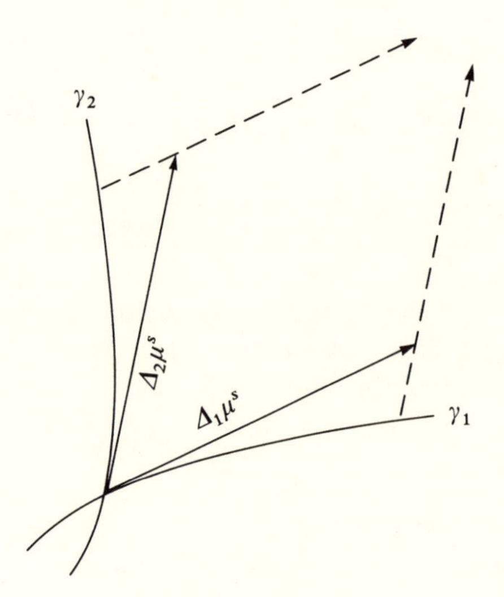

FIGURE 16.1

We will now produce two tensor fields in terms of which we can describe the connection 1-forms.

From eq. (16.2) with $\left\{ {k \atop i\ j} \right\}$ replaced by L^k_{ij} we see that the differences $L^k_{ji} - L^k_{ij}$ are the components of a (1, 2) tensor field skew-symmetric in its covariant indices.

Definition
$T^k_{ij} = L^k_{ji} - L^k_{ij}$ are the coordinate components of *the torsion tensor*, T, of the connection.

For a tensor field of type (r, s) we can form $K^{i_1 \cdots i_r}_{j_1 \cdots j_s, k, l} - K^{i_1 \cdots i_r}_{j_1 \cdots j_s, l, k}$. There is an important set of identities, one for each (r, s), *the Ricci identities*, relating these differences to the torsion tensor and another important tensor field determined by the affine connection. We will derive the one for a vector field, X. Expanding, $X^i_{,j,k}$ using, successively, eq. (16.15) and eq. (16.9) we get

$$
\begin{aligned}
X^i_{,j,k} &= \frac{\partial X^i_{,j}}{\partial \mu^k} + X^p_{,j} L^i_{pk} - X^i_{,q} L^q_{jk} \\
&= \frac{\partial^2 X^i}{\partial \mu^j \, \partial \mu^k} + \frac{\partial}{\partial \mu^k}(X^p L^i_{pj}) + X^p_{,j} L^i_{pk} - X^i_{,q} L^q_{jk} \\
&= \frac{\partial^2 X^i}{\partial \mu^j \, \partial \mu^k} + X^p \frac{\partial}{\partial \mu^k} L^i_{pj} + L^i_{pj}(X^p_{,k} - X^q L^p_{qk}) + L^i_{pk} X^p_{,j} - L^q_{jk} X^i_{,q}
\end{aligned}
$$

Interchanging j and k and subtracting, we get

$$X^i_{,j,k} - X^i_{,k,j} = X^p\left(\frac{\partial}{\partial\mu^k} L^i_{pj} - \frac{\partial}{\partial\mu^j} L^i_{kp}\right)$$
$$- X^q(L^i_{rj}L^r_{qk} - L^i_{rk}L^r_{qj}) - X^i_{,q}(L^q_{jk} - L^q_{kj})$$
$$= X^p\left[\frac{\partial}{\partial\mu^k} L^i_{pj} - \frac{\partial}{\partial\mu^j} L^i_{pk} + (L^i_{rk}L^r_{pj} - L^i_{rj}L^r_{pk})\right]$$
$$+ X^i_{,q}(L^q_{kj} - L^q_{jk})$$

Denoting the coefficient of X^p in the first term on the right by $R^i_{\cdot pkj}$ we have the result

$$X^i_{,j,k} - X^i_{,k,j} = X^p R^i_{\cdot pkj} + X^i_{,q} T^q_{jk} \tag{16.22}$$

Definition

$$R^i_{\cdot pkj} = \frac{\partial}{\partial\mu^k} L^i_{pj} - \frac{\partial}{\partial\mu^j} L^i_{pk} + L^i_{rk}L^r_{pj} - L^i_{rj}L^r_{pk} \tag{16.23}$$

are the coordinate components of *the* (*Riemann*) *curvature tensor*, *R*, *of the connection.*

Note that $R^i_{\cdot pkj}$ is skew-symmetric in its last two covariant indices.

We noted in Section 16.1 that the covariant derivative of a vector field corresponds to a vector-valued 1-form. Similarly, the torsion tensor, T, corresponds to a vector-valued 2-form, $\tilde{T}$, defined by

$$\langle \omega, \tilde{T}(X, Y)\rangle = T(\omega, X, Y)$$

We can write $\tilde{T}(X, Y)$ in terms of a basis $\{e_i\}$ of local vector fields as

$$\tilde{T}(X, Y) = T^k(X, Y)e_k \tag{16.24}$$

This defines the components $T^k(X, Y)$. $T^k\colon (X, Y) \mapsto T^k(X, Y)$ are k ordinary 2-forms. If we write $T = T^k_{ij} e_k \otimes \varepsilon^i \otimes \varepsilon^j$, we find

$$T^k = T^k_{ij}\varepsilon^i \otimes \varepsilon^j \tag{16.25}$$

Then

$$T^k(e_i, e_j) = T^k_{ij} \tag{16.26}$$

and, from (16.24),

$$\tilde{T}(e_i, e_j) = T^k_{ij} e_k \tag{16.27}$$

Definitions
The vector field $\tilde{T}(X, Y)$ is called *the torsion translation* and the T^k are called *the torsion forms* of the connection.

Now $T^k(X, Y) = T^k_{ij} X^i Y^j = L^k_{ji} X^i Y^j - L^k_{ij} X^i Y^j$ when the T^k_{ij} are coordinate components of T. But

$$L^k_{ji} X^i Y^j - L^k_{ij} X^i Y^j = \omega^k_i(X)\, d\mu^i(Y) - \omega^k_i(Y)\, d\mu^i(X) = 2\omega^k_i \wedge d\mu^i(X, Y).$$

So

$$T^k = 2\omega^k_i \wedge d\mu^i \tag{16.28}$$

Equation (16.28) can be thought of as giving the torsion forms in terms of the connection 1-forms. This could be used to define the torsion tensor instead of the way we did it.

We gave the definition of the torsion tensor of a connection in terms of coordinate bases. We can now calculate the components of the torsion tensor in any basis of vector fields by means of the required transformation law. We would expect to then obtain a generalization of (16.28). We will get this by a different method in Section 17.3. The result is

$$T^k = 2(\omega^k_i \wedge \varepsilon^i + d\varepsilon^k) \tag{16.29}$$

Clearly (16.29) reduces to (16.28) for coordinate vector fields.

Again, as we did for the covariant derivative and the torsion tensor we can describe the curvature tensor, which we defined as a (1, 3) tensor, in a sometimes convenient alternative fashion. From the isomorphisms

$$\mathscr{L}(T^*_p, T_p, T_p, T_p; \mathbb{R}) \cong \mathscr{L}(T_p, T_p; \mathscr{L}(T^*_p, T_p; \mathbb{R}))$$

(Theorem 2.12) and $\mathscr{L}(T^*_p, T_p; \mathbb{R}) \cong \mathscr{L}(T_p; \mathscr{L}(T^*_p; \mathbb{R})) \cong \mathscr{L}(T_p, T_p)$, we see that a (1, 3) tensor corresponds to a linear operator-valued (0, 2) tensor. Since $R^i_{\cdot pkj}$ is clearly skew-symmetric in its last two covariant indices, R, the curvature tensor, corresponds to a linear operator-valued 2-form, $\tilde{R}$, defined by

$$\langle \omega, \tilde{R}(X, Y) \cdot Z \rangle = R(\omega, Z, X, Y)$$

We can write $\tilde{R}(X, Y)$ in terms of a basis $\{e_i\}$ of local vector fields as

$$\tilde{R}(X, Y) \cdot e_j = R^i_j(X, Y) e_i \tag{16.30}$$

This defines the components $R^i_j(X, Y)$. $R^i_j\colon (X, Y) \mapsto R^i_j(X, Y)$ are ordinary 2-forms. If we write $R = R^i{}_{\cdot jkl} e_i \otimes \varepsilon^j \otimes \varepsilon^k \otimes \varepsilon^l = e_i \otimes \varepsilon^j \otimes R^i{}_{\cdot jkl} \varepsilon^k \otimes \varepsilon^l$ we find

$$R^i_j = R^i{}_{\cdot jkl} \varepsilon^k \otimes \varepsilon^l \tag{16.31}$$

Then

$$R^i_j(e_p, e_q) = R^i{}_{\cdot jpq} \tag{16.32}$$

and, from (16.30),

$$\tilde{R}(e_k, e_l) \cdot e_j = R^i{}_{\cdot jkl} e_i \tag{16.33}$$

Definitions
The linear operator $\tilde{R}(X, Y)$ on $\mathfrak{X}(M)$ is called *the curvature operator of the connection*, and the R^i_j are called *the curvature forms of the connection.*

A computation analogous to the one for the torsion tensor leads to

$$R^i_j = 2(\omega^i_k \wedge \omega^k_j + d\omega^i_j) \tag{16.34}$$

which gives the curvature forms in terms of the connection forms. Equations (16.29) and (16.34) are called *the (Cartan) structural equations of the affine connection.* Instead of thinking of these equations as giving the torsion and curvature forms in terms of the ε^k and ω^i_j, one can also view them the other way around, as imposing conditions on the 1-forms ε^k and ω^i_j given the torsion and curvature.

PROBLEM 16.11. Derive the Ricci identities (generalizations of eq. (16.22):

$$K^{i_1 \cdots i_r}_{j_1 \cdots j_s, k, l} - K^{i_1 \cdots i_r}_{j_1 \cdots j_s, l, k} = \sum_p^r K^{i_1 \cdots \hat{\imath} \cdots i_r}_{j_1 \cdots \cdots \cdots j_s} R^{i_p}{}_{\cdot ilk} - \sum_q^s K^{i_1 \cdots \cdots \cdots i_r}_{j_1 \cdots \hat{\jmath} \cdots j_s} R^j{}_{\cdot j_q lk} + K^{i_1 \cdots i_r}_{j_1 \cdots j_s, r} T^r_{kl}$$

PROBLEM 16.12. For a manifold with a symmetric (torsionless) connection prove

(i) the 1st Bianchi identity,

$$R^{i}_{\cdot pkj} + R^{i}_{\cdot kjp} + R^{i}_{\cdot jpk} = 0$$

or,

$$\tilde{R}(X, Y)\cdot Z + \tilde{R}(Y, Z)\cdot X + \tilde{R}(Z, X)\cdot Y = 0$$

(ii) the 2nd Bianchi identity,

$$R^{i}_{\cdot jkl,m} + R^{i}_{\cdot jlm,k} + R^{i}_{\cdot jmk,l} = 0$$

or

$$dR^{i}_{j} = 0$$

PROBLEM 16.13. For a pseudo-Riemannian manifold (M, g) define $R_{ijkl} = g_{ip}R^{p}_{\cdot jkl}$. Then (i) $R_{ijkl} = -R_{jikl}$; (ii) $R_{ijkl} = R_{klij}$; and (iii) R_{ijkl} has $(n^2(n^2 - 1))/12$ distinct nonvanishing components. (See Problems 5.6 and 5.7.)

PROBLEM 16.14. For the examples of Problem 16.9 calculate $\hat{R}_{ijkl}$ and $\hat{R}_{ijkl}/\det(\hat{g}_{ij})$.

16.4 The exponential map and normal coordinates

From the existence theorem, Theorem 16.4, we see that through each fixed point, p, of M there are lots of geodesics. This enables us to construct a diffeomorphism between a neighborhood of 0 of T_p and a neighborhood of p, and, as a consequence, a certain useful coordinate system at p.

Recall that Theorem 13.3 in Section 13.2 affirmed the existence of flow boxes for $\dot{\gamma}(u) = X(\gamma(u))$. There is a corresponding result for eq. (16.18); namely, for each $p \in M$ and $v \in T_p$, there exists a neighborhood $\mathscr{W}$ of (p, v) in TM and a $u_0 \in \mathbb{R}$, such that all solutions of (16.18) with initial values in $\mathscr{W}$ are defined for $0 \leq u < u_0$. Hence, in particular, for a fixed point, p, there is a neighborhood $\mathscr{U}$ of 0 in T_p for which all geodesics through p are defined for $0 \leq u < u_0$. Denote the geodesic corresponding to v by γ_v. If $u_0 > 1$ we have a differentiable mapping $\mathscr{U} \to M$ given by $v \mapsto \gamma_v(1)$. If $u_0 \leq 1$ we can make a change of parameter $\bar{u} = au$ with $a > 1/u_0$ and get geodesics with shorter initial vectors, according to Theorem 16.3, and with parameter values going to 1. Thus, in any event, we have a mapping $\exp_p$ from some neighborhood of $0 \in T_p$ to M given by $\exp_p: v \mapsto \gamma_v(1)$. $\exp_p$ is called *the exponential map at p.*

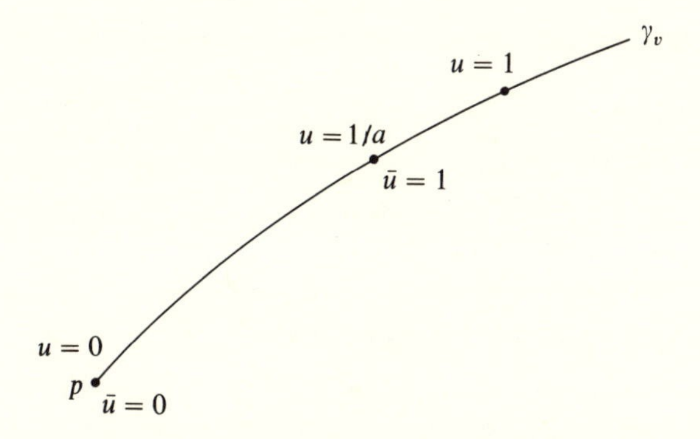

FIGURE 16.2

Theorem 16.7. *For a given $p \in M$ and $v \in T_p$, $\exp_p$ has the property $\exp_p(tv) = \gamma_v(t)$ for all $t \leq 1$.*

Proof. Suppose γ_v has a parameter u. Reparametrize γ_v with $\bar{u} = au$ so that $\bar{u} = 1$ between $u = 0$ and $u = 1$. That is, choose $a > 1$ (Fig. 16.2). Then we get the geodesic $\bar{\gamma}_{v/a}$ with initial vector v/a, by Theorem 16.3, and $\exp_p: v/a \mapsto \bar{\gamma}_{v/a}(1) = \gamma_v(1/a)$. For each $a > 1$ we have such a geodesic, so we have a map $(1/a)v \mapsto \gamma_a(1/a)$ for all $a \geq 1$.

Now let $\{e_i\}$ be a basis of T_p (with $e_i \in \mathscr{U}$). Then for $v \in \mathscr{U}$, $v = v^i e_i$. So we have a map, ϕ, of a neighborhood of $0 \in \mathbb{R}^n$ to a neighborhood of p. In particular, $(u, 0, \ldots, 0) \mapsto ue_1 \mapsto \gamma_{ue_1}(1) = \gamma_{e_1}(u)$, etc. The tangent map of ϕ at $0 \in \mathbb{R}^n$ is nonsingular since the tangents to the coordinate curves at $0 \in \mathbb{R}^n$ map into the linearly independent set $\{e_i\}$. Hence by the inverse function theorem there is a diffeomorphism between a neighborhood of $0 \in \mathbb{R}^n$ and a neighborhood, $\mathscr{V}$, of $p \in M$, and $(\mathscr{V}, \phi^{-1}|_{\mathscr{V}})$ is a local coordinate system at p. In terms of these coordinates, the equations of a geodesic are $\gamma_v^i(u) = v^i u$. $D_u^2 \gamma^k = 0$, and geodesics satisfy $\left(\begin{Bmatrix} k \\ i \quad k \end{Bmatrix} \circ \gamma\right) D_u \gamma D_u \gamma = 0$ (cf., eq. (16.18)). We can summarize these results as follows.

Theorem 16.8. *At each point, p, on a manifold with a connection there is* (i) *a diffeomorphism between a neighborhood $\mathscr{V}$ of p and a neighborhood 0 of T_p. In particular, for each $q \in \mathscr{V}$ there is a geodesic segment starting at p and ending at q;* (ii) *a local coordinate system, called* a normal (*or* Riemannian) *coordinate system in which the geodesics through p are given by $\gamma_v^i(u) = v^i u$.*

From eq. (16.18) we see that for a geodesic in these coordinates $(L_{ij}^k \circ \gamma) D_u \gamma^i D_u \gamma^j = 0$, so if L_{ij}^k is symmetric these are geodesic coordinates (Section 16.3).

Theorem 16.9. *On a pseudo-Riemannian manifold the length of the geodesic* *segment joining the points p and $q = \exp_p(v) = \gamma_v(1)$ is the length of the initial tangent vector, v.*

Proof. *Problem 16.15.*

PROBLEM 16.15. Prove Theorem 16.9.

16.5 Connections on pseudo-Riemannian manifolds

In Section 16.3 we described a certain structure which we imposed on an arbitrary manifold called a connection. The question arises whether one or more such structures can be imposed on a given manifold, i.e., questions of existence and uniqueness. Since every manifold can be given a Riemannian structure, every manifold can be given the connection of Section 16.1. The question of uniqueness is also easily disposed of. From eq. (16.2) it is evident that if L^k_{ij} are the coefficients of a connection, and K^k_{ij} are the components of a tensor, then $L^k_{ij} + K^k_{ij}$ are the coefficients of a connection, and that is all there are.

We see, in particular, that for a pseudo-Riemannian manifold there is a whole class of connections in addition to the Levi–Civita connection defined in terms of the given metric tensor field in Section 16.1. However, it is possible to sort out this particular connection by means of its properties.

Theorem 16.10. *The Levi–Civita connection has the properties*

(1) Its torsion is zero.
(2) If X and Y are both parallel along a curve, γ, then $D_u g(X, Y) \circ \gamma = 0$ (the connection "preserves" or "is compatible with" g).

Proof. (1) is immediate from the definition of T and the symmetry of the Christoffel symbols.

(2) From the definition of $[ij, k]$ and the symmetry of g we get

$$\frac{\partial g_{ik}}{\partial \mu^j} = [ij, k] + [kj, i] = g_{kl}\begin{Bmatrix} l \\ i\ j \end{Bmatrix} + g_{il}\begin{Bmatrix} l \\ k\ j \end{Bmatrix} \tag{16.35}$$

Multiplying both sides of (16.35) by $D_u\gamma^j X^i Y^k$ we get

$$\begin{aligned} X^i Y^k \frac{\partial g_{ik}}{\partial \mu^j} D_u \gamma^j &= Y^k X^i D_u \gamma^j \begin{Bmatrix} l \\ i\ j \end{Bmatrix} g_{kl} + X^i g_{il} Y^k D_u \gamma^i \begin{Bmatrix} l \\ k\ j \end{Bmatrix} \\ &= -Y^k g_{kl} D_u X^l - X^i g_{il} D_u Y^l \end{aligned}$$

using the parallelism of X and Y along γ. Thus, $D_u(g_{ik} X^i Y^k) = 0$ on γ. ■

Theorem 16.11. **(The Ricci lemma)** *On any pseudo-Riemannian manifold with a connection, property (2) of Theorem 16.10 is equivalent to* $g_{ik,j} = 0$.

Proof. Notice first that we have already, for the Levi–Civita connection, from (16.35) that $g_{ik,j} = 0$. Now

$$D_u(g_{ik}X^iY^k) = X^iY^k \frac{\partial g_{ik}}{\partial \mu^j} D_u\gamma^j + Y^k g_{kl} D_u X^l + X^i g_{il} D_u Y^l$$

$$= X^iY^k \frac{\partial g_{ik}}{\partial \mu^j} D_u\gamma^j - Y^k X^i D_u\gamma^j L^l_{ij} g_{kl} - X^i g_{il} Y^k D_u \gamma^j L^l_{kj}$$

if X and Y are parallel along γ. So

$$D_u(g_{ik}(X^iY^k) = X^iY^k D_u\gamma^j \left[\frac{\partial g_{ik}}{\partial \mu^j} - L^l_{ij} g_{kl} - L^l_{kj} g_{il} \right]$$

The conclusion follows from the fact that at any given point X^i and Y^k can be chosen arbitrarily. ■

Now we can show that the Levi–Civita connection is the only connection on a pseudo-Riemannian manifold with properties (1) and (2) of Theorem 16.10; i.e., properties (1) and (2) of Theorem 16.10 actually characterize the Levi–Civita connection among the connections on a pseudo-Riemannian manifold. This is frequently called *the fundamental theorem of Riemannian geometry.*

Theorem 16.12. *If* L^i_{jk} *is a connection on a pseudo-Riemannian manifold and* L^i_{jk} *has properties (1) and (2) of Theorem 16.10, then* $L^i_{jk} = \left\{ {i \atop j\ k} \right\}$.

Proof. From the Ricci lemma, $\partial g_{ik}/\partial \mu^j = L^l_{ij} g_{kl} + L^l_{kj} g_{il}$. Permute the indices and form $\left\{ {i \atop j\ k} \right\}$ on the left side. The right side will reduce to L^i_{jk}. ■

Definition

A connection on a pseudo-Riemannian manifold with property (2) of Theorem 16.10 (or, equivalently with $g_{ik,j} = 0$) is called *a metric connection.*

If we drop the condition that the connection of (M, g) preserves g (property (2) of Theorem 16.10), we can construct the following class of "semimetric" connections.

Theorem 16.13. *If (M, g) is a pseudo-Riemannian manifold and ω is a 1-form, then* $W^i_{jk} = \left\{ \begin{matrix} i \\ j\ k \end{matrix} \right\} + \frac{1}{2}(\delta^i_j \omega_k + \delta^i_k \omega_j - g_{jk} g^{il} \omega_l)$ are the coefficients of a symmetric connection on (M, g). Such a connection is called *a Weyl* (or *conformal*) *connection.*

Proof. Problem 16.17.

Theorem 16.14. *The Weyl connections are the symmetric connections characterized by the property $g_{mn,r} + g_{mn}\omega_r = 0$.*

Proof. Problem 16.19.

Clearly, the Levi–Civita connection is a special case of a Weyl connection.

By Theorem 16.13 we see that with each pair (g, ω) we can associate a connection. This correspondence is not 1–1, for if $g'_{mn} = f g_{mn}$ and

$$\omega'_r = \omega_r - \frac{\partial \ln |f|}{\partial \mu^r}$$

where f is a function on M, then the pair (g', ω') has the same connection as (g, ω). The map $\omega_r \mapsto \omega_r - \partial \ln |f| / \partial \mu^r$; i.e., adding an exact differential, or "gradient" to ω_r is *a gauge transformation,* and $g_{mn} \mapsto f g_{mn}$ is *a conformal transformation.* (See Section 15.1.)

Weyl spaces arose out of an historically important attempt to create a unified field theory (Bergmann, 1976, Chap. XVI). *Projective spaces* are another class of manifolds with a symmetric connection (Schouten, 1954, Chap. VI).

PROBLEM 16.16. Using eq. (16.35) and an orthonormal basis show that on a pseudo-Riemannian manifold the Levi–Civita connection 1-forms satisfy $\lambda^i_j + \lambda^j_i = 0$.

PROBLEM 16.17. Prove Theorem 16.13.

PROBLEM 16.18. On a pseudo-Riemannian manifold the coefficients of a connection are

$$L^l_{ij} = \left\{ \begin{matrix} l \\ i\ j \end{matrix} \right\} + \tfrac{1}{2}\tilde{T}^l_{ij} - \tfrac{1}{2}\tilde{g}^l_{ij}$$

where

$$\tilde{T}^l_{ij} = T^l_{ij} + g^{lk} g_{sj} T^s_{ik} + g^{lk} g_{is} T^s_{jk}$$

and

$$\tilde{g}^l_{ij} = -g^{lk} g_{ij,k} + g^{lk} g_{jk,i} + g^{lk} g_{ki,j}$$

PROBLEM 16.19. Prove Theorem 16.14. (See Problem 16.18.)

PROBLEM 16.20. For a Weyl connection, instead of property (2) of Theorem 16.10, we have that, if X and Y are parallel along a curve, then the ratio of their lengths and the angle between them are preserved.

17

CONNECTIONS ON MANIFOLDS

In Section 16.3 we defined what we mean by a manifold with a (linear or affine) connection. We made the definition in terms of local 1-form fields. There are several other ways of describing connections. We will first present two of these, and then describe the most important structures arising out of a connection on a manifold.

17.1 Connections between tangent spaces

In Section 16.3 we noted that with a connection we can introduce the idea of parallel propagation of vectors along curves. Now we will show that we can go the other way. Hence our definition of a connection and parallel propagation along curves are equivalent. Indeed, the idea of parallel propagation perhaps motivates the term "connection" better than its definition. For on a manifold in general there is no connection between tangent spaces at two different points, but with the introduction of parallel propagation of vectors along a curve we have a mapping $T_{p_0} \to T_p$ between tangent spaces at p_0 and p and thus quite literally a connection between the tangent spaces at different points of M.

Let us go along a given curve, γ, joining p_0 to $\gamma(u) = p$ in a coordinate neighborhood, $\mathscr{U}$, of p_0. Now suppose we have functions, ϕ^i_j, on $\mathscr{U}$ such that

$$X^i_p(\gamma(u)) = \phi^i_j(\gamma(u)) X^j_{p_0} \tag{17.1}$$

for all vectors X_{p_0} in T_{p_0}. Then, assuming adequate differentiability (and slightly abusing our notation by writing $X^i_p(\gamma(u)) = X^i_p(u)$ and $\phi^i_j(\gamma(u)) = \phi^i_j(u)$), we get $D_u X^i_p = (D_u \phi^i_j) X^j_{p_0}$, or

$$D_u X^i_p + f^i_j X^j_p = 0 \tag{17.2}$$

where $f^i_j = -(D_u \phi^i_j)(\phi^i_j)^{-1}$ (matrix multiplication).

Now we want the linear mappings of the tangent spaces to be independent of the choice of parameter of γ, and also of the choice of coordinates. So eq. (17.2) must have these properties. The first condition is satisfied since the f^i_j are linear functions of $D_u\gamma^k$; that is,

$$f^i_j = \ell^i_{jk} D_u \gamma^k \tag{17.3}$$

The second condition requires a certain transformation law for the f^i_j. Namely, from $\bar{X}^i = p^i_j X^j$,

$$D_u \bar{X}^i = \frac{\partial^2 \bar{\mu}^i}{\partial \mu^j \, \partial \mu^k} (D_u \gamma^k) X^j + p^i_j (D_u X^j)$$

and from (17.2),

$$-\bar{f}^i_j \bar{X}^j = \frac{\partial^2 \bar{\mu}^i}{\partial \mu^j \, \partial \mu^k} (D_u \gamma^k) X^j - p^i_j f^j_k X^k$$

or

$$\left(-\bar{f}^i_j p^j_k + p^i_j f^j_k - \frac{\partial^2 \bar{\mu}^i}{\partial \mu^j \, \partial \mu^k} D_u \gamma^j \right) X^k = 0$$

or, since this is to be independent of X^k, and using (17.3), we get

$$\left(\ell^j_{kt} p^i_j - \bar{\ell}^i_{js} p^s_t p^j_k - \frac{\partial^2 \bar{\mu}^i}{\partial \mu^k \, \partial \mu^t} \right) D_u \gamma^t = 0 \tag{17.4}$$

Finally, requiring (17.1) to be valid for all curves, γ, implies that (17.4) must be independent of $D_u \gamma^t$, so

$$\ell^j_{kt} p^i_j = \bar{\ell}^i_{js} p^j_k p^s_t + \frac{\partial^2 \bar{\mu}^i}{\partial \mu^k \, \partial \mu^t} \tag{17.5}$$

Thus, the ℓ^i_{kt} satisfy the same equations (eq. (16.2)) as the Christoffel symbols, $\left\{ \begin{matrix} i \\ k \ t \end{matrix} \right\}$, and the same equations as the coefficients L^i_{kt}, of an affine connection, and as the functions A^i_{kt} defined in Problem 16.4.

17.2 Coordinate-free description of a connection

Definition

A connection at a point, $p \in M$, is a linear map, ∇_p, from T_p to the vector space of derivations from $\mathfrak{X}$ to T_p. That is,

$$\nabla_p \colon v_p \mapsto \nabla_p(v_p)$$

where

$$\nabla_p(v_p) \colon Y \in \mathfrak{X} \mapsto \nabla_p(v_p) Y \in T_p$$

with the following properties, where we write ∇_{v_p} instead of $\nabla_p(v_p)$, replacing the redundant subscript, p, on ∇ with the argument v_p.

1. $\nabla_{av_p+bw_p} = a\nabla_{v_p} + b\nabla_{w_p}$ for $a, b \in \mathbb{R}$.
2. (i) ∇_{v_p} is $\mathbb{R}$-linear.
   (ii) $\nabla_{v_p} fY = (v_p f)Y_p + f(p)\nabla_{v_p} Y$ for $f \in \mathfrak{F}_p$.

With a given vector field, X (instead of v_p), we can extend the above construction.

Definitions
Given a vector field X, *the covariant derivative of Y in the X direction* is the vector field

$$\nabla_X Y: M \to TM$$

given by

$$p \mapsto \nabla_{X(p)} Y$$

The map

$$\nabla: \mathfrak{X} \times \mathfrak{X} \to \mathfrak{X}$$

given by

$$(X, Y) \mapsto \nabla_X Y$$

is *a connection on M*. The (1, 1) tensor field, DY, given by

$$DY(\theta, X) = \nabla_X Y(\theta)$$

where $X \in \mathfrak{X}$ and $\theta \in \mathfrak{X}^*$ is *the covariant derivative of Y*.

(The definition of the covariant derivative of Y in the X direction can be written $\nabla_X Y(p) = \nabla_{X_p} Y$. Compare this with eq. (11.3). We can say we now have a map of two variables, Y and p, described by choosing the variables successively in either order.)

Theorem 17.1. (i) *∇ has properties corresponding to those of ∇_p with f and g in place of a and b in property* (1) *above.* (ii) *The map D defined by $Y \mapsto DY$ has the properties of ∇ in Theorem 16.2.*

Proof. Problem 17.1.

Theorem 17.1(ii) brings us back to the concepts we introduced in Section 16.1, and, in particular, alerts us to a certain notational transgression we have just perpetrated. We defined a connection, ∇, as a map from $\mathfrak{X} \times \mathfrak{X}$ to $\mathfrak{X}$, but recall that in Section 16.1 we already have a map denoted by ∇ from $\mathfrak{X}$ to linear transformations in $\mathfrak{X}$. By Theorem 17.1(ii) this latter map is what we are now calling D. Thus, our definitions in Section 16.1 and here of covariant derivative are the same. The relation between ∇ and D (or ∇ in Section 16.1) is that they are corresponding elements in the isomorphism $\mathscr{L}(V_1, V_2; W) \cong \mathscr{L}(V_2, \mathscr{L}(V_1, W))$. The relation between DY and $\nabla_X Y$ is $\nabla_X Y = C\cdot(XDY)$ where C is the contraction. (See Problem 17.6.)

With regard to notation in the literature, the ∇ of this chapter is just as frequently designated by D, and usually either D or ∇ is used (but not both) for both meanings, the distinction being clear in context.

Now with a basis of vector fields we can proceed as we did in Section 16.1. In particular, if $Y = Y^i e_i$, according to Theorem 17.1(ii) and writing ∇ for D we have

$$\nabla Y = e_i \otimes dY^i + Y^i \nabla e_i \tag{17.6}$$

We define the coefficients, Γ^i_{jk}, of the connection, ∇, by

$$\nabla e_j = \Gamma^i_{jk} e_i \otimes \varepsilon^k \tag{17.7}$$

or $\nabla_{e_k} e_j = \Gamma^l_{jk} e_l$ or $\Gamma^i_{jk} = \nabla e_j(\varepsilon^i, e_k) = \nabla_{e_k} e_j(\varepsilon^i)$. Then from (17.6) and (17.7),

$$\nabla Y = (\langle dY^i, e_j\rangle + Y^k \Gamma^i_{kj}) e_i \otimes \varepsilon^j \tag{17.8}$$

so

$$C\cdot(X\nabla Y) = e_i\langle dY^i, X\rangle + Y^k \Gamma^i_{kj} e_i \langle \varepsilon^j, X\rangle$$

and

$$\nabla_X Y = (XY^i + Y^k \Gamma^i_{kj} X^j) e_i \tag{17.9}$$

Clearly, in a coordinate system, the coordinate components of ∇Y in (17.8) are precisely the functions in (16.8), with $\left\{ \begin{matrix} j \\ i\ k \end{matrix} \right\}$ replaced by Γ^j_{ik}, and the coordinate components of $\nabla_X Y$ in (17.9) are $X^j Y^i_{\ ,j}$ where $Y^i_{\ ,j}$ are the components of ∇Y.

With the Γ^i_{jk} defined here precisely as the A^i_{jk} were defined in Problem 16.4, the transformation law for the Γ^i_{jk} is the same as that for A^i_{lk} (and for $\left\{ \begin{matrix} i \\ j\ k \end{matrix} \right\}$, and for L^i_{jk}, and for ℓ^i_{jk}). Finally, if we define 1-forms ε^i_j by $\varepsilon^i_j = \Gamma^i_{jk}\varepsilon^k$, they will transform just as the λ^i_j of Section 16.1, and the connection 1-forms, ω^i_j, of Section 16.3.

Recall that in Section 16.1, after constructing the covariant derivative, ∇Y, of a vector field, we indicated that the construction can be generalized to obtain the covariant derivative, ∇K, of any tensor field. (We exhibited two specific results in Problem 16.5, and Problem 16.6.) We make a similar generalization now. Note, first, that, while starting with the vector field $\nabla_X Y$, we defined the (1, 1) tensor field ∇Y, we can just as well, starting with $\nabla_X Y$, make the following definition.

Definition
The map $\nabla_X: \mathfrak{X} \to \mathfrak{X}$ given by $Y \mapsto \nabla_X Y$ is called *covariant differentiation in the X direction.*

Theorem 17.2. *∇_X is $\mathbb{R}$-linear on $\mathfrak{X}$ and has the property, for a function, f, and a vector field, Y,*

$$\nabla_X fY = XfY + f\nabla_X Y$$

Proof. Problem 17.3.

We will generalize ∇_X instead of ∇Y. We use the same idea that we used for the Lie derivative.

Theorem 17.3. *Given a vector field, X, there exists a unique derivation, ∇_X, on the algebra of tensor fields on M with the following properties.*

(i) $\nabla_X f = Xf$ *for* $f \in \mathfrak{F}_M$.
(ii) $\nabla_X Y$ *is the covariant derivative of Y in the X direction* (*with respect to the given connection*).
(iii) $\nabla_X\langle\theta, Y\rangle = \langle\theta, \nabla_X Y\rangle + \langle\nabla_X\theta, Y\rangle$ *for* $Y \in \mathfrak{X}$ *and* $\theta \in \mathfrak{X}^*$ ("∇_X *commutes with contractions,*" *see Problem 13.12*).

Proof. ∇_X is defined uniquely on f, X, and θ by (i), (ii) and (iii). For contravariant tensor fields define $\nabla_X K$ locally by

$$\begin{aligned}\nabla_X K^{i_1\cdots i_r} e_{i_1} \otimes \cdots \otimes e_{i_r} = (\nabla_X K^{i_1\cdots i_r} e_{i_1} \otimes \cdots \otimes e_{i_{r-1}}) \otimes e_{i_r} \\ + (K^{i_1\cdots i_r} e_{i_1} \otimes \cdots \otimes e_{i_{r-1}}) \otimes \nabla_X e_{i_r}\end{aligned}$$

and for mixed tensor fields define $\nabla_X K$ locally by

$$\begin{aligned}\nabla_X K^{i_1\cdots i_r}_{j_1\cdots j_s} e_{i_1} \otimes \cdots \otimes \varepsilon^{j_s} = (\nabla_X K^{i_1\cdots i_r}_{j_1\cdots j_s} e_{i_1} \otimes \cdots \otimes \varepsilon^{j_{s-1}}) \otimes \varepsilon^{j_s} \\ + (K^{i_1\cdots i_r}_{j_1\cdots j_s} e_{i_1} \otimes \cdots \otimes \varepsilon^{j_{s-1}}) \otimes \nabla_X \varepsilon^{j_s}\end{aligned}$$

To show $\nabla_X JK = J\nabla_X K + (\nabla_X J)K$, write the local expression for both sides and use the given definitions to get an identity. For any derivation, specification of the operation on the lower-order tensors determines it (uniquely) on the higher-order ones. Finally, existence and uniqueness on each coordinate neighborhood gives existence and uniqueness on the entire manifold (cf., Theorem 12.1). ∎

In terms of a given basis of vector fields, we have, in particular, for a 1-form θ,

$$\nabla_X\theta = \nabla_X\theta_i\varepsilon^i = (\nabla_X\theta_i)\varepsilon^i + \theta_i\nabla_X\varepsilon^i = (X\theta_i)\varepsilon^i + \theta_i\nabla_X\varepsilon^i$$

by Theorem 17.3(i). By Theorem 17.3(iii), $\langle \nabla_X \varepsilon^i, e_j \rangle = -\langle \varepsilon^i, \nabla_X e_j \rangle$, so $\nabla_X \varepsilon^i \cdot e_j = -X^k \langle \varepsilon^i, \nabla_{e_k} e_j \rangle = -X^k \Gamma^i_{jk}$ by the definition of Γ^i_{jk} in equation (17.7). So, finally,

$$\nabla_X \theta = (X\theta_i - \theta_j X^k \Gamma^j_{ik})\varepsilon^i \tag{17.10}$$

Now we define ∇K for any tensor field, K, in terms of $\nabla_X K$ by

$$\nabla K(\theta^1, \ldots, \theta^r, X, X_1, \ldots, X_s) = \nabla_X K(\theta^1, \ldots, \theta^r, X_1, \ldots, X_s)$$

In particular, $\nabla f(X) = \nabla_X f$, $\nabla Y(\theta, X) = \nabla_X Y(\theta)$, and $\nabla \theta(X, X_1) = \nabla_X \theta(X_1)$. From eqs. (17.9) and (17.10), respectively, we get the coordinate components of ∇Y and $\nabla \theta$, already obtained in eqs. (16.9) and (16.14).

Finally, we want to describe parallel propagation of vectors along curves and geodesics (Section 16.2) in terms of the present notation.

If X is a vector field over a curve, γ, we can define $\nabla_X Y$ as before. Then a vector field Y is propagated parallelly along γ if $\nabla_{\dot{\gamma}} Y = 0$ on γ.

If Y is a vector field over a curve, γ, we can extend it locally to a vector field, Z, on a neighborhood of γ. Then $Y = Z \circ \gamma$ and we define

1. $\nabla_{\dot{\gamma}} Y = (\nabla_{\dot{\gamma}} Z) \circ \gamma$, the covariant derivative of Y along γ (cf., Section 16.2).
2. Y is propagated parallelly along γ if $\nabla_{\dot{\gamma}} Y = 0$.
3. γ is a geodesic if $\nabla_{\dot{\gamma}} \dot{\gamma} = 0$.

PROBLEM 17.1. Prove Theorem 17.1.

PROBLEM 17.2. Using eq. (17.9) and the formula of Theorem 12.2(ii) for the differential of 1-forms, show that

$$\nabla_X Y - \nabla_Y X - [X, Y] = 2(d\varepsilon^i + \varepsilon^i_j \wedge \varepsilon^j)(X, Y)e_i \tag{17.11}$$

PROBLEM 17.3. Prove Theorem 17.2.

PROBLEM 17.4. (i) Suppose X_i, $i = 1, \ldots, n$, are independent vector fields on an n-dimensional manifold, M. For any vector fields X and $Y = Y^i X_i$, define $\nabla_X Y$ by $\nabla_X X_i = 0$ and $\nabla_X Y = XY^i X_i$. Then ∇_X satisfies Theorem 17.2. Each X_i is said to be *a parallel vector field* (with respect to this connection).

(ii) We will say a manifold has *a flat connection*, ∇, if it has a coordinate covering on each domain of which the coefficients of ∇ are zero. Show that an affine manifold (see Section 15.3) has a flat connection.

PROBLEM 17.5. Prove that for any three vector fields, X, Y, Z, and any second-order covariant tensor field, g,

$$Z(g(X, Y)) = g(\nabla_Z X, Y) + g(X, \nabla_Z Y) + (\nabla_Z g)(X, Y) \tag{17.12}$$

PROBLEM 17.6. Using contractions we can write the relation between DY and $\nabla_X Y$ as $\nabla_X Y = C\cdot(XDY)$ and we can write eq. (17.12) as $\nabla_Z C_1^1 \cdot C_2^2 \cdot (XYg) = C_1^1 \cdot C_2^2 \cdot \nabla_Z(XYg)$.

PROBLEM 17.7. Show that covariant differentiation satisfies the same formula as Lie differentiation in Theorem 13.12. (Hint: Generalize the second result of Problem 17.6.)

PROBLEM 17.8. Show that the definition of $\nabla_{\dot{\gamma}} Y$ is independent of the vector field Z.

17.3 The torsion and curvature of a connection

In Section 16.3 we introduced and discussed the concepts of the torsion and the curvature of a connection. These tensors were defined in terms of their components in a coordinate basis. We now give an alternative approach which starts with coordinate-free definitions.

Definition
Given a connection, we can define a map

$$\tilde{T}: \mathfrak{X} \times \mathfrak{X} \to \mathfrak{X}$$

by

$$(X, Y) \mapsto \nabla_X Y - \nabla_Y X - [X, Y]$$

That is,

$$\tilde{T}(X, Y) = \nabla_X Y - \nabla_Y X - [X, Y] \tag{17.13}$$

Clearly, $\tilde{T}$ is a vector-valued 2-form.

By Problem 17.2, the components of the vector $\tilde{T}(X, Y)$ with respect to an arbitrary basis are $2(d\varepsilon^i + \varepsilon^i_k \wedge \varepsilon^k)(X, Y)$. In particular, with respect to a coordinate basis its components are $(2\varepsilon^i_k \wedge d\mu^k)(X, Y)$. But these are the components, in a coordinate basis, of the vector $\tilde{T}(X, Y)$ defined in Section 16.3—as we showed in the paragraph just above eq. (16.28). So these vectors are the same. It follows, in particular, that eq. (16.29) is the general form of eq. (16.28). The 2-forms T^k are called *the torsion forms of the connection*, and the (1, 2) tensor, T, defined by $T(\omega, X, Y) = \langle \omega, \tilde{T}(X, Y) \rangle$ is *the torsion tensor of the connection.*

Definition
Given a connection, we can define a map

$$\tilde{R}: \mathfrak{X} \times \mathfrak{X} \to \mathscr{L}(\mathfrak{X}, \mathfrak{X})$$

by

$$(X, Y) \mapsto \nabla_X \nabla_Y - \nabla_Y \nabla_X - \nabla_{[X, Y]}$$

That is,

$$\tilde{R}(X, Y) \cdot Z = \nabla_X \nabla_Y Z - \nabla_Y \nabla_X Z - \nabla_{[X, Y]} Z \tag{17.14}$$

Clearly $\tilde{R}(X, Y)$ is a linear operator on $\mathfrak{X}$, and $\tilde{R}$ is a 2-form. If the components of $\tilde{R}(X, Y)$ in a basis are $R^i_j(X, Y)$ and a (1, 3) tensor, R, is defined by $R(\omega, Z, X, Y) = \langle \omega, \tilde{R}(X, Y) \cdot Z \rangle$, then $R^i_j = R^i_{.jkl} \varepsilon^k \otimes \varepsilon^l$. R is *the (Riemann) curvature tensor of the connection*, and the 2-forms R^i_j are *the curvature forms of the connection.*

We have to show that these $R^i_{.jkl}$ are the $R^i_{.jkl}$ in Section 16.3 when a coordinate basis is used. We write the left side of (17.14) in any basis,

$$\tilde{R}(X, Y) \cdot Z = R^i_{.pkj} X^k Y^j Z^p e_i \tag{17.15}$$

and, in view of our observations just below eq. (17.9), the components of the right side of (17.14) are

$$X^k (Y^j Z^i_{,j})_{,k} - Y^j (X^k Z^i_{,k})_{,j} - [X, Y]^r Z^i_{,r} \tag{17.16}$$

In a coordinate basis, $[X, Y]r = XY^r - YX^r$, so expanding (17.16) we get

$$X^k Y^j [Z^i_{,j,k} - Z^i_{,k,j} + (L^q_{jk} - L^q_{kj}) Z^i_{,q}] \tag{17.17}$$

Equating (17.17) and the coefficient of the right side of (17.15), we see that $R^i_{.pkj}$ satisfy the condition (16.22) by which these coefficients were originally defined.

We now have a rather neat formal description of the curvature tensor of a connection in eq. (17.14). We have some analytical meaning for this concept from the way it was introduced in Section 16.3, namely, as a measure of the noncommutativity of second-order covariant derivatives. This description can also be compared with the noncommutativity relation $L_{[X, Y]} Z = L_X L_Y Z - L_Y L_X Z$ for Lie derivatives. See Problem 13.11. We have, finally, to consider its geometrical significance. This is illustrated by two of its properties.

(1) Let X be a vector field, and let Y be a vector field such that $[X, Y] = 0$, for example, the tangent field of the horizontal curves of a

variation of some curve (Section 15.2) with the given X along it. Then putting $Z = X$ in eq. (17.14) and substituting $\nabla_Y X$ into it from eq. (17.13), we get

$$\tilde{R}(X, Y)\cdot X = \nabla_X(\nabla_X Y - \tilde{T}(X, Y)) - \nabla_Y \nabla_X X$$

If, further, the integral curves of X are geodesics, the last term on the right vanishes, and we have the result that, if X is a vector field whose integral curves are geodesics and Y is a vector field such that $[X, Y] = 0$, then X and Y satisfy

$$\nabla_X \nabla_X Y = \tilde{R}(X, Y)\cdot X + \nabla_X \tilde{T}(X, Y)$$

In particular, if we pick out a geodesic, γ, then on γ, the restriction, Y_γ, of Y satisfies

$$\nabla_{\dot{\gamma}} \nabla_{\dot{\gamma}} Y_\gamma = \tilde{R}(\dot{\gamma}, Y_\gamma)\cdot\dot{\gamma} + \nabla_{\dot{\gamma}} \tilde{T}(\dot{\gamma}. Y_\gamma) \tag{17.18}$$

Equation (17.18) is called *the Jacobi equation*, or *the equation of geodesic deviation.* It is second-order ordinary differential equation. A solution is called *a Jacobi vector field.*

We can interpret (17.18) as a relation between the curvature and the deviation of the geodesics if we consider the case of a Riemannian manifold with the Levi–Civita connection (so $\tilde{T} = 0$) and the geodesics above all passing through a point p on γ. (p cannot be in any coordinate neighborhood in which X and Y are coordinate vector fields, but it can be a limit point of such a neighborhood.)

Let s be the arc-length along γ measured from p, and introduce *the sectional (Riemannian) curvature*

$$K_{xy} = g_{ij} R^i_{.klm} x^j y^k x^l y^m$$

determined by the unit vectors $x = \dot{\gamma}$ and $y = Y_\gamma / \| Y_\gamma \|$. (See Problem 17.10.) Then taking the scalar product of (17.18) with y we get

$$\frac{d^2 \| Y_\gamma \|}{ds^2} + \| Y_\gamma \| g(y, \nabla_x \nabla_x y) + \| Y_\gamma \| K_{xy} = 0$$

One can show that in the limit as $s \to 0$, the middle term drops out (Problem 17.14) and we get

$$\lim_{s \to 0} \frac{1}{\| Y_\gamma \|} \frac{d^2 \| Y_\gamma \|}{ds^2} = -K_{xy} \tag{17.19}$$

Let s^* be the length parameter along the integral curves of Y. Then $s^* = \|Y_\gamma\| v$ + terms of order v^2, where v is a parameter for Y. So

$$\frac{d^2\|Y_\gamma\|}{ds^2} = \lim_{v\to 0} \frac{\partial^2}{\partial s^2}\left(\frac{\partial s^*}{\partial v}\right)$$

and hence along γ near $s = 0$, if $K_{xy} > 0$ the rate of separation of the geodesics decreases, and if $K_{xy} < 0$ the rate of separation of the geodesics increases.

We will see eq. (17.18) (with $\tilde{T} = 0$) again in Section 22.3. There it will be related to an equation of "relative" motion, in which Y_γ is the "relative position vector" of points on neighboring geodesics, subject to a "tidal force."

(2) We have another illustration of the geometrical meaning of the curvature tensor in the special case $R = 0$.

Definition
A vector field X which satisfies $\nabla X = 0$ is called *a parallel vector field.* (Cf., Problem 17.4.)

Theorem 17.4. *On a manifold with a connection, ∇, there exists a neighborhood of p in which $R = 0$ iff there exist n linearly independent parallel vector fields in some neighborhood of p.*

Proof. **If:** At each point, $R(\omega, Z, X, Y) = \langle \omega, \tilde{R}(X, Y)\cdot Z\rangle = 0$ for all ω, Z, X, Y since by eq. (17.14) $R(X_i, X_j)\cdot X_k = 0$ for the parallel vector fields X_i, $i = 1, \ldots, n$.

Only if: We will give a proof for the case when ∇ is the Levi–Civita connection of a pseudo-Riemannian manifold. For the general case see Bishop and Goldberg (1968, pp. 236–237).

Consider two points p and q, and two curves γ_1 and γ_2 joining p and q (Fig. 17.1). Given an initial vector X_p at p, move it parallelly along γ_1 and γ_2 to get X_{q_1} and X_{q_2}, respectively, at q. We will show $R = 0 \Rightarrow X_{q_1} = X_{q_2}$.

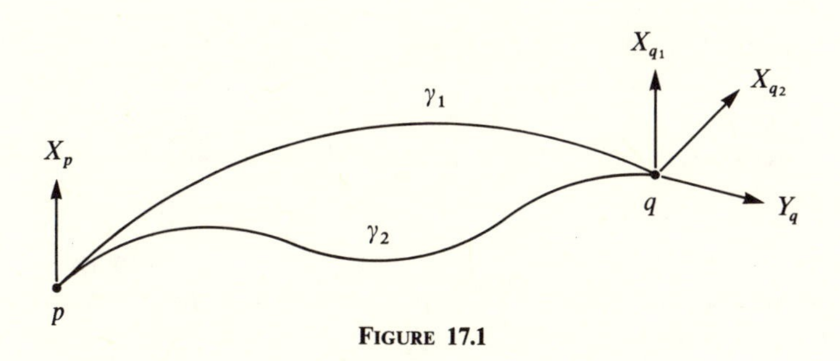

FIGURE 17.1

Parametrize both γ_1 and γ_2 by u going between u_1 and u_2. Imbed γ_1 and γ_2 in a family of curves. Thus, for example, if in coordinates $\gamma_1: u \mapsto \gamma_1^i(u)$ and $\gamma_2: u \mapsto \gamma_2^i(u)$, we can construct the family $f^i(u, t) = \gamma_1^i(u) + t(\gamma_2^i(u) - \gamma_1^i(u))$ on $I = \{(u, t) \in \mathbb{R}^2: u_1 \leq u \leq u_2 \text{ and } 0 = t_1 \leq t \leq t_2 = 1\}$. Now we can also propagate X_p along all the curves of the family and get a vector field, X, on the image of $I - q$. Finally, choose a vector Y_q at q and propagate it back to p along all the curves and get another vector field.

Let v_u be the tangent vector of the $u =$ const. curves, and let v_t be the tangent vector of the $t =$ const. curves. Then

$$
\begin{aligned}
&\nabla_{v_u}(g(X, Y)) - g(X, \nabla_{v_u} Y) \\
&= g(\nabla_{v_u} X, Y) && \text{by eq. (17.12) and Theorem 16.11} \\
&= \int_{u_1}^{u} \frac{\partial}{\partial u} g(\nabla_{v_u} X, Y)\, du && \text{since } \nabla_{v_u} X = 0 \text{ at } u = u_1 \\
&= \int_{u_1}^{u} \nabla_{v_t} g(\nabla_{v_u} X, Y)\, du \\
&= \int_{u_1}^{u} g(\nabla_{v_t} \nabla_{v_u} X, Y)\, du && \text{by eq. (17.12), and since } \nabla_{v_t} Y = 0 \\
&= \int_{u_1}^{u} g((\nabla_{v_u} \nabla_{v_t} X + \tilde{R}(v_t, v_u)\cdot X), Y)\, du && \text{by eq. (17.14)} \\
&= \int_{u_1}^{u} g(\tilde{R}(v_t, v_u)\cdot X, Y)\, du && \text{since } \nabla_{v_t} X = 0
\end{aligned}
$$

In particular, at $u = u_2$ we get

$$\nabla_{v_u}(g(X, Y)) = \int_{u_1}^{u_2} g(\tilde{R}(v_t, v_u)\cdot X, Y)\, du$$

Finally, integrating from t_1 to t_2 we get

$$g(X_{q_2}, Y_q) - g(X_{q_1}, Y_q) = \int_{t_1}^{t_2} \int_{u_1}^{u_2} g(\tilde{R}(v_t, v_u)\cdot X, Y)\, du\, dt$$

and hence, since Y_q is arbitrary, if $R = 0$, then $X_{q_2} = X_{q_1}$. Corresponding to n linearly independent vectors X_p at p we thus get n linearly independent parallel vector fields. ■

Theorem 17.5. *If in Theorem 17.4 ∇ is the Levi–Civita connection of a pseudo-Riemannian manifold, then there is a local coordinate system at p for which*

$$g_{ij} = \begin{cases} \pm 1, & i = j \\ 0, & i \neq j \end{cases}$$

Proof. If X^i are components of a vector field, X then we get a 1-form ω with components $\omega_i = g_{ij}X^j$ and $\omega_{i,k} = g_{ij}X^j{}_{,k}$. If X is a parallel vector field ($\nabla X = 0$) then $\omega_{i,k} = 0$ and hence $d\omega = 0$. We get then, locally, a function f such that $df = \omega$, or $\partial f/\partial \mu^i = \omega_i$.

Let ω^α, and f^α be, respectively, n 1-forms and n functions corresponding to n linearly independent parallel vector fields X_α. Choose new coordinates by

$$\bar{\mu}^\alpha = f^\alpha(\mu^j)$$

Then

$$\bar{\omega}_j^\alpha = \frac{\partial \mu^k}{\partial \bar{\mu}^j}\omega_k^\alpha = \frac{\partial \mu^k}{\partial \bar{\mu}^j}\frac{\partial f^\alpha}{\partial \mu^k} = \delta_j^\alpha$$

Now for two parallel vector fields, X_α and X_β we have $\nabla g(X_\alpha, X_\beta) = 0$ (by eq. (17.12) and the Ricci lemma), or $\partial/\partial\mu^k(g_{rs}X_\alpha^r X_\beta^s) = 0$. But $g_{rs}X_\alpha^r X_\beta^s = g^{rs}\omega_r^\beta\omega_s^\alpha = \bar{g}^{rs}\delta_r^\beta\delta_s^\alpha = \bar{g}^{\beta\alpha}$. Thus, the $\bar{g}^{\beta\alpha}$ are constants. By a linear transformation we get metric coefficients of the kind required. ∎

Theorem 17.6. $R = 0$ *and* $T = 0$ *locally* ⇔ *there exist n parallel coordinate vector fields* ⇔ *there is a coordinate system for which the coefficients of the connection vanish; i.e., the connection is* locally flat. (Chern, 1959, p. 82.)

Proof. The first equivalence comes from eq. (17.13) and the fact (see Section 13.3) that X_i are coordinate vector fields if and only if $[X_i, X_j] = 0$. The second equivalence comes from the definition of the Γ's in eq. (17.7). ∎

PROBLEM 17.9. Show that (i) the properties of the Levi–Civita connection in Theorem 16.10 can be written, respectively,

(1) $[X, Y] = \nabla_X Y - \nabla_Y X$.
(2) $Zg(X, Y) = g(\nabla_Z X, Y) + g(X, \nabla_Z Y)$.

(ii) (1) and (2) together are equivalent to

$$2g(\nabla_X Y, Z) = Xg(Y, Z) + Yg(Z, X) - Zg(X, Y) - g(X, [Y, Z]) + g(Y, [Z, X]) + g(Z, [X, Y])$$

(the Koszul formula).

PROBLEM 17.10. (i) Show that in a pseudo-Riemannian manifold with the Levi–Civita connection,

$$K_{XY} = \frac{g(\tilde{R}(X, Y)\cdot Y, X)}{g(X, X)g(Y, Y) - g(X, Y)^2}$$

depends only on the plane (in T_pM) determined by X and Y at each point. K_{XY} is called *the sectional (Riemannian) curvature of the plane determined by X and Y.*

(ii) Show that R is determined if K_{XY} is known for all pairs X, Y. (See Problem 5.6.)

PROBLEM 17.11. For S^3 with the coordinates and the Riemannian structure of Problem 15.8, compute K_{XY} for pairs of coordinate vectors.

PROBLEM 17.12. The sectional, or Riemannian, curvature, K_{XY}, at a point, p, is the Gaussian curvature of the surface formed by the geodesics through p in the directions of the subspace $\langle\{X, Y\}\rangle$. (Cf., Laugwitz, 1965, pp. 125 ff, or Synge and Schild, 1966, pp. 94 ff.)

PROBLEM 17.13. K_{XY} is independent of X and Y at a point iff

$$R_{ijkl} = \kappa(g_{ik}g_{jl} - g_{il}g_{jk})$$

or

$$\tilde{R}(X, Y)\cdot Z = \kappa(g(Z, Y)X - g(Z, X)Y)$$

and κ is their common value. (See Problem 5.6.)

PROBLEM 17.14. Show $\lim_{s\to 0} \| Y_\gamma \| g(y, \nabla_x \nabla_x y) = 0$ in the derivation of eq. (17.19) (Synge and Schild, p. 97).

PROBLEM 17.15. Prove the last statement in the proof of Theorem 17.4.

17.4 Some geometry of submanifolds

If (P, ϕ) is a submanifold of a pseudo-Riemannian manifold (M, h), then, in general, (P, ϕ) has a natural induced pseudo-Riemannian structure, ϕ^*h. See Section 15.1. In coordinate components the relation between h and $g = \phi^*h$ is given by eq. (15.3). ϕ^*h has the required properties except that in the non-Riemannian case ϕ^*h could be degenerate, for example, on the null curves of a Lorentzian manifold.

We have a criterion for the nature of a submanifold in the following special case

Theorem 17.7. *Suppose (P, ϕ) is a hypersurface of a Lorentzian manifold, (M, h). (i) If ϕ^*h is nondegenerate, then there exists a normal vector, n, at each point such that $h(n, n) < 0$ ($h(n, n) > 0$) iff (P, ϕ) is Riemannian (Lorentzian). (ii) ϕ^*h is degenerate iff there exists a normal vector, n, at each point such that $h(n, n) = 0$.*

Proof. These results are valid at each point according to Problem 5.25. ■

A submanifold, (P, ϕ), if ϕ^*h is nondegenerate, also has a natural induced connection. We can think of P as contained in M, and ϕ as the inclusion map (Section 10.1). Then at each $p \in P$ the tangent space T_pP is a subspace of T_pM. For (P, g) pseudo-Riemannian and, hence, g nondegenerate, we can write $T_pM = T_pP \oplus (T_pP)^\perp$ where $(T_pP)^\perp$ is the orthogonal complement of T_pP and we have the projections $\pi \cdot T_pM = T_pP$ and $\pi^\perp \cdot T_pM = (T_pP)^\perp$.

If X and Y are vector fields on P, we can extend them to M, and if ∇' is a connection on (M, h), then at points of P we have

$$\nabla'_X Y = \pi \cdot \nabla'_X Y + \pi^\perp \cdot \nabla'_X Y \tag{17.20}$$

Theorem 17.8. *The map given by $(X, Y) \mapsto \pi \cdot \nabla'_X Y$ is R-bilinear and $\pi \cdot \nabla'_X fY = (Xf)Y + f\pi \cdot \nabla'_X Y$.*

Proof. Problem 17.16.

Definition
Theorem 17.8 says that $\nabla: \mathfrak{X}(P) \times \mathfrak{X}(P) \to \mathfrak{X}(P)$ given by $(X, Y) \mapsto \pi \cdot \nabla'_X Y$ is a connection, *the induced connection*, and we write $\pi \cdot \nabla'_X Y = \nabla_X Y$.

Theorem 17.9. *If ∇' is the Levi–Civita connection of a pseudo-Riemannian manifold (M, h), then ∇ is the Levi–Civita connection of (P, ϕ^*h).*

Proof. (i) Extend the vector fields X, Y on P to vector fields on M. Then on P,

$$\nabla'_X Y - \nabla'_Y X - [X, Y] = \nabla_X Y + \pi^\perp \cdot \nabla'_X Y - (\nabla_Y X + \pi^\perp \cdot \nabla'_Y X) - [X, Y]$$

so the vanishing of the left side entails the vanishing of

$$\nabla_X Y - \nabla_Y X - [X, Y] \qquad \text{(the tangential field)}$$

and

$$\pi^\perp \cdot \nabla'_X Y - \pi^\perp \cdot \nabla'_Y X \qquad \text{(the normal field)}$$

Thus, we have, in addition to the vanishing of the torsion, the extra result that the map given by $(X, Y) \mapsto \pi^{\perp} \cdot \nabla'_X Y$ is symmetric.

(ii) The condition (2) on (M, h) of Problem 17.9 reduces on P to

$$Xg(Y, Z) = g(\nabla'_X Y, Z) + g(Y, \nabla'_X Z)$$

on P, where X, Y, Z are vector fields on P and $g = \phi^* h$. But $g(\nabla'_X Y, Z) = g(\nabla_X Y + \pi^{\perp} \cdot \nabla'_X Y, Z) = g(\nabla_X Y, Z)$, since $g(\pi^{\perp} \cdot \nabla'_X Y, Z) = 0$. Similarly, for $g(Y, \nabla'_X Z)$, which gives us condition (2) of Problem 17.9 on (P, g). ■

Theorem 17.10. *For (M, h) with the Levi–Civita connection, the map II: $\mathfrak{X}(P) \times \mathfrak{X}(P) \to \mathfrak{X}(P)^{\perp}$ given by $(X, Y) \mapsto \pi^{\perp} \cdot \nabla'_X Y$ is symmetric and bilinear over $\mathfrak{F}_P$.*

Proof. *Problem 17.17.*

Definition

The map, II, of Theorem 17.10 is called *the second fundamental form* or *shape factor*, of P. (Since II is symmetric, the word "form" is a misnomer.)

In terms of the notations ∇ and II, eq. (17.20), for (P, h) with the Levi–Civita connection, is

$$\nabla'_X Y = \nabla_X Y + II(X, Y) \tag{17.21}$$

Equation (17.20) or eq. (17.21) is called *the Gauss formula* or *Gauss decomposition.*

Now we consider another decomposition. If $X \in \mathfrak{X}(P)$ and $Z(p) \in (T_p P)^{\perp}$ for $p \in P$, we can extend these vector fields to M and write

$$\nabla'_X Z = \pi \cdot \nabla'_X Z + \pi^{\perp} \cdot \nabla'_X Z \tag{17.22}$$

as we did before for $X, Y \in \mathfrak{X}(P)$.

Theorem 17.11. *The mapping $\tilde{\mathscr{S}}$: $\mathfrak{X}(P) \times \mathfrak{X}(P)^{\perp} \to \mathfrak{X}(P)$ given by $(X, Z) \to \pi \cdot \nabla'_X Z$ is $\mathfrak{F}_P$ bilinear.*

Proof. Write $\nabla'_X fZ = f\nabla'_X Z + (Xf)Z = f(\pi \cdot \nabla'_X Z + \pi^{\perp} \nabla'_X Z) + (Xf)Z$ and $\nabla'_X fZ = \pi \cdot \nabla'_X fZ + \pi^{\perp} \cdot \nabla'_X fZ$ and compare tangential components. Similarly, for $\nabla'_{fX} Z$. ■

Theorem 17.12. *The map $\nabla^{\perp}$: $\mathfrak{X}(P) \times \mathfrak{X}(P)^{\perp} \to \mathfrak{X}(P)^{\perp}$ given by $(X, Z) \mapsto \pi^{\perp} \cdot \nabla'_X Z$, called* the normal connection, *has the properties of a connection.*

Proof. *Problem 17.18.*

In terms of the notations $\tilde{\mathscr{S}}$ and $\nabla^{\perp}$, eq. (17.22) is

$$\nabla'_X Z = \tilde{\mathscr{S}}(X, Z) + \nabla^{\perp}_X Z \tag{17.23}$$

Equation (17.22) or eq. (17.23) is called *the Weingarten formula* or *Weingarten decomposition.*

Theorem 17.13. *For (M, h) with the Levi–Civita connection, $h(\tilde{\mathscr{S}}(X, Z), Y) = -h(II(X, Y), Z)$ for $X, Y \in \mathfrak{X}(P)$ and $Z \in \mathfrak{X}(P)^{\perp}$.*

Proof. Differentiate $h(Y, Z) = 0$ covariantly with respect to X. Then $h(\nabla'_X Y, Z) + h(Y, \nabla'_X Z) = 0$. Substitute for $\nabla'_X Y$ and $\nabla'_X Z$ from (17.21) and (17.23), respectively, and eliminate two terms by orthogonality. ■

We will illustrate these concepts in the geometry of submanifolds in the special cases where P is a hypersurface in the pseudo-Riemannian manifold (M, h) with the Levi–Civita connection.

1. If Z is unit normal vector field, then differentiating $h(Z, Z) = 1$ we get $h(\nabla'_X Z, Z) = 0$, so $h(\nabla^{\perp}_X Z, Z) = 0$, by (17.23). Since, M is a hypersurface, $\nabla^{\perp}_X Z$ and Z are proportional, so $\nabla^{\perp}_X Z = 0$. Thus, for Z a unit normal vector field, eq. (17.23) reduces to

$$\nabla'_X Z = \tilde{\mathscr{S}}(X, Z) \tag{17.24}$$

For a fixed $Z \in \mathfrak{X}(P)^{\perp}$, the map given by $X \mapsto \tilde{\mathscr{S}}(X, Z)$ is a linear operator on $\mathfrak{X}(P)$, so in this case the map $\mathscr{S}\colon X \mapsto \nabla'_X Z$ is a linear operator on $\mathfrak{X}(P)$. $\mathscr{S}$ is called *the shape operator* (O'Neill, 1983) or the *Weingarten map* (Hicks, 1965; Spivak, 1979). In this case we can write $II(X, Y) = l(X, Y)Z$, and the formula in Theorem 17.13 becomes

$$g(\mathscr{S}(X), Y) = -l(X, Y) \tag{17.25}$$

where $g = \phi^*h$. From this we see that $\mathscr{S}$ is a self-adjoint operator. Moreover, in a basis of T_pP at a point, p, $s_{\alpha\beta} = -l_{\alpha\beta}$, $\alpha, \beta = 1, \ldots, n-1$, where $(s_{\alpha\beta})$ is the matrix of $\mathscr{S}$ in that basis. The algebraic invariants of $\mathscr{S}$ in the special case when P is a surface in the Euclidean space $\mathscr{R}^3_0$ are the Gaussian and mean curvatures of P and the eigenvalues of $\mathscr{S}$ are the principal curvatures.

2. If we specialize further to the case where P is a hypersurface in $\mathscr{R}^n_0$ the classical results of that subject emerge.

(i) Let Z be a unit normal vector field. Then Z has components Z^i, $i = 1, \ldots, n$, with the property $\sum_{i=1}^{n} Z^i Z^i = 1$. Define the map $\phi\colon P \to S^{n-1} \subset \mathscr{R}^n_0$ by $p \mapsto (Z^1(p), \ldots, Z^n(p))$. ϕ is called *the spherical map of*

Gauss. For $X \in \mathfrak{X}(P)$, the components of $\nabla'_X Z$ are $X^j(\partial Z^i/\partial \pi^j)$. Let γ be a curve on P with $[\gamma] = X$. Then the tangent map of ϕ takes $X = [\gamma]$ to $[\phi \circ \gamma]$ and $[\phi \circ \gamma]$ has components

$$D_0(Z^i \circ \gamma) = \frac{\partial Z^i}{\partial \pi^j} D_0 \gamma^j = \frac{\partial Z^i}{\partial \pi^j} X^j$$

which are the components of $\nabla'_X Z$. Thus, $\mathscr{S}$ is the tangent map of the Gauss spherical map.

(ii) To find the components, $l_{\alpha\beta}$, of l in a coordinate system (μ^α) on (P, ϕ) we evaluate $\nabla'_{\partial/\partial\mu^\alpha}(\partial/\partial\mu^\beta)$ and take its normal component. Thus, since

$$\nabla'_{\partial/\partial\mu^\alpha}\left(\frac{\partial}{\partial\mu^\beta}\right) = \nabla'_{\partial/\partial\mu^\alpha}\left(\frac{\partial\phi^k}{\partial\mu^\beta}\frac{\partial}{\partial\pi^k}\right) = \frac{\partial^2\phi^k}{\partial\mu^\alpha\,\partial\mu^\beta}\frac{\partial}{\partial\pi^k}$$

we have

$$l_{\alpha\beta} = h\left(II\left(\frac{\partial}{\partial\mu^\alpha}, \frac{\partial}{\partial\mu^\beta}\right), Z\right) = h\left(\frac{\partial^2\phi^k}{\partial\mu^\alpha\,\partial\mu^\beta}\frac{\partial}{\partial\pi^k}, Z^i\frac{\partial}{\partial\pi^i}\right) = \sum_{k=1}^{n} Z^k \frac{\partial^2\phi^k}{\partial\mu^\alpha\,\partial\mu^\beta}$$

This is the usual classical expression for the components of the second fundamental form (cf., Eisenhart, 1947, p. 215).

(iii) In this case, using coordinate vector fields $\partial/\partial\mu^\alpha$, $\partial/\partial\mu^\beta$ for X, Y, the Gauss formula (17.21) reduces to

$$\frac{\partial^2\phi^i}{\partial\mu^\alpha\,\partial\mu^\beta} = \Gamma^\gamma_{\alpha\beta}\frac{\partial\phi^i}{\partial\mu^\gamma} + l_{\alpha\beta}Z^i \tag{17.26}$$

where Z^i are the components of a unit normal vector. Similarly, using Theorem 17.13, the Weingarten formula (17.24) becomes

$$\frac{\partial Z^i}{\partial\mu^\alpha} = -l_{\alpha\gamma}g^{\gamma\beta}\frac{\partial\phi^i}{\partial\mu^\beta} \tag{17.27}$$

Equations (17.26) and (17.27) are, respectively, the classical equations of Gauss and Weingarten (cf., Eisenhart, 1947, pp. 216–217).

Another important result in the geometry of submanifolds is the relationship between the Riemann curvature tensors of P and M. For M with X, Y, Z tangent to P, we have from (17.14),

$$\tilde{R}'(X, Y)\cdot Z = \nabla'_X\nabla'_Y Z - \nabla'_Y\nabla'_X Z - \nabla'_{[X,Y]}Z$$

Using the Gauss and Weingarten formulas (17.21) and (17.23) on the right side, we get

$$\begin{aligned}\tilde{R}'(X, Y)\cdot Z = \tilde{R}(X, Y)\cdot Z &+ \tilde{\mathscr{S}}(X, II(Y, Z)) - \tilde{\mathscr{S}}(Y, II(X, Z)) \\ &+ II(X, \nabla_Y Z) - II(Y, \nabla_Y Z) - II([X, Y], Z) \\ &+ \nabla_X^{\perp} II(Y, Z) - \nabla_Y^{\perp} II(X, Z) \end{aligned} \tag{17.28}$$

Note that the first three terms on the right are tangent to P, and the last five terms are normal to P. If W is another tangent vector field, then from (17.28), using Theorem 17.13,

$$\begin{aligned} h(\tilde{R}'(X, Y)\cdot Z, W) = h(\tilde{R}(X, Y)\cdot Z, W) &+ h(II(Y, W), II(X, Z)) \\ &- h(II(X, W), II(Y, Z)) \end{aligned} \tag{17.29}$$

Equation (17.29) is called *the Gauss curvature equation.* The fact that the normal part of $\tilde{R}'(X, Y)\cdot Z$ equals the last five terms of (17.28) is called *the Codazzi–Mainardi equation.*

In the special case where (P, ϕ) is a hypersurface in $\mathscr{R}_0^n$, the left side of (17.28) vanishes, and the tangential and normal parts on the right are separately zero, so

$$\tilde{R}(X, Y)\cdot Z = l(X, Z)\mathscr{S}\cdot Y - l(Y, Z)\mathscr{S}\cdot X \tag{17.30}$$

and

$$II(X, \nabla_Y Z) - II(Y, \nabla_X Z) - II([X, Y], Z) + \nabla_X^{\perp} II(Y, Z) - \nabla_Y^{\perp} II(X, Z) = 0 \tag{17.31}$$

For X, Y, Z coordinate fields $\partial/\partial\mu^\alpha$, $\partial/\partial\mu^\beta$, $\partial/\partial\mu^\gamma$, and using eq. (17.15), $\mathscr{S}\cdot X = s_\mu^\lambda\, \partial/\partial\mu^\lambda$, etc., and (17.25) on (17.30), and using (17.7), and $\nabla_X^{\perp} II(Y, Z) = Xl(Y, Z)N$, etc., where N is a unit normal vector field on (17.31), eqs. (17.30) and (17.31) become, respectively,

$$R^{\delta}_{\cdot\gamma\alpha\beta} = l^{\delta}_{\alpha} l_{\beta\gamma} - l^{\delta}_{\beta} l_{\alpha\gamma} \tag{17.32}$$

$$\frac{\partial}{\partial\mu^\alpha} l_{\beta\gamma} + l_{\alpha\delta}\Gamma^{\delta}_{\alpha\beta} = \frac{\partial}{\partial\mu^\beta} l_{\alpha\gamma} + l_{\beta\delta}\Gamma^{\delta}_{\gamma\alpha} \tag{17.33}$$

Equations (17.32) and (17.33) are, respectively, the classical equations of Gauss (*another* "Gauss equation") and Codazzi–Mainardi (cf., Eisenhart, 1947, p. 219).

We conclude this section with a few important observations, without going into details.

(1) Recall that the left side of (17.32), as defined in eq. (16.23), is just a certain complicated combination of the components of g and their first and second derivatives, so the combination of l's on the right side, in terms of which Gauss' definition of the curvature of a surface in $\mathscr{R}_0^3$ can be expressed, depends only on g. Hence Gauss' curvature is an intrinsic property of the surface—Gauss' celebrated *Theorema Egregium* (1827).

(2) The Gauss–Weingarten equations (17.26) and (17.27) can be considered to be an analog of the Frenet–Serret equations of curve theory in $\mathscr{R}_0^3$ (or in $\mathscr{R}_0^n$). They give the derivatives of the tangent and normal vectors in terms of the vectors themselves. For curves in $\mathscr{R}_0^3$, the coefficients, curvature and torsion determine the curve uniquely up to a rigid motion. That is the Fundamental Theorem of Curve Theory. For hypersurfaces in $\mathscr{R}_0^n$, there is an analog, the Fundamental Theorem for Hypersurfaces. That theorem says that the coefficients, the g's and l's, in eqs. (17.26) and (17.27) determine a unique surface in $\mathscr{R}_0^n$ up to a rigid motion. In the case of hypersurfaces, however, since the Gauss–Weingarten equations are partial differential equations (the corresponding Frenet–Serret equations are ordinary differential equations), the coefficients have to satisfy integrability conditions. We saw that (17.32) and (17.33) were necessary conditions for a hypersurface of $\mathscr{R}_0^n$. It turns out that they are also sufficient; that is, they are the integrability conditions.

PROBLEM 17.16. Prove Theorem 17.8.

PROBLEM 17.17. Prove Theorem 17.10.

PROBLEM 17.18. Prove Theorem 17.12.

PROBLEM 17.19. Derive eqs. (17.26) and (17.27).

PROBLEM 17.20. Find $l_{\alpha\beta}$ for S^3 in the local coordinates of Problem 15.8.

PROBLEM 17.21. Verify eqs. (17.32) and (17.33) for S^3 in the local coordinates of Problem 17.20.

PROBLEM 17.22. Show that if X and Y determine a plane in T_pP then the sectional curvatures for P and M are related by

$$K'_{XY} = K_{XY} + \frac{g(II(X, X), II(Y, Y)) - g(II(X, Y), II(X, Y))}{g(X, X)g(Y, Y) - g(X, Y)^2}$$

18

MECHANICS

Our exposition of geometry in Chapters 15, 16 and 17 consisted of starting with a manifold and superimposing a symmetric tensor field of type (0, 2), or prescribing a covariant derivative for pairs of vector fields, and then studying the properties of such abstract structures. A motivation for such generalization and abstraction is that it includes as a special case the classical well-established geometry with which we are familiar.

Starting with a manifold we can superimpose other kinds of tensor fields, or k-dimensional distributions, and then study the resulting structures. The study of manifolds with skew-symmetric tensor fields of type (0, 2) and of manifolds with $n - 1$-dimensional distributions is motivated by the fact that classical mechanics can be described in terms of these structures as a special case.

In classical mechanics, the "motion" of "systems of particles" and "rigid bodies" is described by describing each "position" or "location" by a set of numbers called generalized coordinates. The set of all positions is called *the configuration space*. To get a coordinate-free description we will, in each specific case, model configuration space by a manifold. Then the tangent and cotangent manifolds of the configuration space will turn out to be what are classically called *the state space* and *the phase space*. We will find, in particular, that there is a natural (or "canonical") skew-symmetric tensor field on the cotangent bundle in terms of which we can write the laws of classical mechanics. Again, rather than giving a detailed exposition of this model of classical mechanics at this time, we will generalize, which should enhance both efficiency and insight, and we will point out the relationships to mechanics as we go along.

18.1 Symplectic forms, symplectic mappings, Hamiltonian vector fields, and Poisson brackets

Recall that in Section 11.3 we defined an s-form field (or differential s-form) to be a differentiable map from a manifold, M, to the bundle $\Lambda^s(T^*M)$ of s-forms. If $s = 2$, that is, if ω is a 2-form field, we say it is nondegenerate if $\omega(p)$ is nondegenerate for all $p \in M$.

Definitions

A nondegenerate closed differential 2-form, ω, on M is called *a symplectic* (or *Hamiltonian*) *form*, and the pair (M, ω) is *a symplectic* (or *Hamiltonian*) *manifold.*

It is clear from, the Corollary of Theorem 5.23 that, if M is finite-dimensional, and (M, ω) is symplectic then M must be even-dimensional. Locally we have

$$\omega = \omega_{ij}\, d\mu^i \wedge d\mu^j$$

The requirement of differentiability on M, of course, raises the question of existence of a symplectic form on a given M.

Just as in Chapter 15, when we had the symmetric tensor, g, with the symplectic form, ω, we have a linear map $\omega^\flat: \mathfrak{X} \to \mathfrak{X}^*$ given by $X \mapsto \omega^\flat \cdot X$ where $(\omega^\flat \cdot X) \cdot Y = \omega(X, Y)$. Since ω is nondegenerate we also have the inverse $\omega^\#: \mathfrak{X}^* \to \mathfrak{X}$. In contrast to the symmetric case, where, if M is orientable, with g we can define two volume forms, ω determines a unique $2n$-form,

$$\omega^n = \underbrace{\omega \wedge \cdots \wedge \omega}_{n}$$

on M; that is, a symplectic manifold is oriented.

Now we want to consider mappings between two given symplectic manifolds (M, ω) and (N, θ), or, in particular, mappings from a symplectic manifold to itself.

Definition
A differentiable mapping $\phi: M \to N$ is a *symplectic map* if it "preserves," or "leaves invariant" the symplectic structures.

This descriptive definition is suggestive and convenient, and corresponds to the definition of isometry for pseudo-Riemannian manifolds in Section 15.1. In symbols, $\phi^*\theta = \omega$, or more explicitly, recalling the definition of pull-back, $\phi^*\theta(p) = \phi^* \cdot \theta(\phi(p))$, and the definition of $\phi^* \cdot \theta(\phi(\mathrm{p}))$ by eq. (4.10), ϕ has the property

$$\theta(\phi(p))(\phi_* \cdot v_1, \phi_* \cdot v_2) = \omega(p)(v_1, v_2) \tag{18.1}$$

where $v_i \in T_p(M)$. Compare eq. (15.2).

Theorem 18.1. *If* $\dim M = \dim N$, *then a symplectic map*, ϕ, *is volume-preserving and* ϕ *is a local diffeomorphism.*

Proof. (i) Since $\phi^*\theta = \omega$, $\phi^*\theta \wedge \cdots \wedge \theta = \omega \wedge \cdots \wedge \omega$.
(ii) From (18.1) we have

$$\theta_{ij}(\phi(p)) \frac{\partial \phi^i}{\partial \mu^k} \frac{\partial \phi^j}{\partial \mu^l} = \omega_{kl}(p)$$

and taking det on both sides we get $\det(\partial\phi^i/\partial\mu^k) \neq 0$ since ω and θ are nondegenerate. ■

Definition
X is *a Hamiltonian vector field* on (M, ω) if X preserves ω; i.e., if all the maps, $\{\Theta_u\}$, of the flow, Θ, *of* X preserve the symplectic structure of M.

Theorem 18.2. *X is Hamiltonian* $\Leftrightarrow L_X\omega = 0$.

Proof. This is just a special case of Problem 15.7. ■

Theorem 18.3. *X is Hamiltonian $\Leftrightarrow i_X\omega$ is closed $\Leftrightarrow \omega^b \cdot X$ is closed.*

Proof. Problem 18.1.

Theorem 18.3 says that if X is a Hamiltonian vector field, then there is, locally, a function, H, *a Hamiltonian function* such that $i_X\omega = dH$. Clearly, H is not unique. On the other hand, since ω is nondegenerate, and $\omega^b \cdot \omega^\# = id$, any closed 1-form σ yields a Hamiltonian vector field $\omega^\# \cdot \sigma$. In particular, starting with a function, H we get a Hamiltonian vector field $\omega^\# \cdot dH$.

Definitions
If we are given a function H on M, the vector field $X_H = \frac{1}{2}\omega^\# \cdot dH$ determined by H is called *the Hamiltonian vector field of* H. Then

$$i_{X_H}\omega = 2\omega^b \cdot X_H = dH \tag{18.2}$$

and (M, ω, H) is called *a Hamiltonian system.* (In many of the standard references $i_X\omega = \omega^b \cdot X$. The fact that in our work $i_X\omega = 2\omega^b \cdot X$, which arises from the fact that our $(\omega^b \cdot X) \cdot Y = \omega(X, Y)$ is half of theirs, will lead to some equations which look slightly different from those found in those references.)

Definition
Suppose f is a function and X is a vector field with flow Θ (or, Θ is a 1-parameter group action with infinitesimal generator X). See Section 13.3. If f is constant on integral curves, γ_p, of X, or equivalently, on translations, Θ_u, of Θ, then f is *an integral of* X, or f is *an invariant of* Θ.

It is interesting to compare Hamiltonian vector fields with vector fields defined in Riemannian geometry in Section 15.1. By Theorem 18.2, Hamiltonian vector fields correspond to Killing fields in geometry (but they are much more prevalent.) Coming from functions, they are precisely analogous to gradients. (However, gradients are not, in general, Killing fields.)

Theorem 18.4. *If (M, ω, H) is a Hamiltonian system, then H is an integral of X_H.*

Proof. At each point $D_u(H \circ \gamma) = D_0(H \circ (\gamma \circ \theta_u)) = \langle dH(\gamma(u)), \dot{\gamma}(u) \rangle$ by eq. (9.5). Putting $\dot{\gamma}(u) = X_H(\gamma(u))$ we get

$$D_u(H \circ \gamma) = \langle dH, X_H \rangle(\gamma(u)) = \langle 2\omega^\flat \cdot X_H, X_H \rangle(\gamma(u))$$

by eq. (18.2), But $\langle \omega^\flat \cdot X_H, X_H \rangle = 0$ since ω is skew symmetric. ∎

This result is interpreted mechanically as a statement of conservation of energy. We will examine this more closely in Section 19.2.

As an alternative to imposing a symplectic structure, ω, on M, one can impose a Poisson bracket structure on the differentiable functions, $\mathfrak{F}_M$. Thus, on the one hand, starting with a given ω one can form the following product in $\mathfrak{F}_M$.

Definition
For $f, g \in \mathfrak{F}^M$, *the Poisson bracket of f and g is* $\{f, g\} = 2\omega(X_f, X_g)$.

Theorem 18.5. *Useful alternative descriptions of $\{f, g\}$ are given by*

$$\{f, g\} = L_{X_g} f = X_g f$$

Proof. $L_{X_g} f = X_g f = \langle df, X_g \rangle = 2\omega^\flat \cdot X_f(X_g) = 2\omega(X_f, X_g)$. ∎

Theorem 18.6. *The Poisson bracket has the following three properties.*

(i) *The product $(f, g) \mapsto \{f, g\}$ nakes $\mathfrak{F}_M$ a Lie algebra.* (See Section 11.1.)
(ii) *Each partial map is a derivation.*
(iii) $\{f, g\} = 0 \Leftrightarrow$ *f is constant on integral curves of X_g*
$\Leftrightarrow$ *g is constant on integral curves of X_f*
(*i.e., f is an integral of X_g and g is an integral of X_f*).

Proof. (i) Both the bilinearity and skew-symmetry of the product come almost immediately from the definitions. We get the Jacobi identity from

$$\{f, \{g, h\}\} = L_{X_f} L_{X_g} h = X_f X_g h,$$

$$\{g, \{h, f\}\} = -L_{X_g} L_{X_f} h = -X_g X_f h,$$

and

$$\{h, \{f, g\}\} = L_{X_{\{f,g\}}} h = X_{\{f,g\}} h$$

But $X_{\{f,g\}} = -[X_f, X_g]$ (see the Corollary of Theorem 18.7). So $\{h, \{f, g\}\} = -[X_f, X_g]h$ and adding, we get zero.

(ii) and (iii): *Problem 18.3.* ∎

Now, we can also go in the opposite direction. If on M we have a pairing $\mathfrak{F}_M \times \mathfrak{F}_M \to \mathfrak{F}_M$ with the three properties of Theorem 18.6, then F: $(df, dg) \mapsto \{f, g\}$ defines a 2-vector field on M (recall, Section 11.3, that a 2-vector field is a differentiable map from M to $\Lambda^2(TM) = \bigcup_{p \in M} \Lambda^2(T_p)$.) F is nondegenerate and we have the maps $F^{\#}$: $\mathfrak{X}^* \to \mathfrak{X}$ and $F^{\flat}$: $\mathfrak{X} \to \mathfrak{X}^*$. The corresponding 2-form field, $(v, w) \mapsto (F^{\flat} \cdot v, F^{\flat} \cdot w) \mapsto F(F^{\flat} \cdot v, F^{\flat} \cdot w)$ will be symplectic.

We should notice that we have now defined a variety of related skew-symmetric structures on M which we can compare according to the following listing.

(i) ω, the symplectic structure, gives a pairing of vector fields on M to functions on M
(ii) $\{\ \}$, the Poisson bracket, gives a pairing of functions on M to functions on M
(iii) $[\ ]$, the Lie bracket, gives a pairing of vector fields on M to vector fields on M
(iv) F, defined in the paragraph above, gives a pairing of differential 1-forms on M to functions on M.

One more structure will complete this list.

Definition
For $\sigma, \tau \in \mathfrak{X}^*(M)$, *the Poisson bracket of* σ *and* τ is $\{\sigma, \tau\} = -\frac{1}{2}\omega^{\flat} \cdot [\omega^{\#} \cdot \sigma, \omega^{\#} \cdot \tau]$.

Theorem 18.7. *If* τ *is closed, then* $\{\sigma, \tau\} = \frac{1}{2} L_{\omega^{\#} \cdot \tau} \sigma$.

Proof. Since $X \rfloor \omega = \omega^{\flat} \cdot X$, $\omega^{\#} \cdot \sigma \rfloor \omega = \sigma$. Then $L_{\omega^{\#} \cdot \tau} \sigma = L_{\omega^{\#} \cdot \tau} \omega^{\#} \sigma \rfloor \omega = (L_{\omega^{\#} \cdot \tau} \omega^{\#} \cdot \sigma) \rfloor \omega + \omega^{\#} \cdot \sigma \rfloor L_{\omega^{\#} \cdot \tau} \omega$ by Problem 6.11 and Problem 13.13. But $L_{\omega^{\#} \cdot \tau} \omega = d i_{\omega^{\#} \cdot \tau} \omega + i_{\omega^{\#} \cdot \tau} d\omega = d 2\omega^{\flat} \cdot \omega^{\#} \cdot \tau + i_{\omega^{\#} \cdot \tau} d\omega = 0$ since $d\tau = d\omega = 0$. So, $L_{\omega^{\#} \cdot \tau} \sigma = [\omega^{\#} \cdot \tau, \omega^{\#} \cdot \sigma] \rfloor \omega = \omega^{\flat} \cdot [\omega^{\#} \cdot \tau, \omega^{\#} \cdot \sigma] = -\omega^{\flat} \cdot [\omega^{\#} \cdot \sigma, \omega^{\#} \cdot \tau] = 2\{\sigma, \tau\}$. ■

Corollary. (i) *The relation between the Poisson bracket of functions and the Poisson bracket of differential* 1*-forms is given by*

$$d\{f, g\} = \{df, dg\} \tag{18.3a}$$

i.e., Poisson brackets commute with exterior differentiation.

(ii)

$$X_{\{f, g\}} = -[X_f, X_g] \tag{18.3b}$$

and the Hamiltonian vector fields form a Lie algebra.

PROBLEM 18.1. Prove Theorem 18.3.

PROBLEM 18.2. Prove properties (ii) and (iii) of Theorem 18.6.

PROBLEM 18.3. $\{f, g\} = 0$ iff f is constant on integral curves of X_g iff g is constant on integral curves of X_f.

PROBLEM 18.4. Show that F as specified above defines a nondegenerate 2-vector field, and that the corresponding 2-form field is symplectic.

PROBLEM 18.5. Prove the corollary of Theorem 18.7.

18.2 The Darboux theorem, and the natural symplectic structure of T^*M

There are two key results that pin down to specific concrete cases the abstract structures we have been describing and, in particular, form the bridge to classical mechanics. One is that every symplectic manifold has certain "natural" (or "canonical") local coordinates: "Darboux's theorem for symplectic manifolds." The second is that the cotangent manifold of any manifold has a "natural" (or "canonical") symplectic structure. An important related fact is that the "natural" coordinates on T^*M induced by a coordinate system on M (Section 11.2) are "natural" with respect to its symplectic structure.

Theorem 18.8. *Suppose M has dimension $2n$ and ω is a nondegenerate differential 2-form on M. Then $d\omega = 0$ on M if and only if at each point of M there is a coordinate system with coordinate functions $x^1, \ldots, x^n, y^1, \ldots, y^n$ such that*

$$\omega = dx^1 \wedge dy^1 + \cdots + dx^n \wedge dy^n \tag{18.4}$$

Proof. The one way is obvious. To go the other way, we note, first of all, that on $\mathbb{R}^{2n}$ there is the standard 2-form

$$\hat{\omega}_1 = d\pi^1 \wedge d\pi^{n+1} + \cdots + d\pi^n \wedge d\pi^{2n}$$

where π^i are the standard coordinate functions on $\mathbb{R}^{2n}$. Now, given ω, select a chart $(\mathscr{U}, \mu)$ at p_0 such that the representation $\hat{\omega} = \omega \circ \mu^{-1}$ of ω on $\mu(\mathscr{U})$ has the form of $\hat{\omega}_1$ at $a_0 = \mu(p_0)$. Such a chart exists by Theorem 5.23. The idea of the proof is to find a map, Θ_1, from a neighborhood $\mathscr{V} \subset \mu(\mathscr{U})$ of $a_0 = \mu(p_0)$ in $\mathbb{R}^{2n}$ to $\mathbb{R}^{2n}$ (see Fig. 18.1) such that $(\mu^{-1}(\mathscr{V}), \Theta_1 \circ \mu)$ is the required kind of coordinate system; i.e., a coordinate system in which ω has the form (18.4).

Define a "time-dependent" differential 2-form, Ω, on a neighborhood of $\mu(a_0)$ by

$$\Omega(a, t) = \hat{\omega}(a) + t(\hat{\omega}_1(a) - \hat{\omega}(a))$$

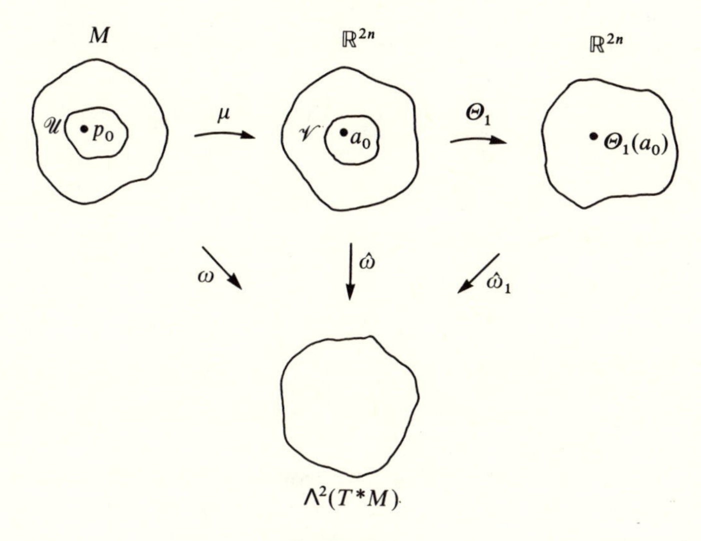

FIGURE 18.1

with $a \in \mu(\mathscr{U})$ and $t \in \mathbb{R}$. Now there is a neighborhood, $\mathscr{V}$, of a_0 for which

(i) $\hat{\omega} - \hat{\omega}_1 = d\sigma$ and $\sigma(a_0) = 0$ since $\hat{\omega}$ and $\hat{\omega}_1$ are closed,
(ii) Ω is nondegenerate for $0 \le t \le 1$ since $\Omega(a_0, t) = \hat{\omega}(a_0)$ is nondegenerate,
(iii) A "time-dependent" vector field, Z, with $Z(a_0, t) = 0$ is defined by $i_Z\Omega = \sigma$ for $0 \le t \le 1$ since Ω is nondegenerate for $0 \le t \le 1$.
(iv) A "flow", Θ, of Z is defined for $0 \le u \le 1$.

(A "flow" in (iv) is an extension of the concept discussed in Section 13.1 according to Θ: $(u, p) \mapsto \Theta(u, p)$ where $\Theta_p(u)$ is such that $\dot{\Theta}_p(u) = Z(\Theta_p(u), u)$. The fact that Θ is defined for $0 \le u \le 1$ in some neighborhood of a_0 comes from an argument like that near the beginning of Section 13.2.)

In order to proceed with the proof we make use of a generalization of the formula of Theorem 13.15. If K is a time-dependent tensor field, and Θ is the flow of a time-dependent vector field Z,

$$K_p^{\#}: (u, t) \in I_1 \times I_2 \mapsto \Theta_u^* K(p, t) \in (T_p)_s^r$$

and when $t = u$,

$$D_u K_p^{\#} = \Theta_u^* \cdot L_Z K(\Theta_u(p), u) + \Theta_u^* \cdot \frac{\partial K}{\partial u}(\Theta_u(p), u) \tag{18.5}$$

Finally, applying (18.5) to our time-dependent 2-form, Ω, we get

$$D_u\Omega_a^\# = \Theta_u^* \cdot L_Z\Omega(\Theta_u(a), u) + \Theta_u^* \cdot \frac{\partial \Omega}{\partial t}(\Theta_u(a), t)$$

for $a \in V$. But, since $d\Omega = 0$, $L_Z\Omega = di_Z\Omega = d\sigma = \hat{\omega} - \hat{\omega}_1$, and $\partial\Omega/\partial t = \hat{\omega}_1 - \hat{\omega}$, so $D_u\Omega_a^\# = 0$. That is, $\Theta_u^* \cdot \Omega(\Theta_a(u), u)$ is constant. Putting $u = 0$ and $u = 1$, we get

$$\hat{\omega}(a) = \Theta_1^* \cdot \hat{\omega}_1(\Theta_a(1)) = \Theta_1^* \cdot \hat{\omega}_1(\Theta_1(a))$$

for $a \in \mathscr{V}$. That is, $\hat{\omega}$ and $\hat{\omega}_1$ are representations of the same form on $\mathscr{V}$ and $\Theta_1(\mathscr{V})$ respectively, so $\Theta_1 \circ \mu$ is the required coordinate map, and $x^i = \pi^i \circ \Theta_1 \circ \mu$ and $y^i = \pi^{n+1} \circ \Theta_1 \circ \mu$ $(i = 1, \ldots, n)$. ■

It is instructive to note that (i) Theorem 18.8 uses the canonical form for an *exterior*-form at a *point*, Theorem 5.23, to get a canonical form for a *differential*-form on a *coordinate neighborhood*, and (ii) Theorem 18.8 says that there is only one symplectic form on a manifold. Contrast these two statements with the geometric situation when M has a pseudo-Riemannian structure.

There is a more general Darboux theorem which gives a local canonical representation for 1-form fields. The result above comes easily from this general theorem as does also a parallel result giving a local canonical representation for contact forms. The proof of the general theorem is in Sternberg (1983, pp. 137 ff). The proof above is given by Abraham and Marsden (1978, p. 175).

Now we go to the other key result mentioned above.

Theorem 18.9. *For any manifold M, there exists a symplectic structure on T^*M.*

Proof. We saw in Theorem 11.18 that T^*M has a natural 1-form field, θ_M. We will show that θ_M is differentiable and $d\theta_M$ is symplectic. This will be done by introducing coordinates.

If $(\mathscr{U}, \mu)$ is a chart on M, then the induced chart, $(\mathscr{V}, \nu)$ on T^*M is obtained by putting $\mathscr{V} = \bigcup_{p \in \mathscr{U}} T_p^*$, $\nu^i = q^i = \mu^i \circ \pi$, $i = 1, \ldots, n$, and $\nu^{n+j} = p_j = \sigma_j \circ \pi$, $j = 1, \ldots, n$, where σ_j is given by

$$\sigma(p) = \sigma_i(p)\, d\mu^i|_p \tag{18.6}$$

for $p \in M$ and $\sigma \in \pi^{-1}(p)$. See Problem 11.9.

With the coordinates q^i, p_j the coordinate vector fields $\partial/\partial q^i, \partial/\partial p_j$ form a basis of vector fields on $\mathscr{V}$. Thus

$$X = X^i \frac{\partial}{\partial q^i} + Y_j \frac{\partial}{\partial p_j} \tag{18.7}$$

for some functions X^i and Y_j. But according to eq. (9.9)

$$\pi_* \cdot \frac{\partial}{\partial q^i} = \frac{\partial \pi^k}{\partial q^i} \frac{\partial}{\partial \mu^k} \qquad \text{and} \qquad \pi_* \cdot \frac{\partial}{\partial p_i} = \frac{\partial \pi^k}{\partial p_i} \frac{\partial}{\partial \mu^k}$$

where π^k are the component functions of π. So $\pi_* \cdot \partial/\partial q^i = \partial/\partial \mu^i$ and $\pi_* \cdot \partial/\partial p_i = 0$, and from eq. (18.7),

$$\pi_* \cdot X = X^i \frac{\partial}{\partial \mu^i} \tag{18.8}$$

By the definition of θ_M in Theorem 11.18

$$\langle \theta_M(\sigma), X \rangle = \langle \sigma, \pi_* \cdot X \rangle \tag{18.9}$$

Substituting from (18.8) and (18.6) into the right side of (18.9), we get

$$\langle \theta_M(\sigma), X(\sigma) \rangle = \sigma_i(p) X^i(\sigma) = p_i(\sigma) X^i(\sigma) \tag{18.10}$$

for $p = \pi(\sigma)$. But $\theta_M(\sigma)$ has the form

$$\theta_M(\sigma) = \xi_i(\sigma)\, dq^i + \eta^j(\sigma)\, dp_j \tag{18.11}$$

so from (18.7),

$$\langle \theta_M(\sigma), X(\sigma) \rangle = \xi_i(\sigma) X^i(\sigma) + \eta^j(\sigma) Y_j(\sigma) \tag{18.12}$$

Comparing (18.10) and (18.12), noting that they are valid for arbitrary choices of X^i and Y_j, we find we must have $\xi_i(\sigma) = p_i(\sigma)$ and $\eta^j(\sigma) = 0$. Hence, from (18.11),

$$\theta_M(\sigma) = p_i(\sigma)\, dq^i \tag{18.13}$$

Since p_i are coordinate functions $\theta_M: \sigma \mapsto \theta_M(\sigma)$ is differentiable and $d\theta = dp_i \wedge dq^i$ is clearly nondegenerate and closed. ■

Corollary. *The natural coordinate system induced on T^*M by a given coordinate system on M, as described in Section 11.2, is a natural (or canonical) coordinate system on the symplectic manifold T^*M in the sense that it is one of the kind guaranteed by Darboux's theorem.*

Theorem 18.10. *If ϕ is a diffeomorphism from M to N, then $\phi^*: T^*N \to T^*M$ takes the natural 1-form, θ_M on T^*M to the natural 1-form, θ_N, on T^*N; i.e., $(\phi^*)^*\theta_M = \theta_N$.*

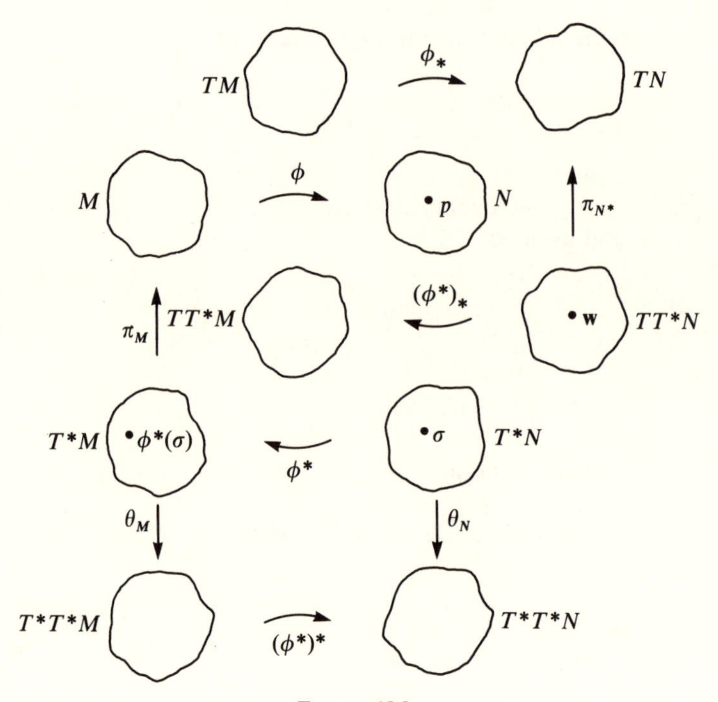

FIGURE 18.2

Proof. The proof is by chasing around pairings with the help of Fig. 18.2. If $\sigma \in T_p^*N$, $\mathbf{w} \in T_\sigma T^*N$, π_M is the projection of T^*M on M, and π_N is the projection of T^*N on N, then

$$\langle (\phi^*)^*\theta_M(\sigma), \mathbf{w}\rangle$$

$$= \langle \theta_M(\phi^*(\sigma)), (\phi^*)_*(\mathbf{w})\rangle \qquad \text{the property of the transpose of } (\phi^*)_*$$

$$= \langle \phi^*(\sigma), (\pi_{M*} \cdot (\phi^*)_*(\mathbf{w}))\rangle \qquad \text{the definition of } \phi$$

$$= \langle \phi^*(\sigma), ((\pi_M \circ \phi^*)_*(\mathbf{w}))\rangle$$

$$= \langle \sigma, (\phi_*((\pi_M \circ \phi^*)_*(\mathbf{w}))\rangle \qquad \text{the property of the transpose of } \phi_*$$

$$= \langle \sigma, ((\phi \circ \pi_M \circ \phi^*)_*(\mathbf{w}))\rangle$$

$$= \langle \sigma, \pi_{N*}(\mathbf{w})\rangle$$

$$= \langle \theta_N(\sigma), \mathbf{w}\rangle \qquad \text{the definition of } \theta_N$$

So $(\phi^*)^*\theta_M = \theta_N$.

Corollary. *If ϕ is any diffeomorphism from M to N, then ϕ^*: $T^*N \to T^*M$ is a symplectic map.*

Proof. *Problem 18.8.*

We can think of a coordinate transformation between overlapping coordinate systems on a manifold as a special case of a mapping of manifolds. A symplectic form on the manifold will have a local representation in each coordinate system.

Definition
In a symplectic manifold a coordinate transformation is *a canonical coordinate transformation* if the overlap map is symplectic.

Theorem 18.11. *If a symplectic form has the Darboux representation in one coordinate system, then it has the Darboux representation in a second coordinate system iff the coordinate transformation is canonical.*

Proof. *Problem 18.9.*

Theorem 18.12. *Two sets of natural (Darboux) coordinates* (q^i, p_i) *and* (Q^i, P_i) *satisfy the following relations at each point of the overlap of their domains.*

$$\frac{\partial p_i}{\partial P_j} = \frac{\partial Q^j}{\partial q^i}, \qquad \frac{\partial p_i}{\partial Q^j} = -\frac{\partial P_j}{\partial q^i}$$

$$\frac{\partial q^i}{\partial P_j} = -\frac{\partial Q^j}{\partial p_i}, \qquad \frac{\partial q^i}{\partial Q^j} = \frac{\partial P_j}{\partial p_i}$$

Proof. *Problem 18.10.*

PROBLEM 18.6. Show that in natural (Darboux) coordinates,

(i) if ω is a symplectic form,

$$i_X\omega = \sum X^i\, dy^i - Y^i\, dx^i$$

where X^i, Y^i are the component functions of X.

(ii) the Poisson bracket of f and g is

$$\{f, g\} = \sum \left(\frac{\partial f}{\partial y^i}\frac{\partial g}{\partial x^i} - \frac{\partial f}{\partial x^i}\frac{\partial g}{\partial y^i}\right)$$

PROBLEM 18.7. Prove Theorem 18.10 by choosing coordinates and showing $P_i\, dQ^i = (\phi^*)^* \cdot p_i\, dq^i$.

PROBLEM 18.8. Prove the corollary of Theorem 18.10.

PROBLEM 18.9. Prove Theorem 18.11.

PROBLEM 18.10. Prove Theorem 18.12. (Hint: use eq. (5.22) with $\begin{pmatrix} I_r & & \\ & -I_s & \\ & & 0 \end{pmatrix}$ replaced by J.)

18.3 Hamilton's equations. Examples of mechanical systems

Theorem 18.13. *In a natural coordinate system, the Hamiltonian vector field, X_H, of a Hamiltonian system (M, ω, H) has the component functions* $\left(\dfrac{\partial H}{\partial y^i}, -\dfrac{\partial H}{\partial x^i}\right)$.

Proof. Write both sides of $i_{X_H}\omega = dH$ in terms of the basic fields dx^i, dy^i. The left side is given by Problem 18.6(i) and

$$dH = \frac{\partial H}{\partial x^i} dx^i + \frac{\partial H}{\partial y^i} dy^i$$

so, equating coefficients, $X_H^i = \partial H/\partial y^i$ and $Y_H^i = -\partial H/\partial x^i$. ■

In applications to mechanics, in the Hamiltonian system (M, ω, H) M is the cotangent manifold of the configuration space, so we write q^i, p_i instead of x^i, y^i for coordinate functions. In these coordinates $\omega = dq^i \wedge dp_i$. The famous *Hamilton canonical equations* are simply the equations in natural coordinates, of the integral curves, γ, of a given Hamiltonian vector field, X_H. Thus, by Theorem 18.13, the Hamilton canonical equations are the equations

$$\frac{dq^i \circ \gamma}{dt} = \frac{\partial H}{\partial p_i}, \qquad \frac{dp_i \circ \gamma}{dt} = -\frac{\partial H}{\partial q^i} \tag{18.14}$$

for the curves, γ.

Let us see how Hamilton's equations describe the behavior of a mechanical system. In mechanics we are concerned with the "motion" of "systems (finite sets) of particles" and/or "rigid bodies."

For the case of systems of particles we assume that, at each instant, we can represent the "position" of the system as a set of n points in $\mathscr{E}_0^3$, a 3-dimensional Euclidean affine space. Recall that by definition, an n-dimensional Euclidean affine space, $\mathscr{E}_0^n$, is an affine space whose vector space, V, is a Euclidean vector space (see Section 5.4(i)). If we choose a point $p_0 \in \mathscr{E}_0^n$, we get a 1–1 correspondence between points of $\mathscr{E}_0^n$ and points of V.

Then choosing an orthonormal basis, $\{e_i\}$, in V we get (rectangular cartesian) coordinates on $\mathscr{E}_0^n$, and, finally, $\mathscr{E}_0^n$ acquires the structure of a Riemannian manifold with rectangular cartesian coordinates by means of the isomorphisms, $e_i \leftrightarrow \partial/\partial\mu^i|_p$, between V and $T_p\mathscr{E}_0^n$ for each $p \in \mathscr{E}_0^n$. Clearly, $\mathscr{E}_0^n$ with this structure is isometric with $\mathscr{R}_0^n$.

We assume that the "motion" of the systems of particles will be determined by "forces" and "constraints," the former describable by a function, a potential, on $\mathscr{E}_0^3$, and the latter determining a submanifold of $\mathbb{R}^{3n}$. The submanifold is called *the configuration space* of the system, and its cotangent manifold is called *the* (*momentum*) *phase space* of the system.

For the case of rigid bodies, we can represent a position of the body as the image of an isometry of $\mathscr{E}_0^3$. In this case *the configuration space* is the set of isometries, and again the cotangent manifold is *the phase space.*

In both cases, the precise way in which forces and constraints determine the motion is assumed to be represented, in local coordinates, by the solutions of Hamilton's equations. The major significance of this model of a mechanical system, in contrast to, for example, the Newtonian model is that, since the induced coordinates in the phase space are always natural, eqs. (18.14) have the same form for *any* choice of coordinates in configuration space.

Consider the following *examples* of simple mechanical systems.

1. THE SIMPLE PLANE PENDULUM. We are given two positive numbers, a length, ℓ, and a mass, m. The position of the ball of the pendulum is represented by a point in $\mathscr{E}_0^3$. Introducing rectangular cartesian coordinates (a, b, c), the configuration space is the circle in $\mathbb{R}^3$,

$$S^1 = \{(a, b, c): a^2 + b^2 = \ell^2, c = 0\}$$

with local coordinate, ϑ, given by $a = \ell \sin \vartheta$, $b = -\ell \cos \vartheta$. Then local coordinates (q, p) of a point, σ, in the cotangent bundle, the phase space, are given by $q = \vartheta \circ \pi$ and $\sigma = p\, d\vartheta$.

The force of gravity is described by a potential function in $\mathscr{E}_0^3$ given by $\Phi(a, b, c) = mgb$ (g = constant). Then on S^1

$$\Phi(\vartheta) = -mg\ell \cos \vartheta \tag{18.15}$$

The kinetic energy is a function on the tangent manifold TS^1 given by

$$\mathsf{T}(\vartheta, v) = \tfrac{1}{2}m\ell^2 v^2 \tag{18.16}$$

where v is the component function of a vector with respect to the coordinate basis $\partial/\partial\vartheta$. Choose $H = \Phi \circ \pi + \mathsf{T}^\flat$ where π is the natural projection of T^*S^1

on S^1 and $\mathsf{T}^\flat$ is the function on T^*S^1 obtained by putting $v = p/m\ell^2$ in (18.16). Then we get, from (18.15) and (18.16),

$$H(q, p) = -mg\ell \cos q + \frac{1}{2m}\frac{p^2}{\ell^2} \tag{18.17}$$

(Strictly speaking, $\Phi(\vartheta)$, $\mathsf{T}(\vartheta, v)$ and $H(q, p)$ should be $\hat{\Phi} \circ \vartheta$, $\hat{\mathsf{T}} \circ (\vartheta, v)$, and $\hat{H} \circ (p, q)$ respectively. We will follow the usual practice of suppressing the extra notation.)

Differentiating (18.17), $\partial H/\partial q = mg\ell \sin q$ and $\partial H/\partial p = p/m\ell^2$ and so Hamilton's equations for the simple plane pendulum are

$$\frac{dq}{dt} = \frac{p}{m\ell^2}, \qquad \frac{dp}{dt} = -mg\ell \sin q$$

2. THE SIMPLE HARMONIC OSCILLATOR. We consider the motion of a particle subject to a central force, of which this example is a special case. Since such a motion occurs in a plane, the position of the particle is represented by a point in $\mathscr{E}_0^3$ where we now use cylindrical coordinates (r, ϑ, z). The central force is described by a potential function

$$\Phi(r, \vartheta, z) = f(r) \tag{18.18}$$

The configuration space is $\mathbb{R}^2$, the set of pairs (r, ϑ). Local coordinates (q^1, q^2, p_1, p_2) of a point, σ, in the cotangent bundle, the phase space, are given by $q^1 = r$, $q^2 = \vartheta$, and $\sigma = p_1\, dr + p_2\, d\vartheta$. Again $H = \Phi \circ \pi + \mathsf{T}^\flat$, and now the kinetic energy, T, is given by

$$\mathsf{T}(v^1, v^2) = \tfrac{1}{2}m[(v^1)^2 + r^2(v^2)^2] \tag{18.19}$$

where $v^1 = dr/dt$ and $v^2 = d\vartheta/dt$. (Note that T is really a function of *four* coordinate functions on the tangent bundle). Putting $v^1 = p_1/m$ and $v^2 = p_2/mr^2 = p_2/m(q^1)^2$ into (18.19) we get from (18.18) and (18.19),

$$H(q^1, q^2, p_1, p_2) = f(q^1) + \frac{1}{2m}\left(p_1^2 + \frac{p_2^2}{(q^1)^2}\right) \tag{18.20}$$

Note that the right side of (18.20) does not actually contain q^2. (Consequently q^2 is called a cyclic coordinate.) Differentiating (18.20) we get Hamilton's equations for this case:

$$\frac{dq^1}{dt} = \frac{p_1}{m}, \qquad \frac{dq^2}{dt} = \frac{p_2}{m(q^1)^2}$$

$$\frac{dp_1}{dt} = -f'(q^1) + \frac{p_2^2}{m(q^1)^3}, \qquad \frac{dp_2}{dt} = 0$$

In particular, we get $p_2 = mr^2\, d\vartheta/dt$ = constant, a well-known result. (Kepler's equal areas in equal times.) For the special case of a simple harmonic oscillator, q^2 is constant, $p_2 = 0$, and $f(r) = \frac{1}{2}kr^2$ where k is a constant, and Hamiltonian's equations reduce to

$$\frac{dq^1}{dt} = \frac{p_1}{m}, \qquad \frac{dp_1}{dt} = -kq^1$$

3. THE DOUBLE PLANE PENDULUM. We are given two lengths ℓ_1 and ℓ_2 and two masses m_1 and m_2. The position of the system is represented by two points in $\mathscr{E}^3_0$. Because of the constraints the position will be determined by two angles ϑ_1 and ϑ_2. The pair $(\vartheta_1, \vartheta_2)$ can be thought of as local coordinates on a certain submanifold of $\mathbb{R}^6$, namely, the 2-dimensional torus. In terms of the natural coordinates x, y, z on $\mathbb{R}^3$,

$$x = (\ell_1 + \ell_2 \sin\vartheta_2)\cos\vartheta_1$$
$$y = (\ell_1 + \ell_2 \sin\vartheta_2)\sin\vartheta_1$$
$$z = \ell_2 \cos\vartheta_2$$

Hence for the double plane pendulum the configuration space is the 2-dimensional torus. Local coordinates (q^1, q^2, p_1, p_2) of a point, σ, in the cotangent bundle, the phase space, are given by $q^1 = \vartheta_1$, $q^2 = \vartheta_2$, and $\sigma = p_1\, d\vartheta_1 + p_2\, d\vartheta_2$. The force is again gravity with a potential Φ defined in $\mathscr{E}^3_0$. In terms of ϑ_1 and ϑ_2, and hence q^1 and q^2, we have

$$\Phi(q^1, q^2) = -m_1 g\ell_1 \cos q^1 - m_2 g(\ell_1 \cos q^1 + \ell_2 \cos q^2) \tag{18.21}$$

and

$$\begin{aligned} \mathsf{T}(v^1, v^2) = {} & \tfrac{1}{2}m_1\ell_1^2(v^1)^2 \\ & + \tfrac{1}{2}m_2[(\ell_1 \sin\theta_1 v^1 + \ell_2 \sin\theta_2 v^2)^2 \\ & + (\ell_1 \cos\theta_1 v^1 + \ell_2 \cos\theta_2 v^2)^2] \end{aligned} \tag{18.22}$$

where $v^1 = d\vartheta_1/dt$ and $v^2 = d\vartheta_2/dt$. (Again, note that T is really a function of *four* coordinates.) Again writing the v's in (18.22) in terms of q's and p's and adding the result to the expression for Φ in (18.21) we get the Hamiltonian function.

4. LAGRANGE'S TOP. In the previous examples we had systems of particles. Now we consider the motion of a rigid body, in the special case in which one point of the body is fixed. For any rigid body moving with a fixed point (the body need not have any symmetry) the set of positions has the structure of a 3-dimensional submanifold of $\mathbb{R}^9$; specifically, the rotations,

SO(3), the orthogonal matrices with determinant equal to 1 in the set of 3 by 3 matrices. That is, the configuration space of a rigid body moving with a fixed point is SO(3). We can choose the "Euler angles" ϑ, φ, χ as local coordinates. The force acting on this system (as in Examples 1 and 3) is gravity, so

$$\Phi(\vartheta, \varphi, \chi) = mg\ell \cos \vartheta \tag{18.23}$$

(We have, among the given properties of this system, a density distribution on $\mathscr{E}_0^3$ which determines m and ℓ). There are standard expressions for T in terms of angular velocity; cf., MacMillan (1960, pp. 180–181), or Arnold (1989, p. 137). Similarly, for angular velocity in terms of Euler angles (op. cit. pp. 184–185, p. 150, respectively). The result is

$$\mathsf{T}(v^1, v^2, v^3) = \tfrac{1}{2}[A(v^1)^2 + A \sin^2 \vartheta (v^3)^2 + C(v^2 + v^3 \cos \vartheta)^2] \tag{18.24}$$

where $v^1 = d\vartheta/dt$, $v^2 = d\varphi/dt$, $v^3 = d\chi/dt$ and A and C are (constant) components of the moments of inertia. Finally, as before, we write the v's in terms of q's and p's to get the Hamiltonian function. In this case we have

$$v^1 = \frac{p_1}{A}$$

$$v^2 + v^3 \cos q^1 = \frac{p_2}{C}$$

$$v^3 \sin^2 q^1 = \frac{p_3 - p_2 \cos q^1}{A}$$

(op. cit. p. 378, p. 151, respectively). Note that, like in Example 2, not all q's will appear explicitly in the expression for $H(q^1, \ldots, p_3)$; p_2, and p_3 are integrals of Hamilton's equations and the system reduces to two equations in two unknowns.

In all the examples above, one step involved expressing the v's in terms of the q's and p's. Neither the fact that this can be done nor the specific form of the relations was justified. The reason for this gap is that, while the concrete physical examples above were introduced at this time to illustrate the considerable previous abstract structure, additional abstract structure, introduced in the next section, is required to fully explain even these simple examples.

PROBLEM 18.11. Show that if H has k cyclic coordinates, then Hamilton's equations reduce to a system of $2(n - k)$ equations in $2(n - k)$ unknowns.

PROBLEM 18.12. Derive Hamilton's equations for the spherical pendulum.

18.4 The Legendre transformation and Lagrangian vector fields

There is another standard model of classical mechanics in terms of the so-called Lagrange equations. These may be derived either directly from Hamilton's equations, or independently starting with *Hamilton's Variational Principle*. If we follow the latter method, we expect, of course, that we can then proceed to derive Hamilton's equations from Lagrange's. To point up this reciprocity, or duality, between Hamilton's and Lagrange's equations we note further that while Hamilton's equations are on the cotangent manifold or (momentum) phase space of the given configuration space, the Lagrange equations are on the tangent manifold, now called *the state space* or *the velocity phase space* of the configuration space. In Hamilton's equations we had a given function, H, *the Hamiltonian* on the cotangent manifold. We proceed now by describing certain structures induced by a given function, L, called *a Lagrangian*, on the tangent manifold.

Let L be a given function on the tangent bundle of a manifold, M. We will fix $p \in M$ and look at only the restriction of L to the vector space T_p. To avoid excessive indexing we will again denote this restriction by L. Let $v \in T_p$, and note that the derivative at v of L, $D_vL\colon T_p \to \mathbb{R}$, a linear function on T_p, is in T_p^*. Let $DL\colon T_p \to T_p^*$ according to $v \mapsto D_vL$ and suppose DL has an inverse. Then given a function L on T_p we can define a function H on T_p^* by

$$H\colon \sigma \mapsto \langle \sigma, DL^{-1}(\sigma)\rangle - L \circ DL^{-1}(\sigma) \tag{18.25}$$

for $\sigma \in T_p^*$. Thus, we have a mapping, $\mathfrak{L}$, from functions on T_p to functions on T_p^*.

Following the standard conventions, we denote the coordinate functions on a coordinate neighborhood of TM by $(q^i, \dot{q}^i)$. Notice that we use the same notation for the first n coordinate functions on TM as on T^*M (see Theorem 18.9), and the notation $\dot{q}^i$ is suggested by the fact that the component functions v^i of v can always be thought of as the derivatives at 0 of the component functions of some curve. In these coordinates

$$DL\colon (\dot{q}^1(v), \ldots, \dot{q}^n(v)) \mapsto \left(\frac{\partial L}{\partial \dot{q}^1}(v), \ldots, \frac{\partial L}{\partial \dot{q}^n}(v)\right) = (p_1(DL\cdot v), \ldots, p_n(DL\cdot v))$$

where (q^i, p_i) are coordinates on T^*M, and (18.25) can be written

$$H(p_1, \ldots, p_n) = p_i\dot{q}^i - L(\dot{q}^i, \ldots, \dot{q}^n) \tag{18.26}$$

where the $\dot{q}^i$ on the right are the solutions of $\partial L/\partial \dot{q}^i = p_i$. (See comment on notation below eq. (18.17).) Moreover, differentiating (18.26) (using $\partial L/\partial \dot{q}^i = p_i$) we have

$$dH = v^i\, dp_i \tag{18.27}$$

Theorem 18.14. *If DL has an inverse, then* $\mathfrak{L}$ is 1–1 and onto.

Proof. Immediate from (18.26).

Definitions
L is a function on TM, then for each $p \in M$ we have a DL, so we have a map, denoted again by DL, from TM to T^*M called *the Legendre transformation of L*, or *the fiber derivative of L*. For sets $\mathscr{U} \subset M$ at which DL has an inverse, $\mathfrak{L}:\mathfrak{F}_{T\mathscr{U}} \to \mathfrak{F}_{T^*\mathscr{U}}$ is called *the Legendre transformation.*

We have been assuming that the given function, L, is such that DL has an inverse. In this case L is called *hyperregular*. We will continue to assume L is hyperregular, though for many of the subsequent results we need only that DL is locally invertible (i.e., L is *regular*).

An important example is given by

$$L(v) = \tfrac{1}{2}g_p(v, v)$$

which we can construct in the case when M has a given pseudo-Riemannian structure, g. Then $D_vL \cdot w = g_p(v, w)$. Thus, at each $p \in M$, DL: $v \to D_vL$ is the linear map $g^\flat$ from T_p to T_p^* induced by g. Further, we can write (18.25) as $H(DL(v)) = \langle DL(v), v\rangle - L(v)$, so, in this case $H(DL(v)) = g_p(v, v) - \tfrac{1}{2}g_p(v, v)$, or $H \circ DL = L$, or $H = L \circ DL^{-1}$. As indicated above, we can make a model of a mechanical system based on L (instead of H). This particular form of L will correspond to a mechanical system in which no forces are acting.

Now, with the Legendre transformation, DL, induced by a given Lagrangian function on TM, we can proceed to build additional structures. Specifically, we have the pull-backs of the natural 1-form field, θ, and the natural symplectic structure, $d\theta$, on T^*M. (We are discarding the subscript M on θ.) We write

$$\theta_L = DL^*\theta \qquad \text{and} \qquad \omega_L = -DL^*\, d\theta$$

Since $d\theta$ is nondegenerate and L is hyperregular, ω_L is nondegenerate. By Theorem 12.4 $\omega_L = -d\theta_L$ so ω_L is closed, and hence ω_L is a symplectic form on TM, and (TM, ω_L) is a symplectic manifold.

Theorem 18.15. *θ_L can also be given by the formula*

$$< \theta_L(v), \mathbf{a}\rangle = \langle DL(v), \pi_* \cdot \mathbf{a}\rangle$$

where $v \in TM$, $\mathbf{a} \in T_vTM$, and π is the natural projection of TM on M. (Compare Theorem 11.18.)

Proof. $\theta_L(v) = DL^*\theta(v) = DL^* \cdot \theta(DL(v))$ by the definition of the pull-back. So

$$\begin{aligned}
\langle \theta_L(v), \mathbf{a} \rangle &= \langle DL^* \cdot \theta(DL(v)), \mathbf{a} \rangle \\
&= \langle \theta(DL(v)), DL_* \cdot \mathbf{a} \rangle \\
&= \langle DL(v), \bar{\pi}_* \cdot DL_* \cdot \mathbf{a} \rangle \quad \text{by the definition of } \theta \\
& \qquad (\bar{\pi} \text{ is the projection of } T^*M \text{ on } M) \\
&= \langle DL(v), (\bar{\pi} \circ DL)_* \cdot \mathbf{a} \rangle \quad \text{by the chain rule} \\
&= \langle DL(v), \pi_* \cdot \mathbf{a} \rangle
\end{aligned}$$

■

We saw, eq. (18.13), that in coordinates (q^i, p_i) on T^*M the natural 1-form, θ, has the form $\theta = p_i \, dq^i$. Therefore, in coordinates $(q^i, \dot{q}^i)$ on TM we have

$$\theta_L = \frac{\partial L}{\partial \dot{q}^i} dq^i \tag{18.28}$$

and

$$\omega_L = \frac{\partial^2 L}{\partial \dot{q}^i \, \partial q^j} dq^i \wedge dq^j + \frac{\partial^2 L}{\partial \dot{q}^i \, \partial \dot{q}^j} d\dot{q}^i \wedge dq^j \tag{18.29}$$

Definitions
The action of L is a map A_L: $TM \to \mathbb{R}$ given by $v \mapsto \langle DL(v), v \rangle$. *The energy of* L is $\mathsf{E} = A_L - L$. *The Lagrangian vector field of* L is the vector field, X_L, on TM such that

$$i_{X_L} \omega_L = d\mathsf{E} \tag{18.30}$$

Note that since ω_L is nondegenerate, there is one and only one X_L. In coordinates

$$A_L(v) = \dot{q}^i \frac{\partial L}{\partial \dot{q}^i} \tag{18.31}$$

(Another comment on terminology. The "action" of L defined above clearly has no relation to the "action" of a 1-parameter group action which was defined in Section 13.2 and which will appear again in Section 19.2. On the other hand, the term "action" as used in mechanics more commonly refers to an integral of what we, following Abraham and Marsden, have called action and is the celebrated quantity in the venerable Maupertius Principle of Least Action; cf., Arnold, 1989, pp. 245–246.)

Now we will write eq. (18.30), the definition of Lagrangian vector field, in coordinate form. We write

$$X_L = Y^k \frac{\partial}{\partial q^k} + Z^l \frac{\partial}{\partial \dot{q}^l}$$

and substitute this and the expression for ω_L in (18.29) into the left side of (18.30). Using the linearity of $i_{X_L}\omega_L$ with respect to X_L and ω_L (and denoting partial derivatives of L by subscripts) we get

$$\begin{aligned} i_{X_L}\omega_L = {} & (Y^k L_{\dot{q}^i q^j}) i_{\partial/\partial q^k}\, dq^i \wedge dq^j + (Y^k L_{\dot{q}^i \dot{q}^j}) i_{\partial/\partial q^k}\, d\dot{q}^i \wedge dq^j \\ & + (Z^l L_{\dot{q}^i q^j}) i_{\partial/\partial \dot{q}^l}\, dq^i \wedge dq^j + (Z^l L_{\dot{q}^i \dot{q}^j}) i_{\partial/\partial \dot{q}^l}\, d\dot{q}^i \wedge dq^j \end{aligned}$$

By Theorem 11.20, $i_{\partial/\partial q^k}\, dq^i \wedge dq^j = \delta^i_k\, dq^j - \delta^j_k\, dq^i$, etc. So

$$i_{X_L}\omega_L = -Y^j L_{\dot{q}^i q^j}\, dq^i + Y^i L_{\dot{q}^i q^j}\, dq^j + Y^i L_{\dot{q}^i \dot{q}^j}\, d\dot{q}^j - Z^j L_{\dot{q}^i \dot{q}^j}\, dq^i \quad (18.32)$$

The right side of (18.30) is

$$dA_L - dL = L_{\dot{q}^i}\, d\dot{q}^i + \dot{q}^i L_{q^j \dot{q}^i}\, dq^j + \dot{q}^i L_{\dot{q}^j \dot{q}^i}\, d\dot{q}^j - (L_{q^i}\, dq^i + L_{\dot{q}^i}\, d\dot{q}^i)$$

Simplifying, we get

$$dE = -L_{q^i}\, dq^i + \dot{q}^i L_{q^j \dot{q}^i}\, dq^j + \dot{q}^i L_{\dot{q}^j \dot{q}^i}\, d\dot{q}^j \quad (18.33)$$

Putting (18.32) and (18.33) into (18.30) and equating coefficients of $d\dot{q}^i$, we get

$$Y^i = \dot{q}^i \quad (18.34)$$

and equating coefficients of dq^i and using (18.34), we get

$$Y^j L_{\dot{q}^i q^j} + Z^j L_{\dot{q}^i \dot{q}^j} = L_{q^i} \quad (18.35)$$

Equations (18.34) and (18.35) together are the coordinate form of eq. (18.30). That is, X_L satisfies (18.30) iff the component functions (Y^i, Z^i) of X_L satisfy (18.34) and (18.35).

Finally, if γ is an integral curve of X_L, then the component functions $q^i \circ \gamma$ and $\dot{q}^i \circ \gamma$ satisfy

$$\frac{dq^i \circ \gamma}{dt} = \hat{Y}^i(q^j \circ \gamma, \dot{q}^k \circ \gamma), \qquad \frac{d\dot{q}i \circ \gamma}{dt} = \hat{Z}^i(q^j \circ \gamma, \dot{q}^k \circ \gamma)$$

Thus, slightly abusing the notation, on an integral curve (18.34) becomes

$$\frac{dq^i}{dt} = \dot{q}^i \tag{18.36}$$

and (18.35) becomes

$$\frac{dq^j}{dt} L_{\dot{q}^i q^j}(q^k, \dot{q}^l) + \frac{d\dot{q}^j}{dt} L_{\dot{q}^i \dot{q}^j}(q^k, \dot{q}^l) = L_{q^i}(q^k, \dot{q}^l)$$

or

$$\frac{d}{dt} L_{\dot{q}^i}(q^k, \dot{q}^l) = L_{q^i}(q^k, \dot{q}^l) \tag{18.37}$$

(See comment on notation below eq. (18.17).) Equation (18.37) with $\dot{q}^i$ given by eq. (18.36) is the Euler–Lagrange equation.

PROBLEM 18.13. With the Lagrangian function, L, in the form $L = \mathsf{T} - \Phi \circ \pi$ obtain the v's in terms of the q's and p's in the four examples in Section 18.3.

PROBLEM 18.14. Make a figure for Theorem 18.15 corresponding to the one accompanying Theorem 18.10.

19

ADDITIONAL TOPICS IN MECHANICS

We will continue to compare the Hamiltonian and Lagrangian formulations of mechanics, and get particular results when we think of configuration space as a pseudo-Riemannian manifold. We will then touch on the important ideas of momentum mappings and conservation laws, and finally on the Hamilton–Jacobi theory for solving the Hamilton canonical equations.

19.1 The configuration space as a pseudo-Riemannian manifold

In Section 18.3 we explained how the motion of a mechanical system can be obtained by means of the Hamilton canonical equations from a given Hamiltonian function on the cotangent space of the configuration space, and in Section 18.4 by means of the Legendre transformation we transferred structures on T^*M to analogs on TM. The correspondence between structures on T^*M and TM is completed by the following relations.

Theorem 19.1. *With the notation of Chapter 18,*

(i) $H = \mathsf{E} \circ DL^{-1}$
(ii) $X_H = DL_* X_L$
(iii) $A_L = \langle \theta, X_H \rangle \circ DL$

Proof. (i) For $\sigma \in T_p^*$,

$$\begin{aligned} \mathsf{E} \circ DL^{-1}(\sigma) &= (A_L - L) \circ DL^{-1}(\sigma) = A_L \circ DL^{-1}(\sigma) - L \circ DL^{-1}(\sigma) \\ &= A_L(v) - L \circ DL^{-1}(\sigma) = \langle DL(v), v \rangle - L \circ DL^{-1}(\sigma) \\ &= \langle \sigma, DL^{-1}(\sigma) \rangle - L \circ DL^{-1}(\sigma) = H(\sigma). \end{aligned}$$

(ii) For $v \in TM$ and $\mathbf{w} \in T_v TM$,

$$\begin{aligned} \omega(DL_* X_L(v), DL_* \mathbf{w}) &= \omega_L(X_L(v), \mathbf{w}) && \text{since } \omega_L \text{ is the pull-back of } \omega \\ &= \tfrac{1}{2} d\mathsf{E}(v)(\mathbf{w}) && \text{by (18.30)} \\ &= \tfrac{1}{2} dH(DL(v))(DL_* \mathbf{w}) && \text{since } d\mathsf{E} \text{ is the pull-back of } dH \\ &= \omega(X_H(DL(v)), DL_* \mathbf{w}) && \text{by (18.2)} \end{aligned}$$

So, since $\mathbf{w}$ is arbitrary and ω is nondegenerate,

$$DL_* X_L(v) = X_H DL(v)$$

(iii) *Problem 19.1* ■

From Theorem 19.1(ii) we see that the integral curves of X_H and X_L go into one another and, hence, the projections of these curves on M are the same. The conclusion for mechanics is that the solutions of the Euler–Lagrange equations give motions of mechanical systems. That is, rather than starting with H on T^*M and solving the Hamilton canonical equations (18.14), we can start with L on TM and solve the Euler–Lagrange equations (18.36) and (18.37).

The standard form of *the Lagrangian*, L, in applications to mechanics is

$$L = \mathsf{T} - \Phi \circ \pi \tag{19.1}$$

where T is *the kinetic energy* and Φ is *the potential energy* of the system or rigid body. T is a function on TM quadratic on each tangent space of M, and Φ is a function on the configuration space, M. The motivation for this form of L comes from the fact that in special cases in terms of rectangular cartesian coordinates on M the Euler–Lagrange equations reduce to Newton's equations, as we will see below.

Thus, from a slightly more abstract point of view, we can always think of the configuration space of a mechanical system (in which forces come from a potential) as a pseudo-Riemannian manifold on which a function, Φ, is prescribed, the pseudo-Riemannian structure coming from the polarization of the quadratic function T (Section 5.4).

In a local coordinate system $(q_i, \dot{q}^i)$ on TM on (M, g) we have $g(v, w)(p) = g_{ij}(q^k)\dot{q}^i(v)\dot{q}^j(w)$, and $g(v, v)(p) = g_{ij}(q^k)\dot{q}^i\dot{q}^j$, so

$$\begin{aligned} D_v g &= \left(\frac{\partial g}{\partial \dot{q}^1}(v), \ldots, \frac{\partial g}{\partial \dot{q}^n}(v)\right) \\ &= 2(g_{ij}\dot{q}^j, \ldots, g_{nj}\dot{q}^j) = 2g^\flat(v) \end{aligned} \tag{19.2}$$

at each point $p \in M$.

Theorem 19.2. *On a pseudo-Riemannian manifold* (M, g) *let* $\mathsf{T}(v) = \frac{1}{2}g(v, v)$. *Then, with* L *given by* (19.1) *we get*

(i) $DL = g^\flat$
(ii) $\langle DL(v), w\rangle = g(v, w)$
(iii) $A_l = 2\mathsf{T}$
(iv) $E = \mathsf{T} + \Phi \circ \pi$

Proof. *Problem 19.2.*

Theorem 19.3. *On a pseudo-Riemannian manifold* (M, g) *let* $\theta_g = (g^\flat)^*\theta$ *and* $\omega_g = -(g^\flat)^*\, d\theta$ where θ is the natural 1-form on T^*M. Then in local coordinates,

$$\theta_g = g_{ij}\dot{q}^i\, dq^j \tag{19.3}$$

$$\omega_g = g_{ij}\, dq^i \wedge d\dot{q}^j + \frac{\partial g_{ij}}{\partial q^k}\dot{q}^i\, dq^j \wedge dq^k \tag{19.4}$$

Proof. $(g^\flat)^* p_j\, dq^j = p_i(g^\flat(v))\, dq^i = g_{ij}\dot{q}^i\, dq^j$, and ω_g is obtained by exterior differentiation. ■

Corollary. *If* (M, g) *has a function* L *on* TM *as in Theorem 19.2 the coordinate expressions for* θ_L *and* ω_L *are given by eqs. (19.3) and (19.4).*

Theorem 19.4. *If* $\phi: M \to N$ *is an isometry, then* $(\phi_*)^*\theta g_N = \theta g_M$. *That is, the 1-form,* θ_g, *is preserved, and* $\phi_*: TM \to TN$ *is symplectic.*

Proof. Since $\phi_* = g_N^\sharp \circ (\phi^*)^{-1} \circ g_M^\flat$,

$$\begin{aligned}
(\phi_*)^*\theta_{g_N} &= (g_N^\sharp \circ (\phi^*)^{-1} \circ g_M^\flat)\theta_{g_N} = (g_M^\flat)^*((\phi^*)^{-1})^*(g_N^\sharp)^*\theta_{g_N} \\
&= (g_M^\flat)^*((\phi^*)^{-1})^*\theta_N \qquad \text{by definition of } \theta_{g_N} \\
&= (g_M^\flat)^*\theta_M \qquad \text{by Theorem 18.10} \\
&= \theta_{g_M} \qquad \text{by definition of } \theta_{g_M}
\end{aligned}$$
■

Theorem 19.5. *With* L *of the form (19.1), the projections* $\gamma_0 = \pi \circ \gamma$ *of the integral curves of* X_L *(i.e., the solutions of the Euler–Lagrange equations) satisfy*

$$\nabla_{\dot{\gamma}_0}\dot{\gamma}_{0\to 0} = -\text{grad}\,\Phi(\gamma_0) \tag{19.5}$$

Proof. With L in the form (19.1), eq. (18.37) is

$$\frac{d}{dt}(g_{ij}\dot{q}^j) = \frac{1}{2}\frac{\partial g_{jk}}{\partial q^i}\dot{q}^j\dot{q}^k - \frac{\partial \Phi}{\partial q^i}$$

Expanding the left side, multiplying both sides by g^{ik} and putting $\dot{q}^i = dq^i/dt$

according to eq. (18.36) we get

$$\ddot{q}^k = g^{lk}\left(\frac{1}{2}\frac{\partial g_{ij}}{\partial q_l} - \frac{\partial g_{li}}{\partial q^j}\right)\dot{q}^i\dot{q}^j - g^{ik}\frac{\partial \Phi}{\partial q^i}$$

$$= \begin{Bmatrix} k \\ i\ j \end{Bmatrix}\dot{q}^i\dot{q}^j - g^{ik}\frac{\partial \Phi}{\partial q^i} \tag{19.6}$$

and this is the coordinate form of (19.5). ∎

Corollaries. (i) *If there are no forces acting so that, in particular,* $\Phi = 0$, *then the projections,* γ_0, *of the integral curves of* X_L *are precisely the geodesics of the Levi–Civita connection of* (M, g).

(ii) *If* M *is pseudo-Euclidean space and* q^i *are rectangular cartesian coordinates, then (19.6) are Newton's equations.*

PROBLEM 19.1. Prove Theorem 19.1(iii).

PROBLEM 19.2. Prove Theorem 19.2.

19.2 The momentum mapping and Noether's theorem

In Chapter 18, after introducing the concept of a Hamiltonian vector field, X, on (M, ω), we focused on the Hamiltonian vector field, X_H, obtained from a given function, H, on (M, ω). Now we switch our point of view, and think of X as coming from a 1-parameter group action. That is, we start with a 1-parameter group action on M that preserves the symplectic structure, *a symplectic action*, and we get a vector field, X_M, the infinitesimal generator of the action. (See Section 13.2.)

Definition
A function, J, on a symplectic manifold, M, is called *a momentum mapping* if

$$i_{X_M}\omega = dJ \tag{19.7}$$

where X_M is the infinitesimal generator of a symplectic action. (Note: this concept is a special case of a more general one which requires more information about Lie groups than we have available.)

The existence of momentum mappings comes from the following result.

Theorem 19.6. *If the natural* 1*-form,* θ, *on* (M, ω) *is invariant under a symplectic action on* M, *then* $i_{X_M}\theta$ *is a momentum mapping.*

Proof. This comes immediately from the identity $di_{X_M}\theta + i_{X_M} d\theta = L_{X_M}\theta$. ∎

The next three theorems give conditions under which the hypothesis of Theorem 19.6 is satisfied, and give formulas for the associated momentum mappings.

Theorem 19.7. (i) *If the manifold of Theorem 19.6 is a cotangent manifold, T^*M, and $\mathscr{A}_M$ is any action on M, then θ is preserved under $\mathscr{A}_{T^*M}$ and $i_{X_{T^*M}}\theta$ is a momentum mapping on T^*M*

(ii)

$$i_{X_{T^*M}}\theta(\sigma_p) = \langle \sigma_p, X_M(p)\rangle \tag{19.8}$$

for $p \in M$ and $\sigma_p \in T_p^*M$.

Proof. (i) The induced action, $\mathscr{A}_{T^*M}$ is symplectic and θ is preserved by Theorem 18.10 and its corollary. Hence, by Theorem 19.6, $i_{X_{T^*M}}\theta$ is a momentum mapping on T^*M.

(ii)

$$\begin{aligned} i_{X_{T^*M}}\theta(\sigma_p) &= \langle \theta(\sigma_p), X_{T^*M}(\sigma_p)\rangle = \langle \sigma_p, \pi_* \circ X_{T^*M}(\sigma_p)\rangle \qquad \text{by definition of } \theta \\ &= \langle \sigma_p, X_M \circ \pi(\sigma_p)\rangle = \langle \sigma_p, X_M(p)\rangle \end{aligned}$$ ∎

EXAMPLE. Let $M = \mathbb{R}^n$ and $\mathscr{A}_M: \mathbb{R} \times \mathbb{R}^n \to \mathbb{R}^n$ by $(u, (a^1, \ldots, a^n)) \mapsto (a^1 + ur^1, \ldots, a^n + ur^n)$ with $r^i \in \mathbb{R}$. Then $X_M: (a^1, \ldots, a^n) \mapsto (r^1, \ldots, r^n)$ and $\langle \sigma, X_M\rangle(a) = p_i r^i$ where $p_i = \sigma_i \circ \pi$ and σ_i are the components of σ. Thus, in this case, the momentum mapping on T^*M gives the X_M component of the momentum.

Theorem 19.8. (i) *If the manifold of Theorem 19.6 is the tangent manifold of a pseudo-Riemannian manifold (M, g), and $\mathscr{A}_M$ is an action on M that preserves g, then θ_g is preserved under $\mathscr{A}_{TM}$ and $i_{X_{TM}}\theta_g$ is a momentum mapping on TM.*

(ii)

$$i_{X_{TM}}\theta_g(v_p) = g(v_p, X_M(p)) \tag{19.9}$$

for $p \in M$ and $v_p \in T_pM$.

Proof. (i) The induced action $\mathscr{A}_{TM}$ is symplectic and θ_g is preserved by Theorem 19.4. Hence, by Theorem 19.6 $i_{X_{TM}}\theta_g$ is a momentum mapping on TM.

(ii) $i_{X_{TM}}\theta_g(v_p)$

$= \langle \theta_g(v_p), X_{TM}(v_p) \rangle$

$= \langle (g^\flat)^*\theta(v_p), X_{TM}(v_p) \rangle$ by the definition of θ_g

$= \langle \theta(g^\flat(v_p)), g^\flat_* X_{TM}(v_p) \rangle$ by the property of transpose

$= \langle g^\flat(v_p), \tilde{\pi}_* \cdot g^\flat_* X_{TM}(v_p) \rangle$ by the definition of θ ($\tilde{\pi}$ is the projection $T^*M \to M$)

$= \langle g^\flat(v_p), \pi_* X_{TM}(v_p) \rangle$ (π is the projection $TM \to M$)

$= \langle g^\flat(v_p), X_M(p) \rangle = g(v_p, X_M(p))$ ■

EXAMPLE. Let$M = \mathbb{R}^3$ and $\mathscr{A}_M$ be an action on $\mathbb{R}^3$ such that for each $u \in \mathbb{R}$, the map $\mathbb{R}^3 \to \mathbb{R}^3$ is a rotation. The instantaneous generator $X_M(a)$ is the velocity of a, and this can be written in the form $r_{ij}a^i$ where (r_{ij}) is a skew-symmetric matrix (Synge and Schild, 1966, section 5.3). If we write

$$(r_{ij}) = \begin{pmatrix} 0 & -r_3 & r_2 \\ r_3 & 0 & -r_1 \\ -r_2 & r_1 & 0 \end{pmatrix}$$

then $X_M(a) = (-r_3a^2 + r_2a^3, r_3a^1 - r_1a^3, -r_2a^1 + r_1a^2) = r \times a$, the ordinary vector product. Hence, $g(v(a), X_M(a)) = g(v(a), r \times a) = g(r \times a, v(a)) = r_i(a \times v(a))^i$, so the momentum mapping on TM gives the X_M component of the angular momentum.

Theorem 19.9. (i) *If the manifold of Theorem 19.6 is a tangent manifold, TM, and* $\mathscr{A}_M$ *is an action on M such that L is preserved under* $\mathscr{A}_{TM}$, *then* θ_L *is preserved under* $\mathscr{A}_{TM}$ *and* $i_{X_{TM}}\theta_L$ *is a momentum mapping on TM.*

(ii)

$$i_{X_{TM}}\theta_L(v_p) = \langle DL(v_p), X_M(p) \rangle \tag{19.10}$$

for $p \in M$ *and* $v_p \in T_pM$.

Proof. Problem 19.3.

In the previous chapter we focused on vector fields coming from Hamiltonian or Lagrangian functions, and so far in this section we have focused on a structure, the momentum mapping, arising from the point of view of a vector field as the infinitesimal generator of an action. From the first point of view, a vector field, through the Hamilton canonical equations or the Euler–Lagrange equations, *determines* a motion of a mechanical system, so that an integral of the vector field (i.e., of the differential equations) gives some information about the motion. From the second point of view, a vector field *arises* from the motion of a mechanical system, and an integral of the vector field is something conserved during the motion. The relation between these two points of view is expressed by *conservation laws.*

Theorem 19.10. *If X_H is a Hamiltonian vector field on M, and X_M is the infinitesimal generator of a symplectic action with moment map, J, on M, then H is an integral of X_M iff J is an integral of X_H.*

Proof. From (19.7) we can write $X_M = X_J$, and the result follows from Theorem 18.6(iii). ∎

Theorem 19.11. **(E. Noether)** *Suppose a 1-parameter group acts on a manifold, M (e.g., the configuration space of a mechanical system). If the Lagrangian, L, is invariant under $\mathscr{A}_{TM}$, then $i_{X_{TM}}\theta_L$ is an integral of X_L, and is given by Eq. (19.10).*

Proof. (i) The fact that L is invariant under $\mathscr{A}_{TM}$ implies that

$$\langle D_v L, w\rangle = \langle D_{\gamma_v(u)} L, \gamma_{w(u)}\rangle$$

for $v, w \in TM$ and γ_v, γ_w are integral curves of X_{TM}. Putting $w = v$ we get that the action of L is invariant under $\mathscr{A}_{TM}$ and hence the energy, E, of L is invariant under $\mathscr{A}_{TM}$, or E is an integral of X_{TM}.

(ii) Let $J = i_{X_{TM}}\theta_L$. Then $dJ = i_{X_{TM}}\omega_L$ since θ_L is invariant according to Theorem 19.9. So X_{TM} has momentum map, J. Also X_L is the Hamiltonian vector field of E. So by part (i) and Theorem 19.10, J is an integral of X_L. ∎

PROBLEM 19.3. Prove Theorem 19.9.

19.3 Hamilton–Jacobi theory

Having formulated a mechanical problem in terms of the Hamilton canonical equations, the solution is obtained by solving these equations. There is an interesting and somewhat effective classical method for this.

We will first put this method into the general context of "the method of characteristics" which we touched on in Section 14.3. We noted there that solutions of certain types of partial differential equations can be obtained from the solution of the system (14.17) of the characteristic vector field of the differential system of the partial differential equation. We can go in the opposite direction.

Theorem 19.12. *If a function, S, of $x^1, \ldots, x^n$ is a solution of*

$$F\left(x^i, \frac{\partial S}{\partial x^i}, S\right) = 0 \tag{19.11}$$

and x^i are solutions of

$$\frac{dx^i}{du} = \frac{\partial F}{\partial p_i}(x^i, p_i, z) \tag{19.12}$$

where $z = S \circ x$ and $p_i = \dfrac{\partial S}{\partial x^i} \circ x$, then z and p_i satisfy

$$\frac{dp_i}{du} = -\left(\frac{\partial F}{\partial x^i}(x^i, p_i, z) + p_i \frac{\partial F}{\partial z}(x^i, p_i, z)\right) \tag{19.13}$$

$$\frac{dz}{du} = p_i \frac{\partial F}{\partial p_i}(x^i, p_i, z) \tag{19.14}$$

Proof. Since S satisfies (19.11), S satisfies

$$\frac{\partial F}{\partial x^i} + \frac{\partial F}{\partial z}\frac{\partial S}{\partial x^i} + \frac{\partial F}{\partial p_j}\frac{\partial^2 S}{\partial x^i\, \partial x^j} = 0$$

or

$$\frac{\partial F}{\partial p_j}\frac{\partial^2 S}{\partial x^i\, \partial x^j} = -\left(\frac{\partial F}{\partial x^i} + \frac{\partial F}{\partial z}\frac{\partial S}{\partial x^i}\right) \tag{19.15}$$

Now, if x^i is a solution of (19.12), then from the definition of p_i,

$$\frac{dp_i}{du} = \frac{\partial^2 S}{\partial x^i\, \partial x^i}\frac{dx^j}{du} = \frac{\partial^2 S}{\partial x^i\, \partial x^i}\frac{\partial F}{\partial p_j}$$

which is precisely the left side of (19.15), and

$$\frac{dz}{du} = \frac{\partial S}{\partial x^i}\frac{dx^i}{du} = \frac{\partial S}{\partial x^i}\frac{\partial F}{\partial p_i}$$

which is eq. (19.14). ■

Corollary. *If F is independent of z, S is a solution of (19.11), and x^i are* *solutions of*

$$\frac{\mathrm{d}x^i}{\mathrm{d}u} = \frac{\partial F}{\partial p_i}(x^i, p_i) \tag{19.16}$$

then

$$\frac{dp_i}{du} = -\frac{\partial F}{\partial x^i}(x^i, p_i) \tag{19.17}$$

Clearly, in different notation, eqs. (19.16) and (19.17) are the Hamilton canonical equations, so the corollary gives us solutions of the Hamilton canonical equations with Hamiltonian function, H, from solutions of the partial differential equation $H(q^i, p_j) = 0$.

EXAMPLE. We saw, in Section 18.3, Example 2, that the Hamiltonian of a particle subject to a central force is

$$H(q^1, q^2, p_1, p_2) = f(q^1) + \frac{1}{2m}\left(p_1^2 + \frac{p_2^2}{(q^1)^2}\right) \tag{18.20}$$

so we want to solve

$$f(q^1) + \frac{1}{2m}\left(\left(\frac{\partial S}{\partial q^1}\right)^2 + \frac{1}{(q^1)^2}\left(\frac{\partial S}{\partial q^2}\right)^2\right) = 0$$

Slightly simplifying the notation by putting $q^1 = r$ and $q^2 = \vartheta$ and looking for solutions S in the form $S(r, \vartheta) = \mathbf{R}(r) + \mathbf{\Theta}(\vartheta)$ we get $(\mathbf{dR}/dr)^2 = -2mf(r) - A^2/r^2$ and $(d\mathbf{\Theta}/d\vartheta)^2 = A^2$ where A^2 is an arbitrary (nonnegative) constant. Then

$$\frac{dr}{du} = \frac{\partial H}{\partial p_1} = \frac{1}{m}p_1 = \frac{1}{m}\frac{\partial S}{\partial r} = \pm\frac{1}{m}\sqrt{-2mf(r) - \frac{A^2}{r^2}}$$

$$\frac{d\vartheta}{du} = \frac{\partial H}{\partial p} = \frac{1}{mr^2}p_2 = \frac{1}{mr^2}\frac{\partial S}{\partial \vartheta} = \pm\frac{A}{mr^2}$$

$$\frac{dp_1}{du} = -\frac{\partial H}{\partial q^1} = -f'(r) + \frac{1}{mr^3}p_2^2 = -f'(r) + \frac{A^2}{mr^3}$$

$$\frac{dp_2}{du} = -\frac{\partial H}{\partial q^2} = 0$$

While in this particular example the problem can be "solved by quadratures," in general, the problem of solving (19.16) could be as formidable as solving the original Hamilton canonial equation. Thus, as an effective method for solving the Hamilton canonical equations, this straightforward application of the general theory for first-order partial differential equations has its limitations.

A variation of this general method, attributed to C. G. J. Jacobi, circumvents the need to solve a system of ordinary equations. We first have to reformulate the partial differential equation (19.11) in the case where the dependent variable is missing.

If F is a function of x^α and p_α, $\alpha = 0, 1, \ldots, n$, we assume we can separate out p_0 in $F(x^\alpha, p_\alpha) = 0$, writing it in the form $p_o + H(x^0, x^i, p_i) = 0$, $i = 1, \ldots, n$. Then we look for functions, S, of x^α such that

$$\frac{\partial S}{\partial x^0} + H\left(x^0, x^i, \frac{\partial S}{\partial x^i}\right) = 0 \tag{19.18}$$

the Hamilton–Jacobi equation.

Definition

A function $S\colon (x^0, x^i, Q^i) \mapsto S(x^0, x^i, Q^i)$ in some neighborhood, $\mathscr{U}$, of $\mathbb{R} \times \mathbb{R}^n \times \mathbb{R}^n$ in which $\partial S/\partial x^i$ satisfy (19.18) and $\det(\partial^2 S/\partial x^i\, \partial Q^j) \neq 0$ is called *a complete integral of (19.18).*

Theorem 19.13. *Suppose we have a complete integral, S, of eq. (19.18), and arbitrary constants, P_i. Let x^i be a solution of*

$$\frac{\partial S}{\partial Q^i} = -P_i \tag{19.19}$$

and let

$$p_i = \frac{\partial S}{\partial x^i} \tag{19.20}$$

where on the right side of (19.20), the arguments, x^i, are the solutions of $\partial S/\partial Q^i = -P_i$. (Hence, x^i and p_i are functions of x^0 and Q^i). Then x^i and p_i are solutions of

$$\frac{dx^i}{dx^0} = \frac{\partial H}{\partial p_i} \tag{19.21}$$

$$\frac{dp_i}{dx^0} = -\frac{\partial H}{\partial x^i} \tag{19.22}$$

(Note, (1) eqs. (19.21) and (19.22) are part of the set of characteristic equations of the Hamilton–Jacobi equation and (2) eqs. (19.21) and (19.22) are the Hamilton canonical equations in slightly different notation.)

Proof. With S a function of x^0, x^i, Q^i we differentiate (19.18) with respect to Q^i and (19.19) with respect to x^0 and we get, respectively,

$$\frac{\partial^2 S}{\partial Q^i\,\partial x^0} + \frac{\partial H}{\partial p_j}\frac{\partial^2 S}{\partial x^j\,\partial Q^i} = 0$$

$$\frac{\partial^2 S}{\partial Q^i\,\partial x^0} + \frac{\partial^2 S}{\partial Q^i\,\partial x^j}\frac{dx^j}{du} = 0$$

which, since $\det(\partial^2 S/\partial Q^i\,\partial x^j) \neq 0$, yields (19.21). Similarly, differentiating (19.18) with respect to x^i and (19.20) with respect to x^0, we get, respectively,

$$\frac{\partial^2 S}{\partial x^0\,\partial x^i} + \frac{\partial H}{\partial x^i} + \frac{\partial H}{\partial p_j}\frac{\partial^2 S}{\partial x^i\,\partial x^j} = 0$$

$$\frac{\partial^2 S}{\partial x^0\,\partial x^i} + \frac{\partial^2 S}{\partial x^i\,\partial x^j}\frac{dx^i}{dx^0} = \frac{dp_i}{dx^0}$$

Combining these, using (19.21) we get (19.22). ∎

EXAMPLE. We will apply the Hamilton–Jacobi method to the problem of the previous example; the motion of a particle with the Hamiltonian (18.20). Now we want to solve

$$\frac{\partial S}{\partial u} + f(r) + \frac{1}{2m}\left[\left(\frac{\partial S}{\partial r}\right)^2 + \frac{1}{r^2}\left(\frac{\partial S}{\partial \vartheta}\right)^2\right] = 0 \tag{19.23}$$

with $S(u, r, \vartheta) = \mathbf{U}(u) + \mathbf{R}(r) + \mathbf{\Theta}(\vartheta)$ we get $d\mathbf{U}/du = -Q^1$, $(d\mathbf{R}/dr)^2 = -2m(f(r) - Q^1) - (Q^2)^2/r$, and $(d\mathbf{\Theta}/d\vartheta)^2 = (Q^2)^2$, so

$$S(u, r, \vartheta, Q^1, Q^2) = -Q^1 u + \int \sqrt{-2m(f(r) - Q^1) - \frac{(Q^2)^2}{r^2}}\,dr + Q^2\vartheta \tag{19.24}$$

is a complete integral of (19.23). Differentiating S with respect to Q^1 and Q^2, and using (19.19), we get, respectively,

$$-u + m\int \frac{1}{I_1}\,dr = -P_1$$

$$\vartheta - Q^2\int \frac{1}{r^2 I_1}\,dr = -P_2$$

where I_1 is the integrand in (19.24). The solutions of these equations for r and ϑ include, when $Q^1 = 0$, the solutions obtained previously using the method of Theorem 19.12.

Finally, the rather formal analytical result of Theorem 19.13 can be given some geometrical content by utilizing the fact that eqs. (18.14) are valid for every canonical (natural) coordinate system ("the Hamilton canonical equations are invariant under canonical coordinate transformations") and introducing the concept of a generating function of a free canonical coordinate transformation.

It follows from Theorem 18.10 that if a canonical coordinate transformation $(q^i, p_i) \mapsto (Q^i, P_i)$ on T^*M comes from a coordinate transformation $(q^i) \mapsto (Q^i)$ on M, then $P_i\, dQ^i = p_i\, dq^i$. Now we want to consider the alternative situation. Specifically, on the graph of $(q^i, p_i) \mapsto (Q^i, P_i)$, i.e., at points in $\mathbb{R}^{2n} \times \mathbb{R}^{2n}$ at which (with slight abuse of notation)

$$\begin{aligned} Q^1 - Q^1(q^1, \ldots, q^n, p_1, \ldots, p_n) &= 0 \\ &\vdots \\ Q^n - Q^n(q^1, \ldots, q^n, p_1, \ldots, p_n) &= 0 \\ P_1 - P_1(q^1, \ldots, q^n, p_1, \ldots, p_n) &= 0 \\ &\vdots \\ P_n - P_n(q^1, \ldots, q^n, p_1, \ldots, p_n) &= 0 \end{aligned}$$

we want to be able to use q^i and Q^i as coordinates. According to the implicit function theorem, the condition we now impose is $\det(\partial Q^i/\partial p_j) \neq 0$.

If (q^i, p_i) and (Q^i, P_i) are two sets of canonical coordinates on T^*M, then on the intersection of their domains, or, equivalently, on the graph in $\mathbb{R}^{2n} \times \mathbb{R}^{2n}$ of the coordinate transformation $d(p_i\, dq^i) = d(P_i\, dQ^i)$, or $d(p_i\, dq^i - P_i\, dQ^i) = 0$, so, locally, $p_i\, dq^i - P_i\, dQ^i = dS(q^i, p_i)$. If now we can take q^i and Q^i as coordinates on the graph, then

$$dS(q^i, Q^i) = p_i\, dq^i - P_i\, dQ^i \tag{19.25}$$

and

$$\frac{\partial S}{\partial q^i} = p_i, \qquad \frac{\partial S}{\partial Q^i} = -P_i \tag{19.26}$$

Definitions
A coordinate transformation with the property $\det(\partial Q^i/\partial p_j) \neq 0$ is called *free*, and a function, S, of q^i and Q^i obtained as above, from a free coordinate transformation is called a *generating function of the transformation.*

We have seen that a free canonical coordinate transformation produces functions of q^i and Q^i. We can go in the opposite direction.

Theorem 19.14. *Every function S, on a $2n$-dimensional submanifold with coordinates q^i and Q^i of $\mathbb{R}^{2n} \times \mathbb{R}^{2n}$ on which $\det(\partial^2 S/\partial Q^i\,\partial q^j) \neq 0$ is a generating function of a free canonical coordinate transformation.*

Proof. Define p_i by $p_i = \partial S/\partial q^i$ (the first equation in (19.26)). Then using our hypothesis we can solve this equation for Q^i as functions of q^i and p_i. Define $P_i(q^i, p_i) = -(\partial S/\partial Q^i)(q^i, Q^i(q^j, p_j))$. Then $p_i\,dq^i - P_i\,dQ^i = (\partial S/\partial q^i)dq^i - (\partial S/\partial Q^i)dQ^i = dS(q^i, Q^i)$, so the transformation is canonical and has generating function, S. From the definition of p_i, $(\partial p_i/\partial Q^j) = (\partial^2 S/\partial q^i\,\partial Q^j)$, so $\det(\partial^2 S/\partial q^i\,\partial Q^j) \neq 0 \Rightarrow \det(\partial Q^i/\partial p_i) \neq 0$ and hence the transformation is free. ■

Theorem 19.15. *Suppose S is a function of q^i and Q^i with $\det(\partial^2 S/\partial q^i\,\partial Q^j) \neq 0$. Then by Theorem 19.14 we get a free canonical coordinate transformation, Ψ, given by (19.26). Suppose S also satisfies*

$$H(q^i, p_j) = K(Q^i) \tag{19.27}$$

where K is a given function of Q^i. (That is, S is a complete integral of (19.27)). Then the free canonical coordinate transformation, Ψ, maps $H(q^i, p_j)$ to $K(Q^i)$.

Proof. Let $\bar{H}$ *be the image of* H *under* Ψ; *i.e.,* $\bar{H} \circ \Psi = H$. Then

$$\begin{aligned} \bar{H}(Q^i(q^i, p_j), P_j(q^i, p_j)) &= H(q^i, p_j) \\ &= H\left(q^i, \frac{\partial S}{\partial q^j}\right) && \text{by the first equation of (19.26)} \\ &= K(Q^i) && \text{by our hypothesis} \end{aligned}$$ ■

Now the process of solving the Hamilton canonical equations for q^i and p_j consists of first solving the Hamilton canonical equations for Q^i and P_j in the new coordinates, which is easy since

$$\frac{dQ^i}{du} = \frac{\partial K}{\partial P_i} = 0, \qquad \frac{dP_j}{du} = -\frac{\partial K}{\partial Q^j} = \text{constants} \tag{19.28}$$

and then finding the equations of transformation, which we get from (19.26) once we have a complete integral of (19.27).

EXAMPLE. We will apply this method to the same problem as in the two previous examples. Now we want to solve

$$f(r) + \frac{1}{2m}\left[\left(\frac{\partial S}{\partial r}\right)^2 + \frac{1}{r^2}\left(\frac{\partial S}{\partial \vartheta}\right)^2\right] = K(Q^1, Q^2) \tag{19.29}$$

With $S(r, \vartheta) = \mathbf{R}(r) + \mathbf{\Theta}(\vartheta)$ we get

$$\left(\frac{d\mathbf{R}}{dr}\right)^2 = -2mf(r) - \frac{(Q^2)^2}{r^2} + 2mK(Q^1, Q^2) \qquad \text{and} \qquad \left(\frac{d\mathbf{\Theta}}{d\vartheta}\right)^2 = (Q^2)^2$$

so

$$S(r, \vartheta, Q^1, Q^2) = \int \sqrt{-2mf(r) - \frac{(Q^2)^2}{r^2} + 2mK(Q^1, Q^2)}\, dr + Q^2\vartheta \tag{19.30}$$

is a complete integral of (19.29). Differentiating S with respect to Q^1 and Q^2, and using the second equation of (19.26), we get, respectively,

$$m\frac{\partial K}{\partial Q^1}\int \frac{1}{I_2}\, dr = -P_1$$

$$-Q^2\int \frac{1}{r^2 I_2}\, dr + m\frac{\partial K}{\partial Q^2}\int \frac{1}{I_2}\, dr + \vartheta = -P_2$$

where I_2 is the integrand in (19.30). Finally, from (19.28),

$$m\frac{\partial K}{\partial Q^1}\int \frac{1}{I_2}\, dr = -P_1 + u\frac{\partial K}{\partial Q^1}$$

$$-Q^2\int \frac{1}{r^2 I_2}\, dr + m\frac{\partial K}{\partial Q^2}\int \frac{1}{I_2}\, dr + \vartheta = -P_2 + u\frac{\partial K}{\partial Q^2}$$

where P_1 and P_2 are constants, which, when $K(Q^1, Q^2) = Q^1$, give precisely the solutions obtained in the previous example. Alternatively, if $\partial K/\partial Q^1 \neq 0$,

$$m\int \frac{1}{I_2}\, dr - u = \frac{p_1}{\partial K/\partial Q^1}$$

$$-Q^2\int \frac{1}{r^2 I_2}\, dr + \vartheta = -P_2 + P_1\frac{\partial K/\partial Q^2}{\partial K/\partial Q^1}$$

which give the same solutions as the previous example except for the designation of the constants.

PROBLEM 19.4. Apply the methods in this section to the other examples in Section 18.3.

20

A SPACETIME

In Chapters 18 and 19 we described models of "the motion of systems of particles and rigid bodies" the subject of classical analytical mechanics. In this chapter and the next we describe a certain model of "spacetime," called Einstein's theory of relativity.* This is usually more specifically called the special theory of relativity, or the theory of special relativity to distinguish it from Einstein's theory of gravitation which we will look at in Chapters 22 through 24.

We could start by simply taking 4-dimensional Minkowski space (Section 15.1) as our model of spacetime, and then describe the consequences of the Lorentz transformations (Section 15.1) for mechanics, electrodynamics, etc. However, to obtain a deeper understanding of some of these apparently not so intuitively obvious consequences, we will take the longer route and develop our model from Galileo's and Newton's notions of space and time.

20.1 Newton's mechanics and Maxwell's electromagnetic theory

Newton's model of "physical space" was a 3-dimensional Euclidean affine space, $\mathscr{E}_0^3$, a 3-dimensional affine space whose vector space has a positive definite symmetric bilinear form, $b \in S^2(V^*)$ (cf., Section 5.4(i)). Hence, in this model we do not have an origin, but we do have straight lines, planes, parallelism, and the distance between two points as the length of the straight line segment joining them.

In Newton's model we have "matter," i.e., "particles" and "bodies." These can be described by associating positive real functions, *mass*, or *mass density*, with certain possible "positions" or "locations" in $\mathscr{E}_0^3$ represented by certain subsets of $\mathbb{R}^n$ (cf., Section 18.3). "Motion" is manifested by the changing of distances between particles. Hence we think of our distances as a function of a single real variable, t. The existence of repetitive changes gives us time units. ("Time is defined so that motion looks simple," Misner, Thorne, and Wheeler, 1973.) In short, we define *a particle in* $\mathscr{E}_0^3$ as a pair, (m, γ), where m is a number and γ is a curve in $\mathscr{E}_0^3$ parametrized by t.

Newton could use $\mathscr{E}_0^3$ as a model of physical space because he used the "fixed stars" in terms of which he could locate a point. In $\mathscr{E}_0^3$ we can "pick

* (And called the relativity theory of Poincaré and Lorentz by Whittaker, 1960.)

a point," and define the velocity of a particle. However, in physical space there is no way to "pick a point," or to talk about the velocity of a particle without reference to "particles" which may be "moving." In order to avoid using fixed stars we introduce frames of reference.

A frame of reference is usually pictured as a moving set of coordinate axes. The essential ingredients of this picture are not the coordinates, but the motion, a concept we associated with matter (particles and bodies), and a concept of rigidity. Thus, a frame of reference is sometimes described as a body in rigid motion (or a rigid body in motion). Most simply, we will define *a frame of reference in* $\mathscr{E}_0^3$ to be a (differentiable) 1-parameter family, $\{\phi_t\}$, of rigid motions, or isometries, of $\mathscr{E}_0^3$. *A point,* $\mathscr{P}$, *of a frame of reference* is an orbit of $\{\phi_t\}$. If μ is any rectangular cartesian coordinate system, then $\mu \circ \phi_t^{-1}$ are *rectangular cartesian coordinates of the frame of reference,* $\{\phi_t\}$. In other words, the coordinates of p with respect to $\{\phi_t\}$ are the coordinates of $p_0 = \phi_t^{-1}(p)$. If v is another rectangular cartesian coordinate system, then $v \circ \phi_t^{-1} = (v \circ \mu^{-1}) \circ \mu \circ \phi_t^{-1}$ and the two sets of coordinate functions $x^i = \mu^i \circ \phi_t^{-1}$ and $y^i = v^i \circ \phi_t^{-1}$ of the frame of reference $\{\phi_t\}$ are related by

$$y^i(p) = a^i_j x^j(p) + b^i \tag{20.1}$$

where the a^i_j are the components of an orthogonal matrix. Finally, our particles in $\mathscr{E}_0^3$ will be associated with frames of reference; i.e., *a particle in a frame of reference,* $\{\phi_t\}$, is a pair $(m, \phi_t^{-1} \circ \gamma)$, and *the velocity in (or, with respect to)* $\{\phi_t\}$ is the velocity of the curve $\phi_t^{-1} \circ \gamma$. The component functions (with respect to the frame $\{\phi_t\}$) are $\gamma^i = x^i \circ \gamma = \mu^i \circ \phi_t^{-1} \circ \gamma$ so $(\phi_t^{-1} \circ \gamma)^{\cdot}(t) = D_t\gamma^i\, \partial/\partial\mu^i$.

If $\{\psi_t\}$ is another frame of reference and v is another rectangular cartesian coordinate system, then for a given point, p, the relation between the coordinates $\mu \circ \phi_t^{-1}$ of p and the coordinates $v \circ \psi_t^{-1}$ of p will depend on t. That is, the coordinate functions $x^i = \mu^i \circ \phi_t^{-1}$ and $y^i = v^i \circ \psi_t^{-1}$ will satisfy relations of the form

$$y^i(p) = f^i(x^j(p), t) \tag{20.2}$$

for all p and t.

By introducing frames of reference we effectively get rid of $\mathscr{E}_0^3$, replacing it by the set of all frames of reference.

Having modeled the basic ingredients in our version of Newtonian mechanics, in order to proceed we have to relate these concepts by "physical laws." That is, we have to throw in some physics, specifically, *Newton's First Law*, or, *The Principle of Inertia.* The effect of this law is essentially to sort out a particular class of frames of reference, so-called *inertial frames.* Newton said (*Principia*, 1687) that there exists a frame of reference (the fixed stars) in $\mathscr{E}_0^3$ with respect to which all "free particles" move in a straight line with constant speed. We translate "free particle" to mean that there are no particles or bodies close enough to it to have any noticeable effect on it. The last phrase in Newton's law is described in our terms by the first of the following definitions.

Definitions

A particle, (m, γ), *moves in a straight line with constant speed with respect to a frame of reference*, $\{\phi_t\}$, if there is a coordinate system, μ, such that $v^i = D_t\gamma^i$ are constant, where $\gamma^i = x^i \circ \gamma = \mu^i \circ \phi_t^{-1} \circ \gamma$. In particular, (m, γ) is *at rest in* $\{\phi_t\}$ if there is a coordinate system such that $v^i = 0$. By eq. (20.1), these definitions are independent of the coordinate system of $\{\phi_t\}$.

Definition

If the functions f^i in eq. (20.2) are given by

$$f^i(x^j, t) = a^i_j x^j - V^i t + b^i \tag{20.3}$$

where a^i_j are the elements of an orthogonal matrix, and V^i and b^i are constants, then (20.2) becomes

$$y^i(p) = a^i_j x^j(p) - V^i t + b^i \tag{20.4}$$

and (20.4) are called *the general Galilean transformations.*

Theorem 20.1. The Newtonian (Galilean) Principle of Relativity. *Suppose* $\{\phi_t\}$ *and* $\{\psi_t\}$ *are two frames of reference and* $\{\phi_t\}$ *is inertial. Then* $\{\psi_t\}$ *is inertial iff* $\{\phi_t\}$ *and* $\{\psi_t\}$ *have coordinates* $x^i = \mu^i \circ \phi_t^{-1}$ *and* $y^i = v^i \circ \psi_t^{-1}$ *which are related by a general Galilean transformation.*

Proof. Problem 20.1.

Another common way of describing (at least in part) the result above is to say the form of the equation $d^2x^i/dt^2 = 0$ is invariant (or the equation is covariant) under Galilean transformations. One can show that the usual form of Newton's Second Law, in the case where the force comes from a potential, is also invariant under Galilean transformations. See Problem 20.2. The important thing about these forms is that they are independent of the transformations themselves; i.e., they do not involve the coefficients of (20.4). (Cf. last paragraph of observation (1), at the end of this section.)

We now turn to a brief account of Maxwell's electromagnetic theory.

Coulomb found that there is an *electric force field*, *E*, produced by stationary charges. Coulomb's law has the same form as Newton's law of gravitation which implies that

$$\operatorname{curl} E = 0, \qquad \operatorname{div} E = \frac{\sigma}{\varepsilon} \tag{20.5}$$

where σ is the charge density, and ε, the permittivity, is a constant whose value depends on the units used.

Ampere found that there is a force on a closed current-carrying wire due to a second closed current-carrying wire. This force is expressed in terms of a vector field, B, *the magnetic induction.* B is proportional to the magnitude of the second current times a certain integral around that wire (the law of Biot and Savart). As a consequence we find

$$\operatorname{div} B = 0, \qquad \operatorname{curl} B = \mu J \tag{20.6}$$

where J is the current density in the second wire, and μ, the permeability, is another constant whose value depends on the units used.

Faraday discovered that there is a current in a wire in the presence of a varying magnetic field, so then a term $-\partial B/\partial t$ must be added to the right side of Coulomb's curl equation in (20.5). Maxwell added $\varepsilon\mu\, \partial E/\partial t$ to the right side of Ampere's curl equation in (20.6) in order to provide for conservation of charge. With these additions the system of equations governing E and B becomes

$$\begin{aligned} \operatorname{curl} E &= -\frac{\partial B}{\partial t}, \qquad \operatorname{div} E = \frac{\sigma}{\varepsilon} \\ \operatorname{div} B &= 0, \qquad \operatorname{curl} B = \mu J + \varepsilon\mu \frac{\partial E}{\partial t} \end{aligned} \tag{20.7}$$

The constants ε and μ appearing in the last two equations of (20.7) come in as a result of the proportionality factors in the Coulomb and Ampere laws. We might expect to be able to eliminate them by choosing proper units, but there are not enough units to be chosen. We can think of Coulomb's law as determining units for charge, and Ampere's law (or Biot–Savart's) as determining units for current, but charge density and current density are related by $J = \sigma V$ where V is the velocity of the charged particles of the current. Thus, we cannot get rid of both ε and μ, and no matter how units are chosen there remains one empirically determined constant in the set of equations.

Now we make two observations about equations (20.7) which evince an important distinction between Newton's mechanics and Maxwell's electromagnetic theory.

(1) In contrast to Newton's mechanics, in Maxwell's equations there is no explicit motion and no involvement of frames of reference. The functions and vector fields in Maxwell's equations are defined on some "medium" and can vary with time. Thus, if we choose for our medium $\mathscr{E}_0^3$, then equations (20.7) are defined on $\mathscr{E}_0^3 \times \mathbb{R}$.

Hertz supposed the medium could be any frame of reference moving uniformly with respect to $\mathscr{E}_0^3$—"the aether contained within ponderable bodies moves with them" (Whittaker, 1960, Vol. I, p. 328)—and he generalized eq. (20.7) to apply to moving media (or moving bodies). If this motion is described by a given vector field, V, the generalized equations have an additional term $-\text{curl } V \times B$ on the left side of the curl E equation, and an additional term, $\varepsilon\mu[\text{curl } V \times E - (\sigma/\varepsilon)V]$ on the left side of the curl B equation. ($V \times B$ and $V \times E$ are the vector products of vector analysis.) See Problem 20.3. Thus, for example, we can imagine a copper wire moving through a static magnetic field in such a way that at points of the wire the variation of B is exactly the same as it was for a stationary wire in a varying magnetic field. Then the field, E, in the wire will be produced by the term curl $V \times B$ in the first equation of the generalized (20.7) instead of $\partial B/\partial t$.

Clearly, by construction, the *form* of the generalized equations is invariant under the transformations (20.4) between two frames of references. However, these equations contain the coefficients of the transformation, specifically, V (cf., comment below Theorem 20.1). V is a velocity with respect to one particular frame of reference, and thus the equations identify a unique frame of reference which should be physically distinguishable.

(2) Maxwell's equations (20.7) have a special property which the generalized system does not have. Take the curl of the last equation of (20.7) and use the first equation to eliminate E. Then each of the components, in rectangular cartesian coordinates, of B satisfies

$$\sum_{j=1}^{3} \frac{\partial^2 B^i}{\partial x^{j^2}} - \varepsilon\mu \frac{\partial^2 B^i}{\partial t^2} = \text{first-order terms} \tag{20.8}$$

The components of E satisfy equations of the same kind.

The solutions of equations of the form (20.8) are waves. Thus, the components of E and B are propagated as waves with speed $1/\sqrt{\varepsilon\mu}$ with respect to the frame $\mathscr{E}_0^3$. The significance of this fact is that having gotten rid of $\mathscr{E}_0^3$ in mechanics by introducing frames of reference, we find it popping up again in electromagnetic theory and having a certain unique importance. While we have been able to do mechanics without worrying about having to do it in some unique frame of reference, electromagnetic theory tells us we should look for a certain unique frame of reference. Still more remarkable is the result that the velocity of propagation $1/\sqrt{\varepsilon\mu}$ turns out to be precisely the empirically measured speed of light!

PROBLEM 20.1. Prove Theorem 20.1. Hint: For the "only if" part show that, for any particle, $(m, \phi_t^{-1} \circ \gamma)$, the assumption that whenever there is a coordinate system, μ, for $\{\phi_t\}$ such that $d\gamma^i/dt$ = constant, then there is a coordinate system, ν, for $\{\psi_t\}$ such that $d\gamma'^i/dt$ = constant implies that the functions f^i in eq. (20.2) satisfy $\partial^2 f^i/\partial x^j \, \partial x^k = 0$. Cf., also, the derivation of the Lorentz transformations in Section 20.3.

PROBLEM 20.2. Prove that Newton's Second Law, in the case where the force comes from a potential, is covariant under Galilean transformations.

PROBLEM 20.3. Derive the form of the Maxwell equations for a moving medium. (See Robertson and Noonan, 1968, Section 2.5.)

20.2 Frames of reference generalized

The fact that $1/\sqrt{\varepsilon\mu}$ is the speed of light leads to the conclusion that light is simply a particular type of electromagnetic radiation, and so various experiments with light were made to clarify the status of Maxwell's equations.

In Section 20.1 we conceived of Maxwell's equations as being defined on $\mathscr{E}_0^3 \times \mathbb{R}$. We could try to generalize by claiming these equations are only valid *locally*—on pieces of $\mathscr{E}_0^3 \times \mathbb{R}$. This permits bodies in relative motion to be at rest with respect to the medium—i.e., bodies drag the medium along with them (there is an ether drag). This possibility was excluded by the phenomenon of aberration of light rays from the stars observed by Bradley in 1728. The only other possibility is that bodies, in particular the earth, move with respect to the medium (there is an ether drift). The Michelson–Morley experiment, 1887, effectively squelched that idea. These experiments (and several others) created a logical impasse. The way in which Einstein resolved this problem amounts, in our terms, to generalizing our concept of frame of reference. In order to motivate this, we first examine the implications of eq. (20.8).

For our purposes it will suffice to examine the special case of eq. (20.8) when the right side is zero. This will occur when $\sigma = J = 0$. We then have precisely the same equation for the components of E. Thus, we are led to consider the basic wave equation

$$\sum_{i=1}^{3} \frac{\partial^2 f}{\partial x^{i2}} - \frac{1}{c} \frac{\partial^2 f}{\partial t^2} = 0 \tag{20.9}$$

Solutions of eq. (20.9) are determined by initial conditions which are propagated on *wave fronts*. A wave front can be thought of as a moving surface in $\mathscr{E}_0^3$, and, in coordinates, as a set of solutions of

$$h(t, x^1, x^2, x^3) = \text{constant} \tag{20.10}$$

for some function h. We will now obtain a necessary condition for h if (20.10) represents a wave front.

A property of a wave front is that it separates regions in which f is constant from those in which a derivative of f is not zero. That is, a wave front consists of points at which the solution of (20.10) may not be analytic. On the other hand, with certain exceptions, the characteristic hypersurfaces, if f and $\partial f/\partial t$ are analytic on a hypersurface they will be propagated as analytic functions in a neighborhood of the hypersurface. Thus, the wave fronts must be characteristic hypersurfaces. The necessary and sufficient conditions that a hypersurface be characteristic is given by the Cauchy–Kowalewski construction. To obtain it for our case we first simplify our notation.

We write eq. (20.9) in the form

$$\sum_{\alpha=0}^{3} \underset{\alpha}{e} \frac{\partial^2 f}{\partial x^{\alpha^2}} = 0 \tag{20.11}$$

where $x^0 = ct$, $\underset{0}{e} = -1$, and $\underset{1}{e} = \underset{2}{e} = \underset{3}{e} = 1$. Now, the hypersurface (20.11) can be imbedded in a family $\{h(x^\lambda) = a\colon a \in \mathbb{R}\}$, and these in turn can be used to construct coordinate transformations

$$y^\lambda = y^\lambda(x^\mu), \qquad \lambda, \mu = 0, 1, 2, 3$$

with $y^0 = h$. By the chain rule,

$$\frac{\partial^2 f}{\partial x^{\alpha^2}} = \frac{\partial^2 f}{\partial y^\lambda\, \partial y^\mu} \frac{\partial y^\lambda}{\partial x^\alpha} \frac{\partial y^\mu}{\partial x^\alpha} + \frac{\partial f}{\partial y^\lambda} \frac{\partial^2 y^\lambda}{\partial x^{\alpha^2}}$$

and substituting this into (20.11) we get

$$\sum_{\alpha=0}^{3} \underset{\alpha}{e} \frac{\partial y^i}{\partial x^\alpha} \frac{\partial y^j}{\partial x^\alpha} \frac{\partial^2 f}{\partial y^i\, \partial y^j} + 2 \sum_{\alpha=0}^{3} \underset{\alpha}{e} \frac{\partial y^i}{\partial x^\alpha} \frac{\partial y^0}{\partial x^\alpha} \frac{\partial^2 f}{\partial y^i\, \partial y^0}$$
$$+ \sum_{\alpha=0}^{3} \underset{\alpha}{e} \left(\frac{\partial y^0}{\partial x^\alpha}\right)^2 \frac{\partial^2 f}{\partial y^{0^2}} + \text{first-order terms} = 0, \qquad i, j = 1, 2, 3 \tag{20.12}$$

Now notice that if we are given f and $\partial f/\partial y^0$ on any hypersurface $y^0(x^\lambda) = \text{constant}$, then all first and second partial derivatives of f can be calculated on this hypersurface except $\partial^2 f/\partial y^{0^2}$, and if $\sum_{\alpha=0}^{3} \underset{\alpha}{e}(\partial y^0/\partial x^\alpha)^2 \neq 0$ we get $\partial^2 f/\partial y^{0^2}$ from eq. (20.12). We can proceed to calculate all the higher-order derivatives of f by differentiating (20.12) and get a power series expansion for f in a neighborhood of $y^0(x^\lambda) = \text{constant}$. Thus, a necessary condition for the function h in eq. (20.10), if eq. (20.10) represents a wave front, is that h satisfies

$$\sum_{\alpha=0}^{3} \underset{\alpha}{e}\left(\frac{\partial h}{\partial x^{\alpha}}\right)^{2} = 0 \tag{20.13}$$

A particular solution of (20.13) is given by

$$h(t, x^1, x^2, x^3) = \sqrt{(x^1)^2 + (x^2)^2 + (x^3)^2} - ct$$

This is a spherical wave front, due to a flash of light at $x^i = t = 0$. At time t this flash will be observed on a sphere of radius ct. These spherical wave solutions suggest a way of interpreting the seemingly contradictory empirical evidence mentioned at the beginning of this section: light is propagated in the same way in any two frames of reference, and yet the "medium" in which Maxwell's equations are valid cannot coincide with both frames.

Suppose points $\mathscr{P}$ and $\mathscr{Q}$ of two relatively moving frames of reference coincide at some instant at which time and place a flash of light is emitted. Then, according to $\mathscr{P}$, t seconds later the light will have reached p and q. But according to $\mathscr{Q}$, after he/she counts t seconds, the light will be at p' and q' (see Fig. 20.1). Thus, according to $\mathscr{Q}$, the light arrived at p in less than t seconds, and arrived at q in more than t seconds. This suggests that in relatively moving frames, time has to be measured differently. In particular, two simultaneous occurrences according to $\mathscr{P}$ are not simultaneous according to $\mathscr{Q}$.

Our original definition of a frame of reference in $\mathscr{E}_0^3$ involved a parameter, t. We used the same parameter for all frames of reference, so, in particular, in coordinate transformations such as (20.2) and (20.4) only

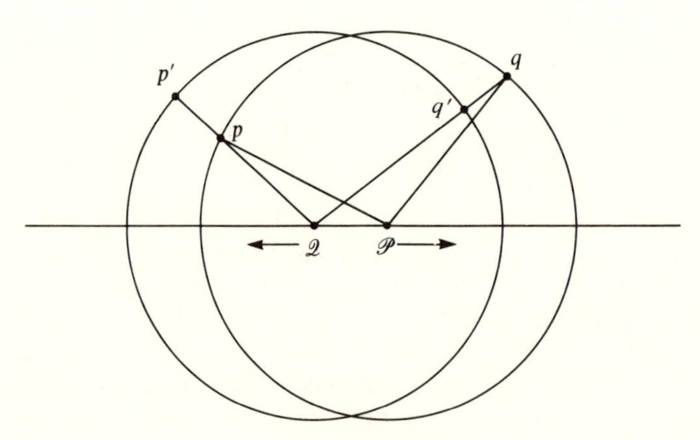

FIGURE 20.1

one "time" appears. We now allow different frames of reference to be parametrized by different times. Now we consider the set of all such *generalized frames*, and in analogy with our statement of Newton's Principle of Inertia we express *Maxwell's law* by the statement: there exists one of these frames of reference with respect to which, in some rectangular cartesian coordinate system, Maxwell's equations are valid. *Einstein's Postulate*, on which his theory of relativity is based, is that there exists a generalized frame of reference for which both Newton's and Maxwell's laws are valid.

The Galilean Principle of Relativity, Theorem 20.1, gave us an entire class of frames of reference, once we had Newton's law. Now, having both Newton's law and Maxwell's law we would like to obtain a class of our generalized frames of reference. In Section 20.1 the class of frames was described by eq. (20.4). Our procedure now will be to try to find transformation laws, corresponding to eq. (20.4), which describe the class of generalized frames of reference in which *both* Newton's law and Maxwell's law are valid.

20.3 The Lorentz transformations

Suppose (x^1, x^2, x^3) are rectangular cartesian coordinates and t is the parameter for a frame of reference $\{\phi_t\}$. Suppose $\{\psi_{t'}\}$ is a second frame of reference with rectangular cartesian coordinates (y^1, y^2, y^3) and parameter t' such that (1) if Newton's law is valid in $\{\phi_t\}$ then Newton's law is valid in $\{\psi_{t'}\}$, and (2) if Maxwell's law is valid in $\{\phi_t\}$ then Maxwell's law is valid in $\{\psi_{t'}\}$. These two conditions will impose a relation between the coordinates and time on the first frame and the coordinates and time on the second frame which we will now obtain. More specifically, we will obtain functions f^λ for which

$$y^\lambda = f^\lambda(x^\alpha) \tag{20.14}$$

where $\alpha, \lambda = 0, 1, 2, 3$ and $x^0 = ct$ and $y^0 = ct'$.

The first condition, the "invariance" of Newton's Law, reads in coordinates,

I. If (t, x^1, x^2, x^3) are related by

$$x^i - x^i_* = v^i(t - t_*) \tag{20.15}$$

in the first frame, then (t', y^1, y^2, y^3) are related by

$$y^i - y^i_* = {}'v^i(t' - t'_*) \tag{20.16}$$

in the second frame. (v^i and ${}'v^i$ are constants.)

Equations (20.15) and (20.16) give the trajectory of a "free particle." v^i and $'v^i$ are its velocity in the respective frames, and the star subscripts denote initial values. (For simplicity we are writing x^i and y^i instead of the more accurate $x^i \circ \gamma(t)$ and $y^i \circ \gamma(t)$.)

The second condition, the "invariance" of Maxwell's Law, implies the "invariance" of wave fronts; i.e.,

II. If h satisfies eq. (20.13) in the first frame, then h', the composition of h and the inverse of (20.14), satisfies

$$\sum_{\alpha=0}^{3} \underset{\alpha}{e}\left(\frac{\partial h'}{\partial y^\alpha}\right)^2 = 0 \tag{20.17}$$

in the second frame.

We can write this condition in an equivalent, more useful fashion.

Recall (Section 14.3), that corresponding to any given first-order partial differential equation there is an equivalent system of characteristic equations (14.17). That is, corresponding to a solution of the partial differential equation there is a family of solutions of (14.17), and conversely. Applying (14.17) to our present case where the first-order partial differential equation is (20.13), f is given by

$$f\left(x^\alpha, \frac{\partial h}{\partial x^\alpha}, h\right) = \sum_{\alpha=0}^{3} \underset{\alpha}{e}\left(\frac{\partial h}{\partial x^\alpha}\right)^2$$

and the first two sets of equations in (14.17) are, respectively,

$$\frac{dx^\alpha}{du} = 2\,\underset{\alpha}{e}\,\frac{\partial h}{\partial x^\alpha}, \qquad \text{no sum on } \alpha$$

and

$$\frac{d}{du}\left(\frac{\partial h}{\partial x^\alpha}\right) = 0$$

So $\partial h/\partial x^\alpha$ are constant along the integral curves of the system, and integrating the first set, choosing $u = (t - t_*)/2$ we get

$$c = -\frac{1}{c}\frac{\partial h}{\partial t} \tag{20.18}$$

and

$$x^i - x^i_* = \frac{\partial h}{\partial x^i}(t - t_*) \tag{20.19}$$

By using (20.13) we can replace (20.18) with

$$\sum_{i=1}^{3}\left(\frac{\partial h}{\partial x^i}\right)^2 = c^2 \tag{20.20}$$

If in eq. (20.19) we fix t, and let x^i_* vary over some initial surface, then eqs. (20.19) are the equations of a wave front moving with velocity $\partial h/\partial x^i$. On the other hand, if we think of the x^i_* as fixed and t varying, then eqs. (20.19) are the equations of *the trajectory of a photon*, or *a ray of light*.

The desired formulation of our second condition, the "invariance" of Maxwell's law, can now be stated as follows.

II′. If (t, x^1, x^2, x^2) are related by eq. (20.19), in which the constant coefficients $\partial h/\partial x^i$ are subject to the condition (20.20), on the first frame, then (t', y^1, y^2, y^3) are related by

$$y^i - y^i_* = \frac{\partial h'}{\partial y^i}(t' - t'_*) \tag{20.21}$$

in which the constant coefficients $\partial h'/\partial y^i$ are subject to

$$\sum_{i=1}^{3}\left(\frac{\partial h'}{\partial y^i}\right)^2 = c^2 \tag{20.22}$$

on the second frame.

The form of the transformation (20.14) can now be obtained by a rather formal application of eqs. (20.15), (20.16), (20.19)–(20.22). Notice first that there is a certain redundancy in these equations. The first condition says that (20.15) implies (20.16) for arbitrary constant coefficients v^i and $'v^i$. The second condition says that a relation (20.19) of the same form as (20.15) implies a relation (20.21) of the same form as (20.16) *with the additional condition on the coefficients imposed by* (*20.20*) *and* (*20.22*). Thus, the set of conditions reduces to (20.15) implies (20.16), and if the coefficients, v^i, of (20.15) satisfy (20.20) then the coefficients, $'v^i$, of (20.16) satisfy (20.22).

To simplify the notation, we introduce $x^0 = ct$ as before, and k^0, a fixed positive number. Further, put $\tau = (x^0 - x^0_*)/k^0$ in place of t, and define constants, k^i, by $v^i/c = k^i/k^0$ $(i = 1, 2, 3)$. Similarly for the y's and the primed symbols. Then (20.15) becomes

$$x^\alpha = x^\alpha_* + k^\alpha \tau \tag{20.23}$$

(20.16) becomes

$$y^\alpha = y_*^\alpha + {}'k^\alpha \tau' \tag{20.24}$$

(20.20) becomes

$$\sum_{\alpha=0}^{3} \underset{\alpha}{e}(k^\alpha)^2 = 0 \tag{20.25}$$

and (20.22) becomes

$$\sum_{\alpha=0}^{3} \underset{\alpha}{e}({}'k^\alpha)^2 = 0 \tag{20.26}$$

Substituting (20.23) into (20.14) and differentiating with respect to τ, we get $dy^\lambda/d\tau = (\partial f^\lambda/\partial x^\alpha)k^\alpha$. From (20.24) (eliminating τ') $dy^i/dy^0 = {}'k^i/{}'k^0$. So

$$\begin{aligned}\frac{\partial f^i}{\partial x^\alpha} k^\alpha = \frac{dy^i}{d\tau} &= \frac{dy^i}{dy^0}\frac{dy^0}{d\tau} \\ &= \frac{{}'k^i}{{}'k^0}\frac{\partial f^0}{\partial x^\alpha} k^\alpha \end{aligned} \tag{20.27}$$

Differentiate both sides of (20.27) again with respect to τ. Since we want (20.14) to be invertible we assume $\det(\partial f^\lambda/\partial x^\alpha) \neq 0$. Then $(\partial f^0/\partial x^\alpha)k^\alpha \neq 0$ for nonvanishing "vectors" k^α and we get

$$\frac{\partial^2 f^i}{\partial x^\alpha\, \partial x^\beta} k^\alpha k^\beta = \frac{\partial f^i}{\partial x^\alpha} k^\alpha \frac{\dfrac{\partial^2 f^0}{\partial x^\alpha\, \partial x^\beta} k^\alpha k^\beta}{\dfrac{\partial f^0}{\partial x^\alpha} k^\alpha} \tag{20.28}$$

Note that eqs. (20.28) are identities in x^α and k^α. Now fix x^α. Since for "vectors" k^α for which $(\partial f^i/\partial x^\alpha)k^\alpha = 0$ we must also have $(\partial^2 f^i/\partial x^\alpha\, dx^\beta)k^\alpha k^\beta = 0$, it follows that $(\partial f^\lambda/\partial x^\alpha)k^\alpha$ divide $(\partial^2 f^\lambda/\partial x^\alpha\, \partial x^\beta)k^\alpha k^\beta$. In other words, there exist functions g_β, of x^α such that

$$\frac{\partial^2 f^\lambda}{\partial x^\alpha\, \partial x^\beta} k^\alpha k^\beta = \left(\frac{\partial f^\lambda}{\partial x^\alpha} k^\alpha\right)(2 g_\beta k^\beta)$$

or, dropping the k's,

$$\frac{\partial^2 f^\lambda}{\partial x^\alpha\,\partial x^\beta} = \frac{\partial f^\lambda}{\partial x^\alpha} g_\beta + \frac{\partial f^\lambda}{\partial x^\beta} g_\alpha \tag{20.29}$$

Now, using (20.27) in (20.26), the latter becomes

$$\left(\sum_{\lambda=0}^{3} \underset{\lambda}{e} \frac{\partial f^\lambda}{\partial x^\alpha} \frac{\partial f^\lambda}{\partial x^\beta}\right) k^\alpha k^\beta = 0 \tag{20.30}$$

But since (20.26) is to follow from (20.25), (20.30) is also to be a consequence of (20.25) and, hence, the coefficients of (20.30) and (20.25) must be proportional; i.e.,

$$\sum_{\lambda=0}^{3} \underset{\lambda}{e} \frac{\partial f^\lambda}{\partial x^\alpha} \frac{\partial f^\lambda}{\partial x^\beta} = \Gamma \underset{\alpha}{e}\, \delta_{\alpha\beta} \tag{20.31}$$

where Γ is a function of the x^α's. Finally, if we differentiate (20.31) and apply (20.29) and (20.31) to the result, we get $g_\alpha = 0$ and $\Gamma =$ constant. Thus, from (20.29), the f^λ are affine functions. See Problem 20.4. That is, (20.14) is

$$y^\lambda = \sum_{\alpha=0}^{3} \underset{\alpha}{e}\, a^\lambda_\alpha x^\alpha + b^\lambda \tag{20.32}$$

Using (20.32) for f^λ, (20.31) becomes

$$\sum_{\lambda=0}^{3} \underset{\lambda}{e}\, a^\lambda_\alpha a^\lambda_\beta = \Gamma \underset{\alpha}{e}\, \delta_{\alpha\beta} \tag{20.33}$$

Now, (20.33) is supposed to be valid for any two sets of rectangular cartesian coordinates for any two frames. In particular, for two sets of rectangular cartesian coordinates with the same origin in the same frame, $t' = t$, and $a^0_i = 0$ so eq. (20.33) reduces to

$$\sum_{i=1}^{3} a^i_j a^i_k = \Gamma\, \delta_{jk} \tag{20.34}$$

But two such rectangular cartesian coordinate systems are related by an orthogonal transformation, i.e., the matrix (a^i_j) is orthogonal, so from (20.34) $\Gamma = 1$ and (20.33) becomes

$$\sum_{\lambda=0}^{3} \underset{\lambda}{e}\, a^\lambda_\alpha a^\lambda_\beta = \underset{\alpha}{e}\, \delta_{\alpha\beta} \tag{20.35}$$

(Note that the matrix $(\underset{\lambda}{e} a_\alpha^\lambda)$, no sum on λ, has the property, described in Section 5.4(i), of the matrix of an orthogonal transformation of a 4-dimensional vector space with an inner product, b, of index 1.)

We have now accomplished the objective stated at the beginning of this section.

Theorem 20.2. *Equations (20.32) with the orthogonality condition (20.35) on the matrix $(\underset{\lambda}{e} a_\alpha^\lambda)$ are the relations imposed by Newton's and Maxwell's laws on the rectangular cartesian coordinates and time in two frames of reference.*

Definitions

Equation (20.32) with the condition (20.35) on the coefficients a_α^λ are called *the Lorentz transformations.* (Cf., Section 15.1 and Problem 20.12.) The frames related by the Lorentz transformations are called *the Lorentzian frames.*

We note in passing that the Lorentz transformations are usually considered the essence of special relativity. Special relativity is frequently described as the theory of the Lorentz transformations. There are shorter "derivations," the shortcuts being achieved (as Einstein did) by invoking homogeneity and/or isotropy of $\mathscr{E}_0^3$ and/or time. Ours, following Fock, is a simplification of a derivation based on geometrical interpretation of the two required conditions (Weyl, 1923).

PROBLEM 20.4. Justify the proportionality argument used in deriving (20.31).

PROBLEM 20.5. Carry out the details for obtaining the Lorentz transformations from eqs. (20.29) and (20.31).

PROBLEM 20.6. The result of this section can be described as the "only if" part of a theorem coresponding to Theorem 20.1. Prove the converse; namely, if the relations between rectangular cartesian coordinates and time in two generalized frames of reference are given by the Lorentz transformations, and Newton's and Maxwell's laws are valid in one frame, then they are valid in the other frame.

20.4 Some properties and forms of the Lorentz transformations

The orthogonality property (20.35) of the coefficient matrices $(\underset{\alpha}{e}\, a_\alpha^\lambda)$ of the Lorentz transformations (20.32) implies that $\det(a_\alpha^\lambda) = \pm 1$. Note that we can also write (20.35) in the form

$$\sum_{\alpha=0}^{3} \underset{\alpha}{e}\, a_\alpha^\lambda a_\alpha^\mu = \overset{\lambda}{e}\, \delta^{\lambda\mu} \qquad (\overset{\lambda}{e} = \underset{\lambda}{e}) \tag{20.36}$$

Finally, we can invert (20.32) to get

$$x^\beta = \sum_{\lambda=0}^{3} \underset{\lambda}{e}\, \tilde{a}^\beta_\lambda y^\lambda - c^\beta \tag{20.37}$$

where $\tilde{a}^\beta_\lambda = a^\lambda_\beta$, and $c^\beta = \sum_{\lambda=0}^{3} \underset{\lambda}{e}\, \tilde{a}^\beta_\lambda b^\lambda$.

We will now show how a Lorentz transformation relating two generalized frames of reference can be described in terms of a "relative velocity" between these two frames.

Consider a particle at rest in the frame $\{\psi_{t'}\}$; i.e., $'v^i \doteq dy^i \circ \gamma/dt' = 0$. In the frame $\{\phi_t\}$ the velocity of this particle is $v^i = dx^i \circ \gamma/dt$, thinking of γ as a function of t. Differentiating the Lorentz transformations (20.37) with respect to t (thinking of y^λ as functions of t) we get

$$c = -a^0_0 \frac{dy^0}{dt} \qquad \text{and} \qquad v^i = -a^0_i \frac{dy^0}{dt}$$

So

$$\frac{v^i}{c} = \frac{a^0_i}{a^0_0} \tag{20.38}$$

Thus, all particles at rest in the frame $\{\psi_{t'}\}$ have the same velocity, v^i, in the frame $\{\phi_t\}$, which fact we can rephrase by saying that $\{\psi_{t'}\}$ *has (relative) velocity* v^i *with respect to* $\{\phi_t\}$.

Because of the symmetry of the relation between two frames, we can, by differentiating the Lorentz transformation (20.32) with respect to t', derive

$$\frac{'v^i}{c} = \frac{a^i_0}{a^0_0} \tag{20.39}$$

where $'v^i$ are the components with respect to $\{\psi_{t'}\}$ of the velocity of a particle fixed in the $\{\phi_t\}$ frame, and can say that $\{\phi_t\}$ has velocity $'v^i$ with respect to $\{\psi_{t'}\}$.

Having introduced the relative velocity, v^i, we can now write the Lorentz transformations (20.32) in terms of this velocity and an orthogonal matrix, (A^i_j). First, from (20.36) putting $\lambda = \mu = 0$ we have

$$(a^0_0)^2 - \sum_{j=1}^{3} (a^0_j)^2 = 1 \tag{20.40}$$

From (20.40) and (20.38) we get

$$a^0_0 = \pm \frac{1}{\sqrt{1 - (v^2/c^2)}} \tag{20.41}$$

where $v^2 = \sum (v^i)^2$; from (20.38) and (20.41) we get

$$a_i^0 = \pm \frac{1}{\sqrt{1 - (v^2/c^2)}} \tag{20.42}$$

and from (20.39) and (20.41) we get

$$a_0^i = \pm \frac{'v^i}{c} \frac{1}{\sqrt{1 - (v^2/c^2)}} \tag{20.43}$$

Next, we put $\lambda = 0$, and $\mu = i$ in (20.36), and we put $\alpha = 0$ and $\beta = i$ in (20.35). We get, respectively (using (20.42) and (20.43)),

$$a_0^0 \, 'v^i = \sum_j v^j a_j^i \tag{20.44}$$

and

$$a_0^0 v^i = \sum_j a_i^j \, 'v^j \tag{20.45}$$

Finally, we define a set of numbers, A_j^i by

$$a_j^i = A_j^i - \frac{a_0^i a_j^0}{1 - a_0^0} \tag{20.46}$$

Then from (20.36) with $\lambda = i$, and $\mu = j$ and (20.44) and (20.45) we conclude that

$$\sum_j A_j^i A_j^k = \delta^{ik} \tag{20.47}$$

and

$$'v^i = -\sum_j A_j^i v^j \tag{20.48}$$

Equations (20.47) and (20.48) are important properties of the A_j^i's.

Now, by substituting the expressions for the a_α^λ's from (20.41), (20.42), (20.43) and (20.46) into the Lorentz transformations, (20.32), we can express them in terms of the relative velocity, v^i, and the orthogonal transformation, A_j^i. We first split (20.32) into

$$ct' = -a_0^0 ct + \sum_j a_j^0 x^j + b^0$$

$$y^i = -a_0^i ct + \sum_j a_j^i x^j + b^i$$

When $b^0 = v^i = 0$ the first equation (noting eq. (20.42)) reduces to $t' = -a_0^0 t$. If we require that $t = t'$ in this case, then we have to use the lower signs in eqs. (20.41)–(20.43). Now, with the substitutions for a_α^λ, and restricting ourselves to the homogeneous transformations, $b^\lambda = 0$, without much loss of generality, we get

$$ct' = \frac{1}{\sqrt{1-(v^2/c^2)}}\,ct - \sum_i \frac{v^i}{c\sqrt{1-(v^2/c^2)}}\,x^i \tag{20.49}$$

$$y^i = \sum_j A_j^i\left[-\frac{v^j}{\sqrt{1-(v^2/c^2)}}\,t + \sum_k\left(\delta^{jk} + \left(\frac{1}{\sqrt{1-(v^2/c^2)}} - 1\right)\frac{v^j v^k}{v^2}\right)x^k\right] \tag{20.50}$$

Finally, with a suitable rotation of coordinates in the frame $\{\psi_{t'}\}$, eq. (20.50) reduces to

$$y^i = -\frac{v^i}{\sqrt{1-(v^2/c^2)}}\,t + \sum_k\left(\delta^{ik} + \left(\frac{1}{\sqrt{1-(v^2/c^2)}} - 1\right)\frac{v^i v^k}{v^2}\right)x^k \tag{20.51}$$

We must make several important observations about the equations we have just obtained.

(1) The determinant of the linear system (20.49), (20.51) is 1. This system has three parameters. These linear transformations are sometimes called *boosts.*

(2) The determinant of the linear system (20.49), (20.50) is ± 1 depending on $\det(A_j^i)$. This has six parameters, and is called *the Lorentz group.* The system (20.49), (20.50) with determinant $+1$ is called *the proper Lorentz group.*

(3) If we add arbitrary constants to the system (20.49), (20.50), then these equations will give all possible relations between a set of rectangular cartesian coordinates and time in one frame of reference and a set of rectangular cartesian coordinates and time in another frame of reference. This set of equations will have 10 parameters and is called *the Poincaré group.* (Cf., the Lorentz transformations, Section 15.1, and Problem 20.12.)

(4) If we make the same rotation of coordinates in $\{\psi_{t'}\}$ that reduced (20.50) to (20.51), then (20.48) reduces to $'v^i = -v^i$. Thus we can say that if $\{\psi_{t'}\}$ has velocity v^i with respect to $\{\phi_t\}$, then $\{\phi_t\}$ has velocity $-v^i$ with respect to $\{\psi_{t'}\}$.

(5) If we replace v^i/c by 0, then the Poincaré group reduces to the general Galilean group, (20.4).

(6) If in (20.49) and (20.51) we choose coordinates in $\{\phi_t\}$ such that the velocity of $\{\psi_{t'}\}$ is in the direction of the $x = x^1$ axis, i.e., $v^2 = v^3 = 0$,

then these equations reduce to

$$\begin{aligned} ct' &= \frac{1}{\sqrt{1-(v^2/c^2)}}ct - \frac{v}{c\sqrt{1-(v^2/c^2)}}x \\ y &= -\frac{v}{\sqrt{1-(v^2/c^2)}}t + \frac{1}{\sqrt{1-(v^2/c^2)}}x \end{aligned} \tag{20.52}$$

If we put $\cosh \vartheta = 1/\sqrt{1-(v^2/c^2)}$, then (20.52) take the form

$$\begin{aligned} ct' &= (\cosh \vartheta)ct - (\sinh \vartheta)x \\ y &= (\sinh \vartheta)ct + (\cosh \vartheta)x \end{aligned} \tag{20.53}$$

We will see in Section 21.1 that ϑ as defined here is precisely the hyperbolic angle described in Section 5.4.

PROBLEM 20.7. Carry out the details of the derivations of eqs. (20.47) and (20.48).

PROBLEM 20.8. Same as Problem 20.7 for eqs. (20.49) and (20.50).

PROBLEM 20.9. Confirm the observations made at the end of the section.

PROBLEM 20.10. Derive the general systems (20.49) and (20.51) starting with the special case (20.52) by writing $x^i = x^i_\parallel + x^i_\perp$ and $y^i = y^i_\parallel + y^i_\perp$ where $x^i_\parallel$ and $y^i_\parallel$ are projections on a line parallel to v^i, and then use (20.52) on $x^i_\parallel$ and $y^i_\parallel$ (cf., Pauli, 1981, p. 10).

20.5 Minkowski spacetime

According to the result of Section 20.3, Theorem 20.2, the coordinates and the parameter values of generalized frames for which both Newton's principle of inertia and Maxwell's equations hold are related by (20.32), in which the coefficients satisfy (20.35). These equations express Einstein's relativity principle, which accommodates both mechanics and electromagnetism (thus "generalizing" the Newton–Galileo principle which applied only to mechanical phenomena).

Not the least important feature of this result is that the quantities and relations appearing can be interpreted as, or identified with, simple structures and relations in a mathematical model conceptually simpler than our model involving frames of reference.

Thus, let (M, g) be a 4-dimensional pseudo-Riemannian manifold with a global coordinate system, μ, on which $g = g_\eta$ has components

$$g_{00} = g\left(\frac{\partial}{\partial\mu^0}, \frac{\partial}{\partial\mu^0}\right) = -1,$$

$$g_{ii} = g\left(\frac{\partial}{\partial\mu^i}, \frac{\partial}{\partial\mu^i}\right) = 1, \qquad i = 1, 2, 3,$$

and

$$g_{\alpha\beta} = g\left(\frac{\partial}{\partial\mu^\alpha}, \frac{\partial}{\partial\mu^\beta}\right) = 0, \qquad \alpha \neq \beta(\alpha, \beta = 0, 1, 2, 3)$$

That is, (M, g_η) is a Lorentzian manifold (Section 15.1) isometric with Minkowski space, $\mathscr{R}^4_1$. Then, from the relations

$$\bar{g}_{\lambda\nu} = \frac{\partial\mu^\alpha}{\partial\bar{\mu}^\lambda}\frac{\partial\mu^\beta}{\partial\bar{\mu}^\nu} g_{\alpha\beta}$$

relating the components of g_η in two coordinate systems, the coordinate transformation

$$\mu^\lambda = \sum_{\alpha=0}^{3} \underset{\alpha}{e}\, a^\lambda_\alpha \bar{\mu}^\alpha + b^\lambda \tag{20.54}$$

with $(\underset{\lambda}{e}\, a^\lambda_\alpha)$ given by (20.35), gives another coordinate system in which g_η has the same values. Thus, the Lorentz transformations can be interpreted as special coordinate transformations on (M, g_η). Coordinatc functions on (M, g_η) related by (20.54) are called *Lorentzian coordinates*. They correspond to the coordinates, x^i, and the parameter values, t, of Lorentzian frames.

Thus, in the presence of a Lorentzian frame (or a Lorentzian coordinate system), we have a 1–1 correspondence between points $p \in (M, g_\eta)$ and pairs $(t, \bar{p})$, where t is a parameter value of the frame and $\bar{p} \in \mathscr{E}^3_0$, and

$$x^i(p) = x^i(\bar{p}),$$

$$x^0(p) = ct \tag{20.55}$$

where on the left the x^α, $\alpha = 0, 1, 2, 3$, are coordinate functions on (M, g_η), and on the right the x^i, $i = 1, 2, 3$ are coordinate functions of the frame.

The tangent space at each point, p, of (M, g_η) is a Lorentzian vector space, and by means of the exponential maps, Section 16.4, tangent vectors go into straight line segments. Hence, properties of vectors in the tangent spaces of (M, g_η) are carried over to properties of lines and line segments in (M, g_η).

Definitions
Choose a timelike vector field, X, on (M, g_η). Then at each point, p, X_p is in one of the nappes, *the future nappe*, of the null cone at T_pM. With this additional structure, (M, g_η) is *time-oriented* and (M, g_η) is *a Minkowski spacetime*. A point in Minkowski spacetime is called *an event*; a timelike curve, λ, is *a world line*; and a pair, (m, λ), where λ is a future-pointing world line, is *a particle in* (M, g_η).

If p is a point on a timelike curve, λ, then in a Lorentz frame having (20.55) we get $x^0(\lambda(u)) = ct$, so t is a function of u. Inverting this function we get

$$\bar{\lambda}(t) = \lambda(u(t)) \tag{20.56}$$

Again from (20.55) $x^i(p) = \mu^i \circ \phi_t^{-1}(\bar{p})$, so that if p is replaced by $\bar{\lambda}(t)$, $\bar{p}$ will be a function, γ, of t. Thus, each timelike curve, λ, in (M, g_η) has a "projection," $\phi_t^{-1} \circ \gamma$, a curve in the frame of reference $\{\phi_t\}$ in $\mathscr{E}_0^3$.

A curve in (M, g_η), in particular a straight line, is either spacelike, timelike or null depending on the corresponding property of its tangent vectors, and, for Lorentzian coordinates, the coordinate curves are mutually orthogonal straight lines. By Theorem 16.9 the length of a line segment joining two points (events) p and q with parameter values 0 and 1 in (M, g_η) is the length of the initial vector in T_pM, and by definition (of distance, Section 15.2) this is the distance, $d(p, q)$, or "separation" between p and q.

Theorem 20.3. *In Lorentzian coordinates, x^α,*

$$d(p, q) = \left| \sum_{\alpha=0}^{3} \underset{\alpha}{e} (x^\alpha(p) - x^\alpha(q))^2 \right|^{\frac{1}{2}} \tag{20.57}$$

Proof. Problem 20.11.

The null lines through p form a (hyper-) cone, given, in Lorentzian coordinates, by

$$\sum_{\alpha=0}^{3} \underset{\alpha}{e} (x^\alpha(p) - x^\alpha(q))^2 = 0 \tag{20.58}$$

Lines through p with

$$\sum_{\alpha=0}^{3} \underset{\alpha}{e} (x^\alpha(p) - x^\alpha(q))^2 < 0 \tag{20.59}$$

correspond, under the exponential map to vectors, v, in T_pM with $g_\eta(v, v) < 0$ and hence are timelike. They can be either future-pointing or past-pointing.

Lines through p with

$$\sum_{\alpha=0}^{3} \underset{\alpha}{e} \, (x^\alpha(p) - x^\alpha(q))^2 > 0 \tag{20.60}$$

correspond to vectors, v, with $g_\eta(v, v) > 0$, and hence are spacelike.

Clearly, the definition (20.57) and the conditions (20.58), (20.59), and (20.60) are independent of the choice of Lorentzian coordinates.

The changing of coordinates in Minkowski spacetime is illustrated by *Minkowski diagrams*, Fig. 20.2. (They will be useful in Chapter 21.) For a given set of coordinates, look at the points of (M, g_η) for which $x^2 = x_*^2$ and $x^3 = x_*^3$. This is a plane. Through each point, p, with coordinates $x_*^\alpha(p)$, we have orthogonal x^0 and x^1 coordinate curves in this plane. We represent these by perpendicular lines on paper in the usual fashion (with the x^0 axis vertical). Points whose coordinates satisfy $-(x^0 - x_*^0)^2 + (x^1 - x_*^1)^2 = 0$, generators of the null cone, are drawn as lines making 45° angles with the coordinate lines. These points are zero distance from p. Points whose coordinates satisfy $-(x^0 - x_*^0)^2 + (x^1 - x_*^1)^2 = \text{constant}$ are drawn as hyperbolas. These points are all equidistant from p.

If we change coordinates by a Lorentz transformation, we get y^0 and

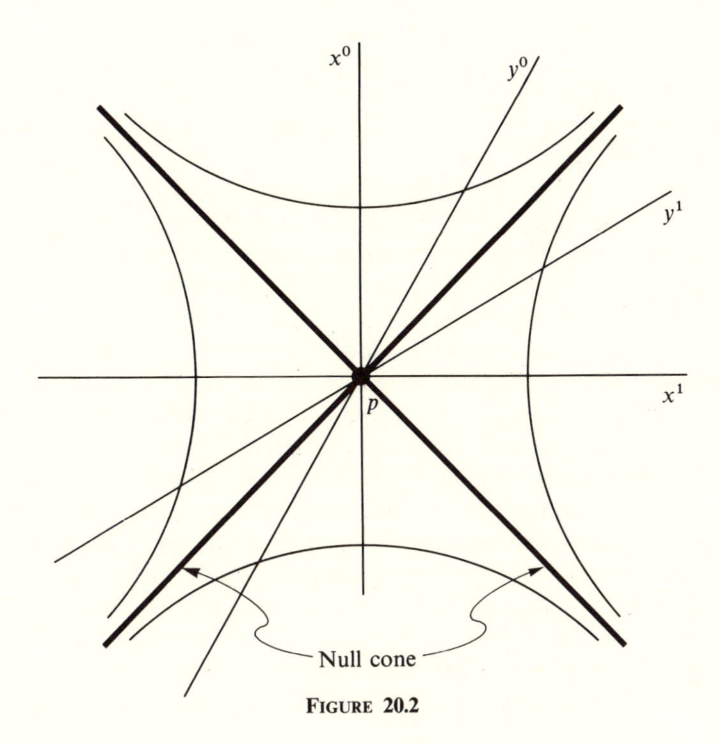

FIGURE 20.2

y^1 coordinate curves which are orthogonal. The null cone and the hyperbolas will look the same. However, from eq. (20.52)

$$x^0 = \frac{v}{c}x^1 + \sqrt{1-(v^2/c^2)}\,y^0 \qquad \text{and} \qquad x^0 = \frac{c}{v}x^1 - \frac{c}{v}\sqrt{1-(v^2/c^2)}\,y^1$$

so the y^α coordinate lines through p have x^α coordinate equations $x^0 = (c/v)x^1 + a^1$ and $x^0 = (v/c)x^1 + a^2$ where a^1 and a^2 are constants. Thus, the new coordinate lines no longer look perpendicular, but they do have the property that they make equal Euclidean angles with the generators of the null cone. We note from this that, in particular, the original representation of the x^0 and x^1 coordinate curves by perpendicular lines on the paper is merely a matter of convenience and has no special significance for Minkowski spacetime.

Going back to some physical interpretation we put $x^0 = ct$ where t is the time coordinate function. Then from (20.19) and (20.20), the null lines on (20.58) correspond to the light rays or the (trajectories of the) photons through p. Putting $\Delta = (\sum_{\alpha=1}^{3} (x^\alpha(p) - x^\alpha(q))^2)^{\frac{1}{2}}$, $t_1 = t(p)$, and $t_2 = t(q)$, eq. (20.59) gives

$$t_2 - t_1 > \frac{\Delta}{c}$$

or

$$t_1 - t_2 > \frac{\Delta}{c}$$

and eq. (20.60) gives

$$-\frac{\Delta}{c} < t_2 - t_1 < \frac{\Delta}{c}$$

In the first case we say the events p and q are *time-ordered* and either *q occurs after p* or *q occurs before p*. In the second case the sign of $t_2 - t_1$ is not independent of coordinates, so there is no way to say which events occurs first. In this case we say the events are *quasisimultaneous*.

Actually, these alternatives reflect a certain symmetry which does not occur in Newtonian mechanics. There, for two different events which do not occur at the same time, it is always possible to find a frame of reference in which they occur at the same place, but for two different events which do not occur at the same place, it is not always possible to find a frame in which they occur at the same time. Now we have the following.

Theorem 20.4. (i) *If p and q are time-ordered, then there exists a frame* (*i.e., coordinates in* (M, g_η)) *in which p and q occur at "the same place."* (ii) *If p*

and q are quasisimultaneous, then there is a frame in which p and q are "simultaneous."

Proof. Write a Lorentz transformation for each event and subtract. If we choose $v^i = (x_2^i - x_1^i)/(t_2 - t_1)$ then $y_2^i = y_1^i$, and if we choose $v^i = \frac{1}{2}c^2(t_2 - t_1)(x_2^i - x_1^i)$ then $t_2' = t_1'$. Note that by our hypotheses, for both choices $v^2 < c^2$, so they can both be made. ∎

If p and q are time-ordered, we can think of p and q as lying on a coordinate curve $y^i = 0$, and $t' =$ constant as consisting of spacelike hyperplanes of simultaneous events, and if p and q are quasi-simultaneous, then we can think of p and q as lying in a coordinate hyperplane $t' = 0$, and $y^i =$ constants as consisting of events occurring at the same place.

To get a better understanding of Minkowski spacetime we will look at its trigonometry. First of all, the hyperbolic hypersurfaces

$$\sum_{\alpha=0}^{3} \underset{\alpha}{e}\,(x^\alpha(p) - x^\alpha(q))^2 = \text{constant}$$

as the set of points whose distance from p is constant are the analogs of spheres.

Spacelike triangles lie in Euclidean submanifolds and hence have the usual properties. We want to consider two other cases.

Theorem 20.5. (A triangle inequality) *If ℓ_1 and ℓ_2 are future-pointing timelike lines intersecting at q, and p on l_1 occurs before q and q occurs before r on l_2, then the line, ℓ_3, through p and r is future-pointing timelike and $d(p, r) \geq d(p, q) + d(q, r)$.*

Proof. The future-pointing line, ℓ_4, through p parallel to ℓ_2 has the same tangent vector as ℓ_2. The tangent vectors of ℓ_4, ℓ_3, and ℓ_2 are in T_pM and their lengths satisfy the corresponding inequality (Problem 5.24). ∎

Theorem 20.6. *Suppose a timelike line, ℓ_1, and a spacelike line, ℓ_2, intersect at q. With p on ℓ_1 and r on ℓ_2 form the triangle pqr. Let ℓ_3 be the line through the points p and r* (see Fig. 20.3).

(i) *If ℓ_3 is spacelike and ℓ_1 and ℓ_2 are orthogonal then*

$$d^2(q, r) = d^2(p, q) + d^2(p, r)$$

(ii) *If ℓ_3 is a null line, then*

$$d(p, q) = d(q, r) \Leftrightarrow \ell_1 \text{ and } \ell_2$$

are orthogonal

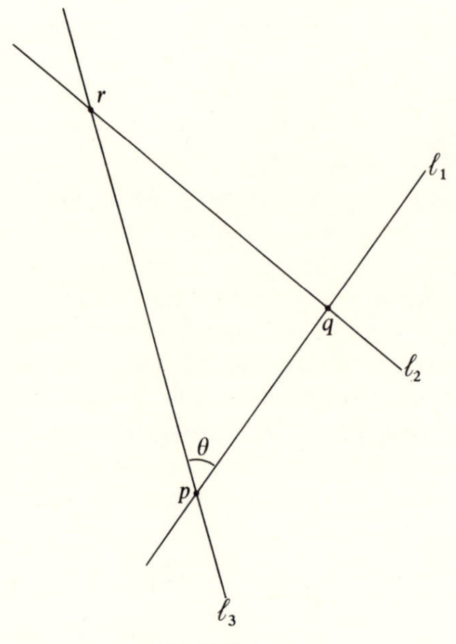

FIGURE 20.3

(iii) *If ℓ_3 is timelike and ℓ_1 and ℓ_3 are both future-pointing, then*

$$d^2(q, r) = 2d(p, q)\, d(p, r) \cosh \vartheta - d^2(p, q) - d^2(p, r)$$

where ϑ is the hyperbolic angle between ℓ_1 and ℓ_3. If further, ℓ_2 and ℓ_1 are orthogonal, then

$$d^2(q, r) = d^2(p, q) - d^2(p, r)$$

and

$$d(p, q) = d(p, r) \cosh \vartheta$$

Proof. The line ℓ_4 through p and parallel to ℓ_2 has the same tangent vector as ℓ_2. Let $\exp_p(v_1) = q$, $\exp_p(v_3) = r$ and $\exp_p(v_4) = r'$ where $d(q, r) = d(p, r')$. Then v_1, v_3, and v_4 satisfy $v_3 = v_1 + v_4$. The results of (i) and (ii) follow by taking the inner product on both sides of this equation. For (iii) we use the relation $\cosh \vartheta = -[g_\eta(v_1, v_3)/\|v_1\| \|v_3\|]$, and when ℓ_1 and ℓ_2 are orthogonal, $g_\eta(v_1, v_3) = -d^2(p, q)$. ∎

Finally, we have three "nonorthogonality properties" for intersecting timelike and null lines.

Theorem 20.7.

(i) *Two lightlike lines are orthogonal iff they are the same.*
(ii) *Two timelike lines are never orthogonal.*
(iii) *A timelike line and a lightlike line are never orthogonal.*

Proof. (i) If ℓ_1 and ℓ_2 are lightlike lines with tangent vectors v_1 and v_2, then $v_1 = a_1 z + w_1$ and $v_2 = a_2 z + w_2$ where z is a unit timelike vector and w_1 and w_2 are spacelike vectors orthogonal to z (cf., Theorem 5.20). Taking scalar products we get $-a_1 a_2 + g_\eta(w_1, w_2)$, $a_1^2 = g_\eta(w_1, w_1)$ and $a_2^2 = g_\eta(w_1, w_2)$, from which we get $[g_\eta(w_1, w_2)]^2 = g_\eta(w_1, w_1) g_\eta(w_2, w_2)$, which implies that $w_1 = a w_2$ and $a_1 = a a_2$ so $v_1 = a v_2$.

(ii) and (iii) *Problem 20.15.*

PROBLEM 20.11. Prove Theorem 20.3.

PROBLEM 20.12. Show that

$$\sum_{\alpha=0}^{3} \underset{\alpha}{e} \, (x^\alpha(p) - x^\alpha(q))^2 = \sum_{\alpha=0}^{3} \underset{\alpha}{e} \, (y^\alpha(p) - y^\alpha(q))^2$$

for two frames and for $p, q \in (M, g_\eta)$ if and only if the coordinates x^α and y^α are related by a Lorentz transformation. (Thus, if we think of the Lorentz transformations as mappings instead of coordinate transformations, then they are precisely the ones that preserve distance, eq. (20.57), the isometries. One can show that the isometries of a pseudo-Riemannian manifold form a group with at most $[n(n + 1)/2]$ parameters, and we see, in particular, that the Lorentz transformations defined in Section 15.1 and the Poincaré group defined in Section 20.4 are just different manifestations of this maximal symmetry.)

PROBLEM 20.13. Show that for a triangle containing two spacelike sides any two of the following implies the third.

(i) The triangle is spacelike.
(ii) The two given sides are perpendicular.
(iii) The Pythagorean theorem is satisfied.

PROBLEM 20.14. The distance between two timelike lines is constant iff they are parallel (cf., Synge, 1965, pp. 44 ff).

PROBLEM 20.15. Prove parts (ii) and (iii) of Theorem 20.7.

21

SOME PHYSICS ON MINKOWSKI SPACETIME

We will sketch some "kinematic" consequences of the Lorentz transformations—kinematic from the viewpoint of $\mathscr{E}_0^3$, though from the viewpoint of spacetime they are just transformations of coordinate intervals. Then we will present formulations of particle dynamics, electromagnetism, and fluid mechanics on (M, g_η).

21.1 Time dilation and the Lorentz–Fitzgerald contraction

In this section we go back again to the consideration of a particle $(m, \phi_t^{-1} \circ \gamma)$ moving with constant velocity. In this case we can describe the situation in terms of two (generalized) frames of reference; the frame $\{\phi_t\}$, and a frame $\{\psi_{t'}\}$ in which the particle is at rest; *a proper frame*, or *rest frame* of the particle. A particle at rest in $\{\phi_t\}$ is called an *observer*, and $\{\phi_t\}$ is the frame of the observer.

Suppose the particle is at the point $\mathscr{P}_j$ in $\{\phi_j\}$ when $t = t_j$, $j = 1, 2$. $(\mathscr{P}_j, t_j)$ determine unique points $\bar{p}_j \in \mathscr{E}_0^3$. Then choosing coordinates in $\{\phi_t\}$ according to observation (6) in Section 20.4, the Lorentz transformations give

$$t_j = \frac{1}{\sqrt{1 - v^2/c^2}}\, t_j' + \frac{v}{c^2\sqrt{1 - v^2/c^2}}\, y(\bar{p}_j) \tag{21.1}$$

(inverting (20.52)).

Now, since $(m, \phi_t^{-1} \circ \gamma)$ is at rest in $\{\psi_{t'}\}$, $\bar{p}_1$ and $\bar{p}_2$ are two $\mathscr{E}_0^3$ points of the same point, $\mathscr{Q}$, of $\{\psi_{t'}\}$ so $y(\bar{p}_1) = y(\bar{p}_2)$. Taking the difference in (21.1) for $\bar{p}_1$ and $\bar{p}_2$ we get

$$t_2 - t_1 = \frac{1}{\sqrt{1 - v^2/c^2}}\, (t_2' - t_1') \tag{21.2}$$

Since $\bar{p}_1$ and $\bar{p}_2$ correspond to $\mathscr{Q}$, $\mathscr{Q}$ has the corresponding parameter values t_1' and t_2'. Hence $t_2' - t_1'$ on the right side of (21.2) is the time interval measured on a clock located at $\mathscr{Q}$.

Finally, for the left side of (21.2), the parameter value t_1 is a parameter value for all points $\mathscr{P}$ of $\{\phi_t\}$. Similarly, for t_2. Hence $t_2 - t_1$ is the time

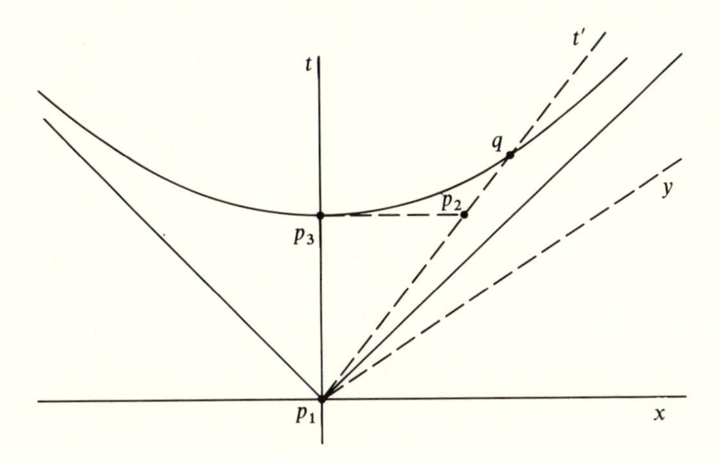

FIGURE 21.1

difference for *any* point in $\{\phi_t\}$. In particular, it is the time interval measured either on a clock at $\mathscr{P}_1$ or on a clock at $\mathscr{P}_2$. Thus, we have obtained the relation between time measurements in the proper frame of a particle and those in the frame of an observer.

The mathematical steps in the above derivation are quite simple; basically, just the algebra of going from the Lorentz transformations to (21.2). It is the physical interpretation which is not so simple.

Another derivation of the phenomenon of time dilation makes use of properties of Minkowski spacetime. Let an observer be represented by the t axis in a Minkowski diagram, Fig. 21.1. (For simplicity take $c = 1$, or $x^0 = t$.) The particle, at rest in $\{\psi_{t'}\}$, can be represented by the t' axis in another set of Lorentzian coordinates. Since the length of the segment p_1p_3 equals the length of the segment p_1q, for any point p_2 on p_1q, $t(p_3) - t(p_1) > t'(p_2) - t'(p_1)$. If p_2 is on the line parallel to the x axis, then $t(p_2) = t(p_3)$, so $t(p_2) - t(p_1) > t'(p_2) - t'(p_1)$, which gives the time dilation.

We can go further. Since the t axis and the line through p_3 and p_2 are orthogonal,

$$t(p_2) - t(p_1) = [t'(p_2) - t'(p_1)] \cosh \vartheta \tag{21.3}$$

by Theorem 20.6(iii). Comparing (21.2) and (21.3) we get the result anticipated at the end of Section 20.4:

Theorem 21.1. *The "angle" introduced in eqs. (20.53) is the hyperbolic angle between the t and t' axes.*

Proof. Immediate.

We can summarize the phenomenon of time dilation by saying that

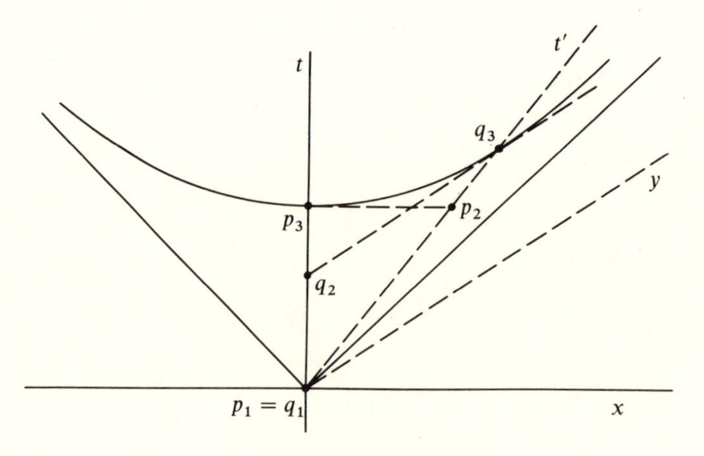

FIGURE 21.2

proper time intervals, time intervals measured by a particle in its own frame (its "proper" frame), are always smaller than those measured by any observer with respect to whom the particle is moving.

In spite of "thought experiments" involving trains and rockets, a completely satisfactory physical feeling for this phenomenon seems somewhat elusive, and it has been marked by apparent "paradoxes." Thus, for example, the principle of relativity expressed by eqs. (20.32) and (20.35) essentially proclaims the equivalence of two frames of a certain class. If the frame of the particle is moving with respect to that of the observer, then the observer frame is moving with respect to that of the particle, so that by the above equivalence, there should be a reciprocity in the measurement of time intervals in the two frames, which, on the surface, our result seems to contradict. A more careful analysis resolves this "paradox" (cf., Fock, 1979, pp. 40–42). While, again, the physical description in terms of moving particles and points in reference frames is quite involved, a simple description of the situation is obtained by means of a Minkowski diagram.

In Fig. 21.2, $p_3 p_2$ is parallel to the x axis and $q_2 q_3$ is parallel to the y axis. If the t axis is the observer and the t' axis is the particle, then $t'(p_2) - t'(p_1) < t(p_2) - t(p_1)$ (as before), but if the t' axis is the observer and the t axis is the particle, then $t(q_2) - t(q_1) < t'(q_2) - t'(q_1)$.

A physical description of the companion consequence of the Lorentz transformations, the Lorentz–Fitzgerald contraction, will be left to the interested reader. Here we will simply note that we can write the Lorentz transformation for any two events p_1 and p_2 as

$$y(p_i) = -\frac{v}{\sqrt{1 - v^2/c^2}}\, t(p_i) + \frac{1}{\sqrt{1 - v^2/c^2}}\, x(p_i)$$

$i = 1, 2$. Then putting $t(p_1) = t(p_2)$ and subtracting we get

$$y(p_2) - y(p_1) = \frac{1}{\sqrt{1 - v^2/c^2}} [x(p_2) - x(p_1)] \tag{21.4}$$

Now, if p_1 and p_2 correspond to the ends of a rod moving in the direction of its length with respect to an observer, (t, x) are the observer's coordinates, and (t', y) are the rod's coordinates, then eq. (21.4) says that the length of the rod as measured in its frame, its *proper* or *rest length*, is always greater than that measured by an observer.

Again, this phenomenon can be described by a Minkowski diagram. In Fig. 21.3 the length of p_1p_3 is greater than the length of p_1p_2 which are respectively the proper and observer lengths of the rod.

The two consequences of the Lorentz transformations which we have been discussing, time dilation and the Lorentz–Fitzgerald contraction, are frequently summarized by the statement that "motion slows down clocks and shortens rods," or, as is commonly stated, clocks *appear* to be slowed down and rods *appear* to be shortened. This description seems to suggest that time and space behave the same, whereas they should more properly be described as behaving oppositely: the time interval measured by the clock in its own frame of reference is smaller than that measured by the observer in his frame, and the length measured by the rod in its own frame of reference is larger than that measured by the observer.

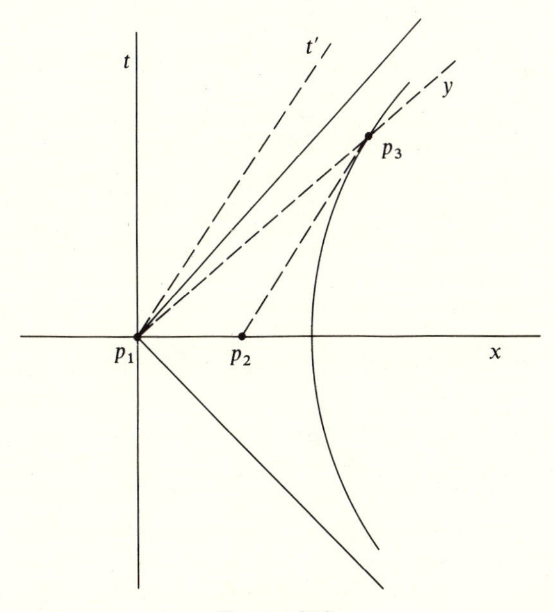

FIGURE 21.3

Another consequence concerns relative velocities, or the addition of velocities. Recall that we have the concepts of the velocity of a particle with respect to a frame (Section 20.1), and the relative velocity of one frame with respect to another, (Section 20.4).

Definition
The relative velocity of a particle, γ_2, with respect to a particle, γ_1, is the velocity of γ_2 in a frame in which γ_1 is at rest.

Suppose that in $\{\phi_t\}$ $\phi_t^{-1} \circ \gamma_1$ has velocity v^i and $\phi_t^{-1} \circ \gamma_2$ has velocity w^i. Let $\{\psi_{t'}\}$ be a frame in which γ_1 is at rest. Then the velocity of $\{\psi_{t'}\}$ with respect to $\{\phi_t\}$ is v^i, and if y^i are rectangular cartesian coordinates in $\{\psi_{t'}\}$ then the relative velocity of γ_2 with respect to γ_1 is dy^i/dt'.

Differentiating the Lorentz transformations (20.51) we get

$$\frac{dy^i}{dt'} = \left[\frac{-v^i}{\sqrt{1 - v^2/c^2}} + \sum_k \left(\delta^{ik} + \left(\frac{1}{\sqrt{1 - v^2/c^2}} - 1\right)\frac{v^i v^k}{v^2}\right) w_k\right] \frac{dt}{dt'}$$

since $dx^k/dt = w^k$ is the velocity of $\phi_t^{-1} \circ \gamma_2$ in $\{\phi_t\}$. Now using the Lorentz transformations (20.49) to get dt/dt' we have the general result

$$\frac{dy^i}{dt'} = \frac{\sqrt{1 - v^2/c^2}}{1 - \sum \frac{v^j w^j}{c^2}} \left(w^i - v^i + \left(1 - \frac{1}{\sqrt{1 - v^2/c^2}}\right)\left(1 - \frac{\sum v^j w^j}{v^2}\right) v^i\right) \tag{21.5}$$

Note that here the v^i are constants, but w^i and dy^i/dt' need not be.

In particular,

(i) If v^i and w^i are parallel, (21.5) becomes

$$\frac{dy^i}{dt'} = \frac{w^i - v^i}{1 - \frac{\sum v^j w^j}{c^2}} \tag{21.6}$$

(ii) If v^i and w^i are perpendicular, (21.5) becomes

$$\frac{dy^i}{dt'} = \sqrt{1 - v^2/c^2}\, w^i - v^i \tag{21.7}$$

We get a somewhat different interpretation of these formulas if we suppose that $\phi_t^{-1} \circ \gamma_1$ has velocity $-v^i$ instead of v^i and again $\{\psi_{t'}\}$ is the frame in which γ_1 is at rest. Then we get (21.5) with v^i replaced by $-\mathrm{v}^i$. If $\{\chi_{t''}\}$ is a frame in which γ_2 is at rest then (21.5) with v^i replaced by $-v^i$

gives the velocity of $\{\chi_{t''}\}$ with respect to $\{\psi_{t'}\}$ in terms of the velocity of, v^i, of $\{\phi_t\}$ with respect to $\{\psi_{t'}\}$ and the velocity, w^i, of $\{\chi_{t''}\}$ with respect to $\{\phi_t\}$. The special case

$$\frac{dy^i}{dt'} = \frac{w^i + v^i}{1 + \dfrac{\sum v^j w^j}{c^2}} \tag{21.8}$$

when v^i and w^i are parallel is Einstein's formula for the addition of velocities.

A third interpretation of the formulas (21.5)–(21.7) is that they give the relation between the components w^i and dy^i/dt' of the velocity of γ_2 in two different Lorentzian frames. Note that when v^i and w^i are small compared with c these formulas all reduce to the usual simple relation.

In the cases which we have been considering, where particles move with constant velocity with respect to a frame, $\{\phi_t\}$, there is a frame in which the particle is at rest. Times and lengths measured in this frame were called proper. Now, if a particle is subject to force, its velocity is not constant in any Lorentzian frame. By generalizing the concept of proper time to include circumstances in which forces are present, we obtain an important invariant.

If λ is a timelike curve in (M, g_η), then its length (Section 15.2) is

$$\int_{u_1}^{u_2} \|\dot{\lambda}(u)\|\, du = \int_{u_1}^{u_2} [-g(\dot{\lambda}(u), \dot{\lambda}(u))]^{\frac{1}{2}}\, du = \int_{u_1}^{u_2} \left[\left(\frac{d\lambda^0}{du}\right)^2 - \sum \left(\frac{d\lambda^i}{du}\right)^2\right]^{\frac{1}{2}} du$$

in Lorentzian coordinates. If λ happens to be a straight line, it will be a coordinate curve in a Lorentzian coordinate system in which the particle is at rest, and using x^0 as the parameter we get

$$x_2^0 - x_1^0 = c(t_2 - t_1) = \int_{u_1}^{u_2} \|\dot{\lambda}(u)\|\, du$$

where $t_2 - t_1$ is the proper time interval. This special case obviously fits into the following general definition.

Definition

$\tau(u) = (1/c) \int_{u_1}^{u} \|\dot{\lambda}(u)\|\, du$ is called *the proper time* of the particle λ (measured from $\lambda(u_1)$).

Theorem 21.2. *If λ is a particle in (M, g_η), then in a Lorentzian coordinate system with associated Lorentzian frame $\{\phi_t\}$, it has a "projection" $\phi_t^{-1} \circ \gamma$*

(Section 20.5). Then

$$\frac{d\tau}{dt} = \sqrt{1 - v^2/c^2} \tag{21.9}$$

where $v = \|(\phi_t^{-1} \circ \gamma)^{\bullet}\|$.

Proof.

$$\frac{d\tau}{dt} = \frac{d\tau}{du}\frac{du}{dt} = \frac{1}{c}\left[\left(\frac{d\lambda^0}{du}\right)^2 - \sum\left(\frac{d\lambda^i}{du}\right)^2\right]^{\frac{1}{2}}\frac{du}{dt}$$

$$= \frac{1}{c}\left[\left(\frac{d\bar{\lambda}^0}{dt}\right)^2 - \sum\left(\frac{d\bar{\lambda}^i}{dt}\right)^2\right]^{\frac{1}{2}} \quad \text{by eq. (20.56)}$$

$$= \frac{1}{c}\left[c^2 - \sum\left(\frac{d\gamma^i}{dt}\right)^2\right]^{\frac{1}{2}} = [1 - v^2/c^2]^{\frac{1}{2}} \qquad \blacksquare$$

Definition
The tangent vector, or velocity, $\dot{\lambda}$, of a particle, λ, parametrized by its proper time is frequently called *the 4-velocity* or, the *relativistic velocity* of the particle.

Theorem 21.3. *The components of the velocity,* $\dot{\lambda}$, *of a particle in terms of the velocity of the associated* $\mathscr{E}_0^3$ *curve in a Lorentz frame are*

$$\left(\frac{c}{\sqrt{1 - v^2/c^2}}, \frac{v^i}{\sqrt{1 - v^2/c^2}}\right) \tag{21.10}$$

Proof. Problem 21.6.

Note that, in contrast to previous velocities we have considered, $(\phi_t^{-1})\circ\gamma$ and $\dot{\lambda}$ are not necessarily constant. However, now the magnitude of $\dot{\lambda}$ is constant, i.e., $g(\dot{\lambda}, \dot{\lambda}) = -c^2$, and $\|\dot{\lambda}\| = c$.

Finally, we make one more definition which we will need in the next section.

Definition
$\mathsf{P} = m\dot{\lambda}$ is *the 4-momentum* or *energy-momentum* of a particle $(m, \dot{\lambda})$.

PROBLEM 21.1. The explanation, using Figure 21.2, of the "reciprocity" between two frames of time dilation depended on the fact that p_2 and q_2 both lie beneath the hyperbola. Prove that that is indeed the case.

PROBLEM 21.2. The formula (21.4), for the Lorentz–Fitzgerald contraction is for the case in which the rod moves in the direction of its length. If, more generally, φ is the angle between the x axis and the rod, then

$$l_r = \frac{\sqrt{1 - (v^2/c^2)\sin^2\varphi}}{\sqrt{1 - v^2/c^2}}\, l_0$$

where l_r is the rest length and l_0 is the length measured by an observer.

PROBLEM 21.3. Construct a figure to explain the "reciprocity" between two frames of the Lorentz–Fitzgerald contraction.

PROBLEM 21.4. Derive eqs. (21.6) and (21.7).

PROBLEM 21.5. Choose coordinates in $\{\phi_t\}$ and $\{\psi_{t'}\}$ so that $v^2 = v^3 = w^2 = w^3 = 0$ where v^i is the velocity of $\{\phi_t\}$ with respect to $\{\psi_{t'}\}$, and w^i is the velocity of $\{\chi_{t''}\}$ with respect to $\{\phi_t\}$. Then (i) $\tanh\vartheta_1 = v_1/c$ and $\tanh\vartheta_2 = w_1/c$ where ϑ_1 is the hyperbolic angle between the t and t' axes and ϑ_2 is the hyperbolic angle between the t and t'' axes, and (ii) the velocity of $\{\chi_{t''}\}$ with respect to $\{\psi_{t'}\}$ is $c\tanh(\vartheta_1 + \vartheta_2)$.

PROBLEM 21.6. Show that Einstein's addition formula respects the limitation of the velocity of one frame with respect to another.

PROBLEM 21.7. Prove Theorem 21.3.

21.2 Particle dynamics on Minkowski spacetime

In classical Newtonian dynamics we introduce the concepts of momentum, energy and force. These concepts should be extendible to relativity. Also in classical dynamics we have conservation laws which we want to still be valid.

For particles, the only kinds of interactions which can be meaningfully formulated in the context of the theory of relativity are collisions and disintegrations since "action at a distance" is excluded. (We will consider interaction between particles and fields in the next section.) In the case of two particles $(m_1, \phi_t^{-1}\circ\gamma_1)$ and $(m_2, \phi_t^{-1}\circ\gamma_2)$ colliding elastically, the classical law of conservation of momentum is

$$m_1(\phi_t^{-\dot{1}}\circ\gamma_2) + m_2(\phi_t^{-\dot{1}}\circ\gamma_2) = \bar{m}_1(\phi_t^{-\dot{1}}\circ\bar{\gamma}_1) + \bar{m}_2(\phi_t^{-\dot{1}}\circ\bar{\gamma}_2) \quad (21.11)$$

(where now, γ_1 and γ_2, and $\bar{\gamma}_1$ and $\bar{\gamma}_2$ are the trajectories before and after collision, respectively, and $\bar{m}_i = m_i$, $i = 1, 2$).

A simple example (Lewis and Tolman, 1909) shows that if formula (21.11) is valid for the frame of reference $\{\phi_t\}$ it will no longer hold if we simply replace ϕ_t by $\psi_{t'}$ in a second frame of reference, $\{\psi_{t'}\}$. In that example, the velocity components satisfy

$$\frac{m_1 v_1^i}{\sqrt{1 - v_1^2/c^2}} + \frac{m_2 v_2^i}{\sqrt{1 - v_2^2/c_2}} = \frac{\bar{m}_1 \bar{v}_1^i}{\sqrt{1 - \bar{v}_1^2/c^2}} + \frac{\bar{m}_2 \bar{v}_2^i}{\sqrt{1 - \bar{v}_2^2/c_2}} \tag{21.12}$$

in both frames. (Here the v's in the denominators are the magnitudes of the corresponding velocities, and, again in this example, $\bar{m}_i = m_i$.)

On the basis of (1) this particular example, and perhaps others (cf., Taylor and Wheeler, 1966), (2) the fact that (21.12) reduces to the classical law for small velocities, and, most importantly, (3) the fact that it has been confirmed experimentally in particle accelerators, eq. (21.12) *is accepted as a physical law*. Along with eq. (21.12) it is sometimes convenient to introduce the following terminology for a particle: $mv^i/\sqrt{1 - v^2/c^2}$ is its *relativistic 3-momentum*, $m/\sqrt{1 - v^2/c^2}$ is its *relativistic mass*, and m is its *rest* or *proper mass*.

The elastic collision of two particles which we have been discussing can be viewed from the standpoint of Minkowski spacetime as the intersection at some point (event), p, in Minkowski spacetime of four particles; (m_1, λ_1), (m_2, λ_2), $(\bar{m}_1, \bar{\lambda}_1)$, and $(\bar{m}_2, \bar{\lambda}_2)$. They have 4-momenta $m_1 \dot{\lambda}_1$, $m_2 \dot{\lambda}_2$, $\bar{m}_1 \dot{\bar{\lambda}}_1$, and $\bar{m}_2 \dot{\bar{\lambda}}_2$ at p. Form the vector $m_1 \dot{\lambda}_1 + m_2 \dot{\lambda}_2 - \bar{m}_1 \dot{\bar{\lambda}}_1 - \bar{m}_2 \dot{\bar{\lambda}}_2$. Now, if any three of the components of a vector at p are zero in all (Lorentzian) coordinate systems, then so is the fourth component. Hence, in view of (21.10), we get, as a consequence of (21.12),

$$\frac{m_1 c}{\sqrt{1 - v_1^2/c^2}} + \frac{m_2 c}{\sqrt{1 - v_2^2/c^2}} = \frac{\bar{m}_1 c}{\sqrt{1 - \bar{v}_1^2/c^2}} + \frac{\bar{m}_2 c}{\sqrt{1 - \bar{v}_2^2/c^2}} \tag{21.13}$$

an additional conservation law for the collision process we have been discussing. This is the source of one of the dramatic results of the theory of relativity.

In order to summarize and generalize in a theorem what we have found above, we first make the following definitions, which we will subsequently have to justify.

Definitions

For a particle, $(m, \phi_t^{-1} \circ \gamma)$, $\mathsf{E} = mc^2/\sqrt{1 - v^2/c^2}$ is called its *total* or *relative energy*, and $\mathsf{E}_0 = mc^2$ is called its *proper* or *rest energy*.

Theorem 21.4. *In any interaction (elastic or inelastic collision, or disintegration) of any number of particles, conservation of relativistic 3-momentum implies both conservation of total energy and 4-momentum.*

Proof. The proof is based as before on the simple observation concerning the vanishing of the components of a vector, and introducing a lot more notation to provide for arbitrary numbers of particles before and after the interaction. ■

To justify our definition of total energy and rest energy we first make another definition.

Definition
The (relativistic) kinetic energy of a particle is

$$\mathsf{T} = \frac{mc^2}{\sqrt{1 - v^2/c^2}} - mc^2 \tag{21.14}$$

This definition is acceptable because:

$$\begin{aligned}(1)\quad \frac{mc^2}{\sqrt{1 - v^2/c^2}} - mc^2 &= mc^2\,\frac{1 - \sqrt{1 - v^2/c^2}}{\sqrt{1 - v^2/c^2}} \\ &= mc^2\,\frac{(1 - \sqrt{1 - v^2/c^2})(1 + \sqrt{1 - v^2/c^2})}{\sqrt{1 - v^2/c^2}(1 + \sqrt{1 - v^2/c^2})} \\ &= \frac{mv^2}{\sqrt{1 - v^2/c^2}(1 + \sqrt{1 - v^2/c^2})}\end{aligned}$$

which becomes the classical Newtonian kinetic energy in the limit as $v/c \to 0$.

(2) In his 1905 paper Einstein found this expression for the kinetic energy of charged particle accelerating in an electromagnetic field. See Problem 21.14 (next section).

Now, in Newtonian mechanics, by conservation of energy, the total kinetic energy of our systems of colliding particles would be constant, since no forces are acting. But because of eq. (21.13)

$$\mathsf{T}_1 + \mathsf{T}_2 + m_1c^2 + m_2c^2 = \frac{m_1c^2}{\sqrt{1 - v_1^2/c^2}} + \frac{m_2c^2}{\sqrt{1 - v_2^2/c^2}}$$

is conserved. To avoid giving up the idea of conservation of energy we think of the quantity conserved as the energy of the system. Then $m_1c^2 + m_2c^2$ will be some kind of energy also. Thus, we call $\mathsf{E} = \sum (m_ic^2/\sqrt{1 - v_i^2/c^2})$ *the total energy of the system* and $\mathsf{E}_0 = \sum m_ic^2$ *the rest energy of the system.*

The fact that, in a given process $\mathsf{E} = \mathsf{E}_0 + \mathsf{T}$ is conserved (constant) suggests that changes of T must be compensated for by changes of the rest mass of the system, and vice versa. Thus, for example if two particles of equal mass, m, collide head-on inelastically, producing a single stationary particle of mass, $\bar{m}$, then the total energy before collision will be $2mc^2/\sqrt{1 - v^2/c^2}$

and after collision it will be $\bar{m}c^2$. So $\bar{m} = 2m/\sqrt{1 - v^2/c^2}$, which is more than just the sum of the two masses.

The idea of the equivalence of mass and energy was suggested by experiments in electromagnetism as far back as 1881 (Whittaker, 1960, p. 51) and it was attractive, esthetically and philosophically, to postulate this equivalence for all kinds of energy in other processes. This has of course been established experimentally, and dramatically utilized. For a derivation for energy other than kinetic energy, see Pauli (181, pp. 121 ff).

Having disposed of the concepts of momentum and energy, there remains only the concept of force to be considered.

Definition

For a particle, λ, in Minkowski spacetime with tangent vector $\dot{\lambda}$, the *relativistic* (*four*, or *Minkowski*) *force*, F, acting on the particle is $\nabla_{\dot{\lambda}}\mathsf{P}$ (cf., Section 17.2).

In a Lorentzian coordinate system with $u^\alpha = d\lambda^\alpha/d\tau$,

$$\mathsf{F} = \frac{d}{d\tau}(mu^\alpha)\frac{\partial}{\partial x^\alpha} = \frac{d}{d\tau}\left(\frac{mc}{\sqrt{1 - v^2/c^2}}\right)\frac{\partial}{\partial x^0} + \frac{d}{d\tau}\left(\frac{mv^i}{\sqrt{1 - v^2/c^2}}\right)\frac{\partial}{\partial x^1} \tag{21.15}$$

Since $u_\alpha u^\alpha = -c^2$ we have $u_\alpha\, du^\alpha/d\tau = 0$ and $u_\alpha\, dmu^\alpha/d\tau = -c^2\, dm/d\tau$. So

$$u_\alpha \frac{d}{d\tau}(mu^\alpha) = -c^2 \frac{d}{d\tau} m \tag{21.16}$$

In particular, we have, from (21.16), that the relativistic force and the relativistic velocity are orthogonal iff the (rest) mass, m, of the particle is constant. Expanding (21.16), and using (21.9) we get

$$\frac{-c}{\sqrt{1 - v^2/c^2}}\frac{d}{d\tau}\frac{mc}{\sqrt{1 - v^2/c^2}} + \frac{v_i}{1 - v^2/c^2}\frac{d}{dt}\left(\frac{mv^i}{\sqrt{1 - v^2/c^2}}\right) = -c^2\frac{d}{d\tau}m \tag{21.17}$$

Using (21.17) in (21.15) in the case m = constant we get

$$\mathsf{F} = \frac{1}{\sqrt{1 - v^2/c^2}}\left(\frac{v_i}{c}\mathsf{F}_N^i\frac{\partial}{\partial x^0} + \mathsf{F}_N^i\frac{\partial}{\partial x^i}\right) \tag{21.18}$$

where

$$\mathsf{F}_N^i = \frac{d}{dt}\left(\frac{mv^i}{\sqrt{1 - v^2/c^2}}\right) \tag{21.19}$$

the components of *the relativistic 3-force* (which approximates the usual Newtonian force when v/c is small.

Finally, we get a relation between the relativistic kinetic energy of a particle and the relativistic 3-force acting on it which generalizes the classical relation. Thus, from $d\mathsf{T}/d\tau = d\mathsf{E}/d\tau - d\mathsf{E}_0/d\tau$ we have

$$\frac{1}{\sqrt{1 - v^2/c^2}} \frac{d\mathsf{T}}{d\tau} = \frac{1}{\sqrt{1 - v^2/c^2}} \frac{d}{d\tau}\left(\frac{mc^2}{\sqrt{1 - v^2/c^2}}\right) - \frac{c^2}{\sqrt{1 - v^2/c^2}} \frac{dm}{d\tau} \tag{21.20}$$

From (21.17) and (21.20),

$$\frac{1}{\sqrt{1 - v^2/c^2}} \frac{d\mathsf{T}}{d\tau} = \frac{v_i}{1 - v^2/c^2} \mathsf{F}^i_N + c^2\left(1 - \frac{1}{\sqrt{1 - v^2/c^2}}\right)\frac{dm}{d\tau}$$

or

$$\frac{d\mathsf{T}}{dt} = v_i \mathsf{F}^i_N - \frac{v^2\sqrt{1 - v^2/c^2}}{\sqrt{1 - v^2/c^2} + 1} \frac{dm}{d\tau} \tag{21.21}$$

Equation (21.21) reduces to the Newtonian relation when $v/c \to 0$ and m is constant. It is sometimes taken to define the relativistic kinetic energy.

PROBLEM 21.8. Describe the motion of a particle if the force acting on it is constant. Show, in particular, that $v < c$.

PROBLEM 21.9. Show that

$$c^2 \frac{d}{dt} \frac{m}{\sqrt{1 - v^2/c^2}} = \mathsf{F}^i_N v_i$$

and if $a^i = dv^i/dt$ is *the 3-acceleration,*

$$\mathsf{F}^i_N = \frac{m}{\sqrt{1 - v^2/c^2}} a^i + \frac{\mathsf{F}^j_N v_j}{c^2} v^i$$

21.3 Electromagnetism on Minkowski spacetime

Recall that in Chapter 20, the requirement to build a relativity theory encompassing both Newton's mechanical laws and electromagnetic theory led to a Minkowski spacetime formulation. In this formulation we saw, in the last section, that Newton's laws had to be slightly modified. It turns out that Maxwell's equations can be formulated in Minkowski spacetime without modification of their classical form. This formulation is in terms of an electromagnetic tensor. There are various motivations for introducing the electromagnetic tensor. We will introduce it by its relation to an electromagnetic force. (For an approach starting with an electromagnetic potential 4-vector see, for example, Rindler, 1966, pp. 120 ff.)

Theorem 21.5. *If a particle, λ, with "charge", e, is subject to a relativistic force,* F, *whose components in a Lorentzian frame are given by*

$$\mathsf{F}_N^i = e(E^i + (v \times B)^i) \tag{21.22}$$

in eq. (21.18), where v is the velocity in the frame, of the curve γ associated with λ, E and B are vector fields in the frame, and $v \times B$ is the cross-product of vector analysis, then a 2-form, $\mathscr{F}$, defined by

$$\mathsf{F} = -e i_{\dot{\lambda}} \mathscr{F} \tag{21.23}$$

on λ has the form

$$\mathscr{F} = -\frac{E_i}{c} dx^0 \wedge dx^i + B_3\, dx^1 \wedge dx^2 - B_2\, dx^1 \wedge dx^3 + B_1\, dx^2 \wedge dx^3 \tag{21.24}$$

In eq. (21.23) $i_{\dot{\lambda}}$ is interior multiplication by $\dot{\lambda}$, and, with abuse of notation, F is the 1-form $(g_\eta)^\flat \cdot \mathsf{F}$.

Proof. If we substitute (21.22) into (21.18) and map the result to a 1-form by $(g_\eta)^\flat$, which we still denote by F, we get

$$\begin{aligned}\mathsf{F} &= \frac{e}{\sqrt{1 - v^2/c^2}}\left(-E_i \frac{v^i}{c} dx^0 + (E_i + (v \times B)_i)\, dx^i\right) \\ &= \frac{e}{\sqrt{1 - v^2/c^2}}\left(-E_i \frac{v^i}{c} dx^0 + E_i\, dx^i + (v_2 B_3 - v_3 B_2)\, dx^1 \right. \\ &\qquad \left. + (v_3 B_1 - v_1 B_3)\, dx^2 + (v_1 B_2 - v_2 B_1)\, dx^3\right)\end{aligned}$$

and computing $i_{\dot{\lambda}}\mathscr{F}$ from (21.24) we get (21.23). ■

We assume (21.22) is valid in every Lorentzian frame. Consequently, $\mathscr{F}$ has the form (21.24) in every frame. The validity of (21.22) is just a matter of the definition of the E^i and the B^i. That is, whatever the force may be, we write it in the form (21.22). (In a different approach, in which E^i and B^i have some prior physical meaning, the 3-force F_{N^i} of eq. (21.22) is called *the Lorentz force*.)

Definition
$\mathscr{F}$ is called *the electromagnetic field tensor.*

It is common practice to arrange the components $\mathscr{F}_{\alpha\beta}$ of $\mathscr{F}$ into a matrix:

$$\mathscr{F}_{\alpha\beta} = \begin{pmatrix} 0 & -E_1/c & -E_2/c & -E_3/c \\ E_1/c & 0 & B_3 & -B_2 \\ E_2/c & -B_3 & 0 & B_1 \\ E_3/c & B_2 & -B_1 & 0 \end{pmatrix}$$

The components of $\mathscr{F}$ in two Lorentzian coordinate systems are, of course, related by the Lorentz transformations. In particular, for the Lorentz transformations (20.52) we get

$$\begin{aligned} E_1' &= E_1 & B_1' &= B_1 \\ E_2' &= \frac{1}{\sqrt{1 - v^2/c^2}}(E_2 - vB_3) & B_2' &= \frac{1}{\sqrt{1 - v^2/c^2}}(B_2 + vE_3) \\ E_3' &= \frac{1}{\sqrt{1 - v^2/c^2}}(E_3 + vB_2) & B_3' &= \frac{1}{\sqrt{1 - v^2/c^2}}(B_3 - vE_2) \end{aligned} \tag{21.25}$$

in which the "primed" frame is moving with velocity v with respect to the "unprimed" frame. It is interesting to note that the first set of these relations also come directly from (21.22). For if the particle is at rest in the "primed" frame then $\mathsf{F}_N^{i'} = eE^{i'}$. But it can be shown (Rindler, 1966, pp. 98, 130) that the relativistic 3-force transforms, in this case, by

$$\mathsf{F}_N^{1'} = \mathsf{F}_N^1 \text{ and } \mathsf{F}_N^{i'} = \frac{1}{\sqrt{1 - v^2/c^2}}\mathsf{F}_N^i, \qquad i = 2, 3$$

Note that in the above discussion we did not use any properties of E or B. Classical electromagnetism requires that E and B satisfy the Maxwell equations (20.7). We can write conditions on $\mathscr{F}$ in Minkowski spacetime which contain the Maxwell equations in Lorentzian coordinates. These conditions are

$$d\mathscr{F} = 0 \tag{21.26}$$

$$d^*\mathscr{F} = \mathscr{S} \tag{21.27}$$

where $\mathscr{S} = i_Z \Omega$ in which

$$Z = c\mu\sigma \frac{\partial}{\partial x^0} + \mu J^i \frac{\partial}{\partial x^i}$$

where μ, σ, and $J = J^i \, \partial/\partial x^i$ are defined in Section 20.1, and

$$\Omega = dx^0 \wedge dx^1 \wedge dx^2 \wedge dx^3$$

the volume form of Minkowski spacetime.

If we write the component expressions for the Maxwell equations, (20.7), in a Lorentzian coordinate system, and put $B^1 = B_1 = \mathscr{F}_{23}$, $B^2 = B_2 = \mathscr{F}_{31}$, etc., then div B is the coefficient of $dx^1 \wedge dx^2 \wedge dx^3$ in $d\mathscr{F}$. With $E^1/c = E_1/c = \mathscr{F}_{01}$, etc., the components of curl $E + \partial B/\partial t$ are the coefficients of the remaining terms of $d\mathscr{F}$. The other two Maxwell equations give $d{*}\mathscr{F} = \mathscr{S}$. This is again obtained quite automatically if we write the E's and B's as $\mathscr{F}$'s, once we write explicit forms of ${*}\mathscr{F}$ and $\mathscr{S}$. These are

$${*}\mathscr{F} = B_i \, dx^0 \wedge dx^i + \frac{E_1}{c} dx^2 \wedge dx^3 - \frac{E_2}{c} dx^1 \wedge dx^3 + \frac{E_3}{c} dx^1 \wedge dx^2 \tag{21.28}$$

$$\mathscr{S} = \frac{\sigma}{c\varepsilon} dx^1 \wedge dx^2 \wedge dx^3 - \mu \, dx^0 \wedge (J_1 \, dx^2 \wedge dx^3 - J_2 \, dx^1 \wedge dx^3 + J_3 \, dx^1 \wedge dx^2) \tag{21.29}$$

(Note that the $*$ operator is not exactly the same as in Section 6.3 where it was defined for a positive definite inner product—the E_i/c terms have opposite sign. Cf. Flanders, 1963, pp. 15–16.) Thus, we obtain eqs. (21.26) and (21.27) as a spacetime formulation of Maxwell's equations.

In terms of components of $\mathscr{F}$, eqs. (21.26) and (21.27) can be written, respectively, as

$$\frac{\partial}{\partial x^\alpha} \varepsilon^{\alpha\beta\gamma\delta} \mathscr{F}_{\beta\delta} = 0 \tag{21.30}$$

where $\varepsilon^{\alpha\beta\gamma\delta}$ is the permutation symbol, Problem 4.15, and

$$\frac{\partial \mathscr{F}^{\alpha\beta}}{\partial x^\beta} = Z^\alpha \tag{21.31}$$

Just as we saw in mechanics, where the relativistic formulation produced an extra invariant in spacetime, eq. (21.13), the relativistic formulation, eqs. (21.26) and (21.27), give us information in classical electromagnetism which cannot be derived in the classical formulation. Specifically, two of the Maxwell equations can be derived from the other two.

Theorem 21.6. *If* div $B = 0$ *in all Lorentzian systems, then* $d\mathscr{F} = 0$, *and consequently* curl $E + \partial B/\partial t = 0$.

Proof. From (21.24),

$$\begin{aligned} d\mathscr{F} &= \frac{\partial E_i/c}{\partial x^\alpha} dx^\alpha \wedge dx^i \wedge dx^0 + \frac{\partial B_1}{\partial x^\alpha} dx^\alpha \wedge dx^2 \wedge dx^3 \\ &\quad + \frac{\partial B_2}{\partial x^\alpha} dx^\alpha \wedge dx^3 \wedge dx^1 + \frac{\partial B_3}{\partial x^\alpha} dx^\alpha \wedge dx^1 \wedge dx^2 \\ &= \frac{\partial E_i/c}{\partial x^\alpha} dx^\alpha \wedge dx^i \wedge dx^0 + \frac{\partial B_1}{\partial x^0} dx^0 \wedge dx^2 \wedge dx^3 \\ &\quad + \frac{\partial B_2}{\partial x^0} dx^0 \wedge dx^3 \wedge dx^1 + \frac{\partial B_3}{\partial x^0} dx^0 \wedge dx^1 \wedge dx^2 \\ &\quad + \frac{\partial B_1}{\partial x^i} dx^i \wedge dx^2 \wedge dx^3 + \frac{\partial B_2}{\partial x^i} dx^i \wedge dx^3 \wedge dx^1 \\ &\quad + \frac{\partial B_3}{\partial x_i} dx^i \wedge dx^1 \wedge dx^2 \end{aligned}$$

By the hypothesis, the sum of the last three terms is zero, so

$$\begin{aligned} d\mathscr{F} = \Bigg(&\frac{\partial E_i/c}{\partial x^\alpha} dx^\alpha \wedge dx^i + \frac{\partial B_1}{\partial x^0} dx^2 \wedge dx^3 \\ &+ \frac{\partial B_2}{\partial x^0} dx^3 \wedge dx^1 + \frac{\partial B_3}{\partial x^0} dx^1 \wedge dx^2 \Bigg) \wedge dx^0 \end{aligned} \tag{21.32}$$

and, consequently, $d\mathscr{F}(\partial/\partial x^i, \partial/\partial x^j, \partial/\partial x^k) = 0$. By our hypothesis this will be valid for all Lorentzian systems, so, since any timelike vector can be written as a sum of spacelike vectors, $d\mathscr{F}$ vanishes on all triples in each tangent space. ■

Theorem 21.7. *If the relation* div $E = \sigma/\varepsilon$ *holds in all Lorentzian systems then* $d * \mathscr{F} = \mathscr{S}$, *and, consequently,* curl $B - \partial E/\partial t = \mu J$.

Proof. From (21.28),

$$d*\mathscr{F} = -\frac{\partial B_i}{\partial x^\alpha} dx^\alpha \wedge dx^i \wedge dx^0 + \frac{\partial E_1/c}{\partial x^0} dx^0 \wedge dx^2 \wedge dx^3$$
$$+ \frac{\partial E_2/c}{\partial x^0} dx^0 \wedge dx^3 \wedge dx^1 + \frac{\partial E_3/c}{\partial x^0} dx^0 \wedge dx^1 \wedge dx^2$$
$$+ \sum \frac{\partial E_i/c}{\partial x^i} dx^1 \wedge dx^2 \wedge dx^3$$

and using (21.29),

$$d*\mathscr{F} - \mathscr{S} = -\frac{\partial B_i}{\partial x^\alpha} dx^\alpha \wedge dx^i \wedge dx^0$$
$$+ \left(\frac{\partial E_i/c}{\partial x^0} dx^2 \wedge dx^3 + \frac{\partial E_2/c}{\partial x^0} dx^3\, dx^1 + \frac{\partial E_3/c}{\partial x^0} dx^1 \wedge dx^2\right) \wedge dx^0$$
$$+ \sum \frac{\partial E_i/c}{\partial x^i} dx^1 \wedge dx^2 \wedge dx^3 - \frac{\sigma}{c\varepsilon} dx^1 \wedge dx^2 \wedge dx^3$$
$$+ \mu(J_1\, dx^2 \wedge dx^3 + J_2\, dx^3 \wedge dx^1 + J_3\, dx^1 \wedge dx^2) \wedge dx^0$$

By the hypothesis, all terms which do not have a factor of dx^0 drop out, and we can finish the argument just as in Theorem 21.6.

Finally, there are two immediate, but important, consequences of eqs. (21.26) and (21.27).

(1) $d\mathscr{F} = 0$ implies that, locally, $\mathscr{F}$ is the exterior derivative of a 1-form, in coordinates, $f\,dt + \omega_i\,dx^i$. Then from (21.24)

$$B = \operatorname{curl} \omega, \qquad E = \operatorname{grad} f - \frac{\partial \omega}{\partial t}$$

(2) $d*\mathscr{F} = \mathscr{S}$ implies conservation of charge. That is, we have $0 = d\mathscr{S} = di_Z W = L_Z W = (\operatorname{div} Z)W$ so $\operatorname{div} Z = 0$.

PROBLEM 21.10. Derive eq. (21.25).

PROBLEM 21.11. Derive eqs. (21.30) and (21.31).

PROBLEM 21.12. If $\mathscr{F}$ satisfies Maxwell's equations, $\sigma = 0$, and curl $J = 0$, then $\mathscr{F}$ is a harmonic form (Section 12.1).

PROBLEM 21.13. Use eq. (21.19) and (21.22) to describe the motion of a particle of mass m and charge e, moving in a plane and subject to a uniform magnetic field, B, acting perpendicular to the plane.

PROBLEM 21.14. Following Einstein, use eqs. (21.21) and (21.25) to derive the formula in Section 21.2 for the relativistic kinetic energy of a particle of charge e accelerated from rest.

21.4 Perfect fluids on Minkowski spacetime

We build on the Newton–Euler–Cauchy laws. In classical continuum mechanics we have a mass density, ρ, and a flow Θ (i.e., a local action, see Section 13.2), which satisfy

$$\text{(i)} \qquad \frac{d}{dt}\int_{\Theta(D,t)} \rho\Omega = 0 \qquad \text{(conservation of mass)} \tag{21.33}$$

and

$$\text{(ii)} \qquad \frac{d}{dt}\int_{\Theta(D,t)} \rho v\Omega = \int_{\Theta(D,t)} \rho f\Omega + \int_{\partial\Theta(D,t)} \sigma i_\nu \Omega \qquad \text{(equation of motion)} \tag{21.34}$$

for D a bounded regular domain of $\mathbb{R}^3$ (Section 12.2). ν is the unit normal vector on $\partial\Theta(D, t)$, v is the velocity of Θ, ρf is the force acting on a unit volume of the fluid by gravity, or an electromagnetic field (if the fluid is charged) or any other external agent, and σ is the internal stress due to pressure, viscosity, etc. The stress vector σ depends on a direction, n, at each point as well as on the point itself. Using (21.34) one can show, following Cauchy, that $\sigma^i(x^j, t, n_j) = \sigma^{ij}(x^k, t)n_j$ i.e., σ is linear in n. Finally, *the Cauchy stress tensor*, σ^{ij}, is symmetric. This property is equivalent to a second equation of motion which involves moments of the terms in (21.34).

Now we write conditions (21.33) and (21.34) as differential equations. We use a "transport theorem,"

$$\frac{d}{dt}\int_{\Theta(D,t)} f\Omega = \int_{\Theta(D,t)} \left(\frac{\partial f}{\partial t} + \operatorname{div} fv\right)\Omega \tag{21.35}$$

Applying (21.35) to (21.33) we get, in rectangular cartesian coordinates,

$$\frac{\partial \rho}{\partial t} + \frac{\partial \rho v^i}{\partial x_i} = 0 \tag{21.36}$$

Applying (21.35) to (21.34), using $\sigma^i = \sigma^{ij}n_j$, the divergence theorem (Theorem 15.5) and rectangular cartesian coordinates again, we get

$$\int_{\Theta(D,t)} \left(\frac{\partial \rho v^i}{\partial t} + \frac{\partial \rho v^i v^j}{\partial x^j} \right) \Omega = \int_{\Theta(D,t)} \rho f^i \Omega + \int_{\Theta(D,t)} \frac{\partial \sigma^{ij}}{\partial x^j} \Omega$$

and with (21.36),

$$\rho \left(\frac{\partial v^i}{\partial t} + v^j \frac{\partial v^i}{\partial x^j} \right) = \rho f^i + \frac{\partial \sigma^{ij}}{\partial x^j} \tag{21.37}$$

or

$$\frac{\partial}{\partial t}(\rho v^i) + \frac{\partial}{\partial x^j}(\rho v^i v^j - \sigma^{ij}) = \rho f^i \tag{21.38}$$

A fluid is a continuous medium in which the stress, σ^{ij}, is a function of the rate of strain,

$$d_{ij} = \frac{1}{2}\left(\frac{\partial v_i}{\partial x^j} + \frac{\partial v_j}{\partial x^i} \right)$$

such that when the latter vanishes an isotropic part remains. We sort out of the large class of possible types of fluids the mathematically simplest.

Definitions
A Newtonian fluid is one for which

$$\sigma_{ij} = -p\,\delta_{ij} - \tfrac{2}{3}\mu tr(d_{ij})\delta_{ij} + 2\mu\, d_{ij}$$

where $\sigma_{ij} = \sigma^{ij}$, p is a function called the (*mean*) *pressure*, and μ is a *coefficient of viscosity*. A *perfect fluid* is a Newtonian fluid in which $\mu = 0$ (nonviscous). Then

$$\sigma_{ij} = -p\,\delta_{ij}$$

and eq. (21.38) becomes

$$\frac{\partial}{\partial t}(\rho v^i) + \frac{\partial}{\partial x^j}(\rho v^i v^j + p\,\delta^{ij}) = \rho f^i \tag{21.39}$$

Now look at (21.36) and (21.39). It is clear that when $f^i = 0$ they can be combined into a single equation $\partial T^{\alpha\beta}/\partial x^\alpha = 0$, $\alpha, \beta = 0, 1, 2, 3$ if we make proper definitions. This suggests that we might be able to write a related spacetime equation for a related spacetime invariant which reduces to this in some limiting case.

We go to Minkowski spacetime, and consider a family of particles each parametrized by proper time. Let U be the vector field which at each point has the value of the 4-velocity $\dot{\lambda}$ of the particle. Form the tensor field $\rho\, U \otimes U/c^2$ with components

$$M^{\alpha\beta} = \rho \frac{U^\alpha U^\beta}{c^2} \tag{21.40}$$

where ρ is *the proper density* (= proper mass/proper volume). $U^0 = c/\sqrt{1 - v^2/c^2}$, $U^i = v^i/\sqrt{1 - v^2/c^2}$ in Lorentzian coordinates, by eq. (21.10), and the density in the given coordinate system will be $\bar{\rho} = \rho/(1 - v^2/c^2)$ (with notation of Section 20.5) since the mass is $m/\sqrt{1 - v^2/c^2}$ and the volume is $\sqrt{1 - v^2/c^2}$ times the proper volume. Hence, in terms of v^i and $\bar{\rho}$ (21.40) becomes

$$M^{ij} = \bar{\rho} \frac{v^i v^j}{c^2} \tag{21.41}$$

$$M^{i0} = \bar{\rho} \frac{v^i}{c} \tag{21.42}$$

$$M^{00} = \bar{\rho} \tag{21.43}$$

From (21.41) and (21.42) for $i = 1, 2, 3$,

$$\begin{aligned} \frac{\partial M^{i\alpha}}{\partial x^\alpha} &= \frac{\partial}{\partial x^j} \frac{\bar{\rho} v^i v^j}{c^2} + \frac{\partial}{\partial x^0} \frac{\bar{\rho} v^i}{c} \\ &= \frac{1}{c^2} \left[\frac{\partial}{\partial t} (\bar{\rho} v^i) + \frac{\partial}{\partial x^j} (\bar{\rho} v^i v^j) \right] \end{aligned} \tag{21.44}$$

and from (21.42) and (21.43),

$$\frac{\partial M^{0\alpha}}{\partial x^\alpha} = \frac{\partial}{\partial x^i} \frac{\bar{\rho} v^i}{c} + \frac{\partial}{\partial x^0} \bar{\rho} = \frac{1}{c} \left(\frac{\partial \bar{\rho}}{\partial t} + \frac{\partial}{\partial x^i} \bar{\rho} v^i \right) \tag{21.45}$$

Thus, comparing (21.44) and (21.39) and comparing (21.45) and (21.36), and identifying ρ and $\bar{\rho}$ in corresponding equations, we have that

$$\frac{\partial M^{\alpha\beta}}{\partial x^\beta} = 0 \tag{21.46}$$

for *a particle flow* or *an incoherent fluid*; i.e., a "fluid" in which there are no internal forces.

To get the proper description in Minkowski spacetime of a fluid with internal stresses we have to add a stress tensor $S^{\alpha\beta}$. For a perfect fluid, $\sigma_{ij} = -p\,\delta_{ij}$, we define

$$S^{\alpha\beta} = \frac{p}{c^2}\left(\frac{U^\alpha U^\beta}{c^2} + g^{\alpha\beta}\right) \tag{21.47}$$

Note that $S^{\alpha\beta}U_\beta = 0$, and in Lorentzian coordinates,

$$S^{ij} = \frac{p}{c^2}\left(\frac{v^i v^j}{c^2(1 - v^2/c^2)} + \delta^{ij}\right) \sim \frac{p}{c^2}\,\delta^{ij} \qquad \text{for small } p \text{ and } v^i$$

$$S^{i0} = \frac{p}{c^2}\,\frac{v^i}{c^2(1 - v^2/c^2)} \qquad \sim 0 \qquad \text{for small } p \text{ and } v^i$$

$$S^{00} = \frac{p}{c^2}\left(\frac{1}{1 - v^2/c^2} - 1\right) \qquad \sim 0 \qquad \text{for small } p \text{ and } v^i$$

In this approximation

$$\frac{\partial}{\partial x^\beta}(M^{\alpha\beta} + S^{\alpha\beta}) = 0 \tag{21.48}$$

reduce to the classical equations (21.36), conservation of mass and (21.39), the equation of motion, for a perfect fluid with no external forces. $M^{\alpha\beta} + S^{\alpha\beta}$ is called *the stress-energy tensor*, or *the energy-momentum tensor of a perfect fluid with no external forces.*

A case in which there are external forces can occur if the fluid is charged. We saw in the last section that a force on a charged particle can be ascribed to an electromagnetic field tensor $\mathscr{F}$ related to the force by eq. (21.23). If instead of a single charged particle we have a "cloud" of charged particles, we change eq. (21.23) to $\mathsf{F} = -\sigma i_U \mathscr{F}$, or, in coordinates,

$$\mathsf{F}^\beta = -\sigma U^\alpha \mathscr{F}_\alpha^{\cdot\beta} \tag{21.49}$$

where σ is the charge density of the fluid and F is the force per unit volume. We replace ρf^i in eq. (21.39) by the components of the 3-force per unit volume. Then by eq. (21.18) $\sqrt{1 - v^2/c^2}\,\mathsf{F}^i = \rho f^i$. To conform to the pattern developed in the previous cases, we would like the effect of ρf^i to be described by a second-order tensor such that the conservation of mass and the equation of motion are expressed as a vanishing "divergence."

Theorem 21.8. *If*

$$E^{\alpha\beta} = \frac{1}{\mu}\mathscr{F}^{\alpha}_{\cdot\nu}\mathscr{F}^{\nu\beta} - \tfrac{1}{4}g^{\alpha\beta}\mathscr{F}^{\lambda\nu}\mathscr{F}_{\lambda\nu})$$

then

$$\frac{\partial}{\partial x^{\beta}} E^{\alpha\beta} = -\sqrt{1 - v^2/c^2}\,\mathsf{F}^{\alpha} = \frac{1}{\mu} Z^{\nu}\mathscr{F}_{\nu}^{\cdot\beta}$$

Proof. Problem 21.17.

For small electromagnetic fields and velocities,

$$\frac{\partial}{\partial x^{\beta}}(M^{\alpha\beta} + S^{\alpha\beta} + E^{\alpha\beta}) = 0 \tag{21.50}$$

reduces to eqs. (21.36) and (21.39) for a charged perfect fluid, and $M^{\alpha\beta} + S^{\alpha\beta} + E^{\alpha\beta}$ is the stress-energy tensor in this case.

PROBLEM 21.15. Derive the "transport theorem," eq. (21.35). Cf., Frankel (1979, p. 60 ff).

PROBLEM 21.16. The form $h^{\alpha}_{\gamma} = g_{\beta\gamma}((U^{\alpha}U^{\beta}/c^2) + g^{\alpha\beta})$ of the tensor appearing in eq. (21.47) is a projection operator; it has the property $h^{\alpha}_{\gamma}h^{\gamma}_{\delta} = h^{\alpha}_{\delta}$, and it maps vectors in each tangent space to their projections in the subspace orthogonal to U.

PROBLEM 21.17. Use Maxwell's equations (21.26) or (21.30) and (21.31), $Z^{\alpha} = \mu\sigma\sqrt{1 - v^2/c^2}\,U^{\alpha}$, and eq. (21.49) to prove Theorem 21.8.

22

EINSTEIN SPACETIMES

In this chapter we will construct Einstein's cosmological models as represented by his field equatons, eq. (22.17) an eq. (22.27). We will include some motivation and justification for these models and one interesting general consequence.

22.1 Gravity, acceleration, and geodesics

To introduce Einstein's cosmological models we go back to the Minkowski spacetime model, and note a couple of ways in which it seems incomplete.

(1) We saw in Problem 21.13 that in Minkowski spacetime we could calculate the trajectory of a charged particle subject to an electromagnetic force field. On the other hand, the existence of a family of Lorentz frames (coordinates), and hence (M, g_η), was predicated on Newton's Principle of Inertia, affirming the existence of a frame in which free particles move on straight lines with constant speed. But in the presence of gravity there really are not any free particles, since every (massive) particle is subject to gravity, so this model excludes gravity. The fact that Einstein's special theory can deal with one of the two macroscopic force fields of physics, but not the other, gives it a certain unsatisfactory incompleteness.

(2) Einstein's principle of relativity, Theorem 20.2, while it extends the Newtonian (Galilean) principle, Theorem 20.1, to electromagnetism, has the same basic limitation as the latter; that is, it is a restricted type of relativity in the sense that physics is the same in every reference frame in a certain restricted class—reference frames moving relative to one another with constant relative velocity. Complete relativity would not require such a restriction. It would also include frames accelerated with respect to one another.

The principle of equivalence is the concept that relates these two deficiencies, and thereby simultaneously corrects both. The result, from the standpoint of the first, is called the theory of gravitation, and from the standpoint of the second is called the theory of general relativity (or, the general theory of relativity).

The principle of equivalence has many, more or less equivalent, formulations. One, roughly, is that gravity is simply a manifestation of using the wrong frame of reference.

Einstein's elevator "thought experiment" is the classical pellucid argument for the principle of equivalence. Thus, suppose a person is enclosed in a (small) elevator in a tall building. If the elevator is falling freely there will be nothing holding him/her to the floor—no gravity. Incidentally, this could not happen if gravitational mass were not the same as inertial mass—otherwise his/her wallet might move relative to him/her. Alternatively, if the elevator is not moving, he/she could interpret the force exerted on him/her by the floor to be due to an upward acceleration.

Another way to look at the principle of equivalence is to consider what happens to equations of motion in different frames of reference. For example, suppose a particle (m, γ) moves according to

$$\frac{d^2\gamma^i}{dt^2} = 0 \tag{22.1}$$

in some frame of reference. In a rotating frame of reference given by

$$x = r\cos(\vartheta + \omega t), \qquad y = r\sin(\vartheta + \omega t), \qquad z = z$$

where ω is a constant, the equations of motion of the particle are

$$\begin{aligned} \frac{d^2r}{dt^2} - r\left(\frac{d\vartheta}{dt}\right)^2 &= \omega^2 r + 2\omega r\frac{d\vartheta}{dt} \\ r\frac{d^2\vartheta}{dt^2} + 2\frac{dr}{dt}\frac{d\vartheta}{dt} &= -2\omega\frac{dr}{dt} \\ \frac{d^2z}{dt^2} &= 0 \end{aligned} \tag{22.2}$$

(where we have written dr/dt for $dr \circ \gamma/dt$, etc.) The left sides are the usual cylindrical coordinate components of acceleration, and on the right sides there are the centrifugal and Coriolis forces per unit mass arising from the rotation. Then, if we go backwards, we can eliminate the centrifugal and Coriolis forces by using the right frame of reference—just like when we let the elevator drop.

If we go to a general frame of reference, then eq. (22.1) will go to a more general form of (22.2). Since a Lorentz frame in $\mathscr{E}_0^3$ corresponds to Lorentz coordinates in (M, g_η), a general frame in $\mathscr{E}_0^3$ will correspond to general coordinates in (M, g_η) so in (M, g_η) going from (22.1) to the generalization of (22.2) corresponds to going from the equation of motion of a particle, λ,

$$\frac{d^2\lambda^\alpha}{d\tau^2} = 0 \tag{22.3}$$

to the equation of motion

$$\frac{d^2\lambda^\alpha}{d\tau^2} = -\begin{Bmatrix} \alpha \\ \beta \ \delta \end{Bmatrix} \frac{d\lambda^\beta}{d\tau}\frac{d\lambda^\delta}{d\tau} \tag{22.4}$$

of λ in general coordinates in (M, g_η). Again we can think of the right side as forces per unit mass. They could include gravity, since the acceleration of gravity is independent of mass. However, if gravity could be included on the right side of eq. (22.4) as suggested by the principle of equivalence as stated above, then we would have a model of gravity in (M, g_η) which cannot be, as we have seen.

A crucial refinement in our discussion of the elevator example is that although in the falling elevator the effects of gravity are small, they are there, and we can only eliminate gravity *completely* at *one point*. In terms of eq. (22.4), this says that there are no coordinates in which contributions to the right side due to gravity completely vanish (gravity is not a fictitious force), though in some coordinate system the right side will vanish at a point and be "small" near the point. This statement has the following two implications.

(1) If eq. (22.4) is to be valid, and the right side cannot be made to vanish, then we can not think of it simply as a coordinate transformation of eq. (22.3) in (M, g_η). But we can think of it as the equation of a geodesic in arbitrary coordinates in a 4-dimensional manifold, so the presence of gravity corresponds to the presence of some nonzero curvature: gravity is a manifestation of the curvature of a 4-dimensional manifold (or conversely).

(2) In a normal coordinate system, physics near the center of coordinates looks approximately like physics without gravity in (M, g_η). In other words, in a neighborhood of each point, physics can be approximated, by means of the exponential map, by physics in the Minkowski spacetime tangent space at the point. In particular, the 4-dimensional manifold of (1) must be Lorentzian (Section 15.1).

22.2 Gravity is a manifestation of curvature

Under the assumption that our 4-dimensional manifold is Lorentzian—the second point above—a simple calculation makes the first point above accessible to experimental verification.

Recall, first (Section 15.1), that in a Lorentzian manifold we have spacelike, timelike, and causal curves. For a future-pointing timelike curve, λ, we define *proper time*, τ, just as before in Section 21.1. Now, let us see what the terms on the right side of (22.4) look like when we go from (M, g_η) to "slightly curved"Lorentzian space and with these meanings of λ and τ.

We assume our manifold has coordinates (x^α) in terms of which $g = (\eta_{\alpha\beta} + \varepsilon h_{\alpha\beta})\, dx^\alpha \otimes dx^\beta$ and ε is small. Let $t = x^0/c$. Writing out the components $\alpha = i = 1, 2, 3$ of (22.4) in terms of t and putting x^i for $\lambda^i = x^i \circ \lambda$

we get

$$\frac{d^2x^i}{dt^2} = -\frac{dx^i}{dt}\frac{d^2t}{d\tau^2}\left(\frac{d\tau}{dt}\right)^2 - \left(\left\{ {i \atop 0\ 0} \right\} + \left\{ {i \atop 0\ 1} \right\}\frac{dx^1}{dt}\bigg/c + \cdots + \left\{ {i \atop 3\ 3} \right\}\left(\frac{dx^3}{dt}\right)^2\bigg/c^2\right)c^2 \tag{22.5}$$

Theorem 22.1. *If $h_{i\alpha}$ are independent of t, then to first order in ε and $(dx^i/dt)/c$, eq. (22.5) reduces to*

$$\frac{d^2x^i}{dt^2} = \frac{\partial}{\partial x^i}\left(\frac{c^2}{2}\,\varepsilon h_{00}\right) \tag{22.6}$$

Proof.

$$\begin{aligned}\left\{ {i \atop 0\ 0} \right\} &= \tfrac{1}{2}g^{i\alpha}\left(\frac{\partial g_{0\alpha}}{\partial x^0} + \frac{\partial g_{\alpha 0}}{\partial x^0} - \frac{\partial g_{00}}{\partial x^\alpha}\right)\\ &= \tfrac{1}{2}(\eta^{i\alpha} + \varepsilon\tilde{h}^{i\alpha})\left(2\frac{\partial}{\partial x^0}(\eta_{i\alpha} + \varepsilon h_{i\alpha}) - \frac{\partial}{\partial x^\alpha}(\eta_{00} + \varepsilon h_{00})\right)\\ &= \tfrac{1}{2}\eta^{i\alpha}\left(2\frac{\partial}{\partial x^0}\varepsilon h_{i\alpha} - \frac{\partial}{\partial x^\alpha}\varepsilon h_{00}\right) \qquad \text{to first order in } \varepsilon\\ &= -\tfrac{1}{2}\varepsilon\frac{\partial}{\partial x^i}h_{00} \qquad \text{if } h_{i\alpha} \text{ are independent of } t\end{aligned} \tag{22.7}$$

Similarly, it is seen that all the $\left\{ {i \atop \alpha\ \beta} \right\}$ have a factor of ε, and in all terms of the term on the right in (22.5), except $\left\{ {i \atop 0\ 0} \right\}$, there is at least one factor of $(dx^i/dt)/c$ so the last term on the right reduces to $-\left\{ {i \atop 0\ 0} \right\}c^2$. Finally, $(dx^i/dt)(d^2t/d\tau^2)$ is of second order in $(dx^i/dt)/c$ and

$$\begin{aligned}\left(\frac{d\tau}{dt}\right)^2 &= -\frac{1}{c^2}(\eta_{\alpha\beta} + \varepsilon h_{\alpha\beta})\frac{dx^\alpha}{dt}\frac{dx^\beta}{dt}\\ &= 1 - \varepsilon h_{00} \qquad \text{to first order}\\ &= -g_{00}\end{aligned} \tag{22.8}$$

so the first term on the right of (22.5) vanishes to first order and eq. (22.5), to that order is

$$\frac{d^2x^i}{dt^2} = -\begin{Bmatrix} i \\ 0\ 0 \end{Bmatrix} c^2 \tag{22.9}$$

Substituting (22.7) into (22.9) we get (22.6). ■

Note that in eq. (22.6) the "force" on the right side has a potential, which is a property of gravity. Writing

$$\Phi = -\frac{c^2}{2}\varepsilon h_{00} \tag{22.10}$$

we get the simple relation

$$-g_{00} = 1 + \frac{2\Phi}{c^2} \tag{22.11}$$

which accounts for the designation, in the literature, of the metric coefficients as "gravitational potentials."

We have not proved that Φ actually is the potential of gravity. However, it has been experimentally confirmed that it has a property in common with gravity—the "redshift" expressed in terms of Φ is what is observed when the values of the gravitational potential are used for Φ. A derivation of the relation between redshift and Φ follows. There are others.

Suppose we have two particles, e, (earth) and, s, (sun) and suppose that s emits photons during a time interval Δt, and these photons arrive at e. The time interval during which these photons arrive at e will be the same, Δt. The proper time in s's frame will be $\Delta\tau_s$, and in e's frame it will be $\Delta\tau_e$. The number of waves emitted by s will be the same as the number of waves received by e. Hence

$$\nu_s\,\Delta\tau_s = \nu_e\,\Delta\tau_e \tag{22.12}$$

where ν_s and ν_e are the proper frequencies of the radiation observed by s and e, respectively. Now by (22.8), (to first order)

$$\Delta\tau_s = \sqrt{-g_{00}(s)}\,\Delta t, \qquad \Delta\tau_e = \sqrt{-g_{00}(e)}\,\Delta t$$

From (22.11) and (22.12)

$$\nu_s\sqrt{1 + \frac{2\Phi_s}{c^2}} = \nu_e\sqrt{1 + \frac{2\Phi_e}{c^2}}$$

or

$$\nu_s = \nu_e \left(1 + \frac{2(\Phi_e - \Phi_s)/c^2}{1 + 2(\Phi_s/c^2)}\right)^{\frac{1}{2}}$$

or, to first order,

$$\frac{\nu_s - \nu_e}{\nu_e} = \frac{\Phi_e - \Phi_s}{c^2} \tag{22.13}$$

Note that $\Phi_e > \Phi_s$ since force $= -\text{grad}\,\Phi$ points toward s, so $\nu_s > \nu_e$, a shift toward the red. Equation (22.13) is the redshift formula. The value of the left side of (22.13) has been determined by a variety of experiments. If the values of the gravitational potential are used for Φ_e and Φ_s on the right, (22.13) is satisfied.

PROBLEM 22.1. Discuss the redshift formula. Give a derivation using conservation of energy and/or give some numerical values for the quantities in (22.13). (See Weinberg, 1972, pp. 79–86.)

22.3 The field equation in empty space

We would like to pin down the manifold which according to our assumptions in Section 22.1 describes gravity. First of all, according to the second point at the end of Section 22.1, the manifold must be Lorentzian. More specifically, we will restrict ourselves to the following class.

Definitions
A spacetime is a connected, 4-dimensional, oriented, time-oriented Lorentzian manifold with the Levi–Civita connection. Thus, at each point, p, the tangent space T_p is a Lorentzian vector space, and each future-pointing timelike vector $v \in T_p$ has *a local rest space*, the subspace of T_p orthogonal to v.

Spacetimes have been studied extensively in recent years (cf., Hawking and Ellis, 1973; Beem and Ehrlich, 1981). Causality concepts have been defined in terms of which one examines the behavior of nonspacelike curves, and the occurrence of singularities. We will refer to the general results of this theory only in the context of the specific spacetimes we will be describing.

The result of Section 22.2 suggests that with any spacetime we can associate a "force" which has at leat some of the properties of gravity. Different spacetimes will give different forces, so, since gravity should be determined uniquely by the distribution of mass in the universe, we can not expect more than one of these manifolds to actually correspond to the force of gravity. Thus, we need to impose some conditions on our manifolds. We can get a clue by looking at the conditions that actually *characterize* the gravitational field (not merely noting properties possessed by the gravitational field, as we have been doing). The conditions that characterize, i.e., determine, Newtonian gravity are that for points in *empty space* it has a potential that satisfies Laplace's equation $\Delta\Phi = 0$ and boundary conditions.

Since the curvature, R, of its connection is one of the main invariants of a pseudo-Riemannian manifold, it would seem reasonable to try to impose some condition in R. Clearly $R = 0$ will not do, for that characterizes flat space, which, on our space, gives only Minkowski spacetime. A weaker condition can be obtained by setting a contraction of R equal to zero. Note that there is no problem of choice of a particular contraction since $C_1^1 \cdot R = 0$ and $C_3^1 \cdot R = -C_2^1 \cdot R$.

Definition
The contraction

$$\mathrm{Ric} = C_2^1 \cdot R$$

of R is a (0, 2) tensor field called *the Ricci curvature tensor.*

In terms of a basis

$$\mathrm{Ric} = R^{\alpha}_{\cdot\beta\alpha\delta}\varepsilon^{\beta} \otimes \varepsilon^{\delta} \tag{22.14}$$

Its values, in terms of the values of R, are

$$\mathrm{Ric}(X, Y) = \sum_{\alpha} R(\varepsilon^{\alpha}, X, e_{\alpha}, Y) \tag{22.15}$$

(see eq. (4.16)) and its components are

$$\mathrm{Ric}_{\beta\delta} = R^{\alpha}_{\cdot\beta\alpha\delta} \tag{22.16}$$

Hence, in these terms, the suggestion above for a condition on R amounts to

$$\mathrm{Ric} = 0 \tag{22.17}$$

Theorem 22.2. Ric *has the following properties.*

(i) Ric *is symmetric.*
(ii) $\mathrm{Ric}(X, Y) = \sum_{\alpha} g_{\alpha\alpha} g(\tilde{R}(e_{\alpha}, X)\cdot Y, e_{\alpha})$ *where* $\{e_{\alpha}\}$ *is any orthonormal basis and* $g_{\alpha\alpha} = \pm 1$.
(iii) *If* u *is a unit vector,* $\mathrm{Ric}(u, u)$ *is* $\pm$ *the sum of the sectional curvatures of the planes containing* u.

Proof. Problem 22.5.

Now we will show that on our spacetime with the proposed condition $\text{Ric} = 0$, we can identify a function Φ with the property $\Delta\Phi = 0$—the quantity we associated with gravity in regions of empty space.

In our spacetime let X be the tangent field of a family of forward-facing timelike geodesics. We consider the special (but important—see Section 24.2) case in which X has orthogonal hypersurfaces.

Suppose F is a vector field on any one of these hypersurfaces with a potential, Φ. In coordinates, $F^i = g^{ij}(\partial\Phi/\partial\mu^j)$. For two nearby points, p and q, we can write $F^i(q) - F^i(p) = (\partial F^i/\partial\mu^j)\Delta\mu^j$ + higher order terms. If Y is the tangent field of a family of coordinate curves with parameter v on these hypersurfaces, then at each point along a given geodesic, γ, we have the vector

$$\lim_{\Delta v \to 0} \frac{F^i(q) - F^i(p)}{\Delta v} = \frac{\partial F^i}{\partial \mu^j} Y^j_\gamma$$

Going to the invariant form of the right side we get the vector field $g^{ik}\Phi_{,j,k}Y^j_\gamma$. Finally, suppose we think of this vector field as a mass-independent "relative force" with potential Φ. Then, if Y_γ is the "relative position vector" of points on neighboring geodesics, we expect it to satisfy the equation of motion

$$(\nabla_{\dot\gamma}\nabla_{\dot\gamma}Y_\gamma)^i = g^{ik}\Phi_{,j,k}Y^j_\gamma \tag{22.18}$$

Theorem 22.3. *If Φ satisfies (22.18), then*

$$\text{Ric}(\dot\gamma, \dot\gamma) = g^{ij}\Phi_{,i,j} \tag{22.19}$$

Proof. The (1, 1) tensor $g^{ik}\Phi_{,j,k}$ is a linear operator, $\mathscr{T}$, on the local rest space of γ at a point of γ, so by (22.18) $\mathscr{T}: Y_\gamma \mapsto \nabla_{\dot\gamma}\nabla_{\dot\gamma}Y_\gamma$. But Y_γ also satisfies the Jacobi equation (17.18) (with $\tilde{T} = 0$) *so* $\mathscr{T}: Y_\gamma \mapsto \tilde{R}(\dot\gamma, Y_\gamma)\cdot\dot\gamma$. The trace of $\mathscr{T}$ is $\text{tr}\,\mathscr{T} = \sum_{i=1}^3 \mathscr{T}(e_i, e_i)$ where e_i is any orthonormal basis of the local rest space. By the original representation of $\mathscr{T}$, $\text{tr}\,\mathscr{T} = g^{ij}\Phi_{,i,j}$, and by the last representation $\text{tr}\,\mathscr{T} = \sum_{i=1}^3 g(\tilde{R}(\dot\gamma, e_i)\cdot\dot\gamma, e_i) = \text{Ric}(\dot\gamma, \dot\gamma)$ by Theorem 22.2 (ii). ■

From Theorem 22.3 it is clear that $\text{Ric} = 0$ implies that $\Delta\Phi = 0$ independently of the choice of X. Conversely, if $\Delta\Phi = 0$ for all choices of X, then so is $\text{Ric}(\dot\gamma, \dot\gamma)$, which implies that $\text{Ric} = 0$ (Problem 22.6).

We can write out Ric in terms of the Christoffel symbols (cf., eq. (16.23)) and thence in terms of the metric coefficients. So eq. (22.17), $\text{Ric} = 0$, is a set of partial differential equations for the g's. Specifically, they are 10 second-order quasilinear partial differential equations for the 10 functions

$g_{\alpha\beta}$. Thus, from a certain point of view we have reduced out problem of finding the relation between gravity and the geometry of spacetime in empty space to solving a system of partial differential equations. While the theory of partial differential equations gives us, in this case, certain important qualitative information, and some local existence and uniqueness theorems (see Adler, Bazin, and Schiffer, 1975, Chap. 7) we get physically interesting solutions by trying to model certain simple physical situations. We will do this in the next chapter.

Once we have a solution of Ric = 0, and hence, an empty space solution for a spacetime, the motion of a particle is governed by eq. (22.4), the "equation of motion" with which we started. We have to be a little careful, though. In the approximation in Section 22.2 the potential, Φ, introduced into the right side and appearing in the redshift formula was supposed to correspond to the gravity of mass other than that of the particle itself. A massive particle could alter the geometry in such a way that it is no longer a geodesic. Hence, we interpret eq. (22.4) as the equation of motion of *a test particle*—a particle of "vanishingly small" mass.

From Ric we can form two more important invariants, which we will need in the next section and in Chapter 24.

Definitions
The *scalar curvature* is $S = g^{\beta\delta}\mathrm{Ric}_{\beta\delta}$. The *Einstein tensor* is $G = \mathrm{Ric} - \frac{1}{2}Sg$, or in components $G_{\alpha\beta} = \mathrm{Ric}_{\alpha\beta} - \frac{1}{2}Sg_{\alpha\beta}$.

Clearly G is symmetric. It has two other important properties.

Theorem 22.4. *g satisfies $G = 0$ iff g satisfies* Ric = 0.

Proof. If g satisfies Ric = 0, then g satisfies $S = 0$, so g satisfies $G = 0$. Conversely, if g satisfies $G = 0$, then g satisfies $g^{\alpha\beta}G_{\alpha\beta} = 0$, But $g^{\alpha\beta}G_{\alpha\beta} = S - \frac{1}{2}g^{\alpha\beta}g_{\alpha\beta}S = -S$, so g satisfies Ric = 0. ■

Theorem 22.5. *The divergence of G vanishes.*

Proof. From the second Bianchi identity (Problem 16.12),

$$R^{\alpha\eta}{}_{\cdot\cdot\beta\gamma,\delta} + R^{\alpha\eta}{}_{\cdot\cdot\gamma\delta,\beta} + R^{\alpha\eta}{}_{\cdot\cdot\delta\beta,\gamma} = 0$$

Put $\beta = \alpha$ and $\gamma = \eta$. Then $S_{,\delta} + R^{\alpha\eta}{}_{\cdot\cdot\eta\delta,\alpha} + R^{\alpha\eta}{}_{\cdot\cdot\delta\alpha,\eta} = 0$, or

$$S_{,\delta} - 2\,\mathrm{Ric}^{\alpha}{}_{\cdot\delta,\alpha} = 0$$

Then

$$g^{\delta\beta}S_{,\delta} - 2\,\mathrm{Ric}^{\alpha\beta}{}_{,\alpha} = 0 \qquad \text{or} \qquad (g^{\alpha\beta}S - 2\,\mathrm{Ric}^{\alpha\beta})_{,\alpha} = 0$$ ■

PROBLEM 22.2. Show that a spacetime has a timelike Killing field, X, iff it has coordinates such that $\partial g_{\alpha\beta}/\partial\mu^0 = 0$. Such a spacetime is called *stationary*.

PROBLEM 22.3. Show that a timelike vector field, U, on a spacetime has orthogonal hypersurfaces *rest spaces*, iff $\operatorname{curl} U(X, Y) = 0$ for all X, Y orthogonal to U. Such a vector field is called *irrotational*. (See definition in Section 15.1 and note that for vector fields X, Y, Z, $\operatorname{curl} X(Y, Z) = g(\nabla_Y X, Z) - g(Y, \nabla_Z X)$).

PROBLEM 22.4. Show that $C^1_1 \cdot R = 0$ and $C^1_3 \cdot R = -C^1_2 \cdot R$.

PROBLEM 22.5. Prove Theorem 22.2.

PROBLEM 22.6. Prove that if $\operatorname{Ric}(\dot{\gamma}, \dot{\gamma}) = 0$ for all timelike geodesic vector fields, then $\operatorname{Ric} = 0$ (cf., Sachs and Wu, 1977, p. 114).

22.4 Einstein's field equation (*Sitz. der Preuss Acad. Wissen.*, 1917)

In Section 21.4 we obtained several symmetric second-order tensors, *stress-energy*, or *energy-momentum* tensors, each of which is supposed to describe some type of matter/energy content of Minkowski spacetime. In the previous section, in setting up the field equation for empty space, no mention was made of any matter/energy distribution of spacetime. Recall that there we were led to $\operatorname{Ric} = 0$ by the argument that in a certain limiting sense this condition reduces to $\Delta\Phi = 0$, which is satisfied by gravity at points of empty space. In Newtonian theory, in regions filled with matter, we have to satisfy $\Delta\Phi = 4\pi\mathbf{g}\rho$, Poisson's equation with gravitational constant $\mathbf{g}$. The obvious corresponding generalization of $\operatorname{Ric} = 0$ is to replace the right side of $\operatorname{Ric} = 0$ with something representing the matter/energy distribution of the region. It would have to be a second-order symmetric tensor, and the energy-momentum tensors we defined in Section 21.4 have these properties. We now assume that these tensors represent the mass/energy content of an arbitrary spacetime, and that, as before, they are divergenceless. This suggests using G on the left rather than Ric, and assuming $G = aT$ where T stands for any stress-energy tensor including those defined in Section 21.4.

We could try to generalize. Thus, for example,

$$G = aT + bg \tag{22.20}$$

where a and b are constants, has all the properties we have required. Some justification for stopping with this much generality is given by the theorem that if T is a second-order symmetric divergenceless tensor whose components are linear in the second-order partial derivatives of the components of g with coefficients which are functions of, at most, the components of g and their first derivatives, then T must have the form $T = c_1 G + c_2 g$ where c_1 and c_2 are constants.* Thus we can, alternatively, either postulate the relation or equivalently postulate that every energy-momentum tensor has the properties of this theorem. Finally, although eq. (22.20) has been used when $-b = \Lambda$, *the cosmological constant*, is not zero, cf. Section 24.3, there are problems with physical interpretations of solutions when $T = 0$, so we are back to

$$G = aT \tag{22.21}$$

Now, in lieu of empirical verification of this postulate, perhaps the main argument for its validity is to show, as we have done in previous situations, that in the "Newtonian limit" we get the standard classical results—in this case the Poisson equation. In the process we will evaluate the constant a.

First note that from

$$\mathrm{Ric}_{\alpha\beta} - \tfrac{1}{2} g_{\alpha\beta} S = a T_{\alpha\beta}$$

we get

$$\mathrm{Ric}_{\alpha\beta} = a(T_{\alpha\beta} - \tfrac{1}{2} g_{\alpha\beta} T^{\lambda}_{\lambda}) \tag{22.22}$$

expressing Ric in terms of the stress-energy tensor. We will use (22.22) in our subsequent calculations.

Now let $g_{\alpha\beta} = \eta_{\alpha\beta} + \varepsilon h_{\alpha\beta}$ as we did in Section 22.2 and find the expansion of $\mathrm{Ric}_{\alpha\beta}$ to first order in ε. $\mathrm{Ric}_{\alpha\beta}$ is written in terms of the Christoffel symbols using eq. (16.23). To first order,

$$\mathrm{Ric}_{\alpha\beta} = \frac{\partial}{\partial x^{\gamma}} \begin{Bmatrix} \gamma \\ \alpha \ \ \beta \end{Bmatrix} - \frac{\partial^2}{\partial x^{\alpha}\, \partial x^{\beta}} \ln \sqrt{|\det(g_{\alpha\beta})|} \tag{22.23}$$

since products of the Christoffel symbols are all of second order. To first order,

$$|\det(g_{\alpha\beta})| = 1 + \varepsilon \sum_{\alpha} h_{\alpha\alpha}$$

and

$$\ln |\det(g_{\alpha\beta})| = \varepsilon \sum_{\alpha} h_{\alpha\alpha} \tag{22.24}$$

* This result was proved in a long paper by E. Cartan in 1922 and in 4 dimensions, more recently, by Lovelock (1972). Other results from which this can be easily derived are in Weyl, 1922, Appendix II and Weinberg, 1972, p. 113.

We can make the same calculations for $\left\{\begin{matrix} i \\ \alpha\ \beta \end{matrix}\right\}$ that we did in Section 22.2 for $\left\{\begin{matrix} i \\ 0\ 0 \end{matrix}\right\}$, so

$$\left\{\begin{matrix} i \\ \alpha\ \beta \end{matrix}\right\} = \frac{\varepsilon}{2}\left(\frac{\partial h_{\beta i}}{\partial x^\alpha} + \frac{\partial h_{i\alpha}}{\partial x^\beta} - \frac{\partial h_{\alpha\beta}}{\partial x^i}\right) \tag{22.25}$$

to first order in ε.

Now from (22.23), (22.24), and (22.25) and assuming, as before, that the g's are indeepndent of x^0,

$$\mathrm{Ric}_{00} = -\frac{\varepsilon}{2}\sum_i \frac{\partial^2 h_{00}}{(\partial x^i)^2} = -\frac{1}{2}\sum_i \frac{\partial^2 g_{00}}{(\partial x^i)^2}$$

Now consider the case of an incoherent fluid, eq. (21.40). Then

$$T_{00} = g_{0\alpha} g_{0\beta} T^{\alpha\beta} = T^{00} = \rho$$

to first order in ε, v^i/c and ρ. Similarly $T^\lambda_\lambda = g_{\alpha\beta}T^{\alpha\beta} = -T^{00}$ to the same approximation. From (22.22)

$$\mathrm{Ric}_{00} = a(T_{00} - \tfrac{1}{2} g_{00} T^\lambda_\lambda)$$

so

$$-\tfrac{1}{2}\sum_i \frac{\partial^2 g_{00}}{(\partial x^i)^2} = a\tfrac{1}{2}T^{00}$$

or

$$\sum_i \frac{\partial^2 g_{00}}{(\partial x^i)^2} = -aT^{00} \tag{22.26}$$

Putting $T^{00} = \rho$, putting $g_{00} + 1 = -2\Phi/c^2$ (eq. (22.11)), and using $\sum_i \partial^2\Phi/(\partial x^i)^2 = 4\pi\mathbf{g}\rho$, Poisson's equation with gravitational constant **g**, we get from (22.26)

$$a = \frac{8\pi\mathbf{g}}{c^2}$$

Further, with $h_{\alpha\beta} = 0$, $\alpha \neq \beta$ and $h_{ii} = h_{00}$, all the other equations of (22.22) are either satisfied identically or with (22.26). (Problem 22.7.) The *Einstein field equation* is

$$G = \frac{8\pi \mathbf{g}}{c^2} T \tag{22.27}$$

In the derivation of Einstein's field equation we assumed that T is divergenceless. We could, alternatively, have assumed Einstein's equation from which it follows that T is divergenceless. We insert here an interesting consequence of this property of T.

Combining all the cases we considered in Section 21.4, we have

$$T^{\alpha\beta} = \left(\rho + \frac{p}{c^2}\right)\frac{U^\alpha U^\beta}{c^2} + \frac{p}{c^2} g^{\alpha\beta} + \frac{1}{\mu}[\mathscr{F}^{\alpha}_{\cdot\nu}\mathscr{F}^{\nu\beta} - \tfrac{1}{4} g^{\alpha\beta}\mathscr{F}^{\lambda\mu}\mathscr{F}_{\lambda\mu}]$$

Then, taking the divergence, we get, using Theorem 21.8,

$$U^\alpha\left[\left(\rho + \frac{p}{c^2}\right)U^\beta\right]_{,\beta} + \left[\left(\rho + \frac{p}{c^2}\right)U^\beta\right]U^\alpha_{;\beta} + g^{\alpha\beta}p_{,\beta} + \frac{1}{\mu}Z^\beta \mathscr{F}_{\beta}^{\cdot\alpha} = 0 \tag{22.28}$$

Multiplying by U_α we get

$$-c^2\left[\left(\rho + \frac{p}{c^2}\right)U^\beta\right]_{,\beta} + U^\beta p_{,\beta} + \frac{1}{\mu}Z^\beta U_\alpha \mathscr{F}_{\beta}^{\cdot\alpha} = 0$$

Then substituting this into (22.28), and using $Z^\beta = \mu\sigma_0 U^\beta$, (Problem 21.17)

$$U^\alpha\left[\frac{1}{c^2}(U^\beta p_{,\beta} + \sigma_0 U^\beta U_\gamma \mathscr{F}_{\beta}^{\cdot\gamma})\right] + \left(\rho + \frac{p}{c^2}\right)U^\beta U^\alpha_{;\beta}$$
$$+ g^{\alpha\beta}p_{,\beta} + \sigma_0 U^\beta \mathscr{F}_{\beta}^{\cdot\alpha} = 0$$

or

$$\left(\rho + \frac{p}{c^2}\right)U^\beta U^\alpha_{;\beta} = \left(\frac{U^\alpha U^\beta}{c^2} + g^{\alpha\beta}\right)[-p_{,\beta} - \sigma_0 \mathscr{F}_{\gamma\beta}U^\gamma] \tag{22.29}$$

From Problem 21.16, eq. (22.29) says that the acceleration of a curve, γ, with tangent vector U is in the local rest space of U, and depends only on the components of pressure and electromagnetic forces in that space. In particular, γ is a geodesic iff the sum of these forces vanishes.

Equation (22.29) is frequently called an equation of motion, and in the sense indicated above it is a consequence of the field equation. However, it seems that there is less here than meets the eye. Like eq. (22.4), eq. (22.29) can be interpreted as the equation of motion of a "test particle"—one of "vanishingly small" mass. The motion of a massive particle is indeed determined by the field equation, but that is a much more complicated matter, cf., Misner, Thorne, and Wheeler (1973, pp. 471–480) and Bergmann (1976, Chap. XV).

PROBLEM 22.7. Show that for an incoherent fluid, Einstein's equation, when linearized as we did, is satisfied with $h_{\alpha\beta} = 0$ for $\alpha \neq \beta$ and $h_{ii} = h_{00}$.

23

SPACETIMES NEAR AN ISOLATED STAR

A solution of the field equations gives the geometry, i.e., the metric tensor, of the spacetime model—at least locally. The geometry is supposed to be "due to," or "determined by" a distribution of mass (energy). Thus, a solution will depend on an assumed mass distribution in the spacetime model.

We get simple models for the geometry near a massive star far from other stars if we assume that we can neglect all other mass in the universe. This is the case of a model being given by solutions of $G = 0$, the metric tensor at points not occupied by mass due to a mass distribution elsewhere.

In a model of spacetime of this kind, one would expect spacetime to have certain *general* geometrical properties. Hence, we *assume* these general properties. This has the effect of limiting to a certain class our search for solution, and simplifies our problem.

23.1 Schwarzschild's exterior solution

We look for a spacetime which is a solution of $G = 0$ with the properties:

1. It is *static*: it has a timelike Killing vector field, X, orthogonal to a family of spacelike hypersurfaces.
2. It is spherically symmetric about one of the integral curves, λ_0, of X; λ_0 is the world line of a particle, (m, λ_0), which represents the star.
3. It approaches Minkowski spacetime far enough away from λ_0.

Theorem 23.1. *Property* (1) *implies that there exists a coordinate system with a coordinate function t such that the* $g_{\alpha\beta}$ *are independent of t and* $t =$ *constant are the orthogonal hypersurfaces, so* $g_{0i} = 0$, $i = 1, 2, 3$.

Proof. Problem 23.1.

Now suppose that at each point where λ_0 intersects a $t =$ constant hypersurface we construct the geodesics on that hypersurface and go out a distance s on each. The assumption (2) of spherical symmetry says that the set of points $\{t = \text{constant},\ s = \text{constant}\}$ forms an ordinary Euclidean 2-sphere and hence it has a metric tensor of the form $r^2(d\vartheta^2 + \sin^2 \vartheta\, d\varphi^2)$. (Here and in the following formulas we write expressions like $d\vartheta \otimes d\vartheta$,

$d\varphi \otimes d\varphi$, etc., in the usual way as $d\vartheta^2$, $d\varphi^2$, etc.) Moreover, this 2-sphere is normal to the geodesics, which can be chosen to be coordinate curves. If r' is the coordinate along the geodesics then with these coordinates our spacetime has a metric tensor of the form $-A'c^2\,dt^2 + B'\,dr'^2 + r^2(d\vartheta^2 + \sin^2\vartheta\,d\varphi^2)$ and by spherical symmetry A', B', and r^2 are functions of only r'. Finally, in terms of r we get a metric tensor, g, of the form

$$-A(r)c^2\,dt^2 + B(r)\,dr^2 + r^2(d\vartheta^2 + \sin^2\vartheta\,d\varphi^2) \tag{23.1}$$

To summarize, we have inferred, from the given mass distribution in spacetime—namely, a single mass point—that spacetime should have certain "symmetries", properties 1–3 above. Using the first two of these we have inferred that such a spacetime must have local coordinates in which g is represented by eq. (23.1). Notice that we are left with only two functions of r available to satisfy the 10 equations $G_{\alpha\beta} = 0$. Thus, that a solution *exists* is remarkable—though uniqueness (which we will get by using the third property) is not surprising.

(There is another result on existence and uniqueness, Birkhoff's theorem, (Hawking and Ellis, 1973, Appendix B) which turns things around. It turns out that by assuming only the property of spherical symmetry we get *uniqueness*.)

Now with the form of g given by (23.1) in coordinates $(t, r, \vartheta, \varphi)$ we will show we can satisfy $G_{\alpha\beta} = 0$ and we will find the form of A and B. Since $G_{\beta\delta} = 0$ is equivalent to $\mathrm{Ric}_{\beta\delta} = 0$, we want to show we can satisfy the latter equation when we write the components

$$\mathrm{Ric}_{\beta\delta} = \frac{\partial}{\partial x^\alpha}\begin{Bmatrix}\alpha\\ \beta\ \delta\end{Bmatrix} - \frac{\partial}{\partial x^\delta}\begin{Bmatrix}\alpha\\ \beta\ \alpha\end{Bmatrix} + \begin{Bmatrix}\alpha\\ \varepsilon\ \alpha\end{Bmatrix}\begin{Bmatrix}\varepsilon\\ \beta\ \delta\end{Bmatrix} - \begin{Bmatrix}\alpha\\ \varepsilon\ \delta\end{Bmatrix}\begin{Bmatrix}\varepsilon\\ \beta\ \alpha\end{Bmatrix} \tag{23.2}$$

in terms of A and B. Thus, we first need to express the Christoffel symbols in terms of A and B. A straightforward procedure makes use of the definition of the Christoffel symbols in Section 16.1.

Since we will also be wanting the equations of the geodesics, an interesting indirect approach will "kill two birds with one stone" as well as considerably shorten the calculations. (This method is given in Adler, Bazin, and Schiffer, 1975. Other, more geometrical methods are given by Frankel, 1979 and by O'Neill, 1983).

To get the equations of the geodesics, we first note that they are the extremals of the energy function defined in Section 15.2. This fact was essentially demonstrated in Theorem 16.5. Now we employ the useful result from the calculus of variations that the extremals of a function of the form $\int_{u_1}^{u_2} f(x^\alpha, d\lambda^\alpha/du)\,du$ can be obtained directly from f. Specifically, a curve, λ, with parameter u is an extremal of $\int_{u_1}^{u_2} f(x^\alpha, y^\alpha)\,du$ iff it satisfies

$$\frac{d}{du}\left(\frac{\partial f}{\partial y^\alpha}\right) = \frac{\partial f}{\partial x^\alpha} \tag{23.3}$$

when $x^\alpha = \lambda^\alpha$ and $y^\alpha = \dot{\lambda}^\alpha$. These are the Euler–Lagrange equations. (To be precise, certain minimum differentiability properties have to be assumed for f). In our case, for the energy function, **E**, $f(x^\alpha, y^\alpha) = g(\dot{\lambda}, \dot{\lambda})$. For nonnull curves we can use the arc length parameter, s, and with g given by eq. (23.1) with $A = e^\xi$ and $B = e^\eta$, and with $x^0 = ct$, $x^1 = r$, $x^2 = \vartheta$, $x^3 = \varphi$,

$$f\left(x^\alpha, \frac{d\lambda^\alpha}{ds}\right) = -e^\xi c^2\left(\frac{dt}{ds}\right)^2 + e^\eta\left(\frac{dr}{ds}\right)^2 + r^2\left[\left(\frac{d\vartheta}{ds}\right)^2 + \sin^2\vartheta\left(\frac{d\varphi}{ds}\right)^2\right] \quad (23.4)$$

Substituting (23.4) into (23.3) we get for $\alpha = 0, 1, 2, 3$, respectively, with $\eta' = d\eta/d\tau$ and $\xi' = d\xi/dr$,

$$\frac{d}{ds}\left(e^\xi c^2 \frac{dt}{ds}\right) = 0 \quad (23.5)$$

$$\frac{d}{ds}\left(2e^\eta \frac{dr}{ds}\right) = -e^\xi \xi' c^2\left(\frac{dt}{ds}\right)^2 + e^\eta \eta'\left(\frac{dr}{ds}\right)^2 + 2r\left[\left(\frac{d\vartheta}{ds}\right)^2 + \sin^2\vartheta\left(\frac{d\varphi}{ds}\right)^2\right] \quad (23.6)$$

$$\frac{d}{ds}\left(r^2 \frac{d\vartheta}{ds}\right) = r^2 \sin\vartheta \cos\vartheta\left(\frac{d\varphi}{ds}\right)^2 \quad (23.7)$$

$$\frac{d}{ds}\left(r^2 \sin^2\vartheta \frac{d\varphi}{ds}\right) = 0 \quad 23.8)$$

Equations (23.5)–(23.8) are the equations of the (nonnull) geodesics of a spacetime with a metric of the form (23.1).

Now compare eqs. (23.5)–(23.8) with the components of

$$\frac{d^2\lambda^\alpha}{ds^2} + \begin{Bmatrix} \alpha \\ \beta\ \gamma \end{Bmatrix} \frac{d\lambda^\beta}{ds}\frac{d\lambda^\gamma}{ds} = 0$$

We get

$$\begin{Bmatrix} 0 \\ 1\ 0 \end{Bmatrix} = \begin{Bmatrix} 0 \\ 0\ 1 \end{Bmatrix} = \tfrac{1}{2}\xi', \qquad \text{from (23.5)}$$

$$\left.\begin{aligned} &\begin{Bmatrix} 1 \\ 0\ 0 \end{Bmatrix} = \tfrac{1}{2}\xi' e^{\xi-\eta}, \qquad \begin{Bmatrix} 1 \\ 1\ 1 \end{Bmatrix} = \tfrac{1}{2}\eta', \\ &\begin{Bmatrix} 1 \\ 2\ 2 \end{Bmatrix} = -re^{-\eta}, \qquad \begin{Bmatrix} 1 \\ 3\ 3 \end{Bmatrix} = -r\sin^2\vartheta\, e^{-\eta}, \end{aligned}\right\} \quad \text{from (23.6)}$$

$$\begin{Bmatrix} 2 \\ 1\ 2 \end{Bmatrix} = \begin{Bmatrix} 2 \\ 2\ 1 \end{Bmatrix} = \frac{1}{r}, \qquad \begin{Bmatrix} 2 \\ 3\ 3 \end{Bmatrix} = -\sin\vartheta\cos\vartheta, \qquad \text{from (23.7)}$$

and

$$\begin{Bmatrix} 3 \\ 1\ 3 \end{Bmatrix} = \begin{Bmatrix} 3 \\ 3\ 1 \end{Bmatrix} = \frac{1}{r}, \quad \begin{Bmatrix} 3 \\ 2\ 3 \end{Bmatrix} = \begin{Bmatrix} 3 \\ 3\ 2 \end{Bmatrix} = \cot\vartheta, \qquad \text{from (23.8)}$$

and all other Christoffel symbols are zero.

Now we substitute these Christoffel symbols into (23.2) using, to simplify the second and third terms of (23.2),

$$\begin{Bmatrix} \alpha \\ \beta\ \alpha \end{Bmatrix} = \frac{\partial}{\partial x^\beta} \ln \sqrt{|\det(g_{\alpha\beta})|}$$

and similarly for $\begin{Bmatrix} \alpha \\ \varepsilon\ \alpha \end{Bmatrix}$, (Problem 16.1), and we get the components, $\mathrm{Ric}_{\beta\delta}$, of the Ricci tensor for a spacetime with the metric (23.1), namely,

$$\frac{2\,\mathrm{Ric}_{00}}{e^{\xi-\eta}} = \xi'' + \tfrac{1}{2}(\xi')^2 - \tfrac{1}{2}\eta'\xi' + \frac{2\xi'}{r} \tag{23.9}$$

$$-2\,\mathrm{Ric}_{11} = \xi'' + \tfrac{1}{2}(\xi')^2 - \tfrac{1}{2}\eta'\xi' - \frac{2\eta'}{r} \tag{23.10}$$

$$-\mathrm{Ric}_{22} = (e^{-\eta}r)' - 1 + e^{-\eta}r\frac{\eta' + \xi'}{2} \tag{23.11}$$

$$\mathrm{Ric}_{33} = \mathrm{Ric}_{22}\sin^2\vartheta \tag{23.12}$$

and the remaining six components are identically zero.

Setting Ric = 0, from (23.9) and (23.10) we get $\eta' + \xi' = 0$. Using the condition (3) with which we started, that this spacetime → Minkowski spacetime for $r \to \infty$, we must have $\eta = -\xi$. Then from (23.11),

$$e^\xi = 1 - \frac{2m}{r} \qquad \text{and} \qquad e^\eta = \frac{1}{1 - (2m/r)}$$

where $2m$ is a constant of integration. Clearly, we must have $r > 2m$. Putting $A = e^\xi$ and $B = e^\eta$ into (23.1) we get "*the Schwarzschild line element*" for $r > 2m$,

$$-\left(1 - \frac{2m}{r}\right)c^2\,dt^2 + \frac{dr^2}{1 - (2m/r)} + r^2(d\vartheta^2 + \sin^2\vartheta\,d\varphi^2) \tag{23.13}$$

Finally, we can express the constant of integration, $2m$, in terms of the mass M of the star, or particle, of the model. We suppose that the quantity

$\Phi = -(c^2/2) \in h_{00} = -(c^2/2)(g_{00} + 1)$ found in the small curvature and velocity case in Section 22.2 (eq. (22.10)) *does* correspond to the Newtonian gravitational potential of M. Then if $\mathbf{g}$ is the gravitational constant, $\Phi = -(\mathbf{g}M/r)$

$$\frac{c^2}{2}(g_{00} + 1) = \frac{\mathbf{g}M}{r}$$

But according to (23.13), $g_{00} = -1 + (2m/r)$, so $(c^2/2)(2m/r) = (\mathbf{g}M/r)$ or

$$2m = \frac{2\mathbf{g}M}{c^2} \tag{23.14}$$

which is called *the Schwarzschild radius.*

It was noted above that the formula (23.13) for the Schwarzschild line element was derived under the condition that $r > 2m$. We get an idea of the magnitude of the Schwarzschild radius, $2m$, by putting $(2\mathbf{g}/c^2) = 1.48 \times 10^{-28}$ cm/g. Then $2m = 1.48 \times 10^{-28}M$. For the sun, $M = 2 \times 10^{33}$ g, so in this case the radius is approximately 3×10^5 cm. Thus, for an "ordinary" star, the Schwarzschild radius is well inside the star, so $r > 2m$ is, for practical purposes, no restriction since we have already restricted ourselves to the exterior of the star. (We will make a model of the interior of certain types of stars in the next chapter.)

The trajectory of a test particle in Schwarzschild spacetime must satisfy the geodesic equations (23.5)–(23.8) with $e^{\xi} = e^{-\eta} = 1 - (2m/r)$. Also, from eq. (23.4), for a timelike geodesic,

$$-\left(1 - \frac{2m}{r}\right)c^2\left(\frac{dt}{ds}\right)^2 + \frac{1}{1 - \dfrac{2m}{r}}\left(\frac{dr}{ds}\right)^2 + r^2\left[\left(\frac{d\vartheta}{ds}\right)^2 + \sin^2\vartheta\left(\frac{d\varphi}{ds}\right)^2\right] = -1 \tag{23.15}$$

In Newtonian mechanics, a particle in a central force field moves in a plane. This suggests that we try a solution with $\vartheta = \pi/2$, the equatorial plane. This will satisfy eq. (23.7). Then eq. (23.5) integrates, with proper time $\tau = s/c$, to

$$\left(1 - \frac{2m}{r}\right)\frac{dt}{d\tau} = l = \text{constant} \tag{23.16}$$

and eq. (23.8) integrates to

$$r^2\frac{d\varphi}{d\tau} = h = \text{constant} \tag{23.17}$$

and, with (23.16) and (23.17), eq. (23.15) reduces to

$$\frac{1}{2}\left(\frac{dr}{d\tau}\right)^2 - \frac{mc^2}{r} + \frac{h^2}{2r^2} - \frac{mh^2}{r^3} = \frac{(l^2 - 1)c^2}{2} \tag{23.18}$$

Comparing these results with the Newtonian case of planetary motion we see that eq. (23.17) is precisely the area integral, Kepler's second law, with h the angular momentum of the planet, and, with $mc^2/r = \mathbf{g}M/r$, eq. (23.18) is precisely the energy integral with the extra term mh^2/r^3 on the left. There are five qualitatively different classes of (trajectories of) particles in the exterior of an isolated star depending on the relative magnitudes of m, h, and l (Problem 23.4).

A photon is supposed to be a null geodesic, i.e., a solution of

$$\frac{d^2\lambda^\alpha}{du^2} + \begin{Bmatrix} \alpha \\ \beta \ \gamma \end{Bmatrix} \frac{d\lambda^\beta}{du}\frac{d\lambda^\gamma}{du} = 0 \tag{23.19}$$

with

$$g_{\alpha\beta}\frac{d\lambda^\alpha}{du}\frac{d\lambda^\beta}{du} = 0 \tag{23.20}$$

With f the same now as it was for a particle, except that the parameter is now u instead of s, eqs. (23.5)–(23.8) with s replaced by u describe the trajectory of a photon. In addition, in this case (23.4) is

$$-\left(1 - \frac{2m}{r}\right)c^2\left(\frac{dt}{du}\right)^2 + \frac{1}{1 - (2m/r)}\left(\frac{dr}{du}\right)^2 + r^2\left[\left(\frac{d\theta}{du}\right)^2 + \sin^2\vartheta\left(\frac{d\varphi}{du}\right)^2\right] = 0 \tag{23.21}$$

Again we put $\vartheta = \pi/2$. From (23.5) we get (23.16) with τ replaced by u, and from (23.8) we get (23.17) with τ replaced by u, and putting these in (23.21) we get

$$c^2l^2 - \left(\frac{dr}{du}\right)^2 = \left(1 - \frac{2m}{r}\right)\frac{h^2}{r^2} \tag{23.22}$$

or, using (23.17) with τ replaced by u,

$$\frac{c^2l^2}{h^2} - \frac{1}{r^2}\left(\frac{dr}{d\varphi}\right)^2 = \frac{1}{r^2}\left(1 - \frac{2m}{r}\right) \tag{23.23}$$

so that for photons, in contrast to the case for particles, the family of trajectories can be written in terms of a single parameter, c^2l^2/h^2.

PROBLEM 23.1. Prove Theorem 23.1. (Hint: The local flow $\Theta: I \times \mathscr{U} \to M$ of X induces a diffeomorphism of a neighborhood $\mathscr{V} \subset \mathscr{U}$ of a point of an orthogonal hypersurface, S, with $I \times S$, and Θ_t are local isometries of M. (See O'Neill, 1983, p. 361.)

PROBLEM 23.2. Show that if we allow A and B to be functions of t, as well as of r, in our initial form of the metric tensor, eq. (23.1), the vanishing of Ric forces us back to their independence of t.

PROBLEM 23.3. Show that $\mathrm{Ric}_{\alpha\beta} = 0$ for $\alpha \neq \beta$, for the metric tensor (23.1).

PROBLEM 23.4. Examine the possible cases of the trajectories of particles and photons in the exterior of a star in Schwarzschild's solution (cf., Robertson and Noonan, 1968, p. 242 ff, or O'Neill, 1983, p. 374 ff).

23.2 Two applications of Schwarzschild's solution

Einstein proposed three tests of general relativity. One was to predict the redshift of light from the sun or other star, which we discussed in Section (22.2). The other two are applications of Schwarzschild's solution.

(i) The precession of the orbit of Mercury

In terms of the parameter φ instead of τ, eq. (23.18) becomes

$$\frac{h^2}{2r^4}\left(\frac{dr}{d\varphi}\right)^2 - \frac{mc^2}{r} + \frac{h^2}{2r^2} - \frac{mh^2}{r^3} = \frac{(l^2-1)c^2}{2}$$

Then, putting $r = 1/\omega$ and differentiating with respect to φ, we get either

$$\frac{d\omega}{d\varphi} = 0 \tag{23.24}$$

or

$$\frac{d^2\omega}{d\varphi^2} + \omega = \frac{mc^2}{h^2} + 3m\omega^2 \tag{23.25}$$

The trajectories of (23.24) are $r = \text{const}$. To find the trajectories described by (23.25), we solve (23.25) by a perturbation method, since $3m\omega^2$ is small compared with mc^2/h^2. Hence we write (23.25) as

$$\frac{d^2\omega}{d\varphi^2} + \omega = C + \frac{\varepsilon}{C}\omega^2$$

where $C = mc^2/h^2$ and $\varepsilon = 3m^2c^2/h^2$ and we look for a solution in the form $\omega(\varphi, \varepsilon) = \omega_0 + \omega_1\varepsilon + \omega_2\varepsilon^2 + \cdots$.

Then

$$\frac{d^2\omega_0}{d\varphi^2} + \omega_0 = C \tag{23.26}$$

$$\frac{d^2\omega_1}{d\varphi^2} + \omega_1 = \frac{\omega_0^2}{C} \tag{23.27}$$

From (23.26), $\omega_0 = C + D\cos(\varphi + \delta)$, where D and δ are arbitrary constants, and putting this into (23.27) with $\delta = 0$ and solving we get a particular solution,

$$\omega_1 = C + \frac{D^2}{2C} + D\varphi\sin\varphi - \frac{D^2}{6C}\cos 2\vartheta$$

To first order

$$\frac{1}{r} = \omega = \omega_0 + \varepsilon\omega_1 = C + \varepsilon\left(C + \frac{D^2}{2C}\right) + D\cos\varphi + \varepsilon\left(D\varphi\sin\varphi - \frac{D^2}{6C}\cos 2\varphi\right)$$

or

$$\frac{1}{r} = C + D\cos(\varphi - \varepsilon\varphi) + \varepsilon\left(C + \frac{D^2}{2C} - \frac{D^2}{6C}\cos 2\varphi\right)$$

Now if ε were zero, the trajectory would be a conic. The effect of the last term is to perturb the radial distance periodically. The ε in the middle term destroys the periodicity of the orbit. The maximum value of this term (giving a minimum for $r = 1/\omega$) occurs now not for φ a multiple of 2π, but for $\varphi - \varepsilon\varphi$ a multiple of 2π. Thus, two successive minima of r occur at $\varphi_1 - \varepsilon\varphi_1 = 2\pi n$ and $\varphi_2 - \varepsilon\varphi_2 = 2\pi n + 2\pi$ or $(\varphi_2 - \varphi_1)(1 - \varepsilon) = 2\pi$, or the change in angle is approximately $\Delta\varphi = 2\pi(1 + \varepsilon)$.

Finally, putting $\varepsilon = 3m^2c^2/h^2$, and using the Newtonian approximation $2m = 2\mathbf{g}M/c^2$, we get an advance in the perihelion of the orbit of $6\pi(\mathbf{g}^2M^2/h^2c^2)$ radians. All these quantities can be determined for the case of Mercury moving in the sun's gravitational field—we get an advance of 42–43 seconds of arc/century. Cf., Problem 23.6.

(ii) The bending of a light ray passing the sun

Proceeding as before in (i) starting with (23.23) instead of (23.18),

$$\frac{d^2\omega}{d\varphi^2} + \omega = 3m\omega^2 \tag{23.28}$$

instead of (23.25). $3m\omega^2$ is evidently small compared with ω, so putting $\varepsilon = 3m$, and $\omega(\varphi, \varepsilon) = \omega_0 + \omega_1\varepsilon + \cdots$, we get

$$\frac{d^2\omega_0}{d\varphi^2} + \omega_0 = 0 \tag{23.29}$$

$$\frac{d^2\omega_1}{d\varphi^2} + \omega_1 = \omega_0^2 \tag{23.30}$$

Equation (23.29) gives

$$\omega_0 = D \cos \varphi \tag{23.31}$$

and putting this into (23.30) we get

$$\omega_1 = \tfrac{2}{3}D^2 - \tfrac{1}{3}B^2 \cos^2 \varphi$$

So, to first order, the trajectory is

$$\frac{1}{r} = D \cos \varphi + \frac{\varepsilon D^2}{3}(2 - \cos^2 \varphi).$$

(Fig. 23.1). Note that if $\varepsilon = 0$, then the trajectory is a straight line. With the extra term we get a curve symmetric about $\varphi = 0$ and as r goes to infinity, φ goes to a solution of

$$0 = D \cos \varphi + \frac{\varepsilon D^2}{3}(2 - \cos^2 \varphi)$$

from which, to first order in ε,

$$\cos \varphi = -\frac{2\varepsilon}{3} D = -2mD$$

so $\varphi = \pm(\pi/2 + 2mD)$ and the angle between the asymptotes is $4mD = (4\mathbf{g}M/c^2)D$. From (23.31). $1/D$ is the minimum value of r in the solution of (23.29). For a light ray which just grazes the sun this gives about 2 seconds of arc.

PROBLEM 23.5. Solve eq. (23.25) without the second-degree term on the right to get the Newtonian orbits

$$r = \frac{h^2/mc^2}{1 + (Ah^2/mc^2) \cos \varphi}$$

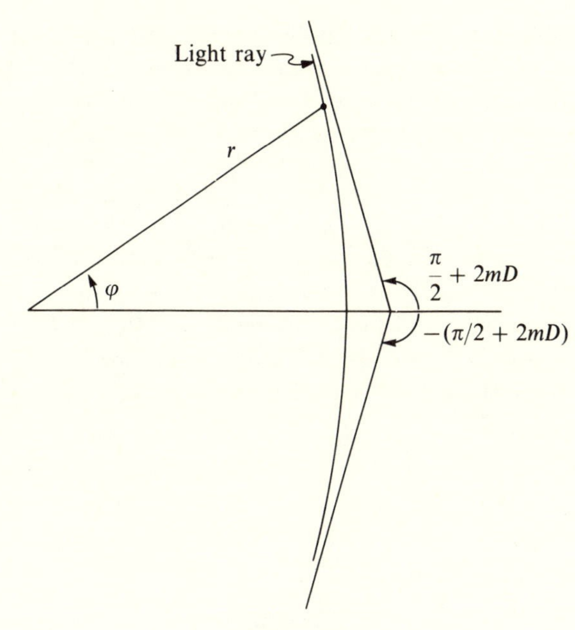

FIGURE 23.1

where $A \geq 0$ is a constant of integration. Thus, the orbits are ellipses with eccentricity Ah^2/mc^2 (Kepler's first law), and from this equation h^2 is obtained from the geometry (eccentricity and principal axes) of the orbit.

PROBLEM 23.6. Look up astronomical data needed in the formulas given for the perihelion advance and the bending of light rays and confirm the values given in our applications.

23.3 The Kruskal extension of Schwarzschild spacetime

In Section 23.1 we obtained a Ricci flat spacetime for $r > 2m$, where m is a small positive number. The metric tensor (23.13) gives a Ricci flat spacetime for $0 < r < 2m$, also. We can show this most easily by putting $A = -e^{\xi}$ and $B = -e^{\eta}$ instead of the substitution above eq. (23.4). Then $\mathrm{Ric}_{22} = 0$ becomes $(e^{-\eta}r)' + 1 = 0$, so $e^{-\eta} = -1 + 2m/r = e^{\xi}$, and we get the same form as before.

Equations (23.5), (23.7), and (23.8) are the same as before, and eq. (23.15) is still valid for a timelike geodesic, so a test particle still has the equations (23.16)–(23.18), and a photon has the equations (23.16) and (23.17) with u in place of τ and eq. (23.22).

Thus, we have a spacetime on $0 < r < 2m$ and a spacetime on $r > 2m$. Alternatively, we can say we have a spacetime on $r > 0$ with a singularity at $r = 2m$. As we have seen in Section 23.1, we have a physical interpretation only for r considerably greater than $2m$. Nevertheless, the occurrence of a singularity provoked considerabl interest in its significance. With that in mind, and with a certain prescience, investigators asked what would happen if there *were* an isolated body small and massive enough so that it lay inside the Schwarzschild radius. That there actually *are* such bodies, the end states of gravitational collapse, or "black holes," has been generally accepted only very recently. We will see that while $r = 2m$ is not a "singularity" (of a slightly modified spacetime), it *does* have an important physical significance.

To see what happens near $r = 2m$ we can restrict ourselves to examining the behavior of particles and photons moving radially; that is, with ϑ and φ constant and, hence, $h = 0$. The equations governing the radial motion of photons are, from eq. (23.16), $(1 - 2m/r)\, dt/du = l$, and from eq. (23.22), $(dr/du)^2 = c^2 l^2$, and from these two we get

$$\pm c \frac{dt}{dr} = \frac{1}{1 - (2m/r)} \tag{23.32}$$

so

$$\pm ct = r + 2m \ln|r - 2m| + \text{constant} \tag{23.33}$$

From Fig. 23.2 it is evident that for $r > 2m$ no timelike curve (in particular, no particle) can approach $r = 2m$ in finite time, t. In physical

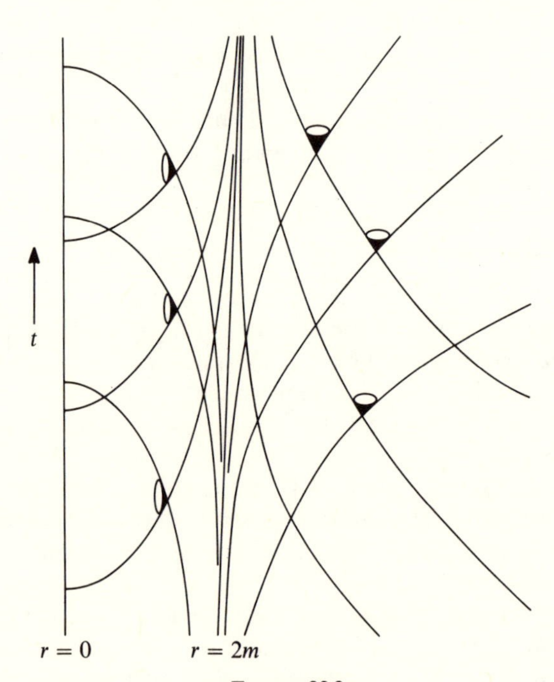

FIGURE 23.2

terms, for an observer at rest with respect to the small massive isolated body and watching a particle fall toward the body, the particle will never pass $r = 2m$. (Also, if in the discussion of the redshift at the end of Section 22.2, s and e are, respectively, the particle and the observer, the relation

$$\frac{\nu_s - \nu_e}{\nu_e} = \frac{\sqrt{-g_{00}(s)} - \sqrt{-g_{00}(e)}}{\sqrt{-g_{00}(s)}}$$

shows that the observer will see a dramatic red shift of the light coming from the particle as it approaches $r = 2m$.) This result can also be gotten analytically from the differential equation for r as a function of t obtained from (23.16) and (23.18). From that equation we get for timelike curves that also for $r < 2m$ as $r \to 2m$, $t \to \pm\infty$. Since for $r < 2m$ the r coordinate curves are timelike and the t coordinate curves are spacelike, in this region a particle travels between a spacelike coordinate value, t, of plus or minus infinity and a finite value during a coordinate time interval $0 < r < 2m$.

Equation (23.16) gives us a clue that the fact that for an observer at rest with respect to the body a particle never crosses $r = 2m$ could be a consequence of what happens to the observer's time more than of what happens to the particle. Hence, instead of examining r as a function of t as we did above, we go to eq. (23.18) and examine r as a function of τ. A straightforward integration for $h = 0$ shows that for all l, there are solutions of (23.18) for which r decreases continuously with increasing τ and which reach $r = 0$ at a finite τ. In particular, there is no problem at $r = 2m$.

Moreover, if we compute $\det(g_{\alpha\beta})$, the sectional curvatures, and other invariants of g and R, we find no problem at $r = 2m$, all of which suggests trying a coordinate change.

Using the minus sign in eq. (23.33) we get the incoming null curves, and clearly these will be straightened out if we put

$$c\bar{t} = ct + r + 2m \ln|r - 2m|, \qquad \bar{r} = r \tag{23.34}$$

(Fig. 23.3). With these coordinates, (23.13) becomes

$$-\left(1 - \frac{2m}{\bar{r}}\right)c^2\, d\bar{t}^2 + 2\, d\bar{t}\, d\bar{r} + \bar{r}^2(d\vartheta^2 + \sin^2\vartheta\, d\varphi^2) \tag{23.35}$$

The singularity is gone. The null curves are $\bar{t}$-constant and the solutions of

$$\frac{d\bar{t}}{d\bar{r}} = \frac{2}{c[1 - (2m/\bar{r})]} \tag{23.36}$$

(cf., eq. (23.32)).

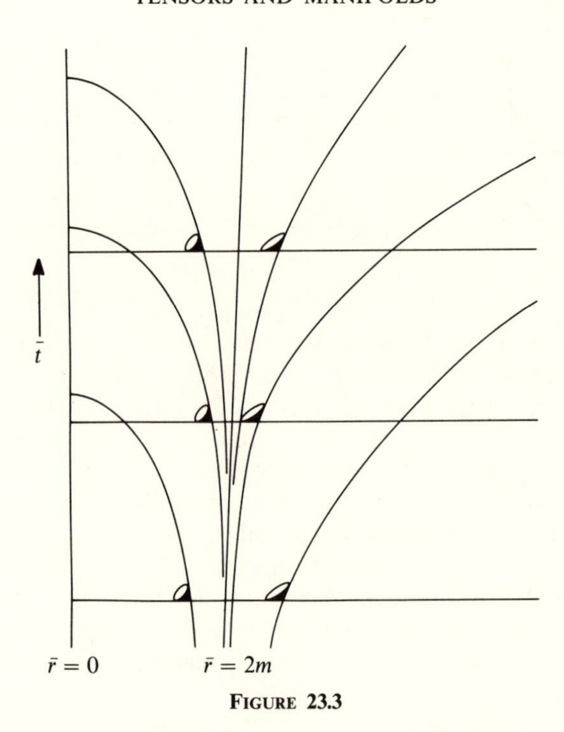

FIGURE 23.3

Equation (23.35) can be considered to be a coordinate representation of the metric tensor of a spacetime into which we have isometrically imbedded the two pieces of Schwarzschild spacetime. Similarly, by choosing the plus sign in eq. (23.33) instead of the minus sign, the outgoing null curves can be straightened out and we get a spacetime without a singularity and containing the two Schwarzschild pieces. These two spacetimes are called *Eddington–Finkelstein spacetimes.* For more details, including their motivation for a further extension, *the Kruskal* (or *Kruskal–Szekeres*) *spacetime,* see either Hawking and Ellis (1973, pp. 150–156), or Misner, Thorne, and Wheeler (1973, pp. 828–832).

The Eddington–Finkelstein spacetimes have the undesirable feature that $r = 2m$ is a null curve. We can avoid that if we can find coordinates (or an isometry) for which the metric tensor has the form

$$F(\bar{t}, \bar{r})(-d\bar{t}^2 + d\bar{r}^2) + G(\bar{t}, \bar{r})(d\vartheta^2 + \sin^2 \vartheta \, d\varphi^2) \tag{23.37}$$

Kruskal found a transformation $\psi\colon (\bar{t}, \bar{r}) \mapsto (\psi_1(\bar{t}, \bar{r}), \psi_2(\bar{t}, \bar{r})) = (t, r)$ from a region of $\mathbb{R}^2$ symmetric about (0, 0) with $\psi(\bar{t}, \bar{r}) = \psi(-\bar{t}, -\bar{r})$ onto $\mathbb{R} \times \mathbb{R}^+$, and such that $F(\bar{t}, \bar{r}) = f^2(\psi_2(\bar{t}, \bar{r})) = f^2(r)$ and $G(\bar{t}, \bar{r}) = \psi_2^2(\bar{t}, \bar{r}) = r^2$. A fairly straightforward approach is to solve the partial differential equations

$$g_{\alpha\beta} = \frac{\partial \bar{\mu}_\gamma}{\partial \mu^\alpha} \frac{\partial \bar{\mu}_\delta}{\partial \mu^\beta} \bar{g}_{\gamma\delta}$$

for ψ^{-1} (locally). (See Adler, Bazin and Schiffer, 1975, pp. 226–230.) The result is

$$\bar{t} = \bar{t}_{\mathrm{I}}(t, r) = \sqrt{\frac{r}{2m} - 1}\, e^{r/4m} \sinh \frac{ct}{4m}$$

$$\bar{r} = \bar{r}_{\mathrm{I}}(t, r) = \sqrt{\frac{r}{2m} - 1}\, e^{r/4m} \cosh \frac{ct}{4m}$$

for $r \geq 2m$

$$\bar{t} = \bar{t}_{\mathrm{II}}(t, r) = \sqrt{1 - \frac{r}{2m}}\, e^{r/m} \cosh \frac{ct}{4m}$$

$$\bar{r} = \bar{r}_{\mathrm{II}}(t, r) = \sqrt{1 - \frac{r}{2m}}\, e^{r/4m} \sinh \frac{ct}{4m}$$

for $r < 2m$, and

$$f^2(r) = \frac{32m^3}{r} e^{-r/2m}$$

for $r > 0$. Note that the region $-\infty < t < \infty, r > 0$ maps onto points of the $\bar{t}, \bar{r}$ plane with $\bar{t} + \bar{r} > 0$, and $\bar{t} + \bar{r} = 0$ corresponds to $t = -\infty$, and $\bar{t} = \bar{r}$ corresponds to $t = +\infty$.

By the symmetry with respect to the origin,

$$\bar{t} = \bar{t}_{\mathrm{I'}}(t, r) = -\bar{t}_{\mathrm{I}}(t, r)$$

$$\bar{r} = \bar{r}_{\mathrm{I'}}(t, r) = -\bar{r}_{\mathrm{I}}(t, r)$$

for $r \geq 2m$, and

$$\bar{t} = \bar{t}_{\mathrm{II'}}(t, r) = -\bar{t}_{\mathrm{II}}(t, r)$$

$$\bar{r} = \bar{r}_{\mathrm{II'}}(t, r) = -\bar{r}_{\mathrm{II}}(t, r)$$

for $r < 2m$ with the same f, is also a solution.

The entire Kruskal extension is shown in Fig. 23.4. The $r =$ const. curves are hyperbolas and the $t =$ const. curves are straight lines in the $(\bar{t}, \bar{r})$ plane. From (23.27) we see that the null lines in a $\vartheta =$ const., $\varphi =$ const. section have slope ± 1. A ray of light or a particle starting at p in Region I, "the Normal Schwarzschild region," heading for $r = 0$ has to pass $t = \infty$ as we

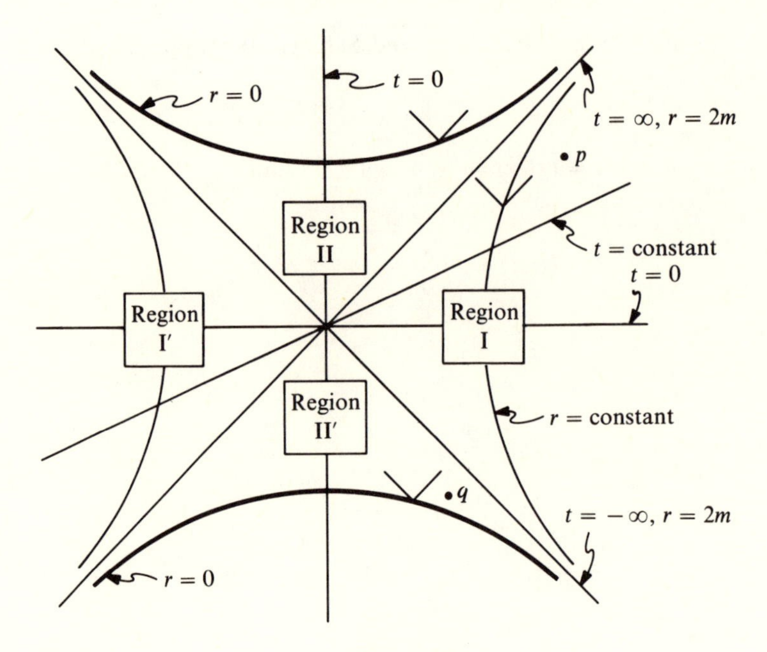

FIGURE 23.4

have already seen. We make the additional observation that every ray of light, or particle, starting at a point q inside the Schwarzschild radius has to end up in either Region I or I′, but a ray of light or particle going into Region I from point q will never be seen by an observer in Region I, since it has to go through $t = -\infty$.

PROBLEM 23.7. Show that the sectional curvatures of Schwarzschild spacetime are all proportional to $1/r^3$. (But they are not all the same.)

23.4 The field of a rotating star

In recent years, generalizations of the Schwarzschild–Kruskal solution have been found. These give fields due to charged massive bodies, rotating massive bodies, and charged rotating massive bodies (Kerr–Newman geometry; Misner, Thorne, and Wheeler, 1973, p. 877 ff).

The Kerr solution for a rotating massive body has a coordinate representation for the metric tensor (in Boyer and Lindquist coordinates),

$$-\left(1 - \frac{2mr}{r^2 + a^2 \cos^2 \vartheta}\right) c^2\, dt^2 + \frac{r^2 + a^2 \cos^2 \vartheta}{r^2 - 2mr + a^2}\, dr^2 + (r^2 + a^2 \cos^2 \vartheta)\, d\vartheta^2$$

$$- 2\frac{2mar \sin^2 \vartheta}{r^2 + a^2 \cos^2 \vartheta}\, c\, dt\, d\varphi + \left[r^2 + a^2 + \frac{2ma^2 r \sin^2 \vartheta}{r^2 + a^2 \cos^2 \vartheta}\right] \sin^2 \vartheta\, d\varphi^2 \quad (23.38)$$

One can verify that this satisfies $\text{Ric} = 0$. We can note also that (1) when $a = 0$ (23.38) reduces to the Schwarzschild solution (23.13); (2) (23.38) is time-independent and axially symmetric; (3) if the signs of a and φ are simultaneously changed, (23.38) does not change, which suggests that a corresponds to a rotation; and (4) the cross-product term $dt\,d\varphi$ which appears corresponds to what we get with a rotating coordinate system in flat space.

We will skip (1) a derivation of (23.38) and (2) further justification for identifying the parameter a with a rotation. Both involve considerable calculations, the latter involving a comparison with an approximate solution (see Adler, Bazin, and Schiffer, 1975, Chap. 7). We will content ourselves with observing a sort of generalization of the Schwarzschild radius.

For $a^2 > m^2$ the coefficients in (23.38) have no singularities, so this spacetime could model a rotating star. For $a^2 < m^2$ a singularity occurs at

$$r^2 - 2mr + a^2 = 0 \tag{23.39}$$

This represents two hyperspheres

$$r_+ = m + \sqrt{m^2 - a^2} \qquad \text{and} \qquad r_- = m - \sqrt{m^2 - a^2}$$

both with $r < 2m$, so (23.38) could still model a rotating star. But, as in the Schwarzschild case, it can also model a black hole. The Kerr solution can be extended to a spacetime which contains it without the singularity, but, like in the Schwarzschild case, the hypersurfaces where the singularity occurred, the two hyperspheres, still have physical significance.

Note, first of all, that the hypersurface

$$r^2 - 2mr + a^2 \cos^2 \vartheta = 0 \tag{23.40}$$

where the coefficient of dt^2 vanishes is now, in contrast to the Schwarzschild metric, (23.13), different from (23.39). At the Schwarzschild radius several things happened:

(i) The t coordinate curves changed from timelike to spacelike.
(ii) The redshift became infinite.
(iii) Photons and particles could only travel inward.

Now, clearly the first two phenomena occur on (23.40). We will show that now the third phenomenon occurs on (23.39) and another interesting phenomenon occurs on (23.40).

Theorem 23.2. *At each point of the hypersurface $r^2 - 2mr + a^2 = 0$ the tangent space of the hypersurface has a one-dimensional null subspace and all* *the other vectors are spacelike; i.e., the tangent space of* $r^2 - 2mr + a^2 = 0$ *is tangent to the null cone.*

Proof. Let $f(r) = r^2 - 2mr + a^2$. Then grad f is a normal vector to $f(r) = 0$. In coordinates,

$$g(\text{grad } f, \text{grad } f) = g^{ij} \frac{\partial f}{\partial \mu^i} \frac{\partial f}{\partial \mu^j}$$

so in our case

$$g(\text{grad } f, \text{grad } f) = \frac{r^2 - 2mr + a^2}{r^2 + a^2 \cos^2 \vartheta} (2r - 2m)^2$$

and on $f(r) = 0$ this vanishes. Thus, we have a normal vector which satisfies the criterion of Problem 5.25 for our conclusion. ■

Corollary. *A particle or photon can cross (23.29) in only one direction.*

Now consider the hypersurface (23.40). It has two parts,

$$r = m \pm \sqrt{m^2 - a^2 \cos^2 \vartheta}$$

We will denote the outer part by r_∞.

Definitions
r_∞ is called *the stationary* (or, *static*) *limit surface.* r_+ is called *an event horizon*, and the region between them is called *the ergosphere* (Fig. 23.5).

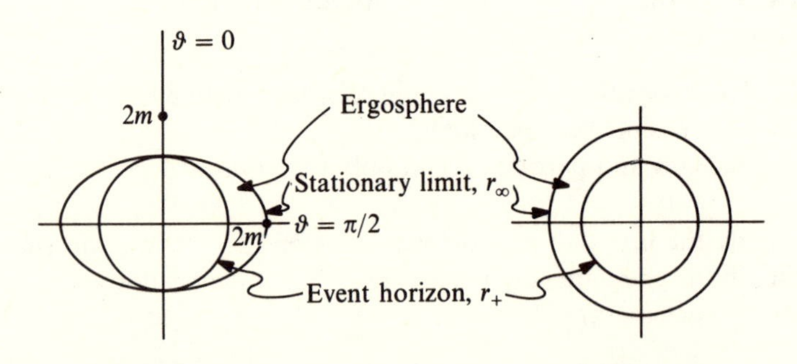

FIGURE 23.5

Consider curves in this spacetime parametrized by t in the coordinate slice r and ϑ constant. They will have tangent vectors $\partial/\partial t + (d\varphi/dt)\,\partial/\partial\varphi$. They can be visualized as coming out of the paper from the t and ϑ constant slice and lying on circular cylinders centered at the origin.

Theorem 23.3. *At each point of the ergosphere there are timelike and null curves parametrized by t in the coordinate slice r and ϑ constant, and for all such curves $d\varphi/dt > 0$.*

Proof. The condition the tangent vectors, v, of our curves have to satisfy is $g(v, v) < 0$ for timelike and $g(v, v) = 0$ for null curves. In our case this condition is

$$g\left(\frac{\partial}{\partial t} + \frac{d\varphi}{dt}\frac{\partial}{\partial\varphi}, \frac{\partial}{\partial t} + \frac{d\varphi}{dt}\frac{\partial}{\partial\varphi}\right) = g_{00} + 2g_{03}\frac{d\varphi}{dt} + g_{33}\left(\frac{d\varphi}{dt}\right)^2 \tag{23.41}$$

is negative or zero, respectively. The discriminant, $g_{03}^2 - g_{00}g_{33}$, of the quadratic (23.41) is positive for $r > r_+$, Problem 23.8. Since $g_{33} > 0$, the parabola $y = g_{00} + 2g_{03}x + g_{33}x^2$ opens upward, so the required condition is satisfied. For $r > r_\infty$, $g_{00} < 0$ so there are both negative and positive possible values for $d\varphi/dt$. In particular, we can have $d\varphi/dt = 0$. However, for $r < r_\infty$, since $g_{00} > 0$, all possible values of $d\varphi/dt$ must be positive. In particular, $d\varphi/dt$ cannot be zero. ■

In physical terms, the fact that in the ergosphere $d\varphi/dt$ cannot be zero is interpreted to mean that particles in the ergosphere cannot be at rest with respect to the rotating body, but are dragged along with its rotation. There is a further argument that this phenomenon could account, at least in part, for the source of the large energy detected in some stellar observations.

PROBLEM 23.8. Show that the discriminant of (23.41) is positive for $r > r_+$.

24

NONEMPTY SPACETIMES

We will obtain two types of solutions of Einstein's equation, (22.27); a model of the interior of a star, and a class of models of the universe as a whole, a class of cosmologies.

24.1 Schwarzschild's interior solution

To make a model of the interior of a star, we start with Einstein's equation in the form (22.22),

$$\mathrm{Ric}_{\alpha\beta} = a(T_{\alpha\beta} - \tfrac{1}{2}T^{\lambda}_{\lambda}g_{\alpha\beta}) \tag{24.1}$$

with $a = 8\pi g/c^2$.

We assume our medium is a perfect fluid with no external forces. Then from eqs. (21.40) and (21.47),

$$T_{\alpha\beta} = \rho\,\frac{U_\alpha U_\beta}{c^2} + \frac{p}{c^2}\left(\frac{U_\alpha U_\beta}{c^2} + g_{\alpha\beta}\right) \tag{24.2}$$

Recall that the velocity field, U, by definition, Section 21.4, is the field of a particle flow, so, since for a particle, λ, $g(\dot{\lambda}, \dot{\lambda}) = -c^2$, we get $g_{\alpha\beta}U^\alpha U^\beta = -c^2$, and from (24.2), $T_{\alpha\beta}U^\beta = -\rho U_\alpha$.

Now we can take U as a coordinate vector field, $\partial/\partial x^0$. Then $U^i = 0$, $i = 1, 2, 3$, the fluid particles are at rest with respect to such a coordinate system, or the coordinates are "comoving." Further, in this case we have $g_{00}(U^0)^2 = -c^2$ and $U_0^2 = -g_{00}c^2$, so

$$T_{00} = -\rho g_{00}, \qquad T^0_0 = -\rho, \qquad T^1_1 = T^2_2 = T^3_3 = \frac{p}{c^2}$$

and

$$T^i_\alpha = 0, \qquad \alpha \neq i, \qquad \alpha = 0, 1, 2, 3, \qquad i = 1, 2, 3$$

For the model of a star we assume our stress-energy tensor, T, is spherically symmetric, so the resulting spacetime would be expected to be spherically symmetric also, as in Schwarzschild's exterior solution.

We assume further that the velocity field, U, is irrotational so that, by Problem 22.3 it has orthogonal hypersurfaces, rest spaces.

Finally, as we noted in Problem 23.2 we do not need the additional assumption that U is a Killing field, i.e., $g_{\alpha\beta}$ are independent of t, but just as before it will turn out that U will have to be one, so we start with a metric tensor of the form (23.1), as for Schwarzschild's exterior solution:

$$g = -e^{\xi}c^2\,dt^2 + e^{\eta}\,dr^2 + r^2(d\vartheta^2 + \sin^2\vartheta\,d\varphi^2) \tag{24.3}$$

where ξ and η are functions of only r.

The components of Ric will be given by eqs. (23.9)–(23.12) just as before. By the orthogonality, $U_i = 0$, and by the results below eq. (24.2),

$$T_{00} - \tfrac{1}{2}g_{00}T^{\lambda}_{\ \lambda} = \frac{e^{\xi}}{2}\left(\rho + \frac{3p}{c^2}\right) \tag{24.4}$$

$$T_{11} - \tfrac{1}{2}g_{11}T^{\lambda}_{\ \lambda} = \frac{e^{\eta}}{2}\left(\rho - \frac{p}{c^2}\right) \tag{24.5}$$

$$T_{22} - \tfrac{1}{2}g_{22}T^{\lambda}_{\ \lambda} = \frac{1}{2}\left(\rho - \frac{p}{c^2}\right)r^2 \tag{24.6}$$

With eqs. (24.4)–(24.6) and the components of Ric given by eqs. (23.9)–(23.11), the $(0, 0)$, $(1, 1)$, and $(2, 2)$ components of the field equation (24.1) are, respectively,

$$\frac{e^{\xi-\eta}}{2}\left[\xi'' + \tfrac{1}{2}(\xi')^2 - \tfrac{1}{2}\xi'\eta' + \frac{2\xi'}{r}\right] = a\frac{e^{\xi}}{2}\left(\rho + \frac{3p}{c^2}\right) \tag{24.7}$$

$$-\frac{1}{2}\left[\xi'' + \tfrac{1}{2}(\xi')^2 - \tfrac{1}{2}\xi'\eta' - \frac{2\eta'}{r}\right] = a\frac{e^{\eta}}{2}\left(\rho - \frac{p}{c^2}\right) \tag{24.8}$$

$$1 - (re^{-\eta})' - re^{-\eta}\frac{\xi' + \eta'}{r} = \frac{a}{2}\left(\rho - \frac{p}{c^2}\right)r^2 \tag{24.9}$$

We can combine these conditions in various ways. Thus, adding $e^{\xi-\eta}$ times the $(1, 1)$ component and the $(0, 0)$ component, we get

$$a\left(\rho + \frac{p}{c^2}\right) = e^{-\eta}\frac{\eta' + \xi'}{r} \tag{24.10}$$

If we solve (24.10) and the $(2, 2)$ component for ρ and p, the density and pressure, we get

$$a\rho = \frac{1}{r^2} - e^{-\eta}\left(\frac{1}{r^2} - \frac{\eta'}{r}\right) \tag{24.11}$$

and

$$a\frac{p}{c^2} = -\frac{1}{r^2} + e^{-\eta}\left(\frac{1}{r^2} + \frac{\xi'}{r}\right) \tag{24.12}$$

and we see, in particular, that ρ, and p are functions of only r. To eqs. (24.10)–(24.12) we can adjoint the condition $T^{\alpha\beta}{}_{,\beta} = 0$. From eq. (22.29) we get

$$\frac{p'}{c^2} = -\frac{\xi'}{2}\left(\rho + \frac{p}{c^2}\right) \tag{24.13}$$

At this point, in order to proceeds, we realize that just as in ordinary fluid mechanics we need another piece of physical information, namely, *an equation of state* for the medium. If we know ρ as a function of p, then we can integrate (24.13) and get

$$-\frac{\xi}{2} = \int \frac{dp}{c^2\left(\rho + \dfrac{p}{c^2}\right)} + \text{const.} \tag{24.14}$$

With ρ as a function of p and ξ as a function of p, we can, in principle, find p as a function of r from (24.13) and then find ρ, ξ, and η.

Alternatively, if we know ρ as a function of r we can integrate eq. (24.11) to get

$$e^{-\eta} = 1 - \frac{1}{r}\int_0^r \rho\tilde{r}^2\, d\tilde{r} \tag{24.15}$$

assuming e^{η} does not go to zero with r. Then we can find ξ and p from the remaining equations.

In order to proceed, we assume ρ is constant. (This may not be very realistic, but it allows us to go ahead and illustrate a development which could be carried out with more technical difficulties under a more realistic assumption.) Now eq. (24.14) becomes

$$-\frac{\xi}{2} = \ln\left(\rho + \frac{p}{c^2}\right) + \text{const.}$$

and using eq. (24.10) we get

$$Cre^{-\xi/2} = e^{-\eta}(\eta' + \xi') \tag{24.16}$$

where C is an arbitrary constant. With ρ constant eq. (24.15) becomes

$$e^{-\eta} = 1 - \frac{r^2}{r_0^2} \tag{24.17}$$

where $r_0^2 = 3/a\rho$. Finally, substituting (24.17) into (24.16) we get a first-order linear differential equation for $e^{\xi/2}$ with the solution

$$e^{\xi/2} = A - B\left(1 - \frac{r^2}{r_0^2}\right)^{1/2} \tag{24.18}$$

where $A = (C/2)r_0^2$ and B is another arbitrary constant. In eqs. (24.17) and (24.18) we have finally arrived at the form of the coefficients of *the Schwarzschild interior line element*. Note that these functions of r are defined only for $0 \leq r < r_0$. With $a = 8\pi\mathbf{g}/c^2$,

$$r_0 = \left(\frac{3}{8\pi\mathbf{g}}\right)^{1/2} c\,\frac{1}{\rho^{1/2}} \approx 0.40 \times 10^{14}\,\frac{1}{\rho^{1/2}} \quad \text{cm}$$

Now we have two spherically symmetrical models of spacetime: Schwarzschild's exterior solution due to a point mass, M, and Schwarzschild's interior solution due to a mass/energy distribution with ρ constant. We can use these two solutions to construct a third model, one in which $\rho = \text{const.} > 0$ for $r \leq \hat{r}$ and $\rho = 0$ (and hence $T = 0$) for $r > \hat{r}$. This could be thought of as a model of a star of "radius" $\hat{r}$. If $\hat{r} < r_0$, then for points $r \leq \hat{r}$ we can use the interior solution we have just found. For $r > \hat{r}$ we assume that the line element of the star is the same as that of some equivalent point mass, M, such that $2\mathbf{g}M/c^2 < \hat{r}$; that is, such that $\hat{r}$ is greater than the Schwarzschild radius of the point mass.

It would seem desirable that this metric tensor field or, equivalently, the functions ξ and η, described differently on different parts of spacetime, should be continuous on the common boundary of these parts. Similarly for the fourth function, p. It turns out that we can achieve this by making suitable choices of the available constants. Thus, equating e^{η} given by (24.17) with the expression for e^{η} in Section 23.1 at $r = \hat{r}$ we get

$$M = \tfrac{4}{3}\pi\hat{r}^3\rho \tag{24.19}$$

That is, given ρ and $\hat{r}$, we must choose a point mass M given by (24.19). To make p continuous on the boundary we assume that it goes to zero as r goes to $\hat{r}$ from the interior. Then $\rho a/C = e^{-\xi/2}$ at $r = \hat{r}$, or $e^{\xi/2} = \frac{2}{3}A$ on $r = \hat{r}$.

Substituting this into (24.18) we get

$$A = 3B\left(1 - \frac{\hat{r}^2}{r_0^2}\right)^{1/2} \tag{24.20}$$

Finally, equating e^ξ given by (24.18) using (24.20) with the expression for e^ξ given in Section 23.1 at $r = \hat{r}$, we get $B = 1/2$. With these values of A and B we have, on the interior,

$$e^\xi = \frac{1}{4}\left[3\left(1 - \frac{\hat{r}^2}{r_0^2}\right)^{1/2} - \left(1 - \frac{r^2}{r_0^2}\right)^{1/2}\right]^2 \tag{24.21}$$

Now that we have a model, we can try to see what we can get out of it. In particular, since we have made certain simplifying assumptions, it is somewhat reassuring to be able to derive the following interesting and credible result. Note, first of all, that since r is not a distance, the factor $\frac{4}{3}\pi\hat{r}^3$ in (24.19) is not a volume. We saw, in Section 15.1, that in terms of coordinates r, ϑ, φ the volume form of a $t =$ const. hypersurface is

$$\sqrt{|\det(g_{ij})|}\, dr \wedge d\vartheta \wedge d\varphi = \left(1 - \frac{r^2}{r_0^2}\right)^{1/2} r^2 \sin\vartheta\, dr \wedge d\vartheta \wedge d\varphi$$

so the volume of the piece between 0 and $\hat{r}$ is

$$\int_0^\pi \int_0^{2\pi} \int_0^{\hat{r}} \left(1 - \frac{r^2}{r_0^2}\right)^{1/2} r^2 \sin\vartheta\, dr\, d\vartheta\, d\varphi = 2\pi r_0^3\left[\sin^{-1}\frac{\hat{r}}{r_0} - \frac{\hat{r}}{r_0}\left(1 - \frac{\hat{r}^2}{r_0^2}\right)^{1/2}\right]$$

and if we expand the terms in the brackets in power series we get finally,

$$\mathsf{V} = \tfrac{4}{3}\pi\hat{r}^3\left[1 + \frac{3}{10}\left(\frac{\hat{r}}{r_0}\right)^2 + \text{higher-order terms}\right] \tag{24.22}$$

If we can indeed interpret ρ as a physical mass density, then the mass of our star is $\rho\mathsf{V}$, which is larger than M. The interesting result, which we find by using the first-order correction for the volume in eq. (24.22), is that

$$\rho\mathsf{V} - M = \frac{\Phi}{c^2} \tag{24.23}$$

where Φ is the potential energy of the mass due to gravity, and Φ/c^2 is the mass corresponding to the potential energy lost, or kinetic energy acquired, if the sphere were to collapse to a point. Cf., Problem 24.4.

PROBLEM 24.1. Show that for our star of "radius" $\hat{r}$, the pressure as a function of r is given by

$$\frac{p}{\rho c^2} = \frac{2}{3 - u^{1/2}} - 1$$

where

$$u = \frac{(r_0/\hat{r})^2 - (r/\hat{r})^2}{(r_0/\hat{r})^2 - 1}$$

PROBLEM 24.2. From Problem 24.1 as well as from eq. (24.21) we see that u must be less than 9. Show $(r_0/\hat{r})^2 = (c^2/2\mathbf{g})(\hat{r}/M)$ and hence deduce the relation

$$\frac{9}{4}\frac{\mathbf{g}}{c^2} M < \hat{r}$$

between M and $\hat{r}$.

PROBLEM 24.3. Fill in the details of the derivation of eq. (24.22).

PROBLEM 24.4. Derive eq. (24.23), using the fact that the gravitational energy loss due to "spherical packing" is $-(16/15)\pi^2\mathbf{g}\rho^2\hat{r}^5$ for a sphere of radius $\hat{r}$. (See Adler, Bazin, and Schiffer, 1975, p. 475.)

24.2 The form of the Friedmann–Robertson–Walker metric tensor and its properties

To obtain the form of the Friedmann–Robertson–Walker metric tensor we make assumptions about the stress-energy tensor, T.

(i) As in Section 24.1, we assume that the medium is a perfect fluid with no external forces, so that T has the form (24.2). Again we take the velocity field U as a coordinate vector field, $\partial/\partial x^0$, so $U^i = 0$ $(i = 1, 2, 3)$ and we get the conditions on the components of T as before.

(ii) We assume $U = -c^2 \operatorname{grad} x^0$. With this assumption, on each curve, λ, of U,

$$\frac{dx^0 \circ \lambda}{d\tau} = U^0 = g(U, \operatorname{grad} x^0) = -\frac{g(U, U)}{c^2} = 1$$

so we can choose x^0 to be the proper time for all particles, and we say *U is proper-time synchronizable.*

Theorem 24.1. $U = -c^2 \operatorname{grad} x^0 \Rightarrow$ *the integral curves of U are geodesics and they have orthogonal hypersurfaces* (*rest spaces*).

Proof. $U = -c^2 \operatorname{grad} x^0 \Rightarrow \operatorname{curl} U = 0$ ($g_{\alpha\beta}U^\alpha$ is exact implies $g_{\alpha\beta}U^\alpha$ is closed). So $\operatorname{curl} U(X, Y) = 0$ for all X and Y orthogonal to U and U has orthogonal hypersurfaces by Problem 22.3. Further $\operatorname{curl} U(U, Z) = g(\nabla_U U, Z) - g(\nabla_Z U, U) = g(\nabla_U, U, Z)$ by eq. (17.12), so $\operatorname{curl} U(U, Z) = 0$ for all $Z \Rightarrow \nabla_U U = 0$. ■

Corollary. *With U chosen as a coordinate vector field,* $\partial/\partial x^0$, *and its rest spaces,* $x^0 = constant$, *chosen as coordinate hypersurfaces, the metric tensor has the form*

$$g = -(dx^0)^2 + g_{ij}\, dx^i\, dx^j \qquad (i, j = 1, 2, 3) \tag{24.24}$$

(iii) We assume T *is spatially isotropic*: at each point of each rest space, $x^0 =$ constant, T is invariant under orthogonal transformations of its tangent space. On the basis of this assumption, we infer that *spacetime, M, is spatially isotropic with respect to U*: at each point there is a set of local isometries which keep the point fixed, and which, given arbitrary unit vectors X and Y in the local rest space at the point, contains an element, ϕ, such that $\phi_* X = Y$ and $\phi_* U = U$. This inference can be given credentials as a form of "Mach's Principle."

From spatial isotropy of M we see that for any two tangent planes $\langle\{X, Y\}\rangle$ and $\langle\{X', Y'\}\rangle$ in the tangent space of the rest space at a point there is an isometry ϕ such that ϕ_* takes $\langle\{X, Y\}\rangle$ into $\langle\{X', Y'\}\rangle$. Under that isometry the surfaces of Problem 17.12 will be isometric and hence the sectional curvatures will be the same. That is, spatial isotropy implies that all sectional curvatures of a rest space at a point are the same.

Now, by means of two more standard geometrical results, spatial isotropy leads to a more specific form of eq. (24.24).

Theorem 24.2. (Schur) *If at each point in some region of a pseudo-Riemannian manifold,* (M, g), *of dimension* $n > 2$ *the sectional curvatures are all the same, then their common value,* κ, *is constant in the region.* (*If* κ *is constant on* (M, g), *then* (M, g) *is called* a space of constant curvature.)

Proof. By Problem 17.13, at each point we have $R_{ijkl} = \kappa(g_{ik}g_{jl} - g_{il}g_{jk})$. By the Ricci Lemma, Theorem 16.11, $R_{ijkl,m} = \kappa_{,m}(g_{ik}g_{jl} - g_{il}g_{jk})$. By the second Bianchi identity, Problem 16.12,

$$\kappa_{,m}(g_{ik}g_{jl} - g_{il}g_{jk}) + \kappa_{,k}(g_{il}g_{jm} - g_{im}g_{jl}) + \kappa_{,l}(g_{im}g_{jk} - g_{ik}g_{jm}) = 0$$

and multiplying this by $g^{ik}g^{jl}$ we get $(n-1)(n-2)\kappa_{,m} = 0$, so κ is constant. ■

Theorem 24.3. *If (M, g) is (locally) a space of constant curvature, κ, then there exists coordinates $(\mathscr{U}, \mu)$ in terms of which g has components*

$$g_{ij} = \frac{\eta_{ij}\, d\mu^i\, d\mu^j}{\left(1 + \frac{\kappa}{4}\eta_{ij}\mu^i\mu^j\right)^2} \tag{24.25}$$

where $\eta_{ij} = \begin{cases} \pm 1 & i = j \\ 0 & i \neq j \end{cases}$

Proof. (1) If (M, g) has coordinates for which g has the form (24.25) where κ is a constant, then direct calculation gives $R_{ijkl} = \kappa(g_{ijk}g_{jl} - g_{il}g_{jk})$ and by Problem 17.13 (M, g) is a space of constant curvature, κ.

(2) We will outline a proof of the following statement. If (M, g) has constant curvature, κ, in a neighborhood $\mathscr{U}$ of p and coordinates $(\mathscr{U}, \mu)$, and $(M, \bar{g})$ has constant curvature, κ, in a neighborhood $\bar{\mathscr{U}}$ of p and coordinates $(\bar{\mathscr{U}}, \bar{\mu})$, then there is a coordinate transformation $\bar{\mu} \circ \mu^{-1}$ such that

$$g_{ij} = \bar{g}_{pq}\frac{\partial \bar{\mu}^p}{\partial \mu^i}\frac{\partial \bar{\mu}^q}{\partial \mu^j} \tag{24.26}$$

That is, locally, there is only one space of constant curvature, κ.

(a) Given g_{ij} and $\bar{g}_{pq}$ we have to show there are functions $\bar{\mu}^i$ of μ^j satisfying eq. (24.26).

(b) Functions $\bar{\mu}^i$ of μ^j and p^i_j of μ^k satisfy

$$g_{ij} = \bar{g}_{pq}\frac{\partial \bar{\mu}^p}{\partial \mu^i}\frac{\partial \bar{\mu}^q}{\partial \mu^j}, \qquad p^i_j = \frac{\partial \bar{\mu}^i}{\partial \mu^j} \tag{24.27}$$

if and only if $\bar{\mu}^i$ satisfy eq. (24.26).

(c) Let

$$F^i_{jk}(\mu^l, \bar{\mu}^m, p^r_s) = \overline{\begin{Bmatrix} t \\ j\ k \end{Bmatrix}} p^i_t - \begin{Bmatrix} i \\ u\ v \end{Bmatrix} p^u_j p^v_k \qquad \text{and} \qquad f^i_j(\mu^l, \bar{\mu}^m, p^r_s) = p^i_j$$

Then functions $\bar{\mu}^i$ of μ^j and p^i_j of μ^k satisfy

$$\frac{\partial p^i_j}{\partial \mu^k} = F^i_{jk}(\mu^k, \bar{\mu}^m, p^r_s), \qquad \frac{\partial \bar{\mu}^i}{\partial \mu^j} = f^i_j(\mu^k, \bar{\mu}^m, p^r_s) \tag{24.28}$$

if and only if they satisfy (24.27).

(d) Equations (24.28) are of the form of eqs. (14.1). Applying the integrability conditions, eqs. (14.3), to these we get

$$p^r_i R^i_{.jkl} - p^s_j p^t_k p^u_l \bar{R}^r_{.stu} = 0 \tag{24.29}$$

(e) By our hypothesis, $R_{ijkl} = \kappa(g_{ik}g_{jl} - g_{il}g_{jk})$ and

$$\bar{R}_{rstu} = \kappa(g_{rt}g_{su} - g_{ru}g_{st}),$$

and these satisfy eq. (24.29).

Parts (1) and (2) together give us the required result. ■

If we change coordinates by $x^i = \sqrt{|\kappa|}\mu^i$ for $\kappa \neq 0$, then the right side of eq. (24.25) becomes

$$\frac{1}{|\kappa|}\frac{\eta_{ij}\,dx^i\,dx^j}{\left(1+\frac{\kappa_0}{4}\eta_{ij}x^ix^j\right)^2}, \qquad \text{where } \kappa_0 = \pm 1,$$

so the metric tensor of each rest space can be written in this form. But κ can change from one rest space to another; that is, κ can be a function of t. If we finally write $1/|\kappa| = \mathsf{R}^2(t)$, and transform to "spherical" coordinates, r, ϑ, φ, and insert the result in eq. (24.24), we get the Friedmann–Robertson–Walker metric tensor in the form

$$g = -c^2\,dt^2 + \mathsf{R}^2(t)\frac{dr^2 + r^2(d\vartheta^2 + \sin^2\vartheta\,d\varphi^2)}{\left(1+\frac{\kappa_0}{4}r^2\right)^2} \tag{24.30}$$

where $\kappa_0 = 0, \pm 1$. (The case $\kappa_0 = 0$ generalizes the case $\kappa = 0$ by inserting a factor of a function of t. It is sometimes called *a simple cosmological spacetime.*)

The Friedmann–Robertson–Walker metric tensor has a couple of interesting properties.

(1) The metric (24.30) gives rise to a redshift in terms of $\mathsf{R}(t)$. Thus, suppose light travels from a star λ_1 to a star λ_2. Since its path is a null geodesic, along it (from 24.30)

$$dl^2 = \frac{dr^2 + r^2(d\vartheta^2 + \sin^2\vartheta\,d\varphi^2)}{\left(1+\frac{\kappa_0}{4}r^2\right)^2} = \frac{c^2\,dt^2}{\mathsf{R}^2}$$

Integrating along this geodesic we have

$$\int_{t_1}^{t_2}\frac{c}{\mathsf{R}}\,dt = l(t_2) - l(t_1) \tag{24.31}$$

During the time interval Δt_1 the star λ_1 emits light observed by λ_2 during the time interval Δt_2. So we also have

$$\int_{t_1+\Delta t_1}^{t_2+\Delta t_2} \frac{c}{\mathsf{R}}\, dt = l(t_2 + \Delta t_2) - l(t_1 + \Delta t_1) \tag{24.32}$$

But the right sides of (24.31) and (24.32) are equal since we are using "comoving" coordinates. So

$$\int_{t_1}^{t_1+\Delta t_1} \frac{1}{\mathsf{R}}\, dt = \int_{t_2}^{t_2+\Delta t_2} \frac{1}{\mathsf{R}}\, dt$$

and, if Δt_1 and Δt_2 are small compared to $t_2 - t_1$,

$$\frac{\Delta t_2}{\mathsf{R}(t_2)} = \frac{\Delta t_1}{\mathsf{R}(t_1)} \tag{24.33}$$

If Δt_1 is the period of some radiation emitted at λ_1, then Δt_2 will be the observed period at λ_2. In terms of frequencies ν_1 and ν_2, (24.33) becomes

$$\frac{\nu_1 - \nu_2}{\nu_2} = \frac{\mathsf{R}(t_2)}{\mathsf{R}(t_1)} - 1 \tag{24.34}$$

Thus, a redshift corresponds to an increasing R—the universe is expanding.

(2) The metric (24.30) gives rise to Hubble's law. Thus, suppose $t_2 - t_1$ is small, and we expand $1/\mathsf{R}$ in a power series around $t = t_2$. Then substituting the result into (24.31) and (24.34), we get respectively,

$$l(t_2) - l(t_1) = h + \frac{1}{2}\frac{\mathsf{R}_2'}{c} h^2 + 0(h^3) \tag{24.35}$$

$$\frac{\nu_1 - \nu_2}{\nu_2} = \frac{\mathsf{R}_2'}{c} h + \frac{\mathsf{R}_2}{c^2}\left[\frac{(\mathsf{R}_2')^2}{\mathsf{R}_2} - \frac{\mathsf{R}_2''}{2}\right] h^2 + 0(h^3) \tag{24.36}$$

where $h = c(t_2 - t_1)/\mathsf{R}_2$. Eliminating h between these two equations by solving (24.35) for h and substituting into (24.36), we obtain

$$c\,\frac{\nu_1 - \nu_2}{\nu_2} = \mathsf{R}_2'\,\Delta l + \frac{\mathsf{R}_2'^2 (\Delta l)^2}{2c}\left(1 - \frac{\mathsf{R}_2''\mathsf{R}_2}{\mathsf{R}_2'^2}\right) + 0((\Delta l)^3)$$

where $\Delta l = l(t_2) - l(t_1)$. Finally, putting $H = \mathsf{R}'/\mathsf{R}$, the "Hubble constant" and $q = -\mathsf{R}''\mathsf{R}/\mathsf{R}'^2$, the "deceleration parameter," we have

$$c\frac{\nu_1 - \nu_2}{\nu_2} = H_2(\mathsf{R}_2\,\Delta l) + \frac{1}{2c}H_2^2(1 + q_2)(\mathsf{R}_2\,\Delta l)^2 + 0((\mathsf{R}_2\,\Delta l)^3) \quad (24.37)$$

$\mathsf{R}_2\,\Delta l$ can be considered an approximation for the physical distance between λ_1 and λ_2, so (24.37) gives the relative change of the frequency of the radiation in terms of the distance between λ_1 and λ_2. In particular, if we use only the first term on the right we get Hubble's law, if H is approximately constant.

PROBLEM 25.4. Prove that the converse of Theorem 24.1 is valid locally.

PROBLEM 24.6. Fill in some of the details of the proof of Theorem 24.3.

24.3 Friedmann–Robertson–Walker spacetimes

We will look for solutions of eq. (24.1) with $\mathrm{Ric}_{\alpha\beta}$ determined by eq. (24.30) and $T_{\alpha\beta}$ with properties (i), (ii), and (iii) in Section 24.2.

If we put $\mathsf{Q} = [1 + (\kappa_0/4)r^2]^{-2}$, then the Christoffel symbols for (24.30) are

$$\begin{Bmatrix} 0 \\ 1\ 1 \end{Bmatrix} = \mathsf{R}\mathsf{R}'\mathsf{Q}, \quad \begin{Bmatrix} 0 \\ 2\ 2 \end{Bmatrix} = \mathsf{R}\mathsf{R}'\mathsf{Q}r^2, \quad \begin{Bmatrix} 0 \\ 3\ 3 \end{Bmatrix} = \mathsf{R}\mathsf{R}'\mathsf{Q}r^2\sin^2\vartheta$$

$$\begin{Bmatrix} 1 \\ 0\ 1 \end{Bmatrix} = \frac{\mathsf{R}'}{\mathsf{R}}, \quad \begin{Bmatrix} 1 \\ 1\ 1 \end{Bmatrix} = \frac{1}{2}\frac{\mathsf{Q}'}{\mathsf{Q}}, \quad \begin{Bmatrix} 1 \\ 2\ 2 \end{Bmatrix} = -\left(\frac{\mathsf{Q}'}{2\mathsf{Q}} + \frac{1}{r}\right)r^2,$$

$$\begin{Bmatrix} 1 \\ 3\ 3 \end{Bmatrix} = -\left(\frac{\mathsf{Q}'}{2\mathsf{Q}} + \frac{1}{r}\right)r^2\sin^2\vartheta$$

$$\begin{Bmatrix} 2 \\ 0\ 2 \end{Bmatrix} = \frac{\mathsf{R}'}{\mathsf{R}}, \quad \begin{Bmatrix} 2 \\ 1\ 2 \end{Bmatrix} = \frac{\mathsf{Q}'}{2\mathsf{Q}} + \frac{1}{r}, \quad \begin{Bmatrix} 2 \\ 3\ 3 \end{Bmatrix} = -\sin\vartheta\cos\vartheta$$

$$\begin{Bmatrix} 3 \\ 0\ 3 \end{Bmatrix} = \frac{\mathsf{R}'}{\mathsf{R}}, \quad \begin{Bmatrix} 3 \\ 1\ 3 \end{Bmatrix} = \frac{\mathsf{Q}'}{2\mathsf{Q}} + \frac{1}{r}, \quad \begin{Bmatrix} 3 \\ 2\ 3 \end{Bmatrix} = \cot\vartheta$$

We saw that property (i) requires that $T_1^1 = T_2^2 = T_3^3$ so that eq. (24.1) in the form

$$\mathrm{Ric}_\beta^\alpha = a(T_\beta^\alpha - \tfrac{1}{2}T_\lambda^\lambda\delta_\beta^\alpha) \quad (24.38)$$

has only two independent components. From the Christoffel symbols we calculate

$$\mathrm{Ric}_{00} = -3\,\frac{\mathsf{R}\mathsf{R}''}{c^2\mathsf{R}^2}$$

$$\mathrm{Ric}_{11} = \mathsf{Q}\left(\frac{\mathsf{R}\mathsf{R}'' + 2\mathsf{R}'^2}{c^2} + 2\kappa_0\right)$$

so eq. (24.38) yields

$$3\,\frac{\mathsf{R}''}{c^2\mathsf{R}} = -\frac{a}{2}\left(\rho + \frac{3p}{c^2}\right) \tag{24.39}$$

$$\frac{\mathsf{R}\mathsf{R}'' + 2\mathsf{R}'^2}{c^2\mathsf{R}^2} + \frac{2\kappa_0}{\mathsf{R}^2} = \frac{a}{2}\left(\rho - \frac{p}{c^2}\right) \tag{24.40}$$

Forming linear combinations of (24.39) and (24.40) we get

$$\frac{3\mathsf{R}'^2}{c^2\mathsf{R}^2} + \frac{3\kappa_0}{\mathsf{R}^2} = a\rho \tag{24.41}$$

$$\frac{2\mathsf{R}''}{c^2\mathsf{R}} + \frac{\mathsf{R}'^2}{c^2\mathsf{R}^2} + \frac{\kappa_0}{\mathsf{R}^2} = -a\,\frac{p}{c^2} \tag{24.42}$$

Finally, differentiating (24.41) and eliminating R'' between the result and (24.39), we get

$$(\rho\mathsf{R}^3)' + \frac{p}{c^2}(\mathsf{R}^3)' = 0 \tag{24.43}$$

In looking for solutions of eqs. (24.39)–(24.43), an obvious, and historically important, case to consider is the static case, $\mathsf{R} = \text{constant}$. From (24.39) and (24.41) we get $0 = \rho + 3p/c^2$ and $3\kappa_0/\mathsf{R}^2 = a\rho$. If ρ and p are to be nonnegative, these conditions imply that $p = \rho = \kappa_0 = 0$, an empty flat universe.

We can get something more interesting if we add the cosmological constant, Λ, which is what Einstein did—reluctantly. It turns out that now we must have $\kappa_0 = 1$ *and* $\mathsf{R} = 2/[a(\rho + p/c^2)]$. With $a = 8\pi\mathbf{g}/c^2$ one gets R, "the radius of the universe" of the order of 10^{10} light years.

The static models were proposed before the redshift was discovered. Hence the fact that they predict a zero redshift is a serious flaw.

An early nonstatic solution of Einstein's equation was obtained by de Sitter (1917) by including the cosmological constant. With notable prescience he assumed R'/R = constant. The equations then require $\kappa_0 = 0$ (a simple cosmological spacetime), and again, unfortunately, $\rho = p = 0$.

Now we look for solutions of eqs. (24.39)–(24.43) with $p = 0$, incoherent dust. Eliminating ρ between (24.39) and (24.41) we get

$$2\mathsf{R}\mathsf{R}'' + \mathsf{R}'^2 + c^2\kappa_0 = 0 \tag{24.44}$$

which is integrated to

$$\mathsf{R}'^2 + c^2\kappa_0 = \frac{D_0 c^2}{\mathsf{R}} \tag{24.45}$$

where D_0 is an arbitrary constant. Putting (24.45) into (24.44) we get

$$\mathsf{R}'' = -\frac{D_0 c^2}{2\mathsf{R}^2} \tag{24.46}$$

and putting this into (24.39) (with $p = 0$) we get

$$D_0 = \frac{a}{3}\rho\mathsf{R}^3 \tag{24.47}$$

This is consistent with eq. (24.39), which in this case, for $\rho \neq 0$, requires $\mathsf{R}'' < 0$, so $D_0 > 0$, and is also consistent with eq. (24.43), which gives, in this case, $\mathsf{R}^3\rho$ = const.

Finally, for the deceleration parameter, q, $\mathsf{R}'' < 0 \Rightarrow q > 0$. Moreover, from eqs. (24.45) and (24.46) we get

$$\frac{\kappa_0 \mathsf{R}}{D_0} = \frac{2q - 1}{2q}$$

so $q < \frac{1}{2}$ for $\kappa_0 = -1$, $q = \frac{1}{2}$ for $\kappa_0 = 0$, and $q > \frac{1}{2}$ for $\kappa_0 = 1$.

Now consider the cases $\kappa_0 = 1$, $\kappa_0 = 0$, and $\kappa_0 = -1$ (see Fig. 24.1).

(i) If $\kappa_0 = 1$, and we introduce a parameter ω by

$$ct = \frac{D_0}{2}(2\omega - \sin 2\omega),$$

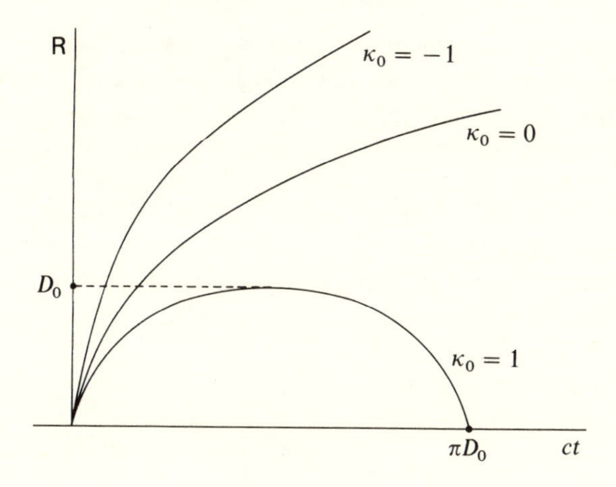

FIGURE 24.1

then eq. (24.25) has a solution of the form

$$\begin{cases} ct = \dfrac{D_0}{2}(2\omega - \sin 2\omega) \\ \mathsf{R} = \dfrac{D_0}{2}(1 - \cos 2\omega) \end{cases}$$

This is a cycloid. The slope is infinite at $t = 0$, which cc ld correspond to the "big band." However, $p = 0$ is not a good assumption for small t.

(ii) If $\kappa_0 = 0$, then integrating (24.45), with $\mathsf{R} = 0$ when $t = 0$,

$$\mathsf{R} = (\tfrac{3}{2}c)^{3/2} D_0^{1/3} t^{2/3}$$

Again we get a big band at $t = 0$. This solution was obtained by A. Friedmann in 1922. Nowadays it is called *Einstein–de Sitter spacetime*. It is another example of a simple cosmological spacetime.

(iii) If $\kappa_0 = -1$, and we introduce a parameter ω by $ct = (D_0/2)(\sinh 2\omega - 2\omega)$, then eq. (24.45) has a solution of the form

$$\begin{cases} ct = \dfrac{D_0}{2}(\sinh 2\omega - 2\omega) \\ \mathsf{R} = \dfrac{D_0}{2}(\cosh 2\omega - 1) \end{cases}$$

and as $t \to \infty$, $\mathsf{R}' \to c$.

The quantities ρ, H, and q in the equations above are subject to astronomical measurement. One might hope that such information could sort out of the three cases, the model that best corresponds to reality. As we have seen, a precise enough determination of q alone would select one of these cases. Alternatively, if we knew H and ρ with enough accuracy we could determine the choice of κ_0 from eq. (24.41). Unfortunately, the range within which all these quantities are presently known is too wide to make a definite determination. This is currently an area of active investigation (cf., Loh and Spillar, 1986).

PROBLEM 24.7. Show that for $\kappa_0 \neq 0$, D_0 is obtained from H and q by

$$D_0 = \frac{2q}{|2q-1|^{3/2}} \frac{c}{H}$$

and for $\kappa_0 = 0$, ρ and H are related by $a\rho = (3/c^2)H^2$.

REFERENCES

Abraham, R., and Marsden, J. E. (1978) *Foundations of Mechanics*, The Benjamin/Cummings Publishing Company.

Adler, R., Bazin, M., and Schiffer, M. (1975) *Introduction to General Relativity*, McGraw-Hill Book Company.

Aris, R. (1962) *Vectors, Tensors and the Basic Equations of Fluid Mechanics*, Prentice-Hall.

Arnold, V. I. (1989) *Mathematical Methods of Classical Mechanics*, Springer-Verlag.

Auslander, L., and MacKenzie, R. E. (1977) *Introduction to Differentiable Manifolds*, Dover Publications.

Beem, J. K., and Ehrlich, P. E. (1981) *Global Lorentzian Geometry*, Marcel Dekker.

Bergmann, G. (1976) *Introduction to the Theory of Relativity*, Dover Publications.

Bishop, R. L., and Goldberg, S. I. (1968) *Tensor Analysis on Manifolds*, The Macmillan Company.

Boothby, W. M. (1975) *An Introduction to Differentiable Manifolds and Riemannian Geometry*, Academic Press.

Brickell, F., and Clark, R. S. (1970) *Differentiable Manifolds, An Introduction*, Van Nostrand Reinhold Company.

Cartan, E. (1922) Sur les équations de la gravitation d'Einstein, *Journal de Mathématiques Pures et Appliquées*, 141–203.

Chern, S. S. (1959) *Differentiable Manifolds*, Class Notes, University of Chicago.

Chevalley, C. (1946) *Theory of Lie Groups I*, Princeton University Press.

Choquet-Bruhat, Y., and DeWitt-Morette, C. (1982) *Analysis, Manifolds, and Physics*, North-Holland Publishing Company.

Coburn, N. (1955) *Vector and Tensor Analysis*, The Macmillan Company.

Courant, R., and Hilbert, D. (1966) *Methods of Mathematical Physics* Vol II, Interscience Publishers.

Dieudonné, J. (1969) *Foundations of Modern Analysis*, Academic Press.

Einstein, A. (1923a) On the electrodynamics of moving bodies Annalen der Physik, Vol 17, 1905, in *The Principle of Relativity*, Dover Publications.

Einstein, A. (1923b) Cosmological considerations on the general theory of relativity Preussischen Akademie der Wissenschaften, Sitzungsberichte, 1917, in *The Principle of Relativity*, Dover Publications.

Eisenhart, L. P. (1947) *An Introduction to Differential Geometry with the Use of the Tensor Calculus*, Princeton University Press.

Eisenhart, L. P. (1949) *Riemannian Geometry*, Princeton University Press.

Flanders, H. (1963) *Differential Forms with Applications to the Physical Sciences*, Academic Press.

Fock, V. (1979) *The Theory of Space, Time and Gravitation* (2nd revised edition), W. H. Freeman and Company.

Frankel, T. (1979) *Gravitational Curvature, An Introduction to Einstein's Theory*, W. H. Freeman and Company.

Greub, W. (1981) *Linear Algebra* (4th edition), Springer-Verlag.

Guillemin, V., and Sternberg, S. (1984) *Symplectic Techniques in Physics*, Cambridge University Press.

Hawking, S. W., and Ellis, G. F. R. (1973) *The Large Scale Structure of Space–Time*, Cambridge University Press.

Hicks, N. J. (1965) *Notes on Differential Geometry*, D. Van Nostrand Company.

Kaplan, W. (1959) *Advanced Calculus*, Addison-Wesley Publishing Company.

Kobayashi, S., and Nomizu, K. (1963) *Foundations of Differential Geometry*, Vol. I, Interscience Publishers.

Lang, S. (1965) *Algebra*, Addison-Wesley Publishing Company.

Laugwitz, D. (1965) *Differential and Riemannian Geometry*, Academic Press.

Lewis, G. N., and Tolman, R. C. (1909) The principle of relativity and non-Newtonian mechanics, *Philosophical Magazine*, **18**, 510–523.

Loh, E. D., and Spillar, E. J. (1986) A measurement of the mass density of the universe, *The Astrophysical Journal*, **307**, L1–L4.

Lovelock D. (1972) The four-dimensionality of space and the Einstein tensor, *Journal of Mathematics and Physics*, **13**, No. 6.

MacMillan, W. D. (1960) *Dynamics of Rigid Bodies*, Dover Publications.

Marsden, J. E. (1981) *Lectures on Geometric Methods in Mathematical Physics*, Society for Industrial and Applied Mathematics.

Misner, C. W., Thorne, K. S., and Wheeler, J. A. (1973) *Gravitation*, W. H. Freeman and Company.

O'Neill, B. (1983) *Semi-Riemannian Geometry with Applications to Relativity*, Academic Press.

Pauli, W. (1981) *Theory of Relativity*, Dover Publications.

Rindler, W. (1966) *Special Relativity* (2nd edition), Oliver and Boyd.

Robertson, H. P., and Noonan, T. W. (1968) *Relativity and Cosmology*, W. B. Saunders Company.

Sachs, R. K., and Wu, H. (1977) *General Relativity for Mathematicians*, Springer-Verlag.

Schouten, J. A. (1954) *Ricci-Calculus, An Introduction to Tensor Analysis and its Geometrical Applications* (2nd edition), Springer-Verlag.

Singer, I. M., and Thorpe, J. H. (1967) *Lecture Notes on Elementary Topology and Geometry*, Scott, Foresman and Company.

Spivak, M. (1965) *Calculus on Manifolds*, W. A. Benjamin.

Spivak, M. (1979) *A Comprehensive Introduction to Differential Geometry* (Vol. II), Publish or Perish.

Sternberg, S. (1983) *Lectures on Differential Geometry* (2nd edition), Chelsea Publishing Company.

Synge, J. L. (1965) *Relativity: The Special Theory* (2nd edition), North-Holland Publishing Company.

Synge, J. L., and Schild, A. (1966) *Tensor Calculus*, University of Toronto Press.

Taylor, E. F., and Wheeler, J. A. (1966) *Spacetime Physics*, W. H. Freeman and Company.

Warner, F. W. (1971) *Foundations of Differentiable Manifolds and Lie Groups*, Scott, Foresman and Company.

Weinberg, S. (1972) *Gravitation and Cosmology: Principles and Applications of the General Theory of Relativity*, John Wiley and Sons.

Weyl, H. (1922) *Space–Time–Matter* (4th edition), Dover Publications.

Weyl, H. (1923) Mathematische Analyse Des Raumproblems in *Das Kontinuum und andere Mongraphien*, Chelsea Publishing Co.

Whitney, H. (1957) *Geometric Integration Theory*, Princeton University Press.

Whittaker, E. T. (1960) *A History of Theories of Aether and Electricity, Vol II, The Modern Theories, 1900–1926*, Harper and Brothers.

Willmore, T. J. (1959) *An Introduction to Differential Geometry*, Oxford University Press.

Woll, J. W., Jr. (1966) *Functions of Several Variables*, Harcourt, Brace and World.

INDEX